THE COMPLETE WORKS OF YUTAKA TANIYAMA

[新版]

谷山 豊全集

日本評論社

1950年(昭和25年)4月

1957年(昭和32年)横浜市・山下公園にて

1955年(昭和30年)9月,代数的整数論に関する国際会議の折,
会場の第一生命ビル屋上にて.最後列右から2人目が谷山.

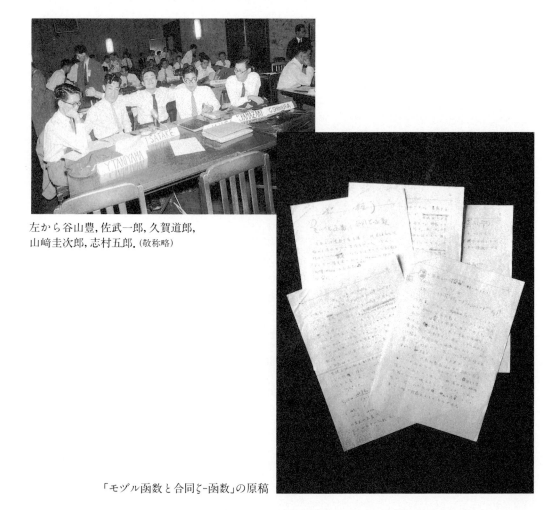

左から谷山豊,佐武一郎,久賀道郎,
山崎圭次郎,志村五郎.(敬称略)

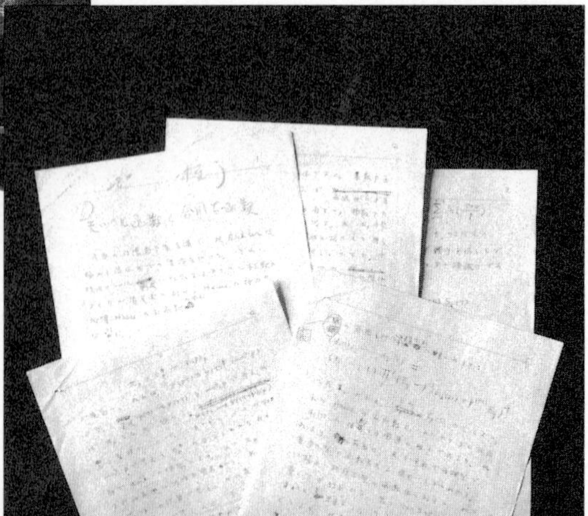

「モヅル函数と合同ζ-函数」の原稿

東大理学部数学科卒業式の日に（昭和 28 年 3 月）同級生一同
前列右から 4 人目が彌永昌吉，左端が谷山豊，後列左から 2 人目が杉浦光夫

東大理学部数学科卒業式の日に彌永昌吉宅にて　有志とともに
中列右から 3 人目が彌永昌吉，前列右端が谷山豊，後列右から 4 人目が杉浦光夫

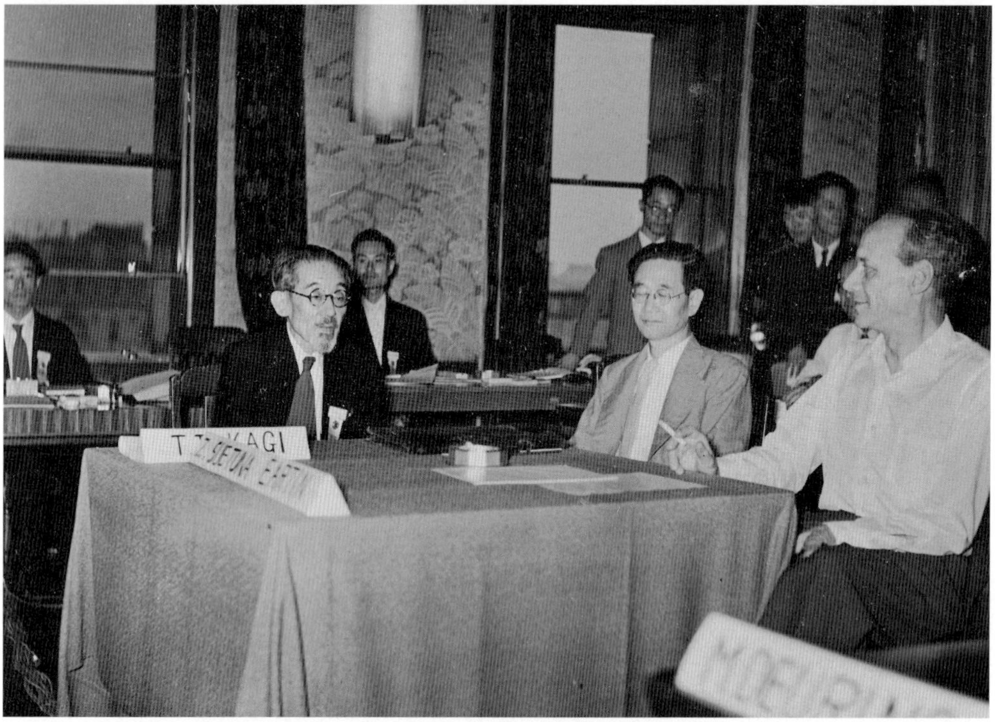

1955年（昭和30年）9月9日，東京第一生命保険相互会社会議室にて
左から高木貞治，末綱恕一，E.Artin

同上．A.Weilによる講演

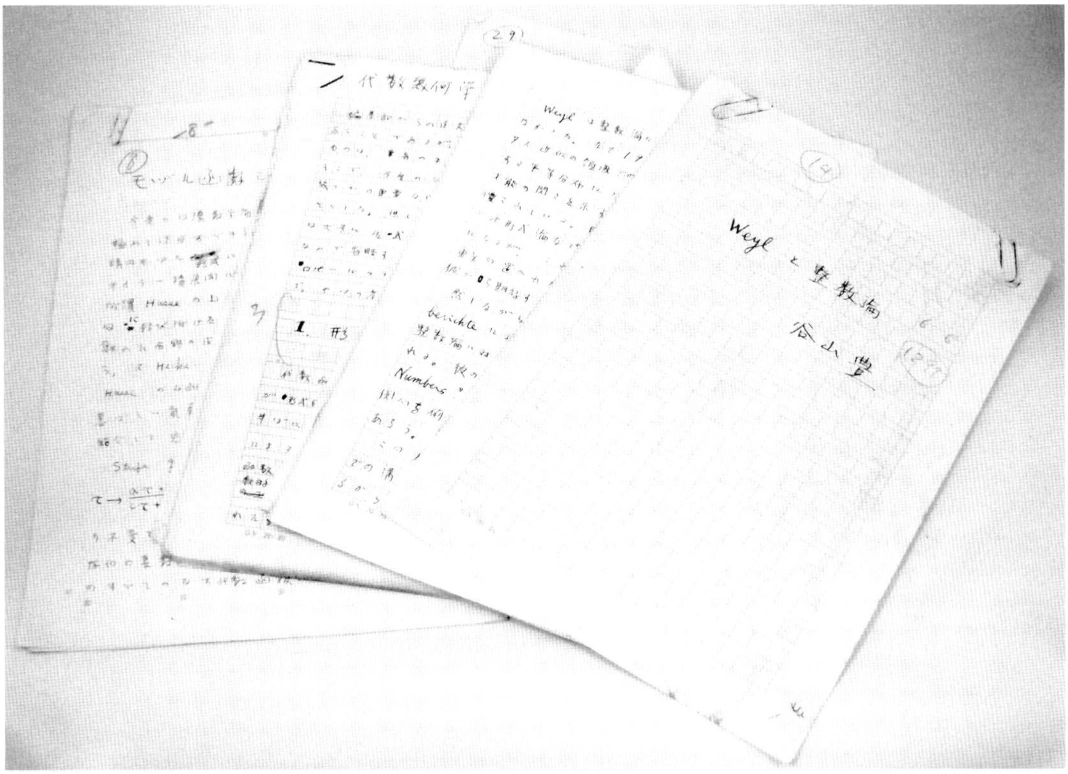

（上）左から「モヅル函数と合同ζ‐函数」（179ページ），「代数幾何学と整数論」（193ページ），「Weyl と整数論」（208ページ）の原稿

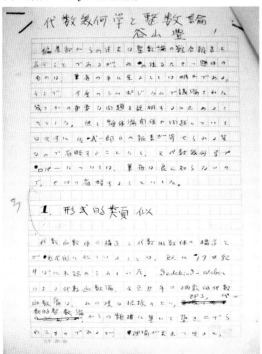

（下（左））「代数幾何学と整数論」

（下（右））「Weyl と整数論」

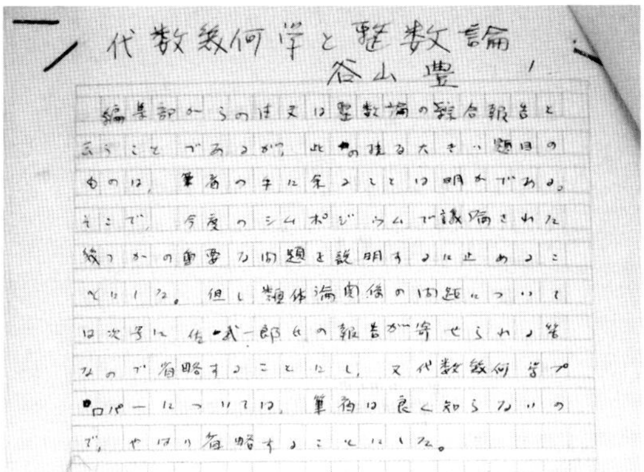

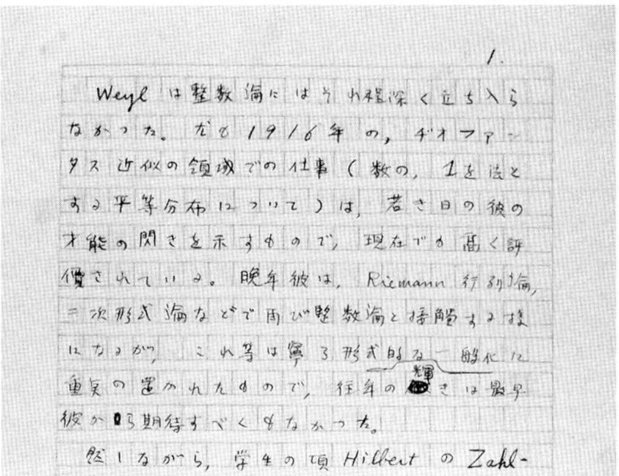

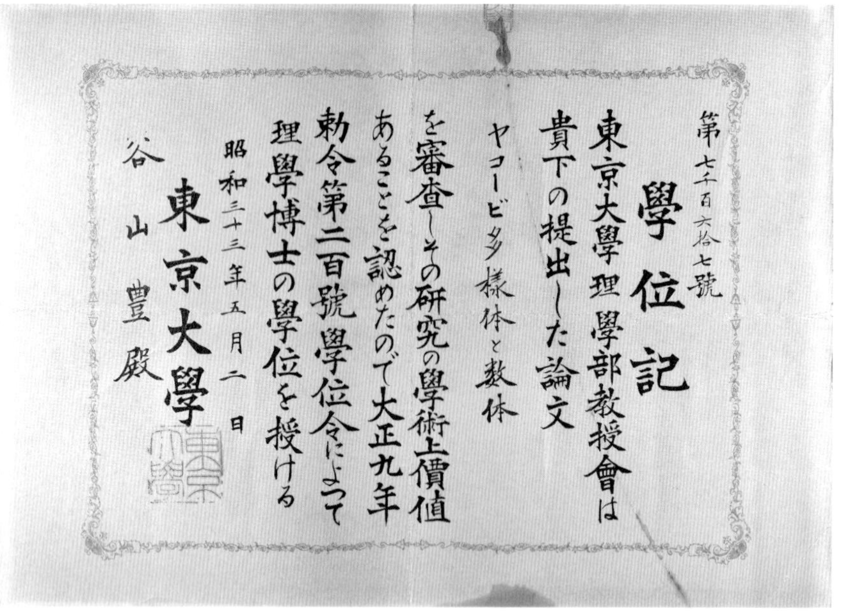

第七千百六拾七號

學位記

東京大學理學部教授會は
貴下の提出した論文
ヤコービ多様体と数体
を審査しその研究の學術上價値
あることを認めたので大正九年
勅令第二百號學位令によって
理學博士の學位を授ける

昭和三十三年五月二日

東京大學

谷山豊殿

（上）学位記（東京大学より　昭和 33 年 5 月 2 日）
（下（左））谷山の蔵書『プラトンの自叙傳』（青木巌訳，昭和 22 年 5 月，和田堀書店）カバー
（下（右））谷山による『プラトンの自叙傳』への書き込み（337 ページ）

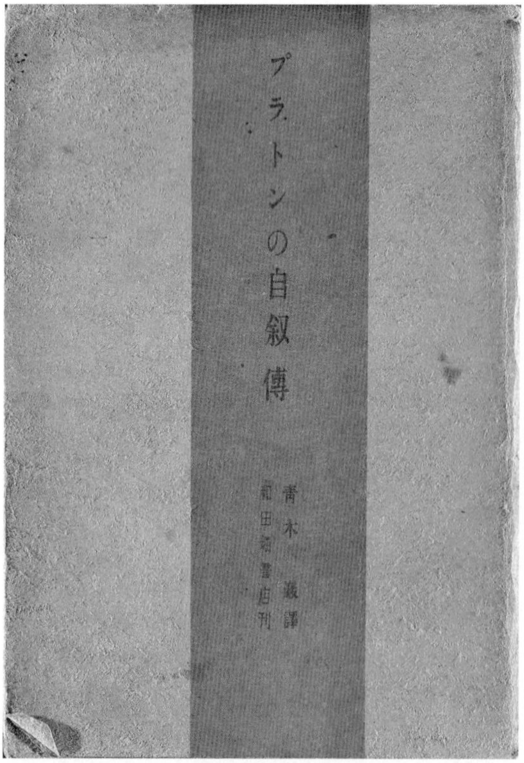

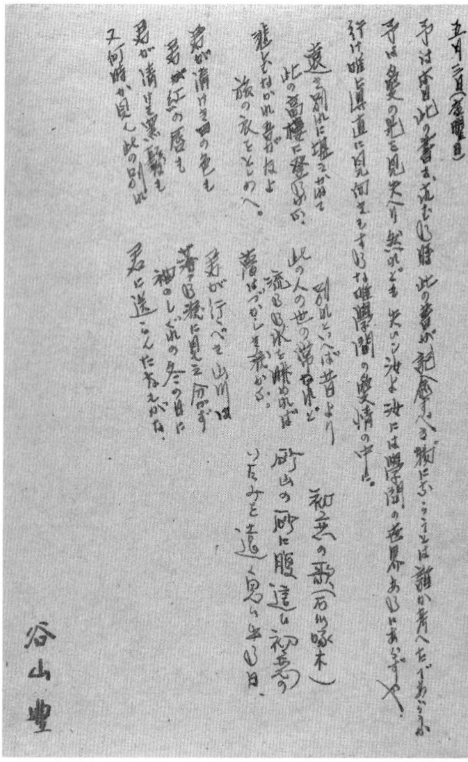

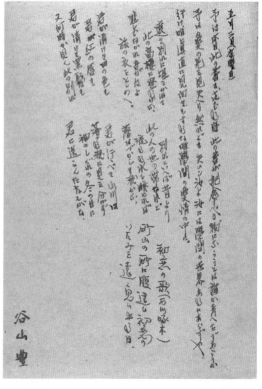

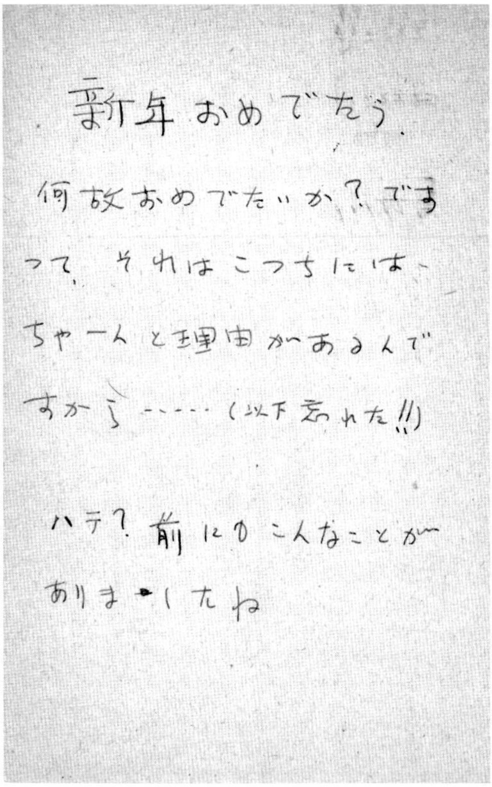
新年おめでたう。

何故おめでたいか？ですって、それはこっちにはちゃーんと理由があるんですから……（以下忘れた!!）

ハテ？前にもこんなことがありましたね

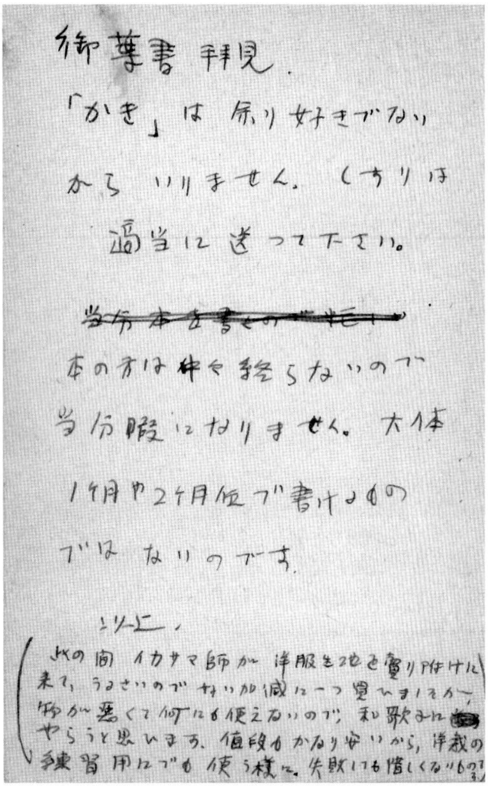
御葉書拝見。

「かき」は余り好きでないからいりません。くすりは適当に送って下さい。

~~当分本を書くので~~

本の方は中々終らないので当分暇になりません。大体1ヶ月や2ヶ月位で書けるものではないのです。

ミツヒコ、

此の間イカサマ師が洋服生地を売りつけに来て、うるさいのでいい加減に一つ買いましたが、物が悪くて何にも使えないので、和歌子にやろうと思います。値段もかなり安いから、洋裁の練習用にでも使う様に。失敗しても惜しくないもので。

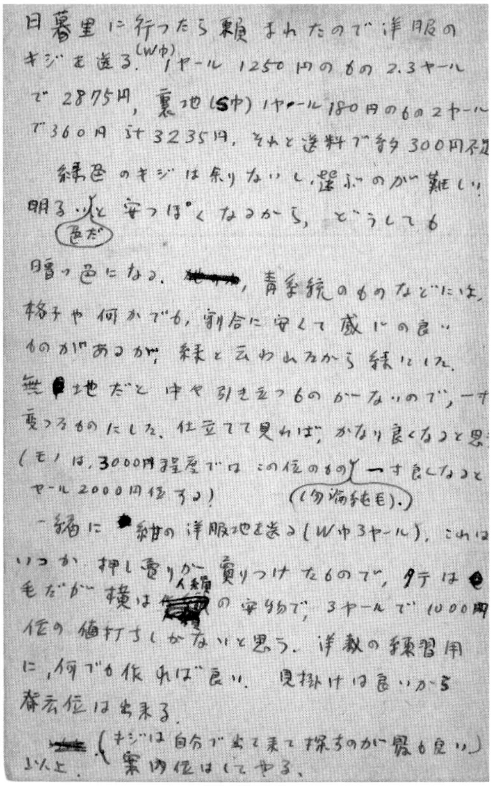
日暮里に行ったら頼まれたので洋服のキジを送る。1ヤール（W巾）1250円のもの2.3ヤールで2875円、裏地（S巾）1ヤール180円のもの2ヤールで360円 計3235円、それと送料で約300円程

緑色のキジは余りないし、選ぶのが難しい。明るい色だと安っぽくなるから、どうしても暗い色になる。青系統のものなどには、格子や何かでも、割合に安くて感じの良いものがあるが、緑と云われるから緑にした。

無地だと中々引き立つものがないので、一寸変ったものにした。仕立てて見れば、かなり良くなると思う。（モノは、3000円程度ではこの位のもの（一寸良くなるとヤール2000円位する）（勿論純毛）。

一緒に紺の洋服地を送る（W巾3ヤール）。これはいつか押し売りが売りつけたもので、タテは毛だが横は人絹の安物で、3ヤールで1000円位の値打ちしかないと思う。洋裁の練習用に、何でも作れば良い。見掛けは良いから普段位は出来る。

（キジは自分で出て来て探すのが最も良い）以上、案内位はしてやる。

（上）成田正雄への書簡
（昭和31年1月1日，310ページ）
（中）谷山欣隆への書簡
（昭和31年10月17日，315ページ）
（下）谷山和歌子への書簡
（昭和32年10月30日，317ページ）

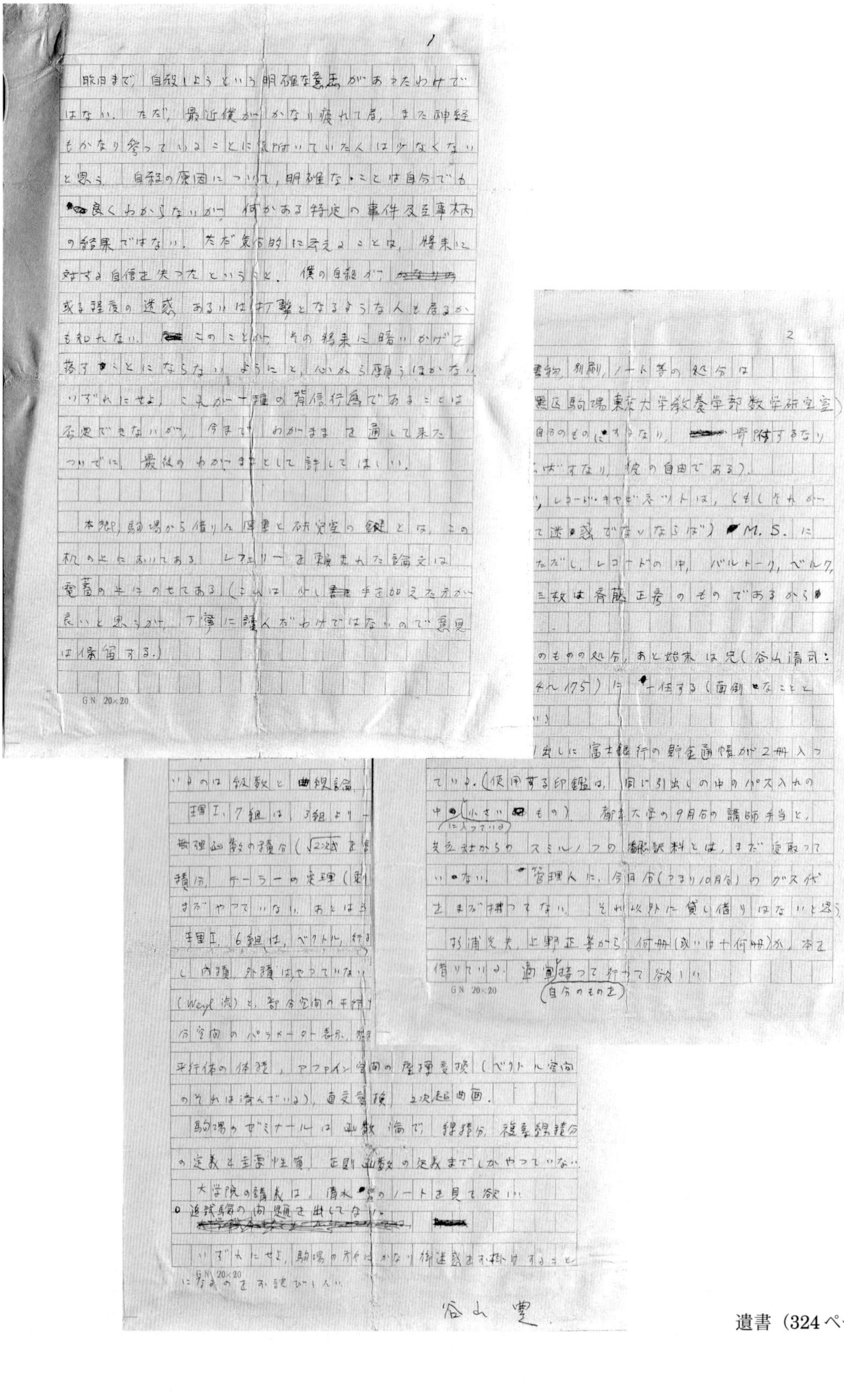
1

昨日まで，自殺しようという明確な意志があったわけではない．ただ，最近僕がかなり疲れて居り，また神経もかなり参っていることに気付いていた人は少なくないと思う．自殺の原因について，明確なことは自分でも良くわからないが，何かある特定の事件乃至事柄の結果ではない．ただ気分的に云えることは，将来に対する自信を失ったということ．僕の自殺が，或る程度の迷惑あるいは打撃となるような人も居るかも知れない．このことが，その人の将来に暗いかげを落すことにならないようにと，心から願うほかない．いずれにせよ，これが一種の背信行為であることは否定できないが，今までわがままを通して来たついでに，最後のわがままとして許してほしい．

本郷，駒場から借りた図書と研究室の鍵とは，この机の上においてある．レフェリーを頼まれた論文は電蓄の上にのせてある（これは少し手を加えた方が良いと思うが，丁寧に読んだわけではないので意見は保留する．）

GN 20×20

2

書物，別刷，ノート等の処分は
黒区駒場東京大学教養学部数学研究室）
自分のものにするなり，寄附するなり
ばすなり，彼の自由である）．
，レコード・キャビネットは，（もしそれが
て迷惑でないならば）M. S. に
ただし，レコードの中，バルトーク，ベルク，
三枚は斉藤正彦のものであるから
のものの処分，あと始末は兄（谷山清司：
4～175）に一任する（面倒なことと
い）
引出しに富士銀行の貯金通帳が2冊入っている．（使用する印鑑は，同じ引出しの中のパス入れの中（に入っている）（小さいもの）．都立大学の9月分の講師手当と，共立社からのスミルノフの翻訳料とは，まだ受取っていない．管理人に，今月分（つまり10月分）のガス代をまだ払ってない．それ以外に貸し借りはないと思う．

杉浦光夫，上野正等から何冊（或いは十何冊）か，本を借りている．適宜（自分のものを）持って行って欲しい．

GN 20×20

いるのは級数と曲線論
理I，7組は，3組より
無理函数の積分（$\sqrt{2次式}$ を含
積分，テーラーの定理（剰
まだやっていない．あとは
理I，6組は，ベクトル，行列
し内積，外積はやっていない
（Weyl流）と，部分空間の平行
分空間のパラメーター表示，
平行体の体積，アファイン空間の座標変換（ベクトル空間のそれは済んでいる），直交変換，2次超曲面．

駒場のゼミナールは函数論で，線積分，複素線積分の定義と主要性質，正則函数の定義までしかやっていない．

大学院の講義は，清水君のノートを見て欲しい．

○追試験の問題を出してない．

いずれにせよ，駒場の方々にかなり御迷惑をお掛けすることになるのをお詫びしたい．

GN 20×20

谷山豊．

遺書（324ページ）

目　　次
CONTENTS

第三部

PARTIII

第四部

PARTIV

第五部
PARTV

装幀=駒井佑二
口絵(3, 4, 6, 8頁)撮影=河野裕昭

第一部

Jacobian Varieties and Number Fields

A mimeographed note, University of Tokyo (September, 1955)

1. The classical theory of complex multiplication solves the problem of construction of abelian extensions of imaginary quadratic fields. Apart from this classical theory, E. Hecke has treated successfully the problem of construction of unramified abelian extensions of certain imaginary biquadratic fields by means of singular values of Hilbert modular functions.

In the following, I shall show that all these theories may be considered as special cases of what I might call "general theory of complex multiplication". This is in close connection with Hasse conjecture on zeta-functions of algebraic curves, and of abelian varieties, over an algebraic number field. In fact, this conjecture can be proved in case we can establish this general theory of complex multiplication. This includes the results of M. Deuring on elliptic curves, and some of results of A. Weil on the curve defined by the equation $ax^n+by^m+c=0$, in case n, m are different prime numbers.

2. Let C be a non-singular curve of genus g, J a jacobian variety of C, φ a canonical mapping of C into J, $\mathcal{A}(J)$ the ring of endomorphisms μ of J and $\{\omega_1,\cdots,\omega_g\}$ a basis of invariant linear differentials on J. We assume all these are defined over an algebraic number field k. We assume now, that

(1) $\mathcal{A}(J)=R$ is isomorphic to the principal order of a certain algebraic number field k_1 of degree $2g$.

We shall identify R with this order. On the other hand, $\delta\mu$ will mean the differential of the endomorphism $\mu\in R$. Now the basis $\{\omega_1,\cdots,\omega_g\}$ can be so chosen that

(2) $$\delta\mu(\omega_i)=\mu^{\sigma_i}\omega_i \qquad (i=1,\cdots,g)$$

for all $\mu\in R$, where σ_i are isomorphisms of k_1 into the field of complex numbers, which are independent of the choice of μ. We assume, for simplicity, that k is absolutely normal and these ω_i are rational over k.

Let $\bar{k}_1$ be the smallest absolutely normal field containing k_1, G the Galois group of $\bar{k}_1/\boldsymbol{Q}$, and G_1 the subgroup of G corresponding to k_1. Let G_{00} be the subgroup of G, consisting of all the elements σ for which $\{\mu^{\sigma_1\sigma},\cdots,\mu^{\sigma_g\sigma}\}$ is a permutation of $\{\mu^{\sigma_1},\cdots,\mu^{\sigma_g}\}$ for every $\mu\in R$. Denote with k^{**} the subfield corresponding to G_{00}. Let G_0 be a subgroup of G_{00}, with which the following decomposition takes place:

(3) $$\sum_{i=1}^{g} G_1\sigma_i = \sum_{j=1}^{s} G_1\tau_j G_0,$$

where $s=g/[G_0\colon 1]$ and τ_j are some of σ_i. We denote by k^* the subfield of k_1 corresponding to this group G_0.

Let $\mathfrak{b}$ be an ideal in k^* (which may then be considered also as an ideal in $\bar{k}_1$). We put now

(4) $$\Gamma(\mathfrak{b}) = N_{\bar{k}_1/k_1}(\mathfrak{b}^{\tau_1^{-1}} \cdots \mathfrak{b}^{\tau_s^{-1}})$$

and denote with H the group of ideals $\mathfrak{b}$ of k^* such that $\Gamma(\mathfrak{b})$ are principal ideals in k_1.

On the other hand, there are a finite number of curves $C_1, \cdots, C_\delta$ having J as a jacobian variety. Let $(m^{(i)})$ be the modules of C_i and (M) the Chow coordinates of $(m^{(1)})+\cdots+(m^{(\delta)})$. Under our assumption (1), we can prove that (M) are algebraic number. Under one more additional assumption

(5) k_1 is relatively normal over all imaginary subfields k_c of it, and there is no automorphism $\sigma \neq 1$ of k_1 over any k_c with the property

$$\sum_{i=1}^{g} G_1\sigma\sigma_i = \sum_{i=1}^{g} G_1\sigma_i,$$

we can prove

THEOREM 1. *$k^*(M)$ is the class field over k^* for the ideal group H.*

This contains the results of the classical theory of complex multiplication for elliptic invariant j as well as those of the theory of Hecke. We can also prove the irreducibility of "class-equations" under certain additional conditions.

3. Let $\mathfrak{a}$ be an ideal in R. Then we call a point b on J a proper $\mathfrak{a}$ Teil-point, if $\alpha b=0$ is equivalent to $\alpha \in \mathfrak{a}$. Let E be the group of roots of unity in k_1. Then, we denote with $E(b)$ the zero cycle $\sum_{\varepsilon \in E}(\varepsilon b)$ on J. We denote with $k(b)_0$ the smallest field containing k, over which the cycle $E(b)$ is rational.

We take a system of ideal numbers in k^* and denote the ideal number of $\mathfrak{b}$ with $\hat{\beta}$. Then we can define the correspondence Γ also for $\hat{\beta}$ in the same manner as (4). Then $\Gamma(\hat{\beta})$ is determined by $\mathfrak{b}$ up to a root of unity factor. We denote with $H_{\mathfrak{a}}$ the group of ideals of k^*, whose ideal numbers $\hat{\beta}$ satisfy

(6) $$\varepsilon'\Gamma(\hat{\beta}) \equiv \varepsilon \pmod{\mathfrak{a}},$$

where ε', ε are roots of unity and $\varepsilon \in E$. Then, we can prove

THEOREM 2. *If J is defined over $k^*(M)$, then $k^*(M, b)_0$ is the class field over k^* for the ideal group $H_{\mathfrak{a}}$.*

4. Now consider in general, an abelian variety A of dimension g, defined over an algebraic number field k, besides the jacobian variety of hitherto considered. $\mathfrak{P}$ being a prime ideal in k, we can reduce mod $\mathfrak{P}$, after Shimura, all A (or J), C, μ's, and obtain $\tilde{A}$ (or $\tilde{J}$), $\tilde{C}$, $\tilde{\mu}$'s. Then, for almost all $\mathfrak{P}$, $\tilde{A}$ is an abelian variety, $\tilde{J}$ is a jacobian variety of a non-singlar curve $\tilde{C}$ and $\tilde{\mu}$'s are endomorphisms of $\tilde{A}$ (or $\tilde{J}$). We call these $\mathfrak{P}$'s "non-exceptional".

Zeta-function $\zeta_C(s)$ for C is defined as follows:

(9) $$\zeta_C(s)=\prod_{\mathfrak{P}}{}' \zeta_{\mathfrak{P}}(s),$$

where $\zeta_{\mathfrak{P}}(s)$ denotes the congruence-zeta-function for C mod $\mathfrak{P}$ and Π' means the product of all non-exceptional $\mathfrak{P}$'s. Zeta-functions $\zeta_A(s)$ for A is defined in a similar manner. Now we can prove

THEOREM 3. *If $\mathcal{A}_0(A)$ (or $\mathcal{A}_0(J)$) contains a subfield of degree $2g$, then we have*

$$\zeta_C(s)=\Phi(s)\frac{\zeta_k(s)\zeta_k(s-1)}{\prod_{i=1}^{2g} L(s-1/2, \chi_i)}.$$

$$\zeta_A(s)=\Psi(s)\frac{\zeta_k(s)\zeta_k(s-g)\prod_{\nu=1}^{g}\Big(\prod_{i_1,\cdots,i_{2\nu}} L(s-g, \chi_{i_1}\cdots\chi_{i_{2\nu}})\Big)}{\prod_{\mu=0}^{g-1}\Big(\prod_{j_1,\cdots,j_{2\mu+1}} L(s-\mu-1/2, \chi_{j_1}\cdots\chi_{j_{2\mu+1}})\Big)}$$

where $\zeta_k(s)$ is the zeta-function for k, χ_i are "Grössencharaktere" of Hecke in k, $L(s, \chi_{i_1}\cdots\chi_{i_\lambda})$ are L-functions with characters $\chi_{i_1}\cdots\chi_{i_\lambda}$, and $\Phi(s)$, $\Psi(s)$ are meromorphic functions of s with simple forms.

($\mathcal{A}_0(A)$ means the constant extension of the algebra $\mathcal{A}(A)$ over the ring of integers to the field of rational numbers).

5. The proof is based on the following fact. Let x be a generic point of A over k. We denote with $\pi_{\mathfrak{P}}$ the endomorphism of A for which

$$\tilde{\pi}_{\mathfrak{P}}\tilde{x}=\tilde{x}^{N\mathfrak{P}}$$

holds, where $N\mathfrak{P}=q$ is the absolute norm of $\mathfrak{P}$, and $\tilde{x}$ a generic point of $\tilde{A}$ (mod $\mathfrak{P}$). Then we have the following ideal decomposition in k,

(10) $$(\pi_{\mathfrak{P}})=N_{k/k_1}(\mathfrak{P}^{\sigma_1^{-1}}\cdots\mathfrak{P}^{\sigma_g^{-1}}).$$

Especially, under the assumption of th. 1, th. 2, we have

(11) $$(\pi_{\mathfrak{P}})=\Gamma(\mathfrak{p})^f,$$

where $\mathfrak{p}^f=N_{k/k^*}\mathfrak{P}$,

(12) $$\pi_{\mathfrak{P}}=\varepsilon'\Gamma(\hat{\varpi})^f,$$

where ε' is a root of unity and $(\hat{\varpi})=\mathfrak{p}$.

Note. Theorems 1, 2 depend on the theory of moduli of algebraic curves. Prof. Weil has kindly remarked me that the existing theory of moduli is not on a sound foundation, but our results may be based on another ground.

*Jacobian Varieties and Number Fields**)

Mimeographed notes (the International Symposium on Algebraic Number Theory, Tokyo-Nikko, September, 1955)

Contents

(Note. The assumptions and the explanations of notations used after page 36 are made in pages 35-36.)

Introduction. The twelfth of the problems proposed by Hilbert in his famous lecture in Paris in 1900 raises a question how to construct the abelian extensions of algebraic number fields by means of special values of analytic functions of one or several variables in some significant connexion with the given field, pursuing the idea of the so-called "Jugendtraum" of Kronecker. This problem is solved by the classical theory of complex multiplication in case where the given field is an imaginary quadratic field. Besides, E. Hecke [1], [2] treated successfully this problem and constructed unramified abelian extensions of certain imaginary biquadratic fields by means of singular values of Hilbert modular functions.

No noteworthy progress has been recorded after Hecke's work, regarding this problem (not elliptic case), until in 1949 M. Deuring [2], [3] has given a purely algebraic foundation of the classical

*) This study was done with the help of subsidy from the ministry of education. (n° 10429)

theory of complex multiplication and suggested a possibility of generalizing his method to the case of ground fields of higher degree.

On the other hand, however, the method of algebraic geometry has been completely renovated in recent years, and it is now possible to interprete the result of Hecke from this point of view. The main purpose of this paper is to develope a theory comprising both classical theory of complex multiplication and the theory of Hecke, by the method of algebraic geometry.

We shall first explain our method in the case treated by Hecke. Let k_r be a real quadratic field, and k_1 an imaginary quadratic extension of k_r: $k_1=k_r(\sqrt{\varDelta})$. We assume that k_1 is not normal over the field $\boldsymbol{Q}$ of rational numbers, and denote with $\bar{k}_1$ the smallest absolutely normal field (of degree 8) containing k_1. We assume furthermore for simplicity's sake that the norm of a fundamental unit of k_r is $+1$. Then there exists a jacobian variety J of hyperelliptic curves C_1, C_2 of genus 2, such that the ring $\mathcal{A}(J)$ of endomorphisms of J is isomorphic to the principal order of k_1. The projective invariants of C_1, C_2 can be expressed by "Thetanullwerte" for arguments T_i where (I, T_i) are Riemann matrices of J. Using these invariants we define the algebraic modules (M) of J. Our curves C_1, C_2, hence also J and all the endomorphisms μ of J can be defined over an algebraic number field k, which can be assumed to be absolutely normal. Let x be a generic point of J over k, $\mathfrak{a}$ an ideal in k_1, and $[\mathfrak{a}]$ the ideal class in k_1 containing $\mathfrak{a}$. Then there is a homomorphism $\lambda_{\mathfrak{a}}$ of J onto an abelian variety $J_{\mathfrak{a}}$ such that $k(\lambda_{\mathfrak{a}}x)=\bigcup_{\alpha\in\mathfrak{a}} k(\alpha x)$. $J_{\mathfrak{a}}$ is also a jacobian variety if and only if $\mathfrak{a}\bar{\mathfrak{a}}=(\alpha)$ belongs to the principal class in k_r in the narrower sense. We denote with C the group of such classes $[\mathfrak{a}]$ in k_1, and with h' the order of C.

$\varDelta'$ being the conjugate of real quadratic number $\varDelta$, we denote with σ_2, σ_3 the elements of the Galois group G of $\bar{k}_1=k_r(\sqrt{\varDelta}, \sqrt{\varDelta'})$ defined by $\sqrt{\varDelta}^{\sigma_2}=\sqrt{\varDelta'}$, $\sqrt{\varDelta'}^{\sigma_2}=\sqrt{\varDelta}$ and $\sqrt{\varDelta}^{\sigma_3}=\sqrt{\varDelta'}$, $\sqrt{\varDelta'}^{\sigma_3}=\sqrt{\varDelta}$, respectively. We denote also with σ_1 the identity, with σ_0 the complex conjugate automorphism of $\bar{k}_1$, and with G_1 the subgroup of G corresponding to the field k_1. Then σ_0, σ_1, σ_2, σ_3 induce all the 4 isomorphisms of k_1.

Endomorphisms μ induce linear transformations of the space of invariant differentials of J, hence we can represent the ring $\mathcal{A}(J)$ with a basis $\{\omega_1, \omega_2\}$ of this space, by the matrices $S_0(\mu)$ with coefficients in k. Characteristic roots of $S_0(\mu)$ belong to $\bar{k}_1$, and are of the form $\mu_1^{\sigma_1}$, $\mu_1^{\sigma_i}$, $\mu_1\in k_1$, where i must be 2 or 3. We call $\{\sigma_1, \sigma_i\}$ the system of isomorphisms for J. There are just $2h'$ non-isomorphic jacobian varieties, with the ring of endomorphisms isomorphic to the principal order of k_1, and h' of them correspond to the system $\{\sigma_1, \sigma_2\}$, while the other h' correspond to $\{\sigma_1, \sigma_3\}$. These h' jacobians with the same system $\{\sigma_1, \sigma_i\}$ are obtained from one J of them in the form $\lambda_{\mathfrak{a}}J$, $[\mathfrak{a}]\in C$. We define the class equation for the principal order of k_1 by modules of these $2h'$ non-isomorphic jacobians.

Let σ be an automorphism of k, then σ permutes these $2h'$ modules among themselves, and if σ is an automorphism over the real quadratic subfield $k^{***}=\boldsymbol{Q}(\sqrt{\Delta\,\Delta'})$ of $\bar{k}_1$, it permutes h' modules having the same system $\{\sigma_1, \sigma_i\}$ among themselves. This shows that the class equation has rational coefficients and it decomposes in k^{***} into two factors, each of which is of degree h' and corresponds to system $\{\sigma_1, \sigma_i\}$. We call these factors the class equations for systems $\{\sigma_1, \sigma_i\}$.

We consider the class equation for $\{\sigma_1, \sigma_2\}$. Let G_0 be the subgroup of G, consisting of σ's such that $S_0(\mu)^\sigma$ has the same characteristic roots as $S_0(\mu)$, (we see clearly $G_0=\{\sigma_1, \sigma_2\}$,) and let k^* be the subfield of $\bar{k}_1$ corresponding to G_0. k^* is an imaginary biquadratic field containing k^{***}. Denote with $\mathcal{G}$ the Galois group of k over k^*. Then, J^σ is of the form $\lambda_{\mathfrak{a}}J$, and we have $\mu^\sigma\lambda_{\mathfrak{a}}=\lambda_{\mathfrak{a}}\mu$, as $S_0(\mu)^\sigma$ has the same characteristic roots as $S_0(\mu)$. Hence we have a homomorphism of $\mathcal{G}$ into C by $\sigma\to[\mathfrak{a}]$, the kernel of which consists clearly of such σ's that $(M)^\sigma=(M)$, where (M) are modules of J. This shows that the field $k^*(M)$ is abelian over k^*, whose Galois group is isomorphic to a subgroup of C. This is the main result of Hecke in [1].

The paper [2] of Hecke is concerned with the arithmetic nature of class equation (for $\{\sigma_1, \sigma_2\}$). $\mathfrak{P}$ being a prime ideal in k, we can reduce J mod. $\mathfrak{P}$, and obtain, for almost all $\mathfrak{P}$, a jacobian variety $\tilde{J}$ and $\mathcal{A}(J)$ is mapped isomorphically into $\mathcal{A}(\tilde{J})$ by this process of reduction: $\mu\to\tilde{\mu}$. $\tilde{J}$ and all $\tilde{\mu}$'s are defined over the finite field $\tilde{k}$ of $N\mathfrak{P}$ elementes, where N denotes always the absolute norm. Let $\pi_{\mathfrak{P}}$ be the endomorphism of J defined by $\tilde{\pi}_{\mathfrak{P}}\tilde{x}=\tilde{x}^{N\mathfrak{P}}$. Then the isomorphic image $\iota\pi_{\mathfrak{P}}$ of $\pi_{\mathfrak{P}}$ into k_1 has the ideal decomposition $(\iota\pi_{\mathfrak{P}})=N_{k/k_1}(\mathfrak{P}^{\sigma_1^{-1}}\mathfrak{P}^{\sigma_2^{-1}})$ in k_1. The left hand side can be written as $\Gamma(\mathfrak{p})^f$ where $\mathfrak{p}^f=N_{k/k^*}\mathfrak{P}$, and $\Gamma(\mathfrak{p})=N_{\bar{k}/k_1}\mathfrak{p}$ (the prime ideal $\mathfrak{p}$ of k^* being considered as an ideal in $\bar{k}_1$) as we have $G_1\sigma_1+G_1\sigma_2=G_1G_0$. Denote with $\sigma(\mathfrak{p})$ the Frobenius automorphism $\left(\dfrac{k^*(M)/k^*}{\mathfrak{p}}\right)$ of a prime ideal $\mathfrak{p}$ of the first degree in k^*, then we have $\tilde{\lambda}_{\Gamma(\mathfrak{p})}\tilde{J}=\tilde{J}^{N\mathfrak{p}}=\widetilde{J^{\sigma(\mathfrak{p})}}$, and also $\mathcal{A}(\tilde{J})=\widetilde{\mathcal{A}(J)}$. Hence $\sigma(\mathfrak{p})\to[\Gamma(\mathfrak{p})]$ by the above isomorphism of Galois group of $k^*(M)/k^*$ into C. This gives the decomposition law and Artin's reciprocity law for $k^*(M)/k^*$, from which we can conclude that $k^*(M)$ is the class field over k^*, for the ideal group $H=\{\mathfrak{b} \mid \Gamma(\mathfrak{b}) \sim 1 \text{ in } k_1\}$. If moreover the class number of k_1 is odd, then every class in C contains an ideal of the form $\Gamma(\mathfrak{p})$, as Hecke showed, hence $\sigma(\mathfrak{p}) \rightarrow [\Gamma(\mathfrak{p})]$ is onto, and $[k^*(M):k^*]=h'$, which means the irreducibility of class equation for $\{\sigma_1, o_2\}$.

These are the main results proved by Hecke, excluding however, no less interesting results about Hilbert modular functions.

But the effectiveness of our method of algebraic geometry becomes clearer, if we consider the field of "Teil-points" of J, which was not treated by Hecke. Indeed, under the assumption that we can find a J defined over $k^*(M)$, we can prove that the field of "Teil-points" of J is a class field over k^*, for a certain explicitly defined ideal group, following an idea similar to the one used by Deuring [2] in elliptic case.

In §6 of this paper, I shall give generalizations of these results of Hecke, under certain conditions on the ground fields. Then, in §7, I shall prove the result concerning the field of Teil-points. But these constructions of abelian extensions do not seem to be sufficient to obtain all the abelian extensions over k^*. The main reason for this is that the results are expressed, not with $\mathfrak{p}$ alone, but with the "mean value" $\Gamma(\mathfrak{p})$ of $\mathfrak{p}$. At any rate, we obtain rather wide range of abelian extensions over k^*, if J is defined over $k^*(M)$. But the problem of construction of J defined over $k^*(M)$, in general case, is not so easy, and we must leave it open.

The same method should be also applicable to abelian varieties, not only jacobians, as shall be seen in §§6, 7, if we could only define algebraic moduli of simple abelian varieties A of dimension g, whose ring $\mathcal{A}(A)$ contains a totally real field k_r of degree g. The word "algebraic" means that the definition of moduli must be compatible with specialization of abelian varieties. It is expected that these algebraic moduli may be defined as the values of Hilbert modular functions for the field k_r, belonging to a subfield of finite index. The situation of this subfield will depend on the structure of unit group of k_r.

From the above study, we can obtain an arithmetic characterization of endomorphisms $\pi_{\mathfrak{P}}$, hence of zero points of the congruence zeta function $\zeta_{\mathfrak{P}}(u)$ of the curve $\tilde{C}$, reduced mod. $\mathfrak{P}$ from C, of which J is a jacobian. Thus we are in a position to prove the Hasse conjecture for zeta function of our curve C, in some singular cases. On this subject, we have some known results in special cases. A. Weil [8] proved namely that the zeta function of a curve defined by

$$ax^n+by^m+c=0 \qquad (*)$$

over an algebraic number field k (containing some roots of unity), can be expressed by the zeta function of the field k and L-functions of Hecke with "Grössencharaktere". M. Deuring [4] proved analogous results for elliptic curves with complex multiplications.

The curve defined by (*) over a finite field was studied by H. Davenport and H. Hasse [1], who proved the Riemann hypothesis for this curve, and proposed for the first time the problem of the arithmetic characterization of $\pi_{\mathfrak{P}}$ for arbitrary curves. They expected that this characterization may be obtained in connection with the construction of class fields by means of "Teilwerte" of abelian functions, the possibility of which had been already seen in the first proof of Riemann hypothesis for elliptic curves, by H. Hasse [1].

In § 5 of this paper, I shall show that the Hasse conjecture is true for curves C having sufficiently many complex multiplications, and in Appendix I shall prove an analogous result for an abelian variety A of dimension g, whose ring $\mathcal{A}(A)$ contains an order of a field of degree $2g$. This results contain that of Deuring [1], and of Weil [8] in case n, m are different prime numbers. Our proof is based on class field theory, and shows that the above expectation of Davenport and Hasse was not wrong.

§§ 1, 2, 3 and 5 of this paper contain some lemmas and propositions, necessary in the proof of above main theorems. In § 1 we shall show that the reduction of jacobian or abelian varieties works well for almost all $\mathfrak{P}$, (Prop. 1). In § 1 we consider the abstract ground field, as this becomes necessary for the proof of Lemma 33, § 6. This problem of reduction of abelian varieties was treated also, in equal characteristic case, by Nèron [1] and Matsusaka [1] for different purposes. Then we shall define the reduction of differentials on non-singular varieties, and shall prove the compatibility of the reduction with the anti-representation of $\mathcal{A}(A)$ by means of a basis of invariant differentials on A (Prop. 2). The main purpose of § 2 is to study the properties of homomorphism $\lambda_{\mathfrak{a}}$. Then we shall prove that the centre of the algebra $\mathcal{A}_0(A)$ is either a totally real field or a totally imaginary quadratic extension of a totally real field, if $\mathcal{A}_0(A)$ is simple (Lemma 16). Using this lemma, we shall give a necessary condition for $\lambda_{\mathfrak{a}}J$ to be also a jacobian (Lemma 17). The sufficient condition for it (in case $g=2$ or 3), will be given later in § 5 (Lemma 30). The proof of these two lemmas gives more generally, for an abelian variety A having a positive divisor X with $\det|E(X)|=1$, a condition for the existence of a positive divisor Y on $\lambda_{\mathfrak{a}}A$ such that $\det|E(Y)|=1$. This formulation should become necessary if we could define the modulus of A. Lemma 19 in § 2 can be replaced by Lemmas 31, 32 in § 5, for our purpose to prove the finiteness of non equivalent curves having isomorphic jacobian varieties. In § 3, the ideal decomposition of $(\iota\pi_{\mathfrak{P}})$ will be given (Lemma 25 and Prop. 3). Then a sufficient condition for our assumptions in the main theorem (Th. 2, § 6) will be given (Lemma 26). Lemma 25 contains also an arithmetic proof of the property of the system $\{\sigma_1, \cdots, \sigma_g\}$ of isomorphisms for J, which is usually proved using the theory of Riemann matrix (cf. Lefschetz [1]).

But the theory of Riemann matrix becomes necessary in some essential steps. I could not prove, namely, Lemma 29 (the isogeny of g dimensional A having isomorphic R of degree $2g$ and the same system $\{\sigma_1, \cdots, \sigma_g\}$), without using Riemann matrices. This lemma and Lemmas 30, 31, 32 above stated, will be proved in § 5, using the theory of Riemann matrix. The proof of existence of abelian varieties A having a given subring R of degree $2g$ of $\mathcal{A}(A)$, and a given system $\{\sigma_1, \cdots, \sigma_g\}$, requires also this theory, at least at present. However, as the criterion for abelian varieties to be jacobians is not known in general, I did not quote this existence theorem, but assumed simply the existence of jacobian variety with given properties.

(As the field of definition of varieties considered in this paper is clear from the context, we sometimes omit to refer to it. Especially, we omit the reference of $\mathfrak{o}$ in the notation of specialization ring $[x \xrightarrow{\mathfrak{o}} y]$, used by Shimura [1]).

Note. As stated above, our theorems 2, 3 depend on the theory of moduli of algebraic curves. Prof. Weil has kindly remarked me, that the existing theory of moduli is not on a sound foundation, but our results may be based on another ground.

§ 1. Preliminaries from reduction theory

The algebraic varieties considered in this paper, are always supposed to be in some projective space S^N or in a product $S^N \times \cdots \times S^M$ of them. Except for this, we use mainly the terminology as in Weil's books [1], [2], [3], and we quote the last book as [VA] throughout the paper. In this §, we also use the notion and terminology in Shimura [1], which we quote as [S]. But we use somewhat different notation. Thus, if a variety V is defined over an algebraic number field k, V is a $\mathfrak{p}$-complete $\mathfrak{p}$-variety $V_{\mathfrak{p}}$ (cf. [S], p. 162), and for almost all $\mathfrak{p}$, $V_{\mathfrak{p}}$ is $\mathfrak{p}$-simple ([S], Theorem 26), where $\mathfrak{p}$ denotes a prime ideal in k (as $V \subset S^N$ or $\subset S^N \times \cdots \times S^M$). "Almost all" means "with a finite number of exceptions" throughout this paper.

First we quote the following

Lemma 1. Let V be a complete variety of dimension n without

$(n-1)$-dimensional singularities, defined over a field k, and let $\xi_1, \cdots, \xi_r$ be a set of functions on V defined over k such that the divisors $(\xi_i)_0$ of zeros of $\xi_i (i=1, \cdots, r)$ have no point in common, then we have an algebraic relation of the following type between monomials $m_j(\xi)$ of ξ_i's of degree ≥ 1,

$$1=\sum a_j m_j(\xi)$$

with a_j in k. (cf. Weil [5], Cor. 1 of Th. 2 and Th. 3).

Now let $\mathfrak{K}$ be the field of functions on V defined over k and let $\xi_{\mu, i}$ be a finite set of functions in $\mathfrak{K}$ such that $\bigcap_\mu (\xi_{\mu, i})_0=\phi$ for each i and $\bigcap_i (1/\xi_{\mu, i})_0=\phi$ for each μ, ϕ denoting an empty set. Then, denoting by $m_j^{(i)}(\xi)$ monomials of $\xi_{\mu, i}$ of degree $\geqq 1$ with fixed i, and $m_j^{(\mu)}(1/\xi)$ having the similar meanings, we have relations of the type

$$1=\sum a_{ji} m_j^{(i)}(\xi) \text{ and } 1=\sum a_{j\mu} m_j^{(\mu)}(1/\xi),$$

with a_{ji}, $a_{j\mu}$ in k. Let v be a non-archimedean valuation of $\mathfrak{K}$, not necessarily trivial on k, such as $v(a_{ji})=v(a_{j\mu})=0$ for all j, i, μ; then from these relations we have $\sup_i v(\xi_{\mu, i}) \prec 0$ for each μ and $\sup_\mu v(1/\xi_{\mu i}) \prec 0$ for each i, that is, $\inf_\mu v(\xi_{\mu, i}) \succ 0$, and hence $\inf_\mu \sup_i v(\xi_{\mu i})=0$.

Let $\mathfrak{V}$ be the set of normalized discrete valuations on k, and, with a finite number of elements a_i in k, we denote by $\mathfrak{V}(a)$ the set of all v such as $v(a_i)=0$ for each a_i, and by an open set in $\mathfrak{V}$ we understand an intersection of a finite number of such $\mathfrak{V}(a)$'s, and if no a_i is 0, we call the open set proper. Then we have clearly

LEMMA 2. Let $\xi_{\mu i}$ be as above, then there is a proper open set $\mathfrak{V}'$ in $\mathfrak{V}$ such that, for each non-archimedean valuation in $\mathfrak{K}$, coinciding on k with one in $\mathfrak{V}'$, we have

$$\inf_\mu \sup_i v(\xi_{\mu i})=0.$$

(This Lemma 2 is a special case of "principe de transport" of A. Weil.)

We note that the theorems of [S], $n^0 6$ hold for any field k if one replaces "almost all $\mathfrak{p}$" by "$\mathfrak{p}$ in some proper open set". Thus V is $\mathfrak{p}$-simple if $\mathfrak{p}$ is in some proper open set. Of course, it may happen that the empty set is also a proper open set, and then these theorems lose their meaning after such modification.

We now consider the reduction of V by a discrete valuation $\mathfrak{p}$ in k. We use the symbol $\sim$ throughout this paper to show the object derived from one of the same kind by reduction mod. $\mathfrak{p}$. For example, let x be a point, φ a function, f a mapping on V, then, $\tilde{V}$ is a variety, and $\tilde{x}$ a point, $\tilde{\varphi}$ a function, $\tilde{f}$ a mapping on $\tilde{V}$ derived from V, x, φ, f respectively by reduction mod. $\mathfrak{p}$. Of course, we do not always obtain a variety from a variety by reduction mod. $\mathfrak{p}$, but we restrict ourselves throughout in this paper to "regular" cases, where the variety thus obtained remains variety.

LEMMA 3. Let V^n be a non-singular variety in a projective space S^N. Then, there is a proper open set of $\mathfrak{B}$ such that, for any $\mathfrak{p}$ in that set $\tilde{V}$ is a non-singular variety, and so if L is the tangent linear space to V at any point y, $\tilde{L}$ is the tangent linear space to V at $\tilde{y}$.

Proof. First we note that for any $\mathfrak{p}$ in some proper open set, we can use the symbol $\tilde{V}$. By assumption, there are a finite number of homogeneous polynomials $F_i(X)$ of degree e_i in $k[X_0,\cdots,X_N]$ such that, for each point $y=(y_j)$ on V, $F_i(y)=0$ for all i and the rank of matrix $\left(\frac{\partial F_i}{\partial X_j}(y)\right)$ is $N-n$; then L can be defined by the set of equations $\sum_j \frac{\partial F_i}{\partial X_j}(y)X_j=0$ for all i. Let $\{G_\mu\}$ be the set of all the sub-determinants of $(\partial F/\partial X)$ of degree $N-n$, each being homogeneous of degree d_μ, then $\xi_{\mu i}=G_\mu(X)/X_i^{d_\mu}$ are functions on V and clearly satisfy the condition of Lemma 2. For any index k such that $\tilde{y}$ has a representative in the affine space defined by $\tilde{X}_k=1$, and for $\mathfrak{p}$ in some proper open set, all functions $\frac{\partial F_i}{\partial X_j}X_k^{1-e_k}$ belong to the specialization ring $[x\to\tilde{y}]$, x being a generic point of V over k, hence we see by Lemma 2, that the rank of $(\partial F_i/\partial X_j)$ at $\tilde{y}$ is $N-n$ for all $\tilde{y}$ on $\tilde{V}$, for $\mathfrak{p}$ in some proper open set. This shows especially, that the tangent linear space to V at $\tilde{y}$ is defined by $\frac{\partial F_i}{\partial X_j}(\tilde{y})X_j=0$, and this proves the lemma.

Considering affine representatives, we can prove this lemma also for abstract varieties. But, for our purpose it is sufficient to prove

COROLLARY. Let V be a non-singular variety in a product $S^N\times\cdots\times S^M$ of projective spaces. Then the same conclusions hold for V as in the above lemma. (Tangent linear spaces mean in this case those in affine representatives.)

Proof. We can imbed V into a projective space of a higher dimension, everywhere biregularly. This imbedding remains everywhere biregular mod. $\mathfrak{p}$ for all $\mathfrak{p}$. Then we have only to apply our lemma.

LEMMA 4. Let V_1, V_2 be two non-singular varieties and f a mapping of V_1 into V_2, everywhere defined on V_1, then, for $\mathfrak{p}$ in some proper open set, there exists a mapping $\tilde{f}$ of $\tilde{V}_1$ into $\tilde{V}_2$, everywhere defined on $\tilde{V}_1$ such that $\widetilde{f(y)}=\tilde{f}(\tilde{y})$ for any point y on V_1. And the graph of $\tilde{f}$ is obtained from that of f by reduction mod. $\mathfrak{p}$.

Proof. Let S^N, S^M be ambient projective spaces of V_1, V_2 respectively, and let x be a generic point of V_1, over k, and Z the graph of f in $V_1 \times V_2$. Then, for each point y on V_1, the product $Z\cdot(y\times V_2)$ is defined and equal to $y\times f(y)$ by assumption, $f(y)$ being a uniquely determined specialization of $f(x)$ over $x\rightarrow y$. Denoting by L the tangent linear space to V_1 at y, this shows especially that Z and $L\times S^M$ are transversal at $y\times f(y)$ (cf. Weil [1], Prop. 7, VI, § 3, and Prop. 20, V, § 3.)

Restricting $\mathfrak{p}$ to some proper open set, we can assume $\tilde{V}_1, \tilde{V}_2, \tilde{Z}$ to be non-singular and $[\tilde{Z}:\tilde{V}_1]=1$, then we imbed $S^N\times S^M$ in one projective space as above, and denote by L', Y the tangent linear spaces to $V_1\times S^M$ and to Z at $y\times f(y)$ respectively, imbedded in that space. Then we see by the above Lemma 3 that $\tilde{L}'$, $\tilde{Y}$ are tangent linear spaces to $V_1\times S^M, Z$ at $\tilde{y}\times\widetilde{f(y)}$, for $\mathfrak{p}$ in a narrower proper open set. As the transversality are defined in language of tangent linear spaces, the same discussion as in the proof of Lemma 3 shows that, for $\mathfrak{p}$ in still narrower proper open set, Z and L' are transversal at $\tilde{y}\times\widetilde{f(y)}$ for all $\tilde{y}$ on $\tilde{V}_1$. Returning to the original varieties in $S^N\times S^M$, we see that $\tilde{Z}\cdot(\tilde{y}\times\tilde{V}_2)$ are defined and the projection from $\tilde{Z}$ to $\tilde{V}_1$ is regular at $\tilde{y}$, and hence a mapping $\tilde{f}$ is defined by $\tilde{Z}$, which is everywhere defined on $\tilde{V}_1$ and clearly $\tilde{y}\times\tilde{f}(\tilde{y})=\tilde{Z}\cdot(\tilde{y}\times\tilde{V}_2)=\tilde{y}\times\widetilde{f(y)}$ by [S], Th. 17.

From this lemma we have directly the following

COROLLARY. Let V_1, V_2, V_3 be non-singular varieties and let f, g be mappings of V_1 into V_2, V_2 into V_3, everywhere defined on V_1, V_2 respectively. Then, for any $\mathfrak{p}$ such that $\tilde{V}_1, \tilde{V}_2, \tilde{V}_3$ are non-singular varieties and $\tilde{f}, \tilde{g}$ are everywhere defined mappings on $\tilde{V}_1, \tilde{V}_2$, the mapping $\widetilde{g\circ f}$ is defined and equal to $\tilde{g}\circ\tilde{f}$.

Let now C be a non-singular curve of genus g, J its jacobian variety, and φ a canonical mapping of C into J, all defined over k. Then for $\mathfrak{p}$ in some proper open set, we have non-singular $\tilde{C}, \tilde{J}$, and $\tilde{\varphi}$ is a mapping of $\tilde{C}$ into $\tilde{J}$, everywhere defined on $\tilde{C}$, and we have $\widetilde{\varphi(P)}=\tilde{\varphi}(\tilde{P})$ for any point P on C. Let further ψ be the composition law on J, that is, a mapping of $J\times J$ onto J such that $\psi(a\times b)=a+b$, and let ρ be the reflexion on J: $\rho(a)=-a$. Then, for $\mathfrak{p}$ in a narrower proper open set, $\tilde{\psi}$ and $\tilde{\rho}$ are defined and satisfy the condition $\tilde{\psi}(\tilde{a}\times\tilde{b})=\widetilde{a+b}$, $\tilde{\rho}(\tilde{c})=\widetilde{(-c)}$, and we have for such $\mathfrak{p}$, by the above corollary, $\tilde{\psi}(\tilde{a},\tilde{\psi}(\tilde{b},\tilde{c}))=\widetilde{a+b+c}=\tilde{\psi}(\tilde{\psi}(\tilde{a},\tilde{b}),\tilde{c})$, $\tilde{\psi}(\tilde{a},\tilde{e})=\tilde{\psi}(\tilde{e},\tilde{a})=\tilde{a}$, $\tilde{\psi}(\tilde{a},\tilde{\rho}(\tilde{a}))=\tilde{e}$, e being the unity element on J, and $\tilde{\psi}(\tilde{a},\tilde{b})=\widetilde{a+b}=\tilde{\psi}(\tilde{b},\tilde{a})$, thus we see that $\tilde{J}$ is an abelian variety with addition defined by

$\tilde{a}+\tilde{b}=\widetilde{a+b}$, and $\tilde{e}$ is the unity element. Moreover, let C^g be a product of g copies of C, Φ the mapping of C^g onto J defined by $\Phi(P_1\times\cdots\times P_g)=\sum\varphi(P_i)$. Then, for generic $P_1\times\cdots\times P_g$ over k, $x=\Phi(P_1\times\cdots\times P_g)$ is generic on J over k. Φ is invariant by the permutation of factors of C^g and we have $\Phi(C^g)=g!J$. Hence for $\mathfrak{p}$ in a still narrower proper open set, $\tilde{\Phi}$ satisfies $\tilde{\Phi}(\tilde{P}_1\times\cdots\times\tilde{P}_g)=\sum\tilde{\varphi}(\tilde{P}_i)$ for any point on $\tilde{C}^g$, and $\tilde{\Phi}$ is invariant by the permutation of factors of $\tilde{C}^g$. Moreover, we have $\tilde{\Phi}(\tilde{C}_g)=g!\ \tilde{J}$ by [S], Th. 19 and by our Lemma 4. Hence for generic $\tilde{x}=\tilde{\Phi}(\tilde{P}_1\times\cdots\times\tilde{P}_g)$, we have $\tilde{k}(\tilde{P}_1,\cdots,\tilde{P}_g)_s=\tilde{k}(\tilde{x})$. By Néron [1], Lemma 15, chap. 1, then, $\tilde{C}$ is of genus g, $\tilde{J}$ is its jacobian variety and $\tilde{\varphi}$ is a cannonical mapping of $\tilde{C}$ into $\tilde{J}$.

We consider now the behavior of the ring $\mathcal{A}(J)$ of endomorphisms of J by the reduction mod. $\mathfrak{p}$.

Lemma 5. Let A_1, A_2 be abelian varieties and let λ, μ be homomorphisms of A_1 into A_2. Then, for such $\mathfrak{p}$ that $\tilde{A}_1, \tilde{A}_2$ are abelian varieties and that $\tilde{\lambda}, \tilde{\mu}$ are homomorphisms of $\tilde{A}_1$ into $\tilde{A}_2$, $\widetilde{\lambda+\mu}$ is defined and is a homomorphism of $\tilde{A}_1$ into $\tilde{A}_2$ which is equal to $\tilde{\lambda}+\tilde{\mu}$.

Proof. We first note that if $\tilde{\lambda}$ is defined as in Lemma 4, it is a homomorphism because $\tilde{\lambda}(\tilde{e}_1)=\widetilde{\lambda(e_1)}=\tilde{e}_2$. Let ρ be a homomorphism of A_1 into $A_2\times A_2$ defined by $\rho(a)=\lambda a\times\mu a$, and let ψ be the law of addition on A_2. Then $\tilde{\rho}$ can be defined and $\tilde{\rho}(\tilde{a})=\tilde{\lambda}\tilde{a}\times\tilde{\mu}\tilde{a}$ holds for $\mathfrak{p}$ satisfying the condition in the lemma, and by the corollary of Lemm 4, $\widetilde{\psi\circ\rho}$ can be defined and is equal to $\tilde{\psi}\circ\tilde{\rho}$, that is, $(\tilde{\lambda}+\tilde{\mu})(\tilde{a})=\tilde{\psi}\circ\tilde{\rho}(\tilde{a})=\widetilde{\psi\circ\rho}(\tilde{a})=\widetilde{(\lambda+\mu)(a)}$, which proves our lemma.

Let then $\mu_1,\cdots,\mu_r$ be a system of generators of the ring $\mathcal{A}(A)$, of endomorphisms of A, so the above lemma and the corollary of Lemma 4 show that for thoes $\mathfrak{p}$, $\tilde{A}$ is an abelian variety and $\tilde{\mu}_1,\cdots,\tilde{\mu}_r$ are endomorphisms of it, $\tilde{\rho}$ is also an endomorphism of $\tilde{A}$ for any ρ in $\mathcal{A}(A)$, and we have clearly a homomorphism of the ring $\mathcal{A}(A)$ into $\mathcal{A}(\tilde{A})$. Then, by [S], Th. 19 and our Lemma 4, $\nu(\tilde{\rho})=\nu(\rho)$, where $\nu(\rho)$ is the index $[k(x):k(\rho x)]$ if this is finite, where x is a generic point of A over k, and $\nu(\rho)=0$ if this index is not finite. This shows especially $\tilde{\rho}\neq 0$ if $\rho\neq 0$, and the homomorphism $\rho\to\tilde{\rho}$ is in fact an isomorphism.

We should remark here another fact. Let δ_A be the identity automorphism of A, and y a point on A such that $n\delta_A y=0$ for some natural number n prime to the characteristic of A, and let furthermore Δ_n be the graph of $n\delta_A$ on $A\times A$. Then, if we denote by $\mathfrak{g}_n(A)$

the 0-cycle of all the points y on A such that $n\delta_A y=0$, we have $\mathfrak{g}_n=\mathrm{pr}_1(\Delta_n\cdot(A\times e))$ and hence, for $\mathfrak{p}$ considered above, $\widetilde{\mathfrak{g}_n(A)}=\mathrm{pr}_1(\tilde{\Delta}_n\cdot(\tilde{A}\times\tilde{e}))$ by [S], Th. 17, 19, and by our Lemma 4, $\tilde{\Delta}_n$ is the graph of $\widetilde{n\delta_A}=n\delta_{\tilde{A}}$, so we see that $\widetilde{\mathfrak{g}_n(A)}=\mathfrak{g}_n(\tilde{A})$ if n is prime to the characteristic of $\tilde{A}$. As $\mathfrak{g}_n(A)$ and $\mathfrak{g}_n(\tilde{A})$ have the same degree, namely n^{2g} by [VA], Cor. 1 of Th. 33, §IX, and as deg $\mathfrak{g}_n(A)=$deg $\tilde{\mathfrak{g}}_n(A)$, we see that if y is in $\mathfrak{g}_n(A)$ and $y\neq e$ we have $\tilde{y}\neq\tilde{e}$. Thus we have arrived at the first main result of this §:

PROPOSITION 1. Let A be an abelian variety, and $\mathcal{A}(A)$ the ring of endomorphisms of A, both defined over a field k. Then, there is a proper open set in $\mathfrak{P}$ such that, for any $\mathfrak{p}$ in that set, $\tilde{A}$ is also an abelian variety, with addition and unity element derived from those of A by the reduction mod. $\mathfrak{p}$, and furthermore $\widetilde{\mathcal{A}(A)}$ is isomorphically imbedded in the ring $\mathcal{A}(\tilde{A})$. For any natural number n prime to the characteristic of $\tilde{A}$, and for any point $y\neq e$ on A such that $n\delta_A y=e$, we have $n\delta_{\tilde{A}}\tilde{y}=\tilde{e}$ and $\tilde{y}\neq\tilde{e}$. Moreover, if A is a jacobian variety of a non-singular curve C of genus g with the canonical mapping φ, both defined over k, then, for any $\mathfrak{p}$ in a narrower proper open set, $\tilde{A}$ is a jacobian variety of the non-singular curve $\tilde{C}$ with the same genus g, and $\tilde{\varphi}$ is a canonical mapping of $\tilde{C}$ into $\tilde{A}$.

We now study the invariant differentials on A, as are defined and investigated by Tamagawa [1]. As this paper is written in Japanese, we quote here his main theorems rather in detail, restricting ourselves to non-singular projective varieties.

The differential forms are defined as follows (cf. Chevalley [1]). Let V^n be a non-singular variety defined over k in a projective space S^N and let x be a generic point of V over k. We call local parameters on V at a point y, a set of n functions $\tau_1,\cdots,\tau_n$ in the specialization ring $[x\to y]$ at y such that, for affine representative $\boldsymbol{A}^N$ containing y, there exist N polynomials $F_1,\cdots,F_N$ in $k[X_1,\cdots,X_N,T_1,\cdots,T_n]$ with the properties, $F_i(x,t)=0$, $\det|\partial F_i/\partial X_j|(y,u)\neq 0$ where $t_i=\tau_i(x)$, $u_i=\tau_i(y)$. Then, the maximal ideal $\mathfrak{P}(y)$ in the regular local ring $[x\to y]$ is of the form $\mathfrak{P}(y)=(\tau_1-u_1,\cdots,\tau_n-u_n)$. Denoting by $\mathfrak{K}$ the field of functions defined over k on V, $\mathfrak{K}$ is separably algebraic over $k(\tau_1,\cdots,\tau_n)$ and one can extend derivations $\partial/\partial\tau_i$ to $\mathfrak{K}$, and if ξ is in $[x\to y]$, clearly $\partial\xi/\partial\tau_i$ is also in $[x\to y]$. Thus ξ can be written uniquely in the form $\xi=\xi(y)+\sum b_i(\tau_i-u_i)+\gamma$, where γ is in $\mathfrak{P}(y)^2$, and $b_i=\dfrac{\partial\xi}{\partial\tau_i}(y)$.

By a tangent vector at y on V, we understand as usual an Ω-linear mapping L of $[x\to y]$ into the universal domain Ω such that

$$L(\varphi\psi)=\varphi(y)L(\psi)+\psi(y)L(\varphi).$$

Such tangent vectors form a linear space $\mathfrak{x}(y)$ over Ω, and for $\xi=\xi(y)+\sum b_i(\tau_i-u_i)+\gamma$, we have

$$L(\xi)=\sum b_i L(\tau_i)=\sum\frac{\partial\xi}{\partial\tau_i}(y)L(\tau_i),$$

hence each L is uniquely determined by the set $(L(\tau_i))$ and $\mathfrak{x}(y)$ has a dimension n over Ω. We denote by $\mathfrak{d}(y)$ the dual space of $\mathfrak{x}(y)$ and call the element of $\mathfrak{d}(y)$ a covariant vector at y on V. Further, by the vector field X on V we understand a mapping $y\to X(y)\in\mathfrak{x}(y)$, defined on some non-void open set of V, in the sense of Zariski topology. We call a vector field X algebraic if for each function φ in $\mathfrak{K}$, there exists a function ψ on V such that $\psi(y)=X(y)(\varphi)$ in some non-void open set on V, in which case ψ is denoted as $X(\varphi)$. If these ψ are all in $\mathfrak{K}$, X is called rational (or defined) over k. If X is rational over k, $\varphi\to X(\varphi)$ is clearly a derivation in $\mathfrak{K}$. Especially, for local parameters $\tau_1,\cdots,\tau_n$ at y, there exist algebraic vector fields $X_1,\cdots,X_n$ such that $X_i(y)(\tau_j)=\delta_{ij}$ $(=0$ if $i\neq j$, $=1$ if $i=j)$ in some non-void open set. They are rational over k and induce derivations $\partial/\partial\tau_1,\cdots,\partial/\partial\tau_n$ respectively, in $\mathfrak{K}$. We identify these X_i with $\partial/\partial\tau_i$. Now, if two algebraic vector fields coincide in some non-void open set, we consider them to be in the same class and by this equivalence relation we can classify all the algebraic vector fields. In each class lies an X with a maximal domain of definition, and we shall identify all vector fields in that class with this X. Then we call singularities of X all the points not belonging to this maximal domain. Denoting $\varphi_i=X(\tau_i)$ for an algebraic X, we have $X=\sum\varphi_i\dfrac{\partial}{\partial\tau_i}$, where $\tau_1,\cdots,\tau_n$ are local parameters at y. If here φ_i are all in $[x\to y]$, then X is defined at y, and $X(y)(\tau_i)=X(\tau_i)(y)=\varphi_i(y)$. Clearly, X is rational over k if and only if all the φ_i are rational over k.

We define the linear differential ω on V as a mapping $y\to\omega(y)\in\mathfrak{d}(y)$ defined on some non-void open set of V, and we call ω algebraic, if for each algebraic vector field X, there exists a function ψ on V such that $\langle X(y),\omega(y)\rangle=\psi(y)$, where the symbol $\langle\cdots\rangle$ denotes the inner product. ω is called rational (or defined) over k, if here ψ is in $\mathfrak{K}$, for every vector field X over k. We classify algebraic ω's as in the case of vector fields and identify then those ω's with the one having the maximal domain of definition in the same class, and we call singularities of ω those points not belonging to this maximal domain. Especially, for each function ξ in $\mathfrak{K}$, we can define a linear differential $d\xi$ by $L\to L(\xi)$ for each tangent vector L at y, and for each point y at which ξ is defined. For an algebraic vector field $X=\sum\varphi_i\frac{\partial}{\partial\tau_i}$, rational over k, we have clearly $\langle X(y), d\xi(y)\rangle$ $=\sum\varphi_i\frac{\partial\xi}{\partial\tau_i}(y)$ and $d\xi$ is algebraic, and rational over k. Let $\tau_1,\cdots,\tau_n$ be local parameters on some non-void open set, then, for each algebraic ω, we have $\omega(y)=\sum\varphi_i(y)d\tau_i(y)$ with φ_i in $\mathfrak{K}$, on the intersection of that open set and domain of definition of ω, and $\varphi_i(y)=\langle\omega(y), \frac{\partial}{\partial\tau_i}(y)\rangle$ on that intersection. Conversely, $\sum\varphi_i(y)d\tau_i(y)$ with φ_i in $\mathfrak{K}$ is an algebraic linear differential, rational over k.

We consider now the transformation of linear differentials. Let V^n, U^m be non-singular varieties and let f be a mapping of V into U, all defined over k, and let x, z be generic points over k of V, U respectively. Let ψ be a function in the specialization ring $[z\to f(y)]$, then $\varphi=\psi\circ f$ is in $[x\to a]$ and $\psi\to\varphi$ gives a homomorphism $[z\to f(y)]$ onto $[x\to y]$ for each y at which f is defined. Then, for a tangent vector L on V at y, we can define one on U at $f(y)$ by $M(\psi)=L(\psi\circ f)$, and $L\to M=df(L)$ gives a linear mapping df of $\mathfrak{x}(y, V)$ into $\mathfrak{x}(f(y), U)$. We shall denote by δf the dual linear mapping of df; so that $\langle\delta f(Z), L\rangle=\langle Z, df(L)\rangle$ for each Z in $\mathfrak{d}(f(y), U)$. Then δf induces a linear mapping of the space of linear differentials on U into that on V, which we denote also by δf. If η is an algebraic differential on $U, \delta f(\eta)$ is also algebraic, because for a function ψ on U, and for an algebraic vector field X on V, we have $\langle dfX(f(y)), d\psi(f(y))\rangle=(dfX)(\psi)(f(y))=X(\varphi)(y)=\langle X(y), d\varphi(y)\rangle$ where $\varphi=\psi\circ f$ and y is in some non-void open set on V, hence $\delta f(d\psi)=d\varphi$ is algebraic. More generally for $\eta=\sum\psi_i d\tau_i$, we have $\delta f(\eta)=\sum(\psi_i\circ f)d\sigma_i$ with $\sigma_i=\tau_i\circ f$ for $\eta=\sum\psi_i d\tau_i$.

We call a differential η on U to be of the first kind if η is everywhere defined on U. Then, if the mapping f is everywhere defined on $V, \delta f(\eta)=\omega$ is of the first kind on V. In fact, denoting by $\tau_1,\cdots,\tau_m$ local parameters at $f(y)$ on $U, \sigma_i=\tau_i\circ f$ are clearly in $[x\to y]$. Therefore $\delta f(d\tau_i)=d\sigma_i$ is defined at y. As η is defined at $f(y)$, we have $\eta=\sum\psi_i d\tau_i$ with ψ_i in $[z\to f(y)]$. Therefore $\psi_i\circ f$ is in $[x\to y]$ and $\delta f(\eta)$ is defined at y.

We can then define invariant differentials on group varieties, but we restrict ourselves here to abelian varieties. Let therefore A^n be an abelian variety defined over k and denote by T_a the translation on A by a point a. Then, for any tangent vector L at the unity element e on $A, dT_a(L)=L_a$ is a tangent vector at a, and the vector field X_L defined by $a\to L_a$ is invariant by any T_b and is algebraic. Indeed we have $X_L(a)\tau_i=dT_{a-b}X_L(b)\tau_i=X_L(b)(\tau_i\circ T_{a-b})$ in a non-void open set $\mathfrak{o}$ of A, at each point of which $\tau_1,\cdots,\tau_n$ are local parameters,

Now let us fix b and consider $\tau_i\circ T_{a-b}(y)=\tau_i(a-b+y)$ as a function $\rho_i(a, y)$ on $A\times A$. We can choose $\tau_1(a),\cdots\tau_n(a), \tau_1(y),\cdots, \tau_n(y)$ as local parameters on $A\times A$, and define $\partial\rho_i/\partial\tau_j(a), \partial\rho_i/\partial\tau_j(y)$. Then we have $X_L(b)(\tau_i\circ T_{a-b})=\sum_j\frac{\partial\rho_i}{\partial\tau_j(y)}(a, b)X_L(b)(\tau_j)$, where the $\frac{\partial\rho_i}{\partial\tau_j(y)}(a, b)$ are functions on A and the $X_L(b)(\tau_j)$ are constants, which shows our assertion. Conversely, for any invariant vector field X, we have $X=X_L$, where $L=X(e)$. Therefore invariant vector fields form a vector space of dimension n over k. Similarly, let ω_e be a covariant vector at e on A, and $\omega_a=\delta T_{-a}\omega_e$. Then we have a linear differential ω: $a\to\omega_a$, clearly invariant by T_a and ω is algebraic. Indeed let $X_1,\cdots, X_n$ be a basis of invariant vector fields. Then we have $\langle\omega(a), X_j(a)\rangle=\sum_i\varphi_i(a)X_j(\tau_i)(a)$ in the above open set $\mathfrak{o}$, as $\omega(a)$ is of the form $\sum\varphi_i(a)(d\tau_i)(a)$ with abstract functions φ_i. Now as ω and X_j are invariant, we have $\langle\omega(a), X_j(a)\rangle=\langle\omega(e), X_j(e)\rangle=$constant, and further $X_j\tau_i$ are functions on A as X_j are algebraic. We have moreover det $|X_j\tau_i(a)|\neq 0$ by the above choice of X and τ, so we see that the φ_i are functions on A. It is also clear that for every invariant ω one has $\omega_a=\delta T_{-a}\omega_e$ so that invariant ω form a linear space of dimension n over k. Obviously such ω are everywhere defined on A.

Let $\omega_1,\cdots,\omega_n$ be a basis of invariant linear differentials on A, then, for each point $a, \omega_1(a),\cdots,\omega_n(a)$ are linearly independent over k, hence $\omega_1,\cdots,\omega_n$ are linearly independent over the function field $\mathfrak{K}$, and therefore form a basis of all the algebraic linear differentials on A. So every algebraic linear differential η can be written in the form $\eta=\sum\varphi_i\omega_i$, with ω_i in $\mathfrak{K}$. But then, as ω_i are everywhere defined on A, η is of the first kind if and only if every φ_i is everywhere defined on A, that is, φ_i is a constant. So we see that η is of the first kind if and only if η is invariant.

Let now A^n, B^m, C^l be three abelian varieties, anb σ, λ be homomorphisms of A into B, B into C respectively, then, for invariant ω on $C, \delta\lambda(\omega)$ is also invariant on B, and $\delta(\lambda\circ\sigma)(\omega)=\delta\sigma\circ\delta\lambda(\omega)$. But we have also $\delta(\sigma+\tau)(\omega)=\delta\sigma(\omega)+\delta\tau(\omega)$ for two homomorphisms of A into B. For, let μ, ρ, ν be homomorphisms: $A\xrightarrow[\mu]{}A\times A\xrightarrow[\rho]{}B\times B\xrightarrow[\nu]{}B$, given by $\mu(a)=a\times a$, $\rho(a\times b)=\sigma(a)\times\tau(b)$, $\nu(c\times d)=c+d$. Let further σ_1, τ_2 be homomorphisms of $A\times A$ into $B\times B$, given by $\sigma_1(a\times b)=\sigma(a)\times e$, $\tau_2(a\times b)=e\times\tau(b)$, then, we have $\sigma+\tau=\nu\circ\rho\circ\mu$, $\sigma=\nu\circ\sigma_1\circ\mu$, $\tau=\nu\circ\tau_2\circ\mu$ and hence we have only to prove $\delta\mu\circ\delta\rho\circ\delta\nu=\delta\mu\circ(\delta\sigma_1+\delta\tau_2)\circ\delta\nu$. But every invariant differential ω on $B\times B$ is a sum $\omega_1+\omega_2$ of ω_1, ω_2, each of which is an invariant differential on each factor B. We have evidently $\delta\sigma_1(\omega)=\delta\sigma(\omega_1)$, $\delta\tau_2(\omega)=\delta\tau(\omega_2)$ and $\delta\rho(\omega_1)=\delta\sigma(\omega_1)$, $\delta\rho(\omega_2)=\delta\tau(\omega_2)$. Especially $\delta\nu(\eta)$ is invariant on $B\times B$, if η is invariant on B. So we have $\delta\rho\circ\delta\nu(\eta)=(\delta\sigma_1+\delta\sigma_2)\circ\delta\nu(\eta)$, proving our assertion. Thus we can represent the module $\mathcal{H}(A, B)$ of homomorphisms of A into B by means of the invariant linear differentials on A and on B. Namely, let $\eta_1,\cdots,\eta_m$, and $\omega_1,\cdots,\omega_n$ be basis of the linear spaces of invariant differentials on B and on A respectively. So we can write $\delta\sigma(\eta_i)=\sum S_{ij}\omega_j$ with constant S_{ij}. Then, $\sigma\to S(\sigma)=(S_{ij})$ is clearly a representation of $\mathcal{H}(A, B)$. Especially, we have an anti-representation of the ring $\mathcal{A}(A)$ of endomorphisms on A in this manner, which is clearly faithful if the characteristic is 0.

Something more holds in case of jacobians. Let C be a non-singular curve of genus g and J a jacobian variety of C, φ a canonical mapping of C into J, all defined over k, and let C^g be the product of g copies of C, Φ the extension of φ onto C^g and $P_1\times\cdots\times P_g$ a generic point of C^g over k. Then $x=\Phi(P_1\times\cdots\times P_g)$ is generic on J over k. Denote further by $\omega_1,\cdots,\omega_g$ a basis of differentials of the first kind on C, and let $\omega_i^{(k)}$ mean ω_i on the k-th factor C in C^g. Then $\omega_i^{(k)}(i, k=1,\cdots, g)$ form a basis of differentials of the first kind on C^g. Let then η be an invariant linear differential on J, then $\delta\Phi(\eta)=\zeta'$ is of the first kind on C^g as Φ is everywhere defined on C^g. Hence we have $\zeta'=\sum\alpha_{ij}\omega_i^{(j)}$ with constant α_{ij}. But as the graph of Φ is invariant by permutations of factors of C^g, ζ' must be of the form $\sum\alpha_i(\eta_i^{(1)}+\cdots+\eta_i^{(g)})$, hence $\eta\to\zeta=\sum\alpha_i\eta_i$ is a linear mapping of the space of the invariant differentials on J into that of the differentials of the first kind on C. This mapping is one-to-one. Indeed, denote by $\tau_1,\cdots,\tau_g$ local parameters on J at a generic point x. Then $\sigma_1=\tau_1\circ\Phi,\cdots,\sigma_g=\tau_g\circ\Phi$ are also local parameters at $P_1\times\cdots\times P_g$ on C^g, as $C^g\times x$ and the graph of Φ are transversal at $(P_1\times\cdots\times P_g)\times x$. Hence $\delta\Phi(\omega)=\sum(\varphi_i\circ\Phi)d\sigma_i$ can not be zero, unless $\omega=\sum\varphi_i d\tau_i$ is zero. Especially the mapping $\eta\to\zeta$ is onto as both spaces have the same dimension.

In the following the same scheme as the above quoted theory of Tamagawa, we can now define and study the reduction of differentials mod. $\mathfrak{p}$. Let V^n be non-singular. We assume in the rest of this §1 that $\tilde{V}$ is always a non-singular variety, and that x is a generic point of V over k. Then we call local parameters on V at a point $\tilde{y}$ on $\tilde{V}$ a set of n functions $\tau_1,\cdots,\tau_n$ on V, belonging to the specialization ring $[x\to\tilde{y}]$ and, such that $\det\left|\dfrac{\partial F_i}{\partial X_j}\right|(y, u)\neq 0$, where $\tilde{u}_i=\tau_i(\tilde{y})$ and F_i have similar meaning as in the above definition of local parameters in the usual case. Then, $\tilde{\tau}_1,\cdots,\tilde{\tau}_n$ are clearly local parameters at $\tilde{y}$ on $\tilde{V}$. We first prove the following

LEMMA 6. Let V^n be a non-singular variety in a projective space

S^N and we assume $\tilde{V}$ to be non-singular. Then, there exist a finite set of systems of n functions $(\tau^{(i)})=(\tau_1^{(i)},\cdots,\tau_n^{(i)})\,(i=1,\cdots,s)$ on V such that for each point $\tilde{y}$ on $\tilde{V}$, at least one of these $(\tau^{(i)})$ is a system of local parameters on V at $\tilde{y}$.

Proof. By assumption, there is a finite number of polynomials G_i in $k[X_0,\cdots,X_N]$ such that the rank of $(\partial G_i/\partial X_j)$ is $N-n$ everywhere on $\tilde{V}$. Every $\tilde{y}$ has a representative in an affine space, say, defined by $\tilde{X}_0=1$. Let det $\left|\frac{\partial G_i}{\partial X_j}\right|(\tilde{y})\neq 0$ with $j=n+1,\cdots N$, say. Then we can clearly take as local parameters at $\tilde{y}$ n functions $\xi_1=X_1/X_0,\cdots,\xi_n=X_n/X_0$.

Let then $\tau_1,\cdots,\tau_n$ be local parameters on V at $\tilde{y}$, and let ξ be a function defined over k on V belonging to $[x\to\tilde{y}]$. Then, $\partial\xi/\partial\tau_i$ is also in $[x\to\tilde{y}]$. If ξ is a rational function of $\tau_1,\cdots,\tau_n$, we have clearly $\widetilde{\frac{\partial\xi}{\partial\tau_i}}=\frac{\partial\tilde{\xi}}{\partial\tilde{\tau}_i}$. Then, as the field $\tilde{\mathfrak{K}}$ of functions defined over $\tilde{k}$ on $\tilde{V}$ is separably algebraic over $\tilde{k}(\tilde{\tau}_1,\cdots,\tilde{\tau}_n)$ by assumption, the derivation $\frac{\partial}{\partial\tilde{\tau}_i}$ can be extended to $\tilde{\mathfrak{K}}$, and we have $\widetilde{\partial\xi/\partial\tau_i}=\partial\tilde{\xi}/\partial\tilde{\tau}_i$ also for any ξ in $[x\to\tilde{y}]$. We call a tangent vector on V at $\tilde{y}$, a mapping $\mathfrak{C}$ from the ring $[x\to\tilde{y}]$ into the universal domain Ω such that $\mathfrak{C}(\varphi\psi)=\tilde{\varphi}(\tilde{y})\mathfrak{C}(\psi)+\tilde{\psi}(\tilde{y})\mathfrak{C}(\varphi)$ and $\mathfrak{C}(\alpha\varphi+\beta\psi)=\tilde{\alpha}\mathfrak{C}(\varphi)+\tilde{\beta}\mathfrak{C}(\psi)$ for constant α,β in $[x\to\tilde{y}]$. Then we have, as before, for local parameters $\tau_1,\cdots,\tau_n$ on V at $\tilde{y}$, $\mathfrak{C}(\varphi)=\sum\widetilde{\frac{\partial\varphi}{\partial\tau_i}}(\tilde{y})\mathfrak{C}(\tau_i)$. Similarly we call a covariant vector on V at $\tilde{y}$ a $\tilde{k}$-linear mapping of the space of all $\mathfrak{C}$'s into $\tilde{\Omega}$. Furthermore, we say that an algebraic vector field X on V is defined at $\tilde{y}$, if it can be written in the form $X=\sum\varphi_i\frac{\partial}{\partial\tau_i}$ with local parameters $\tau_i,\cdots,\tau_n$ on V at $\tilde{y}$ and with $\varphi_1,\cdots,\varphi_n$ in $[x\to\tilde{y}]$. Clearly this condition does not depend on the choice of $\tau_1,\cdots,\tau_n$. Then, we can define a tangent vector $X(\tilde{y})$ on V at $\tilde{y}$ by $X(\tilde{y})(\xi)=\left(\sum\varphi_i\frac{\partial\xi}{\partial\tau_i}\right)(\tilde{y})$ $=\sum\tilde{\varphi}_i(\tilde{y})\frac{\partial\tilde{\xi}}{\partial\tilde{\tau}_i}(\tilde{y})$ for ξ in $[x\to\tilde{y}]$. As $\frac{\partial}{\partial\tau_i}(\tilde{y})$ form a basis of all tangent vectors at $\tilde{y}$ on V, every tangent vector at $\tilde{y}$ can be obtained as $X(\tilde{y})$ with some algebraic X. We can define from it a vector field $\tilde{X}$ on $\tilde{V}$ by $\tilde{X}(\tilde{y})(\tilde{\xi})=X(\tilde{y})(\xi)=\sum\tilde{\varphi}_i(\tilde{y})\frac{\partial\tilde{\xi}}{\partial\tilde{\tau}_i}(\tilde{y})$. Clearly $\tilde{X}$ is algebraic. This definition of $\tilde{X}$ is also independent from the choice of local parameters $\tau_1,\cdots,\tau_n$. Similarly, for ξ in $[x\to\tilde{y}]$ we define a covariant vector $d\xi(\tilde{y})$ on V at $\tilde{y}$ by $\langle X(\tilde{y}), d\xi(\tilde{y})\rangle = X(\tilde{y})(\xi)$ for all algebraic X defined at $\tilde{y}$, and then define $\widetilde{d\xi}$ by $\langle \tilde{X}(\tilde{y}), \widetilde{d\xi}(\tilde{y})\rangle = \langle X(\tilde{y}), d\xi(\tilde{y})\rangle$. We have clearly $\widetilde{d\xi} = d\tilde{\xi}$. More generally, an algebraic differential $\omega = \sum \psi_i d\tau_i$ is said to be defined at $\tilde{y}$ if $\tau_1, \cdots, \tau_n$ and $\psi_1, \cdots, \psi_n$ are in $[x \to \tilde{y}]$, and for such ω, we can define $\omega(\tilde{y})$ and $\tilde{\omega}$ in the evident manner. We say that this $\tilde{\omega}$ is obtained from ω by the reduction mod. $\mathfrak{p}$. Note that if X or ω is rational over k, then $\tilde{X}$ or $\tilde{\omega}$ is also rational over $\tilde{k}$ respectively.

The compatibility of transformation δf of algebraic differential with the reduction of it: $\widetilde{\delta f(\eta)} = \delta\tilde{f}(\tilde{\eta})$ for every algebraic η, defined on $\tilde{V}$; is seen as follows: We consider only the case where $\tilde{V}, \tilde{U}$, are non-singular and $\tilde{f}$ is everywhere defined on $\tilde{V}$, where f denotes a mapping of V into U. Denote by x, z generic points of V, U over k respectively. If ψ is in the ring $[z \to \widetilde{f(y)}]$, then $\psi\circ f$ is in $[x \to \tilde{y}]$ and $\widetilde{\psi\circ f} = \tilde{\psi}\circ\tilde{f}$. Then, we have $\delta f(d\psi) = d(\psi\circ f)$, hence $\delta f(d\psi)$ is defined at $\tilde{y}$. More precisely, denoting by $\tau_1, \cdots, \tau_n$ local parameters on V at $\tilde{y}$, we have $\delta f(d\psi) = \sum \frac{\partial(\psi f)}{\partial\tau_i} d\tau_i$ and $[\delta f(d\psi)](\tilde{y})$ $= \sum \frac{\partial(\psi\circ f)}{\partial\tau_i}(\tilde{y}) d\tau_i(\tilde{y}) = \sum \frac{\partial(\tilde{\psi}\circ\tilde{f})}{\partial\tilde{\tau}_i}(\tilde{y}) d\tilde{\tau}_i(\tilde{y}) = [\delta\tilde{f}(d\tilde{\psi})](\tilde{y})$ so we have $\widetilde{\delta f(d\psi)} = \delta\tilde{f}(d\tilde{\psi})$. In general, if $\sigma_1, \cdots, \sigma_m$ are local parameters on U at $\widetilde{f(y)}$, we have

$$[\delta f(\eta)](\tilde{y}) = \sum \varphi_i\circ f(\tilde{y})[\delta f(d\sigma_i)](\tilde{y}) = \sum \tilde{\varphi}_i\circ\tilde{f}(\tilde{y})[\delta\tilde{f}(d\tilde{\sigma}_i)](\tilde{y}) = [\delta\tilde{f}(\tilde{\eta})](\tilde{y})$$

for algebraic $\eta = \sum \varphi_i d\sigma_i$, defined at $\widetilde{f(y)}$. Hence we see $\widetilde{\delta f(\eta)} = \delta\tilde{f}(\tilde{\eta})$.

When f is a birational transformation of V onto U, everywhere biregular, and $\tilde{f}$ is also everywhere biregular, then we have also $\widetilde{df(X)} = d\tilde{f}(\tilde{X})$ for every algebraic X on V, defined on $\tilde{V}$. For, in this case, if $\tau_1, \cdots, \tau_n$ are local parameters on V at $\tilde{y}$, $\sigma_1 = \tau_1\circ f^{-1}, \cdots, \sigma_n = \tau_n\circ f^{-1}$ are local parameters on U at $\widetilde{f(y)}$, and we have $df\left(\frac{\partial}{\partial\tau_i}\right) = \frac{\partial}{\partial\sigma_i}$. Hence for algebraic $X = \sum \varphi_i \frac{\partial}{\partial\tau_i}$ we have $df(X) = \sum \varphi_i\circ f^{-1} \frac{\partial}{\partial\sigma_i}$, and $[df(X)](\widetilde{f(y)})(\psi) = \sum \tilde{\varphi}_i(\tilde{y}) \widetilde{\frac{\partial\psi}{\partial\sigma_i}}(\widetilde{f(y)})$ by definition. As the right hand side is equal to $d\tilde{f}(\tilde{X})(\widetilde{f(y)})(\tilde{\psi})$, this shows $\widetilde{df(X)} = d\tilde{f}(\tilde{X})$. Thus we have proved

LEMMA 7. *Let V, U be non-singular varieties and let f be a mapping from V into U, all defined over k. We assume f to be everywhere defined on V. Let furthermore $\mathfrak{p}$ be a discrete valuation* on k such that $\tilde{V}, \tilde{U}$ are non-singular and $\tilde{f}$ is everywhere defined on $\tilde{V}$. Then, for each algebraic linear differential η on U, defined on some non-void open set of $\widetilde{f(V)}, \delta f(\eta)$ is defined on the inverse image of that set, on $\tilde{V}$, and we have $\widetilde{\delta f(\eta)}=\delta\tilde{f}(\tilde{\eta})$. Moreover, if f is everywhere biregular birational transformation and $\tilde{f}$ is also everywhere biregular, then, for each algebraic vector field X on V, defined on some non-void open set on $\tilde{V}, df(X)$ is defined on the image of that open set, on U, and we have $\widetilde{df(X)}=d\tilde{f}(\tilde{X})$.

Let now A be an abelian variety and let $\tilde{A}$ be also an abelian variety. Then, if an invariant vector field X is defined at $\tilde{e}$, then $dT_aX=X$ is defined at $\tilde{a}$ by the above Lemma 7, and therefore X is everywhere defined on $\tilde{A}$. We have also $\tilde{X}=\widetilde{dT_aX}=dT_{\tilde{a}}\tilde{X}$ by this lemma, whence follows that X is invariant on $\tilde{A}$. Similarly, if an invariant linear differential ω is defined at $\tilde{e}$, then ω is everywhere defined on $\tilde{A}$ and $\tilde{\omega}$ is invariant on $\tilde{A}$. Clearly, the mappings $X\to\tilde{X}$ and $\omega\to\tilde{\omega}$ are onto. Then, let λ be a homomorphism of A into an abelian variety B, and let us assume that $\tilde{B}$ is also an abelian variety and that $\tilde{\lambda}$ is a homomorphism. Denote by $\omega_1,\cdots,\omega_n$ and by $\eta_1,\cdots,\eta_m$ bases of invariant differentials on A and on B respectively. Then we have $\delta\lambda(\eta_i)=\sum s_{ij}\omega_j$, so by Lemma 7, $\delta\tilde{\lambda}(\tilde{\eta}_i)=\sum\widetilde{s_{ij}}\tilde{\omega}_j$, if $\tilde{\omega}_i, \tilde{\eta}_j$ are all defined. Especially, let $\omega_1,\cdots,\omega_n$ be a basis, defined at $\tilde{e}$. If we denote by $S(\mu)=(\mu_{ij})$ the matrix of anti-representation of the endomorphism μ of A with $\omega_1,\cdots,\omega_n$, and by $\tilde{S}(\tilde{\mu})$ that of $\tilde{\mu}$ with $\tilde{\omega}_1,\cdots,\tilde{\omega}_n$, so we have $\widetilde{S(\mu)}=\tilde{S}(\tilde{\mu})$. Thus we have obtained the second main result of this §1:

PROPOSITION 2. *Let A^n, B^m be two abelian varieties, $\mathcal{H}(A, B)$ the module of homomorphisms of A into B, and let $\{\omega_1,\cdots,\omega_n\}$ and $\{\eta_1,\cdots,\eta_m\}$ be bases of the spaces of invariant linear differentials on A and on B respectively, and all these be defined over k. Let furthermore $\mathfrak{p}$ be a dicrete valuation of k such that $\tilde{A}, \tilde{B}$ are abelian varieties, $\tilde{\lambda}$ are homomorphisms of $\tilde{A}$ into $\tilde{B}$ for all λ in $\mathcal{H}(A, B)$, and that $\{\tilde{\omega}_1,\cdots,\tilde{\omega}_n\}, \{\tilde{\eta}_1,\cdots,\tilde{\eta}_m\}$ are all defined and form bases of the spaces of differentials of $\tilde{A}, \tilde{B}$ respectively. Then, if we denote $\tilde{S}(\tilde{\lambda}), S(\lambda)$ the matrices of representation of $\mathcal{H}(A, B)$, $\mathcal{H}(\tilde{A}, \tilde{B})$ with $\omega_1,\cdots,\omega_n$, $\eta_1,\cdots,\eta_m$ respectively, we have $\widetilde{S(\lambda)}=\tilde{S}(\tilde{\lambda})$ for any λ in $\mathcal{H}(A, B)$. When B coincides with A and $\eta_1,\cdots,\eta_m$ coincide with $\omega_1,\cdots,\omega_n$, we have the same result for anti-representation $S(\mu)$, $\tilde{S}(\tilde{\mu})$ of the ring $\mathcal{A}(A)$, $\mathcal{A}(\tilde{A})$ of endomorphisms of $A, \tilde{A}$ respectively.*

Note that there exists a proper open set $\mathfrak{V}'$ of $\mathfrak{V}$ such that all $\mathfrak{p}$'s in $\mathfrak{V}'$ satisfy the conditions of this proposition.

The following lemma plays an important role in the later applications of this Proposition 2.

LEMMA 8. Let A, B be two abelian varieties, λ a homomorphism of A onto B, all defined over k, and let x be a generic point of A over k. We assume the characteristic p not to be zero. Then, denoting by $S(\lambda)$ a representation of λ by invariant differentials on A and on B, $S(\lambda)=0$ if and only if $k(x^p)\supset k(\lambda x)$, where $k(x^p)$ means the field $k(x_1^p,\cdots,x_N^p)$, with an affine representative $(x_1,\cdots,x_N)$ of x.

This lemma is a direct consequence of [VA], Lemma 3, §1.

§2. Isogenous abelian varieties

Two abelian varieties A, B of the same dimension g is called isogenous if there exists a homomorphism λ of A onto B. This relation is symmetric by [VA], Th. 27, §VII, and we can classify by it all the abelian varieties into categories. If A, B are isogenous, then the rings $\mathcal{A}_0(A)$ and $\mathcal{A}_0(B)$ are isomorphic, where $\mathcal{A}_0(A)$ means the constant extension of the algebra $\mathcal{A}(A)$ over the ring of rational integers, to the field $\boldsymbol{Q}$ of rational numbers. Each element of $\mathcal{A}_0(A)$ is of degree at most $2g$, and $\mathcal{A}_0(A)$ is a semi-simple algebra. More precisely, if A is isogenous to a product $(A_1\times\cdots\times A_1)\times\cdots\times(A_h\times\cdots\times A_h)$, where A_i are not mutually isogenous and are simple, then, $\mathcal{A}_0(A)$ is isomorphic to the direct product of $\mathcal{A}_0(A_i\times\cdots\times A_i)$, which are respectively isomorphic to the matrix algebra, over skew-fields $\mathcal{A}_0(A_i)$, with degree equal to the number of factors A_i([VA], Th. 29, §VII.) Especially, if $\mathcal{A}_0(A)$ contains a field R_0 of degree $2g$, the above number h must be equal to 1, and $\mathcal{A}_0(A)$ is simple.

As the théorème 9 of [VA], §III is frequently used in the following, we quote it as a lemma.

LEMMA 9. Let A, B be two abelian varieties and let f be a mapping of B into A, all defined over k. Then, f is everywhere defined on B and of the form $\lambda+a$, where λ is a homomorphism of B into A, rational over k and $a=f(e)$ is a constant, rational over k.

The following lemma is important for our purpose.

LEMMA 10. Let R be a subring of $\mathcal{A}(A)$, and let $\mathfrak{a}$ be a left ideal of R. Then, there exist an abelian variety B, and a homomorphism $\lambda_{\mathfrak{a}}$ of A onto B, both defined over a field k, over which A and all endomorphisms in R are defined, with the property: $k(\lambda_{\mathfrak{a}}x) = \bigcup_{\alpha\in\mathfrak{a}} k(\alpha x)$, x being a generic point of A over k. B is determined uniquely by $\mathfrak{a}$ except for an isomorphism, and then also $\lambda_{\mathfrak{a}}$ is uniquely determined by $\mathfrak{a}$ except for automorphisms on B. The kernel of $\lambda_{\mathfrak{a}}$

is a finite group $\mathfrak{g}_{\mathfrak{a}}$ of all points a such that $\alpha a=e$ for all α in $\mathfrak{a}$. Each α in $\mathfrak{a}$ can be written in the form $\alpha=\alpha_0\cdot\lambda_{\mathfrak{a}}$, with a homomorphism α_0 of B onto A. Finally, let R' be another subring containing R and $\mathfrak{a}'$ be the left ideal $R'\mathfrak{a}$, then B' is isomorphic to B and we can choose $\lambda_{\mathfrak{a}}$ as $\lambda_{\mathfrak{a}'}$.

Proof. (The following construction is suggested by G. Shimura.) As all α in R, and the addition on A are rational over k, for any α, β, in R, we have $k((\alpha+\beta)x)\subset k(\alpha x)\cup k(\beta x)$ and $k(\alpha\beta x)\subset k(\beta x)$. We have therefore $\bigcup_{\alpha\in\mathfrak{a}} k(\alpha x)=\bigcup_{i=1}^{r} k(\alpha_i x)$ with generators $\alpha_1,\cdots,\alpha_r$ of $\mathfrak{a}$, and this field being containd in $k(x)$ and finitely generated, $\alpha_1 x\times\cdots\times\alpha_r x$ has a locus B over k in the product $A\times\cdots\times A$. Let then x, y be independent generic points of A over k, then, $\alpha_1 x\times\cdots\times\alpha_r x$ and $\alpha_1 y\times\cdots\times\alpha_r y$ are independent generic points of B over k, whose sum on $A\times\cdots\times A$ belongs to B, being of the form $\alpha_1(x+y)\times\cdots\times\alpha_r(x+y)$, hence B is an abelian variety ([VA], Prop. 3, §IV). As $\bigcup k(\alpha_i x)$ is contained in $k(x)$, there is a homomorphism $\lambda_{\mathfrak{a}}$ of A onto B, defined over k, such that $k(\lambda_{\mathfrak{a}}x)=\bigcup k(\alpha_i x)=\bigcup_{\alpha\in\mathfrak{a}} k(\alpha x)$, by the above lemma. One of such $\lambda_{\mathfrak{a}}$ is clearly given by $\lambda_{\mathfrak{a}}x=\alpha_1 x\times\cdots\times\alpha_r x$. The uniqueness of $\lambda_{\mathfrak{a}}$ is evident by Lemma 9. Then, $\lambda_{\mathfrak{a}}a=e$ means $\alpha_1 a=\cdots=\alpha_r a=e$, that is, $\alpha a=e$ for all α in $\mathfrak{a}$. Finally for every α in $\mathfrak{a}$, $k(\alpha x)$ being contained in $k(\lambda_{\mathfrak{a}}x)$, the mapping $\lambda_{\mathfrak{a}}x\rightarrow\alpha x$ is a homomorphism α_0 of B onto A by Lemma 9, rational over k and $\alpha=\alpha_0\lambda_{\mathfrak{a}}$ as is required. The last assertion is evident by the above construction.

LEMMA 11. Let $\mathfrak{p}$ be a discrete valuation of k such that $\tilde{A}$ can be defined and is an abelian variety and that every endomorphism of A can be reduced to an endomorphism of $\tilde{A}$, then, B is $\mathfrak{p}$-simple, $\tilde{B}$ is abelian and $\tilde{\lambda}_{\mathfrak{a}}$ is a homomorphism of $\tilde{A}$ onto $\tilde{B}$ and can be written as $\lambda_{\tilde{\mathfrak{a}}}$.

Proof. $\widetilde{A\times\cdots\times A}=\tilde{A}\times\cdots\times\tilde{A}$ is an abelian variety by assumption. Let λ_i be a homomorphism of A into $A\times\cdots\times A$ defined by $\lambda_i x=e\times\cdots\times\alpha_i x\times\cdots\times e$; then $\tilde{\lambda}_i$ is a homomorphism of $\tilde{A}$ into $\tilde{A}\times\cdots\times\tilde{A}$ and hence $\tilde{\lambda}_{\mathfrak{a}}=\widetilde{\sum\lambda_i}=\sum\tilde{\lambda}_i$ is also a homomorphism by Lemma 5. But then, $\tilde{\lambda}_{\mathfrak{a}}$ can be written as $\lambda_{\tilde{\mathfrak{a}}}$, hence $\nu(\tilde{\lambda}_{\mathfrak{a}})=\nu(\lambda_{\tilde{\mathfrak{a}}})=\nu(\lambda_{\mathfrak{a}})$. Therefore as Z is $\mathfrak{p}$-simple, B is also $\mathfrak{p}$-simple by [S], Th. 19. Thus we see that $\tilde{\lambda}_{\mathfrak{a}}\tilde{A}=\tilde{B}$, so B is an abelian variety. This proved our lemma. (Z means the graph of $\lambda_{\mathfrak{a}}$.)

Now, we consider mutually isogenous abelian varieties. Let A, B be two abelian varieties of the same dimension g and let λ be a homomorphism of A onto B. Let k be a field over which A, B, λ and all the endomorphisms of A and of B are defined, and let x be a generic point of A over k. Let μ be in $\mathcal{A}_0(A)$, then, for some natural number n, $n\mu$ is in $\mathcal{A}(A)$ and by Lemma 9, $\lambda\cdot n\nu(\lambda)\mu$ can be written as $\rho\cdot\lambda$ with an endomorphism ρ of B; hence, there is a μ^* in $\mathcal{A}_0(B)$, such that $\mu^*\lambda=\lambda\mu$ and the correspondence $\mu\rightarrow\mu^*$ is clearly an isomorphism of $\mathcal{A}_0(A)$ onto $\mathcal{A}_0(B)$. Especially, in the above case $\lambda=\lambda_{\mathfrak{a}}$, for any μ in the right order of $\mathfrak{a}$ in $\mathcal{A}_0(A)$, there is a natural number m such that $m\mu$ is in $\mathcal{A}(A)$ and then, $k(\lambda_{\mathfrak{a}m\mu}x)=k(\lambda_{\mathfrak{a}}m\mu x)$, but as $\mathcal{A}(A)\mathfrak{a}\mu\subset\mathcal{A}(A)\mathfrak{a}$, we have $k(\lambda_{\mathfrak{a}m\mu}x)\supset k(\lambda_{\mathfrak{a}m}x)$ and there is an endomorphism μ^* of B such that $\lambda_{\mathfrak{a}}m\mu=\mu^*\lambda_{\mathfrak{a}}m\delta_A$. This μ^* is clearly just the above determined endomorphism. Thus, $\mathcal{A}(B)$ contains the right order of $\mathfrak{a}$ in $\mathcal{A}(A)$. Then, let $\mathfrak{b}$ be an integral left ideal in $\mathcal{A}_0(A)$ with the right order of $\mathfrak{a}$ as a left order, and let $\mathfrak{b}$ lie in $\mathcal{A}(A)$. Then, denoting by $\beta_1,\cdots,\beta_s$ generators of $\mathfrak{b}$, we have $k(\lambda_{\mathfrak{b}}^*\lambda_{\mathfrak{a}}x)=\bigcup_i k(\beta_i^*\lambda_{\mathfrak{a}}x)=\bigcup_i k(\lambda_{\mathfrak{a}}\beta_i x)=k(\lambda_{\mathfrak{a}\mathfrak{b}}x)$ and hence $\lambda_{\mathfrak{a}\mathfrak{b}}=\varepsilon\lambda_{\mathfrak{b}}^*\lambda_{\mathfrak{a}}$ with an isomorphism ε. We note further, that if $\mathfrak{a}=(\alpha)$ is principal, we have $\lambda_{\mathfrak{a}}=\eta\alpha$ with an isomorphism η.

LEMMA 12. Let $R,\mathfrak{a}$ be as in Lemma 10 and we assume that R is the principal order of a field R_0 of degree d. Then, we have

$$\nu(\lambda_{\mathfrak{a}})=N\mathfrak{a}^{2g/d}$$

where $N\mathfrak{a}$ denotes the absolute norm in the field R_0.

Proof. We have $\nu(n\delta_A)=n^{2g}=N(n)^{2g/d}$ ([VA], Cor. 1 of Th. 33, §IX), and hence $\nu(\gamma)=N(\gamma)^{2g/d}$ for γ in R. We choose an ideal $\mathfrak{b}$ in R such that $(\nu(\lambda_{\mathfrak{a}}),\mathfrak{b})=1$ and $\mathfrak{a}\mathfrak{b}=(\gamma)$ is a principal ideal. Then, $\nu(\lambda_{\mathfrak{b}}^*)\nu(\lambda_{\mathfrak{a}})=\nu(\lambda_{\mathfrak{b}}^*\lambda_{\mathfrak{a}})=\nu(\lambda_{\mathfrak{a}\mathfrak{b}})=\nu(\gamma)=N(\gamma)^{2g/d}=(N\mathfrak{a}N\mathfrak{b})^{2g/d}$ as $\mathfrak{b}$ is contained in the right order of $\mathcal{A}(A)\mathfrak{a}$. Then, as $k(\lambda_{\mathfrak{a}}x)\supset k(\gamma x)$, $\nu(\gamma)$ is divisible by $\nu(\lambda_{\mathfrak{a}})$, so $N(\mathfrak{a})^{2g/d}$ must be divisible by $\nu(\lambda_{\mathfrak{a}})$ by the choice of $\mathfrak{b}$. Similarly $N(\mathfrak{b})^{2g/d}$ is divisible by $\nu(\lambda_{\mathfrak{b}})$, hence we have $\nu(\lambda_{\mathfrak{a}})=(N\mathfrak{a})^{2g/d}$ as asserted.

LEMMA 13. Let A be simple. Let $\mathcal{A}(A)=R$ be commutative and the principal order of the field R_0. Let further $\mathfrak{a},\mathfrak{b}$ be two ideals of R. Then, $\lambda_{\mathfrak{a}}A$ and $\lambda_{\mathfrak{b}}A$ are isomorphic, if and only if $\mathfrak{a}\sim\mathfrak{b}$ in R, that is, $\mathfrak{a}=\mathfrak{b}\alpha$ with α in $\mathcal{A}_0(A)$.

Proof. We first note that, if $k(\beta x)\subset k(\lambda_{\mathfrak{a}}x)$ with some β in R, then, denoting by $\mathfrak{c}$ the ideal $(\mathfrak{a},\beta)$, we have $k(\lambda_{\mathfrak{c}}x)=k(\lambda_{\mathfrak{a}}x)$ and $N\mathfrak{c}=N\mathfrak{a}$ by the above Lemma 12 and hence $\mathfrak{c}=\mathfrak{a}$, that is, β must be in $\mathfrak{a}$. Especially, if $k(\lambda_{\mathfrak{a}}x)=k(\lambda_{\mathfrak{b}}x)$, then $\mathfrak{a}=\mathfrak{b}$. Assume $\mathfrak{a}=\mathfrak{b}\alpha$. Let n be such a number that $n\alpha$ belongs to R. Then, as $n\mathfrak{a}=n\alpha\mathfrak{b}$, $\lambda_{n\mathfrak{a}}A=n\delta\cdot\lambda_{\mathfrak{a}}A$ is isomorphic to $\lambda_{\mathfrak{b}\cdot n\alpha}A=(n\alpha)^*\lambda_{\mathfrak{b}}A$, hence $\lambda_{\mathfrak{a}}A$ is isomorphic to $\lambda_{\mathfrak{b}}A$ as $n\delta,(n\alpha)^*$ are endomorphisms of $\lambda_{\mathfrak{a}}A,\lambda_{\mathfrak{b}}A$ respectively. Conversely, let $\lambda_{\mathfrak{a}}A$ and $\lambda_{\mathfrak{b}}A$ be isomorphic. We choose a number m such that $m\mathfrak{a}\subset\mathfrak{b}$. Then $\lambda_{m\mathfrak{a}}A=m\delta\cdot\lambda_{\mathfrak{a}}A$ is also isomorphic to $\lambda_{\mathfrak{b}}A$ and there is an isomorphism ζ of $\lambda_{m\mathfrak{a}}A$ onto $\lambda_{\mathfrak{b}}A$ such that $k(\zeta\lambda_{m\mathfrak{a}}x)=k(\lambda_{m\mathfrak{a}}x)$. The right hand side being contained in $k(\lambda_{\mathfrak{b}}x)$, there is a homomorphism ρ^* of $\lambda_{\mathfrak{b}}A$ onto $\zeta\lambda_{m\mathfrak{a}}A=\lambda_{\mathfrak{b}}A$, that is, ρ^* is an endomorphism of $\lambda_{\mathfrak{b}}A$. As R is maximal, ρ is in R and $k(\lambda_{\mathfrak{b}\rho}x)=k(\rho^*\lambda_{\mathfrak{b}}x)=k(\zeta\lambda_{m\mathfrak{a}}x)=k(\lambda_{m\mathfrak{a}}x)$ and hence $m\mathfrak{a}=\mathfrak{b}\rho$ by the above remark and $\mathfrak{a}=\mathfrak{b}(\rho/m)$, which proves our lemma.

LEMMA 14. Let $\mathcal{A}(A)$ contain a subring R which is the principal order of a field R_0 of degree $2g$. Let B be an abelian variety of the same dimension g as A and let λ be a homomorphism of A onto B. Then, if $\nu_i(\lambda)=1$, and μ^* is an endomorphism of B, for any μ in R, then λ must be of the form $\lambda_{\mathfrak{a}}$ with an ideal $\mathfrak{a}$ in R.

Proof. Let $\mathfrak{a}$ be an ideal in R. We denote by $\mathfrak{g}(\mathfrak{a})$ the group of points of a in A, such that $\alpha a=e$ for all α in $\mathfrak{a}$. Then, if $\mathfrak{a}\subset\mathfrak{b}$, $\mathfrak{g}(\mathfrak{a})\supset\mathfrak{g}(\mathfrak{b})$. Clearly, $\mathfrak{g}((\mathfrak{a},\mathfrak{b}))=\mathfrak{g}(\mathfrak{a})\cap\mathfrak{g}(\mathfrak{b})$ because any δ in $(\mathfrak{a},\mathfrak{b})$ is of the form $\delta=\mu\alpha+\nu\beta$ with α in $\mathfrak{a}$, β in $\mathfrak{b}$ and μ,ν in R. We have further $\mathfrak{g}(\mathfrak{a}\cap\mathfrak{b})\supset\mathfrak{g}(\mathfrak{a})+\mathfrak{g}(\mathfrak{b})$ and, as $(\mathfrak{a}\cap\mathfrak{b})\cdot(\mathfrak{a},\mathfrak{b})=\mathfrak{a}\cdot\mathfrak{b}$, $\nu_s(\lambda_{(\mathfrak{a}\cap\mathfrak{b})(\mathfrak{a},\mathfrak{b})})=\nu(\lambda_{\mathfrak{a}\mathfrak{b}})$, hence $\nu_s(\lambda_{(\mathfrak{a},\mathfrak{b})})\nu_s(\lambda_{\mathfrak{a}\mathfrak{b}})=\nu_s(\lambda_{\mathfrak{a}})\nu_s(\lambda_{\mathfrak{b}})$, where $\nu_s(\lambda)$ means the separable part of index $[k(x)\colon k(\lambda x)]$, hence the order of $\mathfrak{g}(\mathfrak{a})$ if $\lambda=\lambda_{\mathfrak{a}}$. ($\nu_i(\lambda)$ means the inseparability index of $[k(x):k(\lambda x)]$.) Thus we see $\mathfrak{g}(\mathfrak{a}\cap\mathfrak{b})=\mathfrak{g}(\mathfrak{a})+\mathfrak{g}(\mathfrak{b})$. Now, let $\mathfrak{h}=\{h_i\}$ be a finite set of order finite points of A; and let $\mathfrak{a}(\mathfrak{h})$ denote the set of α in R such that $\alpha h_i=0$ for all h_i in $\mathfrak{h}$. Clearly $\mathfrak{a}(\mathfrak{h})$ is an ideal of R. We write also $R\mathfrak{h}=\bigcup_{\substack{\mu\in R\\ h_i\in\mathfrak{h}}}\mu h_i$. Then, clearly $\mathfrak{g}(\mathfrak{a}(\mathfrak{h}))\supset R\mathfrak{h}$ and $\mathfrak{a}(\mathfrak{h})=\bigcap_{h_i\in\mathfrak{h}}\mathfrak{a}(h_i)$. But, for μ,ν in R, we have $\mu h=\nu h$ if and only if $\mu\equiv\nu \bmod \mathfrak{a}(\mathfrak{h})$; hence Rh has just $N\mathfrak{a}(h)$ elements. As $\mathfrak{g}(\mathfrak{a}(h))$ has also $\nu_s(\mathfrak{a}(h))\leq N\mathfrak{a}(h)$ elements (cf. Lemma 12), we see $\mathfrak{g}(\mathfrak{a}(h))=Rh$. Hence also $\mathfrak{g}(\mathfrak{a}(\mathfrak{h}))=\bigcup\mathfrak{g}(\mathfrak{a}(h_i))=\bigcup Rh_i=R\mathfrak{h}$. Thus if $\mathfrak{h}\neq\mathfrak{g}(\mathfrak{a}(\mathfrak{h}))$, then there exists a μ in R such that $\mu\mathfrak{h}\not\subset\mathfrak{h}$. Let especially $\mathfrak{h}$ be the kernel of λ; then, as by assumption $\lambda\mu=\mu^*\lambda$ with endomorphism μ^* of λA for any μ in R, the kernel of $\lambda\mu=\mu^*\lambda$ must contain $\mathfrak{h}$, that is, $\mu\mathfrak{h}\subset\mathfrak{h}$ for any μ in R, hence $\mathfrak{h}=\mathfrak{g}(\mathfrak{a}(\mathfrak{h}))$. As $\nu_i(\lambda)=1$, this shows $\lambda=\lambda_{\mathfrak{a}(\mathfrak{h})}$, which proves our lemma.

LEMMA 15. Let $\mathcal{A}(A)=R$ itself be the principal order of the field R_0. Then, every endomorphism of $B=\lambda A$ is of the form μ^* with μ in R. Especially, if $\mathcal{A}(B)$ is also the principal order, $\mu\to\mu^*$ gives an isomorphism of $\mathcal{A}(A)$ and $\mathcal{A}(B)$.

Proof. We first remark that if $\mathfrak{B}$ is a subring of R of all μ for which μ^* is in $\mathcal{A}(B)$, then $\mathfrak{B}$ is an order of R. That $\mathfrak{B}$ is really a ring is evident. $\mathfrak{B}$ contains $n\delta_A$ for any n, and further, for each μ, there is a natural number m such that $m\mu^*$ is in $\mathcal{A}(B)$, hence $m\mu$ is in $\mathfrak{B}$. Let μ^* be in $\mathcal{A}(B)$, and μ be not in $\mathcal{A}(A)$. We choose an integer n such that $n\mu$ is in $\mathcal{A}(A)$. Then, as $n\mu^*\lambda=\lambda n\mu$, we have $\nu(n\mu^*)=\nu(n\mu)$. But as μ is not in $\mathcal{A}(A)$ and $\mathcal{A}(A)$ is the principal order, μ is not integral, hence there is a prime number q such that the exponent of q in $\nu(n\mu)$ is smaller than that in $\nu(n\delta)$, which is not greater than $\nu(n\mu^*)$ as μ^* is an endomorphism, and we arrive at a contradiction.

For the following, we must quote some more results from Weil's book [VA]. With any prime number l different from the characteristic of the universal domain, we can attach to every point a of l-th power order on A, a vector a_l with $2g$ l-adic numbers mod. 1 as components, which are called l-adic coordinates of a. With this vector a_l, we can represent faithfully the ring $\mathcal{A}(A)$, and we denote the representation matrix of μ in A as $M_l(\mu)$, whose elements are l-adic integers. The characteristic equation of $M_l(\mu)$ has rational integral coefficients and is the same for all $l \neq p$, and further we have $\det M_l(\mu)=\nu(\mu)$. Further, to every divisor X on A there corresponds a skew-symmetric l-adic integral matrix $E_l(X)$ of degree $2g$, and for two divisors X, Y on $J, E_l(X)=E_l(Y)$ if and only if $X \equiv Y$. Let B be also an abelian variety, and λ a homomorphism of A onto B. Then, taking l-adic coordinate also on B, we can represent the homomorphism λ by a transformation matrix $M_l(\lambda)$ by means of l-adic coordinates of A and B. $M_l(\lambda)$ has also l-adic integral elements and, for each divisor X on B, we have $E_l(\lambda^{-1}(X))={}^tM_l(\lambda)E_l(X)M_l(\lambda)$, where ${}^tM$ means the transposed matrix of M. Especially, if A is a jacobian variety J of a non-singular curve C with canonical mapping φ, there is a divisor Θ on J with generic point $\sum_{i=1}^{g-1}\varphi(P_i)$, P_i denoting independent generic points of C over k. This Θ we call a Riemann divisor. Then, $E_l(\Theta)$ is l-adic unimodular. With the above homomorphism λ and with any divisor X on B, we can define a homomorphism λ'_X of B into J, and for this λ'_X we have $M_l(\lambda'_X)=E_l(\Theta)^{-1}\cdot{}^tM_l(\lambda)E_l(X)$. In case of $B=J$ and $X=\Theta, \lambda$ an endomorphism, we denote λ'_Θ as λ'. Hence, $M_l(\lambda')=E_l(\Theta)^{-1}\cdot{}^tM_l(\lambda)E_l(\Theta)$. Then, denoting by $\sigma(\mu)$ the trace of $M_l(\mu)$, we have $\sigma(\mu\mu')>0$ for any $\mu \neq 0$ in $\mathcal{A}(J)$. Moreover, for a homomorphism λ of J onto B, and for a positive divisor $X \neq 0$ on B, $\sigma(\mu'\lambda'_X\lambda\mu)>0$ with $\mu \neq 0$ in $\mathcal{A}(J)$ ([VA], Th. 25, §VI, Th. 31, §VIII, Th. 34, §IX). Clearly, $\mu \to \mu'$ is an involutive anti-automorphism of $\mathcal{A}(J)$, and $\lambda'_X\lambda$ is invariant by it. Finally, there is also an involutive anti-automorphism $\mu \to \mu'$ of $\mathcal{A}(A)$ for a simple abelian variety A, with the property $\sigma(\mu'\mu)>0$ for $\mu \neq 0$ in $\mathcal{A}(A)$ ([VA], Th. 38, §X). From this, we can define $\mu \to \mu'$ for A which is isogenous to $B\times\cdots\times B$ with simple B. Indeed, as μ is of the form $\begin{pmatrix}\mu_{11} & \cdots & \mu_{1s}\\ & \cdots & \\ \mu_{s1} & \cdots & \mu_{ss}\end{pmatrix}$ in this case, where μ_{ij} are in $\mathcal{A}(B)$, we have only to define $\mu'={}^t\begin{pmatrix}\mu'_{11} & \cdots & \mu'_{1s}\\ & \cdots & \\ \mu'_{s1} & \cdots & \mu'_{ss}\end{pmatrix}$. Thus, we have clearly $\sigma(\mu'\mu)=\sigma(\mu\mu')>0$ for $\mu\neq 0$. In most general case, where $\mathcal{A}(A)$ is a direct sum of simple algebras, we define $\mu\to\mu'$ by the above anti-automorphism in each factor. Then we have also $\sigma(\mu'\mu)=\sigma(\mu\mu')>0$ for $\mu\neq 0$ in $\mathcal{A}(A)$.

When a field R_0 is isomorphic to an algebraic number field of a finite degree, we call conjugates of R_0 or of elements μ in R_0, isomorphic images of R_0 or of μ into the complex number fields, respectively. We call R_0 or μ totally real, totally positive or totally imaginary, according as all its conjugates are real, positive or imaginary, respectively.

LEMMA 16. Let R_0 be a subfield of $\mathcal{A}_0(A)$. We assume that R_0 is invariant as a whole by the anti-automorphism $\mu\to\mu'$ of $\mathcal{A}_0(A)$, and we denote by K the subfield of R_0 consisting of elements fixed by this anti-automorphism. Then, K is totally real, and R_0 is either equal to K or is totally imaginary. In the latter case, $\mu\to\mu'$ induces the complex conjugate automorphism in every conjugate of R_0.

Proof. We assume, there is a conjugate $R_0^{(1)}$ of R_0, such that $\mu'^{(1)}\neq\overline{\mu^{(1)}}$ for some μ in R_0 and we show that this assumption leads to a contradiction. We denote $\overline{\mu^{(1)}}=\mu^{(2)}$, which is also a conjugate of μ. We can assume $|\mu^{(1)}|\geq|\mu'^{(1)}|$, because $\overline{1/\mu^{(1)}}\neq 1/\mu'^{(1)}$ also. We can further assume that $|\mu^{(1)}|\neq|\mu^{(i)}|$ for $i=3,\cdots,2g$. Indeed, we can replace $\mu^{(1)}$ by $\mu^{(1)}+\varepsilon$, where ε is a small rational number such that $|\mu^{(1)}+\varepsilon|\neq|\mu^{(i)}+\varepsilon|$ if $|\mu^{(1)}|\neq|\mu^{(i)}|$, and that $|\mu^{(1)}+\varepsilon|\geq|\mu'^{(1)}+\varepsilon|$. But then, $|\mu^{(i)}+\varepsilon|\neq|\mu^{(j)}+\varepsilon|$ also for $j\neq 1,2$ such that $|\mu^{(1)}|=|\mu^{(j)}|$, as cosine law shows. Then, let $|\mu^{(j)}|/|\mu^{(1)}|=1+\delta_i$ and let $\delta=\inf_{i=3,\cdots 2g}|\delta_i|$. δ is positive by assumption. By Dirichlet's method, we can choose a natural number N such that the arguments ϑ_1, ϑ_2 of $\mu^{(1)N}$, $\mu'^{(1)N}$ are so near to zero that $|\sin\vartheta_i|<\delta\eta(i=1,2)$, where η is a given small positive number. Then, we can choose a natural number t_1 such that $\left|\dfrac{\mu^{(1)N}}{t_1}-1\right|<\delta\eta$. Next, we can choose a rational number t_2 such that $0<\left|\dfrac{\mu'^{(1)N}}{t_1}-t_2\right|<3\delta\eta$ and the argument ϑ_3 of $\left(\dfrac{\mu'^{(1)N}}{t_1}-t_2\right)$ satisfies $|\vartheta_3|<\dfrac{5}{24}\pi$. As we assumed that η is very small, and as $\left|\dfrac{\mu'^{(1)N}}{t_1}\right|\leq 1-\delta+2\delta\eta$, by assumption, and as $|\sin\vartheta_2|<\delta\eta$, we see that the argument of $\left(\dfrac{\mu'^{(1)N}}{t_1}-t_2\right)$ is different from π at most $\pm\dfrac{\pi}{24}$, if η is so small that $\eta<\dfrac{1}{8}\sin\dfrac{\pi}{24}$. We denote $\frac{\mu^N}{t_1}-t_2=\nu$, then, $\frac{\mu^{(1)N}}{t_1}-t_2=\nu^{(1)}$, $\frac{\mu'^{(1)N}}{t_1}-t_2=\nu'^{(1)}$ and the argument ϑ_4 of $\nu^{(1)}\nu'^{(1)}$ satisfies $|\vartheta_4-\pi|<\frac{\pi}{4}$, we have also $|\nu^{(1)}\nu'^{(1)}|<8\delta\eta$, as $\left|\frac{\mu'^{(1)N}}{t_1}\right|<\left|\frac{\mu^{(1)}}{t_1}\right|<1+\delta\eta$, and $|t_2|<1+4\delta\eta$. On the other hand, $|\nu^{(i)}|\geqq\frac{\delta}{4}$, for $i\neq 1, 2$, as $\left|\frac{\mu^{(i)N}}{t_1}-\frac{\mu^{(1)N}}{t_1}\right|>\delta(1-\delta\eta)$ by the choice of δ. Therefore, we have $|\nu^{(i)}\nu'^{(i)}|\geqq\frac{\delta^2}{16}$ if $\nu^{(i)}\nu'^{(i)}\neq\overline{\nu^{(1)}\nu'^{(1)}}, \nu^{(1)}\nu'^{(1)}$. Then, $\left(\frac{1}{\nu^{(1)}\nu'^{(1)}}+\frac{1}{\overline{\nu^{(1)}\nu'^{(1)}}}\right)<\frac{-1}{4\delta\eta}$, as the argument ϑ_5 of $1/\nu^{(1)}\nu'^{(1)}$ satisfies also $|\vartheta_5-\pi|<\frac{\pi}{4}$. As $|1/\nu^{(i)}\nu'^{(i)}|\leqq\frac{16}{\delta^2}$ for another i, we see that, if $\eta<\frac{\delta}{2^8 g}$, then $\sum_i\frac{1}{\nu^{(i)}\nu'^{(i)}}<0$, which contradicts the property $\sigma(1/\nu\cdot 1/\nu')>0$. Thus we have $\mu'^{(i)}=\overline{\mu^{(i)}}$ for any i, which proves our assertion.

COROLLARY. Let R_0 be a subfield of $\mathcal{A}_0(A)$, which is invariant by the anti-automorphism $\mu\to\mu'$ as a whole. Let $\bar{k}_3$ be the smallest normal field (over $\boldsymbol{Q}$) which contains an isomorphic image k_3 of R_0 into the complex number field. Then, σ_0 commutes with all the automorphisms σ of $\bar{k}_3$, where σ_0 means the complex conjugate automorphism of $\bar{k}_3$.

Proof. If $\bar{k}_3$ is real, there is nothing to prove. In another case, k_3 is a totally imaginary quadratic extension of a totally real field. Let $k_3=\boldsymbol{Q}(\mu)$. Then, $\mu^\sigma\to\mu^{\sigma_0\sigma\sigma_0}$ is an automorphism of $k_3^\sigma=Q(\mu^\sigma)$, because σ_0 fixes all k_3^σ as a whole. For any real number ν^σ in k_3, $\nu=(\nu^\sigma)^{\sigma^{-1}}$ is also real. Hence we have $\nu^{\sigma_0\sigma\sigma_0}=\nu^\sigma$, which shows that the automorphism $\mu^\sigma\to\mu^{\sigma_0\sigma\sigma_0}$ fixes all the real numbers in k_3^σ. Thus $\mu^{\sigma_0\sigma\sigma_0}$ must equal to either μ^σ or $\mu^{\sigma\sigma_0}$. But $\mu^{\sigma_0\sigma\sigma_0}=\mu^{\sigma\sigma_0}$ is impossible, as $\mu^{\sigma_0}\neq\mu$ by assumption. As we can choose k_3^σ instead of k_3, this shows $\sigma_0\sigma=\sigma\sigma_0$ for all automorphisms σ of $\bar{k}_3$.

For the later use we need a criterion for a homomorphic image of a jacobian variety to be also jacobian. Here we only give a necessary condition.

LEMMA 17. Let J be a simple jacobian variety. Let $\mathcal{A}(J)=R$ be commutative and be the principal order of the field R_0, and let $\mathfrak{a}$ be an ideal of R. Then, if $\lambda_{\mathfrak{a}}J$ is also a jacobian variety, then $\mathfrak{a}\mathfrak{a}'=(\alpha)$ with a totally positive element α of the field K, consisting of elements of R_0 fixed by the automorphism $\mu\to\mu'$ of R_0.

Proof. We denote $J^*=\lambda_{\mathfrak{a}}J$. Let Θ, Θ^* be Riemann divisors of J, J^* respectively, which certainly exist by assumption. We denote further $\lambda=\lambda_{\mathfrak{a}}$. As $\mathfrak{a}$ is an ideal of the unique maximal order R, $\mathcal{A}(B)$ is also the principal order, so by Lemma 15, $\mu\to\mu^*$ gives an isomorphism of $\mathcal{A}(A)=R$ onto $\mathcal{A}(B)$, and we can write $\mathcal{A}(B)=R^*$. We first assume that $\mathfrak{a}$ is prime to the characteristic p. At any rate, $\lambda'_{\Theta^*}\circ\lambda=\rho$ is an endomorphism of J. Then, for each α in $\mathfrak{a}$, there is a homomorphism α_0 of J^* onto J such that $\alpha=\alpha_0\lambda$ (cf. Lemma 10), thus we see: $M_l(\alpha')=E_l(\Theta)^{-1}\cdot{}^tM_l(\alpha)E_l(\Theta)=E_l(\Theta)^{-1}\cdot{}^tM_l(\lambda)E_l(\Theta^*)E_l(\Theta)^{-1}\cdot{}^tM_l(\alpha_0)E_l(\Theta)=M_l(\lambda'_{\Theta^*})M_l(\alpha'_{0\Theta})$. Hence $\alpha'=\lambda'_{\Theta^*}\cdot\alpha'_{0\Theta}$ and $\lambda'_{\Theta^*}\alpha'_{0\Theta}\lambda'_{\Theta^*}\lambda=\alpha'\rho=\rho\alpha'=\lambda'_{\Theta^*}\lambda\alpha'=\lambda'_{\Theta^*}\alpha'^*\lambda$. Therefore we have $\alpha'^*=\alpha'_{0\Theta}\lambda'_{\Theta^*}$, for each α in $\mathfrak{a}$. Then, for a generic point x^* of J^* over k, $k(\lambda_{\mathfrak{a}'^*}x^*)=\bigcup_{\alpha\in\mathfrak{a}}k(\alpha'^*x^*)\subset k(\lambda'_{\Theta^*}x^*)$. But, for each prime number $l\neq p$, as $E_l(\Theta)$ and $E_l(\Theta^*)$ are unimodular, the determinants of $M_l(\lambda'_{\Theta^*})=E_l(\Theta)^{-1}\cdot{}^tM_l(\lambda_{\mathfrak{a}})E_l(\Theta^*)$ and $M_l(\lambda_{\mathfrak{a}})$ have the same l-factor, and this l-factor is equal to that in $\nu(\lambda_{\mathfrak{a}})=N\mathfrak{a}^{2g/d}=N\mathfrak{a}'^{*2g/d}=\nu(\lambda_{\mathfrak{a}'^*})$, (cf. Lemma 12), hence l-factors in $\nu(\lambda'_{\Theta^*})$ and in $\nu(\lambda_{\mathfrak{a}'^*})$ are equal for all $l\neq p$, therefore $k(\lambda_{\mathfrak{a}'^*}x^*)=k(\lambda'_{\Theta^*}x^*)$ as $\mathfrak{a}$ is prime to p. Here we can choose $x^*=\lambda_{\mathfrak{a}}x$, then, $k(\lambda_{\mathfrak{a}\mathfrak{a}'}x)=k(\lambda_{\mathfrak{a}'^*}\lambda_{\mathfrak{a}}x)=k(\lambda'_{\Theta^*}\lambda_{\mathfrak{a}}x)=k(\rho x)$, therefore as was seen in the proof of Lemma 13, we have $\mathfrak{a}\mathfrak{a}'=(\rho)$. As $\rho=\lambda'_{\Theta^*}\lambda$, so $\rho'=\rho$ and ρ is in K, and $\sigma(\mu'\rho\mu)>0$ for any $\mu\neq0$ in $\mathcal{A}(J)$ as quoted above. But, as was seen in the proof of the above lemma, we can choose μ such that one conjugate of the totally real number $\mu'\mu$ is far greater than the other conjugates. As $\sigma(\mu'\rho\mu)=\sigma(\rho\mu'\mu)>0$, ρ must be then totally positive. This gives our assertion in case $(\mathfrak{a}, p)=1$. If $\mathfrak{a}$ is not prime to p, we can choose $\mathfrak{b}$ in the same class as $\mathfrak{a}$ and prime to p. Then, Lemma 13 assures our conclusion for this $\mathfrak{a}$ also.

COROLLARY. Let J be as in the lemma and let λ be a homomorphism of J onto an abelian variety B. Then, for any positive divisor $X\neq0$ on B, $\lambda'_X\lambda$ is totally positive. Especially, δ'_Y is totally positive for any positive divisor $Y\neq0$ on J.

Proof. This corollary follows immediately from the last part of the above proof.

Finally we consider some properties of Riemann divisor.

LEMMA 18. If a positive divisor X on a jacobian variety J is linearly equivalent to a Riemann divisor Θ, then $X=\Theta$.

Proof. (This proof is due to Hironaka.) By assumption, there is a function ψ on J such that $\Theta+(\psi)=X$, where (ψ) is the divisor of ψ. Let C be a non-singular curve, of which J is a jacobian, and let φ be a canonical mapping, with which Θ is defined, and let J, X, C, φ, ψ be defined over k. Let u be a generic point of J over k, then the product $\varphi(C)\cdot\Theta_{-u}, \varphi(C)\cdot X_{-u}$ are defined and $\varphi(C)\cdot\Theta_{-u}$ is linearly equivalent to $\varphi(C)\cdot X_{-u}$ on $\varphi(C)$. (cf. [VA], Cor. of Th. 3,

§II; Θ_{-u} means the translation $T_{-u}\Theta$). But as the divisor $\varphi(C)\cdot\Theta_{-u}$ is non special on the curve $\varphi(C)$, which is non-singular (cf. [VA], Prop. 16, Prop. 18, §V), this implies first that $\varphi(C)\cdot\Theta_{-u}=\varphi(C)\cdot X_{-u}$, then, that ψ induces a constant function on $\varphi(C)_u$. Let $M, P_1,\cdots, P_g$, $Q_1,\cdots, Q_g$ be $2g+1$ independent generic points of $\varphi(C)$ over k, and we denote $x_i=\sum_{\kappa=1}^{i-1}Q_\kappa+\sum_{\nu=i}^{g}P_\nu$ $(i=1,\cdots, g+1)$. Then x_i are generic points of J over k, hence we have $\psi(M+x_i)=\psi(Q_i+x_i)=\psi(P_i+x_{i+1})=\psi(M+x_{i+1})$ as ψ is constant on $\varphi(C)_{x_i}, \varphi(C)_{x_{i+1}}$. By induction, we have $\psi(M+x_1)=\psi(M+x_{g+1})$. But as x_1 and x_{g+1} are independent generic points of J over $k(M)$, this is impossible unless ψ is constant on J. Thus we see $X=\Theta$.

LEMMA 19. Let X be a positive divisor on a jacobian variety J with Riemann divisor Θ, defined with φ. Then, δ'_X is of the form $\delta'_X=\zeta'\zeta$, with a unit ζ in $\mathcal{A}(J)$, if and only if X is a Riemann divisor corresponding to the same curve as Θ.

Proof. First we assume $\delta'_X=\zeta'\zeta$. Then we have $E_l(X)=E_l(\Theta)^{-1}$ $M_l(\delta'_X)={}^tM_l(\zeta)E_l(\Theta)M_l(\zeta)=E_l(\zeta^{-1}\Theta)$, hence $X\equiv\zeta^{-1}\Theta$ as quoted above. Then, $\zeta X-\Theta$ is linearly equivalent to $\Theta_a-\Theta$ with some point a on J ([VA], Cor. 2 of Th. 32, §VIII), hence $\zeta X=\Theta_a$ by the above lemma. Then, we have only to choose a canonical mapping $\varphi_1=\zeta^{-1}\varphi+b$ instead of φ with a point b such that $(g-1)\zeta b=a$. The converse is evident by the uniqueness (except for automorphisms and additive constants) of canonical mapping.

We call two Riemann divisors Θ_1, Θ_2 equivalent if there is an automorphism ζ of $\mathcal{A}(J)$ and a point a on J such that $\Theta_2=\zeta\Theta_{1a}$. Then, we have

COROLLARY. Let J be as in Lemma 17. Then the number of non-equivalent Riemann divisors on J is at most equal to the index $\gamma=[\eta: \zeta\zeta']$, where η is the group of all totally positive units in R, and $\zeta\zeta'$ is the group of all the units in R of the form $\zeta\zeta'$.

Proof. The automorphism $\mu\to\mu'$ does not depend on the choice of Θ_i, as is seen by Lemma 16. Let Θ_1, Θ_2 be two Riemann divisors on J. Then, $E_l(\Theta_1^{-1})E_l(\Theta_2)=E_l(\delta'_{\Theta_2})$ and δ'_{Θ_2} is a totally positive unit, by Cor. of Lemma 17. Thus we have one-to-one correspondence $\Theta_i\leftrightarrow\delta'_{\Theta_i}$ of $\{\Theta_i\}$ into η. Then, by Lemma 19, Θ_i and Θ_j are equivalent if and only if $\delta'_{\Theta_i}=\zeta'\delta'_{\Theta_j}\zeta$ with a unit ζ, which proves our corollary.

§3. Ideal decomposition of the endomorphism π

First we prove some general lemmas. Let A be an abelian variety defined over a field k. Let $\bar{\bar{k}}$ be an algebraically closed field containing k, over which all the endomorphisms of A are rational, and let

x be a generic point of A over k, such that $k(x)$ and $\bar{\bar{k}}$ are linearly disjoint over k. Let finally σ be an automorphism of k, then, σ can be extended to an isomorphism of $\bar{\bar{k}}(x)$, and x^σ has a locus A^σ over k^σ. A^σ is clearly an abelian variety. For an endomorphism μ of A, we can define an endomorphism $\mu^{\bar{\sigma}}$ of A^σ by $\mu^{\bar{\sigma}}x^\sigma=(\mu x)^\sigma$. Then, $\mu\rightarrow\mu^{\bar{\sigma}}$ is an isomorphism of the ring $\mathcal{A}(A)$ onto $\mathcal{A}(A^\sigma)$. Especially, if σ fixes every element of the field k, σ can be extended to $k(x)$ by $x^\sigma=x$, and then, we have $\mu^{\bar{\sigma}}x=(\mu x)^\sigma$. $\mu\rightarrow\mu^{\bar{\sigma}}$ is an automorphism of $\mathcal{A}(A)$ in this case.

Let $\{\omega_1,\cdots,\omega_g\}$ be a basis of invariant linear differentials of A, rational over k. Let $S_0(\mu)$ be the anti-representation of $\mathcal{A}(A)$ with respect to this basis. Then, all coefficients of $S_0(\mu)$ lie in the smallest field k' containing k, over which all endomorphisms of A are rational. Let furthermore $\mathcal{A}(A)$ contain a subring R, which is isomorphic to an order of an algebraic number field k_1. Denote by $\bar{k}_1$, the smallest normal field containing k_1. Then, if the characteristic of k is zero, $S_0(\mu)$ can be transformed to $S(\mu)$, which are diagonal for all μ in R. Indeed, the minimal equation of $S_0(\mu)$ is irreducible for μ in R, and all diagonal elements of the $S(\mu)$ are algebraic conjugates. Let μ be a primitive element of $R:R_0=\boldsymbol{Q}(\mu)$; and let $S(\mu)=\begin{pmatrix}\mu_1^{\sigma_1} & & \\ & \ddots & \\ & & \mu_1^{\sigma_g}\end{pmatrix}$ where σ_i are isomorphisms of the field $\boldsymbol{Q}(\mu_1)$ and $\sigma_1=$identity. Then, for any ν in R, we have $S(\nu)=\begin{pmatrix}\nu_1^{\sigma_1} & & \\ & \ddots & \\ & & \nu_1^{\sigma_g}\end{pmatrix}$. We fix an isomorphism ι of R into k_1 by $\mu\rightarrow\iota(\mu)=\mu_1$.

LEMMA 20. Let k' be the smallest field containing k, over which all endomorphisms of A are rational. Then k' is a finite algebraic normal extension of k, and $\sigma\rightarrow\bar{\sigma}$ induces an isomorphism of the Galois group k' over k into the group of automorphisms of $\mathcal{A}(A)$.

Proof. As $\mathcal{A}(A)$ has a finite basis, k' is finitely generated. Denote by k_μ the smallest field containing k, over which μ is rational. Then we have clearly $k_{\mu^{\bar{\sigma}}}=k_\mu^\sigma$ for an automorphism σ of $\bar{\bar{k}}$ over k. But, as $k'=\bigcup_{\mu\in\mathcal{A}(A)}k_\mu$, we have $k'^\sigma=\bigcup_{\mu\in\mathcal{A}(A)}k_\mu^\sigma=\bigcup_{\mu\in\mathcal{A}(A)}k_{\mu^{\bar{\sigma}}}=k'$. Thus, k' is invariant by all the automorphisms σ of $\bar{\bar{k}}$ over k. As $\bar{\bar{k}}$ can be taken to have an arbitrarily high transcendency degree, k' must be algebraic and normal over k. The second assertion is evident.

LEMMA 21. Let the characteristic of k be 0, and let σ be an automorphism of the algebraic closure of k. We denote by S_0^σ the anti-representation of the ring $\mathcal{A}(A^\sigma)$ by $\{\omega_1^\sigma,\cdots,\omega_g^\sigma\}$. Then, we have

$S_0^{\sigma}(\mu^{\bar{\sigma}})=[S_0(\mu)]^{\sigma}$. Especially, if σ is an automorphism over k, we have $S_0(\mu^{\bar{\sigma}})=[S_0(\mu)]^{\sigma}$ and moreover, if $R^{\bar{\sigma}}=R$, we have $R(\mu^{\bar{\sigma}})=\begin{pmatrix} \mu_1^{\sigma'\sigma_1} & & \\ & \ddots & \\ & & \mu_0^{\sigma'\sigma_g} \end{pmatrix}$ with an automorphism σ' of the field $\boldsymbol{Q}(\mu_1)$ isomorphic to R_0.

Proof. The first part is evident by definition. As to the second part we remark that, as $R^{\bar{\sigma}}=R$, $\mu\to\mu^{\bar{\sigma}}$ induces an automorphism of $\boldsymbol{Q}(\mu_1)$. If we denote it as σ', the second assertion is also clear.

LEMMA 22. Let A and $S_0(\mu)$, $S(\mu)$ be as in the above lemma. We assume that, for μ generating R, there is no identical automorphism σ of $\boldsymbol{Q}(\mu_1)$ such that

$$\{\mu_1^{\sigma\sigma_1},\cdots,\mu_1^{\sigma\sigma_g}\}=\{\mu_1^{\sigma_1},\cdots,\mu_1^{\sigma_g}\} \qquad \text{(as sets)}.$$

Then, for any two μ,ν, both generating R, $S_0(\mu)$ and $S_0(\nu)$ have the same characteristic roots if and only if $\mu=\nu$.

Proof. "If" part is trivial. Let $S_0(\mu)$ and $S_0(\nu)$, therefore $S(\mu)$ and $S(\nu)$ have the same characteristic roots. Then, $\nu_1=\mu_1^{\sigma_i}$ for some i, and therefore,

$$S(\nu)=\begin{pmatrix} \nu_1^{\sigma_1} & & \\ & \ddots & \\ & & \nu_1^{\sigma_g} \end{pmatrix}=\begin{pmatrix} \mu_1^{\sigma_i\sigma_1} & & \\ & \ddots & \\ & & \mu_1^{\sigma_i\sigma_g} \end{pmatrix}.$$

But as μ,ν generates R_1, $\mu_1\to\nu_1=\mu_1^{\sigma_i}$ is an automorphism of $\boldsymbol{Q}(\mu)$, hence by assumption, σ_i must be the identity. Hence $\nu_1=\mu_1$, and $\nu=\mu$.

The following lemma is fundamental for the purpose of this §.

LEMMA 23. Let A be an abelian variety defined over a field k with $q=p^f$ elements and R a subring of $\mathcal{A}(A)$. We assume that R is the principal order of the field R_0. We denote by π the endomorphism defined by $x\to x^q$, where $x=(x_0,\cdots,x_N)$ is a generic point of A over k and $x^q=(x_0^q,\cdots,x_N^q)$. Let $S^*(\lambda)$ be a representation of a homomorphism λ of A onto an abelian variety $B=\lambda A$ with invariant linear differentials of A and of B, and let finally $\mathfrak{q}$ be the largest ideal in R such that $S^*(\lambda_{\mathfrak{q}^r})=0$ for some natural number r. Then if some power π^h of π belongs to R, a prime ideal $\mathfrak{p}$ of R divides (π^h) if and only if it divides $\mathfrak{q}$.

Proof. For any ideal $\mathfrak{a}$ in R, we have, by Lemma 8, §1, $S^*(\lambda_{\mathfrak{a}})=0$ if and only if $k_R(x^p)\supset k_R(\lambda_{\mathfrak{a}}x)$, and this holds if and only if $k_R(x^p)\supset k_R(\alpha x)$ for all α in $\mathfrak{a}$. Therefore, if two ideals $\mathfrak{q}_1,\mathfrak{q}_2$ satisfy $S^*(\lambda_{\mathfrak{q}_i^r})=0$, then the greatest common divisor $(\mathfrak{q}_1,\mathfrak{q}_2)$ satisfies a similar condition, hence there exists certainly the largest one $\mathfrak{q}$. Clearly this $\mathfrak{q}$ has no multiple factors. Let $\mathfrak{p}_1,\cdots,\mathfrak{p}_s$ be all the different prime divisors of (π^h) in R. Then, there is a natural number r such that for any μ in $\mathfrak{p}_1\cdots\mathfrak{p}_s$, μ^r is divisible by π^h, therefore $k_R(x^p)\supset$

$k_R(\mu^r x)$, that is, $S^*(\lambda_{(\mathfrak{p}_1\cdots\mathfrak{p}_s)^r})=0$. Thus we see that $\mathfrak{q}$ divides $\mathfrak{p}_1\cdots\mathfrak{p}_s$. Next, for any ν in $\mathfrak{q}$, we have $S^*(\nu^r)=0$ with some r, that is, $k_R(x^p)\supset k_R(\nu^r x)$, so $k_R(x^{p^2})\supset k_R((\nu^r x)^p)$. But as $\nu^r x$ is also a generic point of A, and as the isomorphism $\mu\to\mu^*$ defined by $\mu^*\nu^r=\nu^r\mu$ is identical on R, we have similarly $k_R((\nu^r x)^p)\supset k_R(\nu^{2r}x)$, and by induction, we have $k_R(\pi^h x)=k_R(x^{q^h})\supset k_R(\nu^{rfh}x)$, that is, ν^{rfh} is divisible by π^h. This shows that $\mathfrak{q}^{rfh}$ is divisible by $\mathfrak{p}_1\cdots\mathfrak{p}_s$, hence $\mathfrak{q}=\mathfrak{p}_1\cdots\mathfrak{p}_s$, which was to be proved.

COROLLARY. If an anti-representation $S(\mu)$ of $\mathcal{A}(A)$ by invariant differentials on A is diagonal for all elements of R, then the above $\mathfrak{q}$ is the set of all μ in R such that $S(\mu)=0$, and we have $S^*(\lambda_{\mathfrak{q}})=0$.

Proof. In this case, if $S(\mu)\neq 0$, then $S(\mu^r)\neq 0$ for any r. Then, $S(\mu)=0$ means $k(x^p)\supset k(\mu x)$ for all μ in $\mathfrak{q}$, therefore $k(x^p)\supset k(\lambda_{\mathfrak{q}}x)$, hence $S^*(\lambda_{\mathfrak{q}})=0$, by Lemma 8.

We recall some properties of the endomorphism π. Let C be a non-singular curve of genus g, defined over a field k_0, and let J be a jacobian variety of C, φ a canonical mapping of C into J. We can and do assume J and φ to be defined over a finite algebraic extension k of k_0. (cf. Chow [2] or Matsusaka [1]). Let k be a finite field of $q=p^f$ elements, and let $P_1,\cdots,P_g$ be independent generic points of C over k. Then there is a correspondence I in $C\times C$ defined as the locus of $P_1\times P_1^q$ over k, and an endomorphism π of A is defined by $x=\sum\varphi(P_i)\to\pi x=\sum\varphi(I(P_i))=x^q$. Moreover we have $\pi\pi'=\pi'\pi=q\delta_J$ (cf. Weil [2], 2nd part, Cor. 1 of Th. 13 and [VA], n° 48). Furthermore, let $F(X)=\prod_{i=1}^{2g}(X-\varpi_i)$ be the characteristic polynomial of the l-adic representation $M_l(\pi)$ of π, then the congruence zeta function of C over k is of the form $\zeta(s)=[(1-q^{-s})(1-q\cdot q^{-s})]^{-1}\prod_{i=1}^{2g}(1-\varpi_i q^{-s})$ as is seen in [VA], n° 69. The "Riemann hypothesis" for $\zeta(s)$ shows that each ϖ_i has the absolute value $q^{\frac{1}{2}}$ (Weil [2], 2nd part, §IV). If π generates a field over $\boldsymbol{Q}$, therefore, every algebraic conjugate of π has the absolute value $q^{\frac{1}{2}}$.

More generally, let A be an abelian variety, defined over a finite field k with $q=p^f$ elements, and let x be a generic point of A over k. We define the endomorphism π^* by $\pi^*x=x^q$. If A is simple, there is a homomorphism λ of a jacobian variety J of a curve C on A onto A. We assume that λ, J are defined over k, then, we have $\pi^*\lambda=\lambda\pi$, by definition. Therefore, for the minimal polynomial $F(X)$ for π with rational integral coefficients, we have $F(\pi^*)=0$. As A is simple, π^* generates a commutative field. This shows especially, all isomorphic images of π^* into the complex number field have the same absolute value $q^{\frac{1}{2}}$, $\pi^*\pi^{*\prime}=q\delta_A$, as $\pi^{*\prime(i)}=\overline{\pi^{*(i)}}$ for any conjugate $\pi^{*(i)}$

of π^*, by Lemma 16, as the centre of $\mathcal{A}_0(A)$ is invariant by $\mu\to\mu'$ as a whole.

The same conclusion holds for non-simple A. Let $\mathcal{A}(A)=\mathcal{A}(A_1)+\cdots+\mathcal{A}(A_r)$, where $\mathcal{A}(A_i)$ are matrix algebras over skew-fields. Then, we have a decomposition $\pi^*=\pi_1^*+\cdots+\pi_r^*$. Therefore we have only to prove our conclusion in case where $A_1=B\times\cdots\times B$, with simple B. Then, π_1^* is contained in the centre of $\mathcal{A}(A)$, which is isomorphic to the centre of $\mathcal{A}(B)$. Hence we have $\pi_1^*\pi_1^{*\prime}=q\delta_A$, and all the characteristic roots of $M_l(\pi_1^*)$ have the absolute value $q^{\frac{1}{2}}$. Thus the same holds for π^*.

We note that if J, λ are not rational over k, then, we must operate in a finite extension of k, over which λ, J are rational. Thus we have, in general, $\pi^*\pi^{*\prime}=q\varepsilon$, where ε is a root of unity in $\mathcal{A}(A)$. But the absolute values of the characteristic roots of $M_l(\pi^*)$ are $q^{\frac{1}{2}}$ also in this case.

From now on, in the rest of this paper, we assume that k is a finite algebraic number field. Then, for the set $\mathfrak{P}$ of normalized discrete valuations of k_1, considered in §1, "$\mathfrak{p}$ in some proper open set" is equivalent to "for almost all $\mathfrak{p}$". Thus, for almost all $\mathfrak{p}$, the assertions in Prop. 1, Prop. 2 in §1 hold. *Henceforward, we only consider such "good natured $\mathfrak{p}$" or its extension $\mathfrak{P}$ in some overfields of k*, with additional conditions if necessary. Other $\mathfrak{P}$'s will be called exceptional.

Furthermore, we use the following notations throughout the rest of this paper. A is an abelian variety. If A is a jacobian variety of a non-singular curve C of genus g, we denote A with J. φ is a canonical mapping of C into J. A is defined over a field k. C and φ are defined over k_0, which contains k. *We assume that $\mathcal{A}_0(A)$ contains a field R_0 of degree $2g$ and we denote $R=R_0\cap\mathcal{A}(A)$. R is clearly an order of R_0.* We denote by x a generic point of A (or J) over k. k_R means the smallest field containing k, over which all the endomorphisms in R are rational. $S_0(\mu)$ is an anti-representation of $\mathcal{A}(A)$ with a basis $(\omega_1\cdots,\omega_g)$ of invariant differentials of A, where ω_i are all rational over k. Also, $S(\mu)$ is an anti-representation, which is diagonal for μ in R: $S(\mu)=\begin{pmatrix}\mu_1^{\sigma_1} & & \\ & \ddots & \\ & & \mu_1^{\sigma_g}\end{pmatrix}$, ($\sigma_1=$identity e). Then, we identify R with an order of a field $k_1=\iota(R_0)=\boldsymbol{Q}(\nu_1)$ (where ν generates R) by $\mu\to\mu_1=\iota(\mu)$. We denote by $\bar{k}_1$ the smallest normal field containing k_1, and by G its Galois group over $\boldsymbol{Q}$. G_1 denotes the subgroup corresponding to the subfield k_1. Let G_{00} be the subgroup of G of all σ in G such that $\{\mu_1^{\sigma_1},\cdots,\mu_1^{\sigma_g}\}^{\sigma}=\{\mu_1^{\sigma_1},\cdots,\mu_1^{\sigma_g}\}$ (as a set) for all μ in R, and k^{**} the subfield of $\bar{k}_1$, corresponding to G_{00}. σ_i can be considered as an element of G, then, σ_i is determined up to the left multiplication of elements in G_1. Then, we denote by G_0 one of the subgroups of G_{00}, such that the decomposition: $\sum_{i=1}^{g} G_1\sigma_i = \sum_{i=1}^{s} G_1\tau_i G_0$ takes place, where $s=g/(\text{order of } G_0)$. This condition is equivalent to $G_0 \cap \sigma_i^{-1} G_1 \sigma_i = \{e\}$ for $i=1,\cdots g$; we denote by k^* the subfield of $\bar{k}$, corresponding to this group G_0. Clearly we have $\bar{k}_1 \supset k^* \supset k^{**}$. We denote by k_D a normal field, containing $k, \bar{k}_1$ and k_R; then, $S_0(\mu)$ can be transformed into the diagonal form $S(\mu)$ by a matrix in k_D. (We do not assume that k_D is fixed once for all.) We denote by σ_0 the complex conjugate automorphism of the complex number field, and we use the symbol "—" to denote it: $\mu^{\sigma_0} = \bar{\mu}$. Then k_r denotes the subfield of k_1, consisting of the numbers μ_1, such that $\bar{\mu}_1 = \mu_1$. Let $\mathfrak{P}$ be a prime ideal in a finite algebraic extension of k, then the endomorphism defined by $\tilde{x} \to \tilde{x}^{N(\mathfrak{P})}$ (N means always absolute norm) is clearly the centre of $\mathcal{A}(J)$, and hence in $\tilde{R}$, for, if otherwise $\tilde{R}$ and this endomorphism would generate a commutative algebra of degree $> 2g$. Hence we can denote it as $\tilde{\pi}_{\mathfrak{P}}$, with $\pi_{\mathfrak{P}}$ in R. (This $\pi_{\mathfrak{P}}$ can be defined for allmost all $\mathfrak{P}$.)

Let $\mathfrak{a}$ be an ideal in R. Then $\lambda_{\mathfrak{a}}$ means the homomorphism of A into an abelian variety B, defined in Lemma 10, §2.

We do not repeat these assumptions, conventions and explanations of notations in the following.

We can see that k_R contains k^{**}. Indeed, let σ be an isomorphism of $k_R \cup k^{**}$ over k_R. Then, as $\mu^{\bar{\sigma}} = \mu$, we see $S_0(\mu) = S_0(\mu^{\bar{\sigma}}) = S_0(\mu)^{\sigma}$ by Lemma 21, hence we have $\{\mu_1^{\sigma_1}, \cdots, \mu_1^{\sigma_g}\} = \{\mu_1^{\sigma_1}, \cdots, \mu_1^{\sigma_g}\}^{\sigma}$. Thus σ induces on $\bar{k}_1$ an automorphism belonging to G_{00}, hence fixes the elements of k^{**}.

We prove furthermore

LEMMA 24. Let A and R satisfy the conditions in Lemma 22, and let R be invariant as a whole by all the automorphisms $\bar{\sigma}$. Then we have $k_R = k^{**} \cup k$.

Proof. As is seen above, we have $k_R \supset k^{**} \cup k$. Let σ be an isomorphism of k_R over $k^{**} \cup k$. Then, $S_0(\mu^{\bar{\sigma}}) = S_0(\mu)^{\sigma}$. But, as σ fixes the elements of k^{**}, $S_0(\mu)^{\sigma}$ has the same characteristic roots as $S_0(\mu)$. For each μ generating R, we have therefore $\mu^{\bar{\sigma}} = \mu$ by Lemma 22. Hence σ must be identity on k_R.

In the rest of this §, we assume that R is the principal order of R_0.

LEMMA 25. We have the following ideal decomposition in k_1:

$$(\iota\pi_{\mathfrak{P}})=N_{k_D/k_1}(\mathfrak{P}^{\sigma_1^{-1}}\cdots\mathfrak{P}^{\sigma_g^{-1}})$$

where $\mathfrak{P}$ is a prime ideal in k_D, σ_i are supposed to be extended to k_D. $\sigma_1,\cdots,\sigma_g,\sigma_0\sigma_1,\cdots,\sigma_0\sigma_g$ induce all the $2g$ isomorphisms of k_1.

Proof. (By the above assumption, R is the principal order of R_0). We first determine the largest ideal $\mathfrak{q}$ in R, such that each $\nu\in\mathfrak{q}$ satisfies $\widetilde{S(\nu)}=0$, that is, $S(\nu)\equiv 0 \pmod{\mathfrak{P}}$. But this holds if and only if $\nu_i^{\sigma_i}\equiv 0 \pmod{\mathfrak{P}}$, that is $\nu_1\equiv 0 \pmod{\mathfrak{P}^{\sigma_i^{-1}}}$ for $i=1,\cdots,g$. If we denote by $\mathfrak{p}_i$ the prime ideal in k_1, divisible by $\mathfrak{P}^{\sigma_i^{-1}}$, we see therefore $\iota(\mathfrak{q})=\prod'\mathfrak{p}_i$, where $\prod'$ means the product over all different $\mathfrak{p}_i$'s. Thus, by Cor. of Lemma 23, $\mathfrak{p}_1,\cdots,\mathfrak{p}_g$ are all the prime divisors of $\iota(\pi_{\mathfrak{P}})$ in k_1. First we choose as $\mathfrak{P}$ a prime ideal of the first degree in k_D, which is possible as there are infinitely many such $\mathfrak{P}$'s. Then we have $\iota(\pi_{\mathfrak{P}})\cdot\overline{\iota(\pi_{\mathfrak{P}})}=p$ by Lemma 16, §2, and as p decomposes completely in k_1 in this case, we have first $\iota(\pi_{\mathfrak{P}})\neq\overline{\iota(\pi_{\mathfrak{P}})}$, that is, k_1 is totally imaginary by Lemma 16, §2, and next, $\iota(\pi_{\mathfrak{P}})$ must contain just g mutually different prime ideals in k_1, no two of which are complex conjugate. We see therefore that, for above $\mathfrak{p}_i=N_{k_D/k_1}(\mathfrak{P}^{\sigma_i^{-1}})$, $\mathfrak{p}_i\neq\mathfrak{p}_j(i\neq j)$ and $\mathfrak{p}_i\neq\bar{\mathfrak{p}}_j$. This shows, σ_i does not belong to $G_1\sigma_j(i\neq j)$ and to $G_1\sigma_0\sigma_j$. Especially, $\sigma_1,\cdots,\sigma_g$ and $\sigma_0\sigma_1,\cdots,\sigma_0\sigma_g$ induce all the $2g$ isomorphisms of k_1 over $\boldsymbol{Q}$. Hence we have for any prime ideal $\mathfrak{P}$ of arbitrary degree,

$$\prod_{i=1}^{g}N_{k_D/k_1}(\mathfrak{P}^{\sigma_i^{-1}})\cdot\prod_{i=1}^{g}\overline{N_{k_D/k_1}(\mathfrak{P}^{\sigma_i^{-1}})}=N\mathfrak{P}=(q)=\iota(\pi_{\mathfrak{P}})\cdot\overline{\iota(\pi_{\mathfrak{P}})}.$$

As the prime divisors of $\iota\pi_{\mathfrak{P}}$ are $\mathfrak{p}_1,\cdots,\mathfrak{p}_g$, this proves our lemma.

PROPOSITION 3. Let $k^{\#}$ be an algebraic number field of finite degree, which contains k_R. Then we have the following ideal decomposition of $\iota\pi_{\mathfrak{P}}$ in k_1:

$$(\iota\pi_{\mathfrak{P}})=N_{\bar{k}_1/k_1}(\mathfrak{p}^{\tau_1^{-1}}\cdots\ \mathfrak{p}^{\tau_s^{-1}})^f\,,$$

where $\mathfrak{p}^f=N_{k^{\#}/k^*}\mathfrak{P}$.

Proof. We choose k_D so as to contain $k^{\#}$ and denote by $\mathcal{P}$ a prime divisor of $\mathfrak{P}$ in k_D; $N_{k_D/k^{\#}}\mathcal{P}=\mathfrak{P}^h$. Then we have $\pi_{\mathcal{P}}=\pi_{\mathfrak{P}}^h$. But by Lemma 25, we have

$$\iota(\pi_{\mathcal{P}})=N_{k_D/k_1}(\mathcal{P}^{\sigma_1^{-1}}\cdots\mathcal{P}^{\sigma_g^{-1}})=N_{\bar{k}_1/k_1}[\prod_{\substack{\sigma\in G_0\\ i=1,\cdots,s}}(N_{k_D/\bar{k}_1}\mathcal{P})^{\sigma^{-1}\tau_i^{-1}}]$$

as $k_D,\bar{k}_1$ are both normal. The last expression is equal to

$$\prod_{i=1}^{s}N_{\bar{k}_1/k_1}(N_{k_D/k^*}\mathcal{P})^{\tau_i^{-1}}=\prod_{i=1}^{s}N_{\bar{k}_1/k_1}(\mathfrak{p}^{fh\tau_i^{-1}}),$$

which proves our proposition.

We shall write $\Gamma(\mathfrak{a})=N_{\bar{k}_1/k_1}(\mathfrak{a}^{\tau_1^{-1}}\cdots\mathfrak{a}^{\tau_s^{-1}})$ for any ideal $\mathfrak{a}$ in k^{**}. Γ is a homomorphism of the group of ideals of k^* into that of k_1, and induces clearly a homomorphism of ideal class group of k^* into that of k_1, for absolute ideal class group as well as for "Strahlklassengruppe" modulo a natural number. We note that as σ_0 does not belong to G_{00}, k^{**} is, hence k^* is, imaginary, so we have not to distinguish between the absolute class in the "narrower sense" and that in the "wider sense" in k^* or in k_1.

We remark: Let $\mathfrak{P}$ be a prime ideal of the first degree in $\bar{k}_1$, and let $\mathfrak{p}=N_{\bar{k}_1/k^*}\mathfrak{P}$; then $\Gamma(\mathfrak{p})=N_{\bar{k}_1/k_1}(\mathfrak{P}^{\sigma_1^{-1}}\cdots\mathfrak{P}^{\sigma_g^{-1}})$. Hence $\Gamma(\mathfrak{p})$ is independent of the choice of G_0 if $\mathfrak{p}$ decomposes in $\bar{k}_1$ into prime ideals of the first degree.

LEMMA 26. If there is no automorphism σ of k_1 other than identity over some imaginary subfield of k_1 such that $\{\mu_1^{\sigma\sigma_1},\cdots\mu_1^{\sigma\sigma_g}\}=\{\mu_1^{\sigma_1},\cdots,\mu_1^{\sigma_g}\}$, and if k_1 is relatively normal over any imaginary subfield, then, for any prime ideal $\mathfrak{p}$ of the first degree in k^* whose norm does not ramify in k, we have $\Gamma(\mathfrak{p})^\sigma\neq\Gamma(\mathfrak{p})$ for all isomorphisms $\sigma\neq e$ of k_1. On the other hand, if $\Gamma(\mathfrak{p})^\sigma\neq\Gamma(\mathfrak{p})$ for all isomorphisms $\sigma\neq e$ and for some $\mathfrak{p}$, then, by the reduction modulo a prime divisor $\mathfrak{P}$ of $\mathfrak{p}$ in $k^\#$, $\tilde{J}$ is simple, so $\mathcal{A}(\tilde{J})=\tilde{R}$ and $\{\mu_1^{\sigma\sigma_1},\cdots,\mu_1^{\sigma\sigma_g}\}\neq\{\mu_1^{\sigma_1},\cdots,\mu_1^{\sigma_g}\}$ for any automorphism $\sigma\neq e$ of k_1.

Proof. We first prove the first half. If $\Gamma(\mathfrak{p})^\sigma=\Gamma(\mathfrak{p})$ for some $\sigma\neq e$, then, as $\mathfrak{p}$ is of the first degree, $\sum G_0\tau_i^{-1}G_1\sigma=\sum G_0\tau_i^{-1}G_1$, that is, $\sum\sigma^{-1}G_1\tau_iG_0=\sum G_1\tau_iG_0$. Thus, if σ is an automorphism of k_1, then, as σ commutes with G_1, we have $\sum G_1\sigma^{-1}\sigma_i=\sum G_1\sigma_i$, hence $\{\mu_1^{\sigma\sigma_1},\cdots,\mu_1^{\sigma\sigma_g}\}=\{\mu_1^{\sigma_1},\cdots,\mu_1^{\sigma_g}\}$, so that the field k_4 of elements of k_1 fixed by σ must be real by the first assumption. If σ is not an automorphism of k_1 then by the second assumption we see that also in this case k_4 must be real. Then, $\Gamma(\mathfrak{p})$ must be a real ideal. On the other hand, however, as $\mathfrak{p}$ is of the first degree, $\Gamma(\mathfrak{p})\cdot\overline{\Gamma(\mathfrak{p})}=p$, so, $\Gamma(\mathfrak{p})\cdot\overline{\Gamma(\mathfrak{p})}$ has no multiple factor, as the norm of $\mathfrak{p}$ does not ramify in k, hence $\Gamma(\mathfrak{p})\neq\overline{\Gamma(\mathfrak{p})}$, that is, $\Gamma(\mathfrak{p})$ cannot be real, which is a contradiction.

Next, we prove the second half. As $\Gamma(\mathfrak{p})^\sigma\neq\Gamma(\mathfrak{p})$ for all $\sigma\neq e$, $\pi_{\mathfrak{P}}$ cannot belong to any proper subfield of R_0 by the above Proposition 3. As the centre of $\mathcal{A}(\tilde{A})$ contains $\tilde{\pi}_{\mathfrak{P}}$, it contains also $\tilde{R}$, hence we must have $\mathcal{A}(\tilde{J})=\tilde{R}$, as otherwise $\mathcal{A}(\tilde{A})$ should be a matrix algebra with centre of lower degree than $2g$. The last assertion $\{\mu_1^{\sigma\sigma_1},\cdots,\mu_1^{\sigma\sigma_g}\}\neq\{\mu_1^{\sigma_1},\cdots,\mu_1^{\sigma_g}\}$ is easily seen by going backward in the proof of the first half. (Then, the assumption that $\mathfrak{p}$ is of the first degree is unnecessary.)

COROLLARY. If k_1 does not contain any imaginary proper subfield, then, all the conclusions in the above lemma hold.

LEMMA 27. We have

$$|[N_{k_D/k_1}(\beta^{\sigma_1^{-1}}\cdots\beta^{\sigma_g^{-1}})]^\sigma|=|N\beta|^{\frac{1}{2}}$$

for any element β in k_D and for any isomorphism σ of k_1. Especially, we have $|\Gamma(\gamma)^\sigma|=|N\gamma|^{\frac{1}{2}}$ for any element γ in k^* and for any isomorphism σ of k.

Proof. For simplicity, we write here $k_D=k$. If A is simple, then, σ_0 commutes with any automorphism σ of k_1 by the corollary of Lemma 16. Hence

$$[N_{k/k_1}(\beta^{\sigma_1^{-1}}\cdots\beta^{\sigma_g^{-1}})]^\sigma\,\overline{[N_{k/k_1}(\beta^{\sigma_1^{-1}}\cdots\beta^{\sigma_g^{-1}})]^\sigma}$$
$$=([N_{k/k_1}(\beta^{\sigma_1^{-1}}\cdots\beta^{\sigma_g^{-1}})]\,\overline{[N_{k/k_1}(\beta^{\sigma_1^{-1}}\cdots\beta^{\sigma_g^{-1}})]})^\sigma=(N\beta)^\sigma=N\beta.$$

Similarly $\Gamma(\gamma)^\sigma\overline{\Gamma(\gamma)^\sigma}=(N\gamma)^\sigma$. If σ_0 does not commute with some σ, then, A is not simple and $\mathcal{A}_0(A)$ is a matrix algebra over a skew field, as was remarked at the beginning of §2. All $\pi_{\mathfrak{P}}$ is contained in the centre $\mathfrak{R}_0$ of $\mathcal{A}_0(A)$. As $\mathfrak{R}_0$ is invariant as a whole by all the automorphism of $\mathcal{A}_0(A)$, it is a totally imaginary quadratic extension of a totally real field. Indeed, for almost all prime ideal $\mathfrak{P}$ of the first degree in k, we have $(\iota\pi_{\mathfrak{P}})\neq\overline{(\iota\pi_{\mathfrak{P}})}$ as $(\iota\pi_{\mathfrak{P}}\cdot\overline{\iota\pi_{\mathfrak{P}}})=(p)$, hence $\mathfrak{R}_0$ is certainly imaginary. By the corollary of Lemma 25, we see that σ_0 commutes with any automorphism σ of the smallest normal field $\bar{\mathfrak{R}}_0$ (over $\boldsymbol{Q}$) containing $\mathfrak{R}_0$. On the other hand, $\mathfrak{p}_i=N_{k/k_1}\mathfrak{P}^{\sigma_i^{-1}}$ are all different for almost all $\mathfrak{P}$ of the first degree. For any automorphism σ of $\bar{k}_1$ over $\mathfrak{R}_0$, we have $\pi_{\mathfrak{P}}^\sigma=\pi_{\mathfrak{P}}$, hence $\mathfrak{p}_i^\sigma=\mathfrak{p}_j$ for some j. Thus we see that $(N_{k/k_1}\beta^{\sigma_i^{-1}})^\sigma=(N_{k/k_1}\beta^{\sigma_j^{-1}})$, hence $N_{k/k_1}(\beta^{\sigma_1^{-1}}\cdots\beta^{\sigma_g^{-1}})$ is invariant by this σ, so belongs to $\mathfrak{R}_0$. Then, we have only to prove our assertions for all automorphisms σ of $\bar{\mathfrak{R}}_0$, and this can be proved just as above.

§4. Zeta function of a curve

Now we give the first application of the above considerations.

We fix some k_D and denote it with k. We first assume that R is the principal order of R_0. In this § we identity μ and $\iota\mu$, and omit the symbol ι.

For an ideal $\mathfrak{a}$ in R, we call "$\mathfrak{a}$ Teil-point" on A such a point b that $\alpha b=0$ for all α in $\mathfrak{a}$. Such points form a group $\mathfrak{g}(\mathfrak{a})$, whose order is $N\mathfrak{a}$ by Lemma 12, §2. We call an $\mathfrak{a}$ Teil-point proper if it is not $\mathfrak{b}$ Teil-point for any $\mathfrak{b}\supsetneqq\mathfrak{a}$. Then, if b is a proper $\mathfrak{a}$ Teil-point, $\mu b=\nu b$ if and only if $\mu\equiv\nu \bmod \mathfrak{a}$ for any μ,ν in R. Therefore $\mathfrak{g}(\mathfrak{a})$ can be written in the form Rb with this b. Especially, μ, b is also proper if and only if $(\mu,\mathfrak{a})=1$. Hence there are just $\varphi(\mathfrak{a})$ proper $\mathfrak{a}$ Teil-points. Let b be one of these. Then every one of $\varphi(\mathfrak{a})$ proper $\mathfrak{a}$ Teil-points can be written in the form μb with μ in R, $(\mu,\mathfrak{a})=1$.

Let $\mathfrak{a}$ be an ideal in R, prime to 2 and to the discriminant of R_0, and let b be a proper $\mathfrak{a}$ Teil-point. We consider the field extension $k(b)$ over k. For any isomorphism σ of $k(b)$ over k, b^σ is also a proper $\mathfrak{a}$ Teil-point, as σ fixes k, hence also k_R, and so $\bar{\sigma}$ fixes the elements in R. We can write therefore $b^{\sigma}=\mu_{\sigma}b$ with μ_{σ} in R, and, if we denote by $[\mu]$ the residue class in R mod $\mathfrak{a}$, to which μ belongs, $[\mu_{\sigma}]$ is uniquely determined by σ independently of μ_{σ}.

We denote it with $[\sigma]$ also. As μ is rational over k, $b^{\sigma}=\mu b$ is rational over $k(b)$, hence $k(b)$ is normal over k. We denote the Galois group of $k(b)/k$ with $\mathfrak{G}$. Then, for two σ, τ in $\mathfrak{G}$, $b^{\sigma\tau}=(\mu_{\sigma}b)^{\tau}=\mu_{\sigma}\mu_{\tau}b$ as $\mu^{\tau}=\mu$ for μ in R, hence $[\sigma\tau]=[\sigma]\cdot[\tau]$, and $\sigma\rightarrow[\sigma]$ is a homomorphism of $\mathfrak{G}$ into the prime residue class group mod $\mathfrak{a}$ in R. But, if $[\sigma]=[1]$, then $b^{\sigma}=b$ so $\sigma=e$, hence this is in fact an isomorphism. This shows especially that $k(b)$ is abelian over k.

Let now $\mathfrak{P}$ be a prime ideal in k, satisfy the conditions imposed in §3, which is in addition prime to $N\mathfrak{a}$. Then $\tilde{b}$ is a proper $\mathfrak{a}$ Teil-point on $\tilde{A}$. In fact, for any β in R, not belonging to $\mathfrak{a}$, we have $\beta b\neq e$ by assumption, and $(N\mathfrak{a})\cdot\beta b=e$, so $\widetilde{\beta b}=\tilde{\beta}\tilde{b}\neq\tilde{e}$ by Prop. 1, §1 and by $(N\mathfrak{a}, p)=1$. We further impose on $\mathfrak{P}$ an additional condition: $\widetilde{k(b)}=\tilde{k}(\tilde{b})$, which certainly holds for almost all $\mathfrak{P}$. Let $\sigma=\sigma(\mathfrak{P})$ be the Frobenius automorphism of $\mathfrak{P}$ in $\mathfrak{G}$, so we have $\widetilde{b^{\sigma(\mathfrak{P})}}=\tilde{b}^{N\mathfrak{P}}$, and right hand side is equal to $\tilde{\pi}_{\mathfrak{P}}\tilde{b}$ by definition. Hence we have $b^{\sigma(\mathfrak{P})}=\pi_{\mathfrak{P}}b$ by Prop. 1, §1, as $b^{\sigma(\mathfrak{P})}$ and $\pi_{\mathfrak{P}}b$ are also $\mathfrak{a}$ Teil-points. Thus we have proved $[\sigma(\mathfrak{P})]=[\pi_{\mathfrak{P}}]$. For an ideal $\mathfrak{B}=\mathfrak{P}_1\cdots\mathfrak{P}_r$ in k, prime to the exceptional $\mathfrak{P}$'s we see from this, $[\sigma(\mathfrak{B})]=[\pi_{\mathfrak{P}_1}\cdots\pi_{\mathfrak{P}_r}]$, denoting by $\sigma(\mathfrak{B})=\sigma(\mathfrak{P}_1)\cdots\sigma(\mathfrak{P}_r)$ the Artin-symbol of $\mathfrak{B}$; we denote, $\pi_{\mathfrak{B}}=\pi_{\mathfrak{P}_1}\cdots\pi_{\mathfrak{P}_r}$. Then we have, $\sigma(\mathfrak{B})=e$ if and only if $\pi_{\mathfrak{B}}\equiv 1 \pmod{\mathfrak{a}}$. But, as $k(b)/k$ is abelian, $k(b)$ is contained in some "Strahlklassenkörper" mod F of k, and we may and do take F as a natural number divisible by $\mathfrak{a}$. If γ is in the ideal group for this "Strahlklassenkörper", namely, if $\gamma\equiv\eta$ (mod. F), where η is a unit in k, then, γ belongs a fortiori to the ideal group for the abelian extension $k(b)/k$, hence $\sigma((\gamma))=e$ by class field theory. Thus we see that, if $\gamma\equiv\eta \pmod{F}$, then $\pi_{(\gamma)}\equiv 1 \pmod{\mathfrak{a}}$.

Now, for an ideal $\mathfrak{B}$ in k, prime to exceptional prime ideals in k, we define $\chi(\mathfrak{B})=\pi_{\mathfrak{B}}/|\pi_{\mathfrak{B}}|$. As $|\pi_{\mathfrak{B}}|=|N_{\mathfrak{B}}|^{\frac{1}{2}}$, $|N\mathfrak{B}|^{\frac{1}{2}}\chi(\mathfrak{B})$ is a number in k_1. By Lemma 25, §3, we have

$$(|N\mathfrak{B}|^{\frac{1}{2}}\chi(\mathfrak{B}))=N_{k/k_1}(\mathfrak{B}^{\sigma_1^{-1}}\cdots\mathfrak{B}^{\sigma_g^{-1}}).$$

Especially, for a principal ideal (β) in k, we have,

$$\chi((\beta))=\varepsilon(\beta)N_{k/k_1}(\beta^{\sigma_1^{-1}}\cdots\beta^{\sigma_g^{-1}})/|N\beta|^{\frac{1}{2}},$$

where $\varepsilon(\beta)$ is a unit in k_1. But, $\varepsilon(\beta)$ is a root of unity in k_1 as $|\varepsilon(\beta)^{\sigma}|=1$ for any σ in G. Indeed, as $|\pi_{\mathfrak{P}}^{\sigma}|=|N\mathfrak{P}|^{\frac{1}{2}}$ for any σ in G, we have $|\pi_{(\beta)}^{\sigma}|=|N\beta|^{\frac{1}{2}}$, on the one hand and $|[N_{k/k_1}(\beta^{\sigma_1^{-1}}\cdots\beta^{\sigma_g^{-1}})]^{\sigma}=|N\beta|^{\frac{1}{2}}$ on the other hand. Hence, from the relation $\varepsilon(\beta)N_{k/k_1}(\beta^{\sigma_1^{-1}}\cdots\beta^{\sigma_g^{-1}})=\pi_{(\beta)}$

we have the assertion.

Let now $\beta\equiv 1 \pmod F$, then $N_{k/k_1}(\beta^{\sigma_1^{-1}}\cdots\beta^{\sigma_g^{-1}})\equiv 1 \pmod F$, as F is a natural number. Moreover, as was just seen, $\pi_{(\beta)}\equiv 1 \pmod{\mathfrak{a}}$. Hence $\varepsilon(\beta)\equiv 1 \pmod{\mathfrak{a}}$ as F is divisible by $\mathfrak{a}$, so $\varepsilon(\beta)=1$, as $\mathfrak{a}$ is prime to 2 and to the discriminant of k_1. Thus, $\chi(\mathfrak{B})$ is a "Grössencharakter" of Hecke (cf. Hecke [3]), whose module is the product of F and all the exceptional prime divisors in k. In the same way, we see that $\chi_i(\mathfrak{B})=\dfrac{\pi_{\mathfrak{B}}^{\sigma}}{|\pi_{\mathfrak{B}}|}$ is also a "Grössencharakter", for any σ in G. Indeed, $\pi_{(\beta)}\equiv 1 \pmod{\mathfrak{a}}$ implies $\pi_{(\beta)}^{\sigma}\equiv 1 \pmod{\mathfrak{a}^{\sigma}}$. Moreover, F is also divisible by $\mathfrak{a}^{\sigma}$, and $\mathfrak{a}^{\sigma}$ is prime to the discriminant of k_1

We now pass to the case where R is not the principal order. We take an ideal $\mathfrak{b}$ in R, which has the principal order of R_0 as its order. Then, the ring $\mathcal{A}(\lambda_{\mathfrak{b}}A)$ of endomorphisms of $\lambda_{\mathfrak{b}}A$ contains the principal order of the field R_0^*, as was noted before. As $\lambda_{\mathfrak{b}}$ and $\lambda_{\mathfrak{b}}A$ are defined over k, the endomorphism $\pi_{\mathfrak{P}}^*$ of $\lambda_{\mathfrak{b}}A$, defined by $\pi_{\mathfrak{P}}^*\lambda_{\mathfrak{b}}=\lambda_{\mathfrak{b}}\pi_{\mathfrak{P}}$ has the same meaning for $\lambda_{\mathfrak{b}}A$ as $\pi_{\mathfrak{P}}$ has for A. Indeed, as $\tilde{\lambda}_{\mathfrak{b}}$ is rational over $\tilde{k}$, we have, $\tilde{\pi}_{\mathfrak{P}}^*\tilde{\lambda}_{\mathfrak{b}}\tilde{x}=\tilde{\lambda}_{\mathfrak{b}}\tilde{\pi}_{\mathfrak{P}}\tilde{x}=\tilde{\lambda}_{\mathfrak{b}}\tilde{x}^q=(\tilde{\lambda}_{\mathfrak{b}}\tilde{x})^q$. As R_0, R_0^* are fields of degree $2g$, we can transform l-adic matrices $M_l(\mu)$, and $M_l^*(\mu^*)=M_l(\lambda_{\mathfrak{b}})M_l(\mu)M_l(\lambda_{\mathfrak{b}})^{-1}$ into diagonal forms, and we can assume

$$M_l^*(\mu^*)=\begin{pmatrix}\mu_1^{\rho_1} & & \\ & \ddots & \\ & & \mu_1^{\rho_{2g}}\end{pmatrix}=M_l(\mu),$$ where $\rho_1,\cdots,\rho_{2g}$ are all the isomorphisms of k_1. But, as $\chi_i^*(\mathfrak{B})=\dfrac{\pi_{\mathfrak{B}}^{*\rho_i}}{|\pi_{\mathfrak{B}}^*|}$ is a Hecke character, $\chi_i=\dfrac{\pi_{\mathfrak{P}}^{\rho_i}}{|\pi_{\mathfrak{P}}|}$ is also a "Grössencharakter" for every ρ_i. Thus we have proved

PROPOSITION 4. Let A be an abelian variety, let $\mathcal{A}(A)$ contain a ring R, isomorphic to an order of a field k_1 of degree $2g$. Let A and all the endomorphisms in R be defined over an algebraic number field k of a finite degree. We assume that k contains all the conjugates of k_1. Then, the symbol $\chi_i(\mathfrak{B})=\pi_{\mathfrak{B}}^{\rho_i}/|\pi_{\mathfrak{B}}^{\rho_i}|$ is a "Grössencharakter" of Hecke in k, whose module contains all the exceptional prime ideals in k.

When $A=J$ is a jacobian variety of a non-singular curve C, defined over k, this proposition 4 gives a result on ζ-function of C. The ζ-function $\zeta_C(s)$ of C is defined as follows: Denote $\zeta_{\mathfrak{P}}(s)$ the congruence ζ-function of the curve $\tilde{C} \pmod{\mathfrak{P}}$, then,

$$\zeta_C(s)=\prod{}'\zeta_{\mathfrak{P}}(s)$$

where $\prod'$ means that $\mathfrak{P}$ runs through all non-exceptional prime ideals $\mathfrak{P}$ in k. (cf. Weil [8] or Deuring [4]). When a canonical mapping φ of C into J is also rational over k, then, as quoted above, in §3,

$\zeta_{\mathfrak{P}}(s)$ is of the form,

$$\zeta_{\mathfrak{P}}(s)=[(1-q^{-s})(1-q^{1-s})]^{-1}\prod_{i=1}^{2g}(1-\pi_{\mathfrak{P}}^{\rho_i}q^{-s})$$

where $q=N\mathfrak{P}$, and ρ_i runs over all the $2g$ isomorphisms of k_1. Thus we have proved

THEOREM 1. *Let C be a non-singular curve, J a jacobian variety of C, and φ a canonical mapping of C into J. Let $\mathcal{A}(J)$ contain a field R_0 of degree $2g$. We assume that C, J, φ and all endomorphisms in R_0 are defined over an algebraic number field k. Then, the zeta function $\zeta_C(s)$ of C over k has the following form:*

$$\zeta_C(s)=\Phi(s)\zeta(s)\zeta(s-1)\prod_{i=1}^{2g}L\left(s-\frac{1}{2},\ \chi_i\right)^{-1}$$

where $\zeta(s)$ means the zeta function of k, $L(s,\chi_i)$ means the L-function of k with the "Grössencharakter" χ_i and $\Phi(s)$ is a product of rational functions of q^{-s} for a finite numbers of $q=N\mathfrak{P}$.

REMARK. As shall be proved later, if J is simple and $\mathcal{A}_0(J)=R_0$ is a field of degree $2g$, then, C can be defined over an algebraic number field of finite degree, hence there exists always a field k satisfying the assumption of our theorem.

§5. Some Lemmas from analytic theory

As A is defined over an algebraic number field k, we can apply the theory of theta-functions and Riemann matrices. Let Ω be a Riemann matrix of A. Ω is a period-matrix of invariant differentials $\omega_1,\cdots,\omega_g$ of A. Therefore, denoting as before by $S_0(\mu)$ the representation of $\mathcal{A}(A)$ by these invariant differentials $\omega_1,\cdots,\omega_g$, there is a rational integral square matrix $M(\mu)$ of degree $2g$ such that

$$S_0(\mu)\Omega=\Omega M(\mu).$$

If μ is in R and of degree $2g$, then the characteristic equation of $M(\mu)$ is an irreducible equation over $\boldsymbol{Q}$, satisfied by $\iota(\mu)$, so $M(\mu)$ has $2g$ simple characteristic roots $\mu_1^{\sigma_1},\cdots,\mu_1^{\sigma_g},\ \bar{\mu}_1^{\sigma_1},\cdots,\bar{\mu}_1^{\sigma_g}$. Therefore the rank of $M(\mu)-\mu_1^{\sigma_i}I$ is $2g-1$ if μ generates R, where I is the unit matrix.

LEMMA 28. We can choose as Ω a matrix of the following form: The components of $\Omega=(\omega_{ij})$ satisfy the relation $\omega_{ij}=\omega_{1j}^{\sigma_i}$ $(i=1,\cdots g,$ $j=1,\cdots,2g)$, and $\omega_{1,1}\cdots\omega_{1,2g}$ form a basis of an ideal of $\iota(R)$.

Proof. We transform the basis $\omega_1,\cdots,\omega_g$ of differentials so that the representation becomes diagonal for μ in R: $S(\mu)=\begin{pmatrix}\mu_1^{\sigma_1} & & \\ & \ddots & \\ & & \mu_1^{\sigma_g}\end{pmatrix}$,

then Ω is transformed accordingly. After that, we denote $M(\mu)=(m_{ij})$. We then have, by $S(\mu)\Omega=\Omega M(\mu)$,

$$(*)\qquad \sum_j \omega_{ij}(m_{jk}-\mu_1^{\sigma_i}\delta_{jk})=0 \qquad i=1,\cdots,g;\ k=1,\cdots,2g$$

where $\delta_{ij}=1$ if $i=j$, $=0$ if $i\neq j$. By the permutation of columns, and multiplication by a constant factor, we can assume $\omega_{11}=1$. Now let μ be of degree $2g$ in R. Then, the rank of $M(\mu)-\mu_i^{\sigma_i}I$ is $2g-1$, so we see that ω_{1j} are uniquely determined $(j=2,\cdots,2g)$ and belong to $\boldsymbol{Q}(\mu_1)$. Then, as we have, for $i=2,\cdots,g$,

$$\sum_j \omega_{1j}^{\sigma_i}(m_{jk}-\mu_1^{\sigma_i}\delta_{jk})=0,$$

$\{\omega_{11}^{\sigma_i},\cdots,\omega_{12g}^{\sigma_i}\}$ is a solution of the above equation (*) for the index i. This is the unique solution if we set $\omega_{i1}=1$. (If $\omega_{i1}=0$, then we should have two independent solutions of (*) for the index i. Thus $\omega_{i1}\neq 0$. Then we can assume $\omega_{i1}=1$ by multiplication of a constant factor.)

Finally, for each ν in R, we must have $\nu_1\omega_{1k}=\sum\omega_{1j}d_{jk}$ with rational integral d_{jk}, so $\{\omega_{11},\cdots,\omega_{1,2g}\}$ must be a basis of some ideal of $\iota(R)$.

LEMMA 29. Let A^* be another abelian variety, and let $\mathcal{A}(A^*)$ contain a subring R^* isomorphic to R, and R^* admit the same diagonal representation $S(\mu^*)=\begin{pmatrix}\mu_1^{\sigma_1} & & \\ & \ddots & \\ & & \mu_1^{\sigma_g}\end{pmatrix}$ as R. Then A and A^* are isogenous.

Proof. By Lemma 28, A and A^* have Riemann matrices $\Omega=(\omega_{ij})$, $\Omega^*=(\omega_{ij}^*)$ of the type indicated as in that lemma. But, then, as ω_{ij}^* is in k_1 and as $\omega_{11},\cdots,\omega_{1,2g}$ form a basis of an ideal of R, we have $\omega_{1j}^*=\sum_k\omega_{1k}r_{kj}(j=1,\cdots,2g)$ with rational r_{jk}, and therefore $\omega_{ij}^*=\omega_{1j}^{*\sigma_i}=\sum\omega_{1k}^{\sigma_i}r_{kj}=\sum\omega_{jk}r_{kj}$, hence $\Omega^*=\Omega\boldsymbol{R}$ with rational $\boldsymbol{R}=(r_{kj})$, from which follows the isogeny of A and A^* by the theory of theta functions.

We now prove the converse of Lemma 17, §2, in case $g=2$ or 3. For this purpose, we must quote some results of the algebraic theory of theta functions due to A. Weil [4], [6].

Let Ω be a Riemann matrix of an abelian variety A^* of dimension g; then, by $i\Omega=\Omega\boldsymbol{J}$ with $i=\sqrt{-1}$ we obtain a real square matrix $\boldsymbol{J}$ of degree $2g$, called Weil matrix of real Riemann matrix. For each positive divisor X of A^*, there exists a skew symmetric matrix $E(X)$ of degree $2g$, with rational integral coefficients, which is called Riemann form of $\boldsymbol{J}$. Then, $F(X)=E(X)\boldsymbol{J}$ is symmetric and positive definite, and conversely, for any matrix E with these properties, there exists a positive divisor X on A^* such that $E=E(X)$. Especially, in case of jacobian J^*, the matrix $E(\Theta)$ for a Riemann divisor Θ is unimodular. We note that $C={}^tE(X)^{-1}$ is a so-called principal matrix of Ω, hence, we can transform with ${}^tE(\Theta)^{-1}$, (whose elementary divisors are certainly all equal to 1,) Ω isomorphically into (I, T), where I is the unit matrix, $T=(t_{ij})$ is symmetric and its imaginary part is negative definite. This T is a "Thetamodul" of the curve C^* corresponding to the divisor Θ. Let now A_1^*, A_2^* be two abelian varieties and Ω_1, Ω_2 Riemann matrices of A_1^*, A_2^* respectively. Let further λ be a homomorphism of A_1^* onto A_2^* and $S^*(\lambda)$ its representation with differentials. Then, there is a rational integral matrix $M(\lambda)$ such that $S(\lambda)\Omega_1=\Omega_2 M(\lambda)$, hence $M(\lambda)\boldsymbol{J}_1=\boldsymbol{J}_2 M(\lambda)$ for Weil matrices of A_1^*, A_2^*. Conversely, for each rational integral matrix M with this property, there exists a homomorphism λ of A_1^* onto A_2^* such that $M=M(\lambda)$. Thus, if A_1^* is a jacobian variety J^* with Riemann divisor Θ, for any divisor X on A_2^* there exists a homomorphism λ'_X with the matrix $M(\lambda'_X)=E(\Theta)^{-1}\cdot {}^tM(\lambda)E(X)$. Especially, if $A_2^*=J^*$, we denote $\lambda'_\Theta=\lambda'$. Then, we can see easily Trace $M(\lambda'\cdot\lambda)>0$. Therefore, if a subfield R_0^* of $\mathcal{A}_0(J^*)$ is invariant as a whole by the anti-automorphism $\mu\to\mu'$, this anti-automorphism induces the complex conjugate automorphism of all conjugates of R_0^*, as is seen just as in Lemma 16, §2.

To any point b of finite order on A^*, there corresponds a vector b_0 in one-to-one way, which has $2g$ rational numbers mod. 1 as coefficients. Then, $\lambda b=e$ if and only if $M(\lambda)a_0\equiv 0$ (mod. 1), hence $|\det M(\lambda)|$ is equal to $\nu(\lambda)$.

We now prove

LEMMA 30. Let J be simple, and $\mathcal{A}(J)=R$ be the principal order of the field $\mathcal{A}_0(J)=R_0$, and we assume $g=2$ or 3. We denote by K the subfield of R_0 consisting of elements μ such that $\mu'=\mu$. Then, for an ideal $\mathfrak{a}$ in R such that $\mathfrak{a}\mathfrak{a}'=(\alpha)$ with a total-positive element α in K, $J^*=\lambda_{\mathfrak{a}}J$ is also a jacobian variety.

Proof. As α is in K, we have $\alpha'=\alpha$. We fix as before the isomorphism $\mu\to\mu^*$ of R and $R^*=\mathcal{A}(J^*)$ by $\mu^*\lambda=\lambda\mu$. This is really an isomorphism onto by Lemma 15. As α is divisible by $\mathfrak{a}$, we can define by $\alpha=\lambda^*\lambda_{\mathfrak{a}}$ a homomorphism of J^* onto J, which we can write in the form $\lambda_{\mathfrak{a}'^*}$. For simplicity, we denote $\lambda=\lambda_{\mathfrak{a}}$, $\lambda^*=\lambda_{\mathfrak{a}'^*}$. Then, we define a matrix E^* by

$$E^*={}^tM(\lambda)^{-1}E(\Theta)M(\lambda^*)$$

where Θ is a Riemann matrix of J. This E^* is clearly rational, and is skew symmetric, because,

$$E(\Theta)^{-1}\cdot{}^tM(\lambda)\cdot{}^tE^*=E(\Theta)^{-1}\cdot{}^tM(\lambda)\cdot{}^tM(\lambda^*)(-E(\Theta))M(\lambda)^{-1}$$
$$=-M(\alpha')M(\lambda)^{-1}=-M(\lambda^*)=-E(\Theta)^{-1}\cdot{}^tM(\lambda)E^*.$$

(as $\alpha'=\alpha=\lambda^*\lambda$). Then, $F^*=E^*\boldsymbol{J}^*$ is symmetric. For

$${}^tF^*={}^t\boldsymbol{J}^*\cdot{}^tM(\lambda^*)\cdot{}^tE(\Theta)M(\lambda)^{-1}={}^tM(\lambda^*)\cdot{}^t\boldsymbol{J}\cdot{}^tE(\Theta)M(\lambda)^{-1}$$
$$={}^tM(\lambda^*)E(\Theta)\boldsymbol{J}M(\lambda)^{-1}={}^tM(\lambda^*)E(\Theta)M(\lambda)^{-1}\boldsymbol{J}^*=F^*.$$

From this we see also $F^*={}^tM(\lambda)^{-1}FM(\lambda^*)={}^tM(\lambda)^{-1}FM(\alpha)M(\lambda)^{-1}$, hence $FM(\alpha)$ is symmetric. But F is positive definite and $M(\alpha)$ can be transformed by real matrix into a diagonal matrix, whose diagonal elements are then all positive by assumption, hence F^* is also positive definite. We have clearly $|\det E^*|=1$, as $\nu(\lambda)=\nu(\lambda^*)=N\mathfrak{a}$.

Finally, we must prove that E^* is an integral matrix. First we note that, as μ'^* is in $\mathfrak{a}'^*$ for any μ in $\mathfrak{a}$, $\mu'^*=\mu_0^*\cdot\lambda^*$ with a homomorphism μ_0^* of J onto J^*, hence $\lambda^*\cdot\mu_0^*\cdot\lambda^*\cdot\lambda=\lambda^*\cdot\mu'^*\cdot\lambda=\lambda^*\cdot\lambda\cdot\mu'=\mu'\cdot\lambda^*\cdot\lambda$ as $\lambda^*\lambda=\alpha$, and $\lambda^*\cdot\mu_0^*=\mu'$. Then,
${}^tM(\mu_0^*)E^*={}^tM(\lambda^*\mu_0^*)E(\Theta)M(\lambda)^{-1}={}^tM(\mu')E(\Theta)M(\lambda)^{-1}=E(\Theta)M(\mu)M(\lambda)^{-1}$. Here we note that, as μ is in $\mathfrak{a}$, $M(\mu)M(\lambda_\mathfrak{a})^{-1}$ is an integral matrix. On the other hand, by the definition of $\lambda=\lambda_\mathfrak{a}$, if $\mu b=e$ for all μ in $\mathfrak{a}$, then $\lambda b=e$. But $\mu b=e$ if and only if $M(\mu)b_0\equiv 0$ (mod. 1), hence if and only if ${}^tM(\mu_0^*)\cdot{}^tM(\lambda^*)E(\Theta)b_0\equiv 0$ (mod. 1). Then, if ${}^tM(\mu_0^*)$ have a common right integral factor L with $|\det L|>1$, then the number of points b which satisfy $\lambda b=e$, should be greater than that of b^* satisfying $\lambda^*b^*=e^*$, which is impossible as $\nu(\lambda)=\nu(\lambda^*)$. Thus, as ${}^tM(\mu_0^*)E^*=E(\Theta)M(\mu)M(\lambda)^{-1}$ is integral for each μ in $\mathfrak{a}$, we see that E^* is itself integral.

E^* is therefore a Riemann form, with determinant 1, hence we can transform the Riemann matrix of J^* with this E^* into the form

$$\Omega^*=(I,\ T)$$

where T is symmetric and has negative definite imaginary part. But as all "Thetamodul" is Riemann's one if $g=2$ or 3, so Ω^* is a period matrix of some curve of genus g, which proves our lemma.

LEMMA 31. Let J be as in the above lemma, with Riemann divisor Θ. Then, for each positive divisor X with unimodular Riemann form $E(X)$, δ'_X is a totally positive unit in R, and conversely, for each totally positive unit η in R, there exists a positive divisor Y on J with unimodular Riemann form $E(Y)$, such that $\eta=\delta'_Y$.

Proof. As we have $M(\delta'_X)=E(\Theta)^{-1}E(X)$, δ'_X is a unit in R. But, $E(X)\boldsymbol{J}=M(\delta'_X)E(\Theta)\boldsymbol{J}$ and $E(\Theta)\boldsymbol{J}$ are both definite, so all the characteristic roots of $M(\delta'_X)$ must be positive as $M(\delta'_X)$ can be brought to a diagonal form, hence δ'_X is totally positive. Conversely, if we set $E=E(\Theta)M(\eta)$, then, as $\eta'=\eta$ by assumption, E is skew symmetric. Next $E\boldsymbol{J}=M(\eta)E(\Theta)\boldsymbol{J}$ is positive definite as η is totally positive, hence E is a unimodular Riemann form, and there is a Y such that $E=E(Y)$, namely, $\eta=\delta'_Y$.

LEMMA 32. Two Riemann forms $E(X)$ and $E(Y)$ are equivalent if and only if $\delta'_X=\zeta'\delta'_Y\zeta$ with a unit ζ in $\mathcal{A}(J)$.

Note. Here we call two Riemann forms equivalent if there is a unimodular matrix M such that $M=M(\zeta)$ for a unit ζ and $E(X)={}^tME(Y)M$. If so, the "Thetamodul" obtained by them are equal.

Proof. Evident by definition.

REMARK. Lemmas 31, 32 are not used in this paper.

§6. Algebraic and arithmetic nature of "class equation"

In the following §§6–7, we assume that A (or J) is simple, and except for Lemma 33, $\mathcal{A}(A)=R$ is the principal order of R_0. In this §6, we assume furthermore that k_D contains k^*, (and also k_0 in case of J). Therefore φ and C are rational over k_D. We denote $k_D=k$ in this §.

If two A, A^* have isomorphic R, R^*, and admit the same diagonal representation: $S(\mu)=S^*(\mu^*)$, then, they are isogenous, hence $A^*\cong\lambda_{\mathfrak{a}}A$ with an ideal $\mathfrak{a}$ in R (by Lemmas 29, 14, 15), where $\cong$ means the isomorphism of abelian varieties. Therefore, there are just h abelian varieties $A_1,\cdots,A_h$, having isomorphic R and admitting the same $S(\mu)$, where h is the class number of k_1. Indeed, for any ideal $\mathfrak{a}$ in R, $\mu\to\mu^*$, (where $\mu^*\lambda_{\mathfrak{a}}=\lambda_{\mathfrak{a}}\mu$) defines an isomorphism of R onto $\mathcal{A}(\lambda_{\mathfrak{a}}A)$ such that $S(\mu), S^*(\mu^*)$ have the same characteristic roots.

Let σ be an automorphism of k. If there is an automorphism τ of k_1 such that $\sum_{i=1}^{g}G_1\sigma_i\sigma=\tau\cdot\sum_{i=1}^{g}G_1\sigma_i$, then A and A^σ admit the same representation $S(\mu)$ of $\mathcal{A}(A), \mathcal{A}(A^\sigma)$. In fact, we have $S_0^\sigma(\mu^{\bar\sigma})=(S_0(\mu))^\sigma$ by Lemma 21, so $S_0^\sigma(\mu^{\bar\sigma})$ has characteristic roots $(\mu_1^{\sigma_1\sigma}\cdots\mu_1^{\sigma_g\sigma})=(\mu_1^{\tau\sigma_1}\cdots\mu_1^{\tau\sigma_g})$, where S_0^σ have the same meaning as in Lemma 21. If we identify $\mu^{\bar\sigma}$ with $\mu_1^\tau=\iota(\mu^{\bar\sigma})$, then we have an isomorphism $\mu\to\mu^*=\iota^{-1}(\mu_1)$ of R onto $R^{\bar\sigma}$, such that $S(\mu)=S^*(\mu^*)$. Hence this σ permutes the isomorphism-classes of $A_1,\cdots,A_h$. We denote with $[\mathfrak{a}(\sigma)]$ the ideal class of k_1, containing an ideal $\mathfrak{a}(\sigma)$, defined by $A^\sigma\cong\lambda_{\mathfrak{a}(\sigma)}A$. Clearly, $[\mathfrak{a}(\sigma)]$ is determined uniquely by σ (cf. Lemma 13), hence we shall write also $[\mathfrak{a}(\sigma)]=[\sigma]$.

Now, we assume that, *for almost all prime ideal* $\mathfrak{p}$ *of the first degree in* k^*, *we have* $\Gamma(\mathfrak{p})^\sigma\neq\Gamma(\mathfrak{p})$, *for all isomorphisms* $\sigma\neq e$ *of* k_1. (A sufficient condition for this, is given in Lemma 26, §3.) Then, by Lemma 26, the conditions of Lemma 22 are satisfied. Thus, if two matrices $S_0(\mu), S_0(\nu)$ for μ,ν in R have the same characteristic roots, then we have $\mu=\nu$. Especially, let σ be an automorphism of k over k^{**}, and let $A^\sigma=\zeta\lambda_{\mathfrak{a}(\sigma)}A$, where ζ is an isomorphism of $\lambda_{\mathfrak{a}(\sigma)}A$ onto A^σ. Then we have $\mu^{\bar\sigma}\zeta\lambda_{\mathfrak{a}(\sigma)}=\zeta\lambda_{\mathfrak{a}(\sigma)}\mu$. Indeed, $S_0^\sigma(\mu^{\bar\sigma})=S_0(\mu)^\sigma$ has the same characteristic roots as $S_0(\mu)$, as σ is an automorphism over k^{**}. But if we define μ^* by $\mu^*\zeta\lambda_{\mathfrak{a}(\sigma)}=\zeta\lambda_{\mathfrak{a}(\sigma)}\mu$, then $S_0^\sigma(\mu^*)$ has clearly the same characteristic roots as $S_0(\mu)$, hence as $S_0^\sigma(\mu^{\sigma})$. Thus we see $\mu^*=\mu^{\bar\sigma}$. But then, we have $\lambda_{\mathfrak{a}}^{\bar\sigma}\zeta\lambda_{\mathfrak{a}(\sigma)}A\cong\lambda_{\mathfrak{a}^{\bar\sigma}}\zeta\lambda_{\mathfrak{a}(\sigma)}A=\lambda_{\mathfrak{a}}^*\zeta\lambda_{\mathfrak{a}(\sigma)}A\cong\zeta\lambda_{\mathfrak{a}(\sigma)\mathfrak{a}}A$. We have therefore $A^{\tau\sigma}\cong(\lambda_{\mathfrak{a}(\tau)}A^\sigma)\cong\lambda_{\mathfrak{a}(\tau)}^{\bar\sigma}\zeta\lambda_{\mathfrak{a}(\sigma)}A\cong\zeta\lambda_{\mathfrak{a}(\sigma)\mathfrak{a}(\tau)}A$. This shows that, $\sigma\to[\sigma]$ is a homomorphism of the Galois group of k over k^{**}, into the ideal class group of k_1. Denote with $\hat{k}$ the subfield of k, corresponding to the kernel of this homomorphism. Then $\hat{k}$ is relatively abelian over k^{**}, and its Galois group over k^{**} is isomorphic to the subgroup of ideal class group of k_1.

But it is advantageous to consider $\hat{k}$ over k^*, rather than over k^{**}. Let $\mathfrak{p}$ be a prime ideal of the first degree in k^*, which does not ramify in $\hat{k}$ and satisfies the condition $\Gamma(\mathfrak{p})^{\sigma} \neq \Gamma(\mathfrak{p})$. Denote with $\sigma(\mathfrak{p}) = \left(\frac{\hat{k}/k^*}{\mathfrak{p}}\right)$ the Frobenius automorphism of $\mathfrak{p}$. Then, $\widetilde{A^{\sigma(\mathfrak{p})}} = \tilde{A}^{N\mathfrak{p}} = \tilde{A}^p$, considered modulo a prime divisor $\mathfrak{P}$ of $\mathfrak{p}$ in k, hence $\tilde{A}^p = \lambda_{\widetilde{\mathfrak{a}(\sigma(\mathfrak{p}))}}\tilde{A}$. But as all the prime divisors of $(\pi_{\mathfrak{P}})$ in k_1 divide also $\Gamma(\mathfrak{p})$, we have $\tilde{k}(\tilde{x}^p) \supset \tilde{k}(\lambda_{\widetilde{\Gamma(\mathfrak{p})}}\tilde{x})$ by the Cor. of Lemma 23, §3. But as we have $[\tilde{k}(\tilde{x}) : \tilde{k}(\tilde{x}^p)] = p^g = N_{\widetilde{\Gamma(\mathfrak{p})}} = [\tilde{k}(\tilde{x}) : \tilde{k}(\lambda_{\widetilde{\Gamma(\mathfrak{p})}}\tilde{x})]$, so $\tilde{k}(\tilde{x}^p) = \tilde{k}(\lambda_{\widetilde{\Gamma(\mathfrak{p})}}\tilde{x})$, that is $\lambda_{\widetilde{\Gamma(\mathfrak{p})}}\tilde{A}$ is isomorphic to $\tilde{A}^p$, hence to $\lambda_{\widetilde{\mathfrak{a}(\sigma(\mathfrak{p}))}}\tilde{A}$. This means first, that $\widetilde{\Gamma(\mathfrak{p})}$ belongs to the same class as $\widetilde{\mathfrak{a}(\sigma(\mathfrak{p}))}$, and then, that $[\Gamma(\mathfrak{p})] = [\mathfrak{a}(\sigma(\mathfrak{p}))]$, since $\mathcal{A}(\tilde{A}) = \tilde{R}$. Thus we have, $[\sigma(\mathfrak{p})] = [\Gamma(\mathfrak{p})]$, that is, $\Gamma(\mathfrak{p})^f \sim 1$ in k_1 if and only if $\sigma(\mathfrak{p})^f = e$ (identity), which proves the following

PROPOSITION 5. Let A be a simple abelian variety such that $\mathcal{A}(A)$ is isomorphic to the principal order of a field k_1 of degree $2g$, and let the condition $\Gamma(\mathfrak{p})^{\sigma} \neq \Gamma(\mathfrak{p})$ be satisfied. Let $\hat{k}$ be the subfield of k, which corresponds to the subgroup of the Galois group of k over k^*, consisting of all automorphisms σ such that $A^{\sigma} \cong A$. Then $\hat{k}$ is the class field over k^*, for the ideal group H, where $H = \{\mathfrak{b} \mid \Gamma(\mathfrak{b}) \sim 1 \text{ in } k_1\}$.

But we cannot go further, unless we confine ourselves to the case of jacobian.

We can attach, following van der Waerden [1], to each projective curve D' of genus g in the same universal domain Ω, a set of quantities $(m) = (m_1, \cdots, m_n)$, called moduli of D', such that two curves are birationally equivalent if and only if they have the same moduli These moduli (m) depend rationally on D', that is, $(m)^{\sigma}$ is the moduli of D'^{σ}, for any automorphism σ of Ω. If $\tilde{Q}(m)$ is perfect, we can find a non-singular curve D with moduli (m), defined over the field $\tilde{Q}(m)$, where $\tilde{Q}$ is the prime field (cf. Deuring [1]). In our case, with the complex number field as the universal domain, we have the following results. If two curves C, C^* have the same Riemann matrix Ω and equivalent Riemann forms $E(\Theta)$, $E(\Theta^*)$, where Θ, Θ^* denote the Riemann divisors of C, C^* respectively, then they have the same "Thetamodul" T, as T is obtained using these Riemann forms. Especially, if we can take Θ^* equal to Θ, then, by Torelli's theorem, C and C^* are birationally equivalent. Thus, for one jacobian variety J, there are only a finite number δ of non-equivalent curves

$C_1,\cdots C_\delta$, having J as their jacobian, and we have $\delta\leq\gamma$, where $\gamma=[(\eta):(\bar{\zeta}\zeta)]$ is the index of the group $(\bar{\zeta}\zeta)$ in the group (η) of all the totally positive units in k_r. ($(\bar{\zeta}\zeta)$ consists of the units in k_r, of the form $\bar{\zeta}\zeta$ with a unit ζ in k_1.) (cf. Cor. of Lemma 17, and Lemma 19, or Lemmas 31, 32.) Let now $(m^{(1)}),\cdots,(m^{(\delta)})$ be the moduli of $C_1,\cdots,C_\delta$ respectively, which can be considered as coordinates of points in an affine space of dimension n. Then we denote by (M) the Chow-coordinates of these δ points $(m^{(1)}),\cdots,(m^{(\delta)})$, and call (M) the moduli of J. If two J, J^* have the same moduli, then we have first $\delta=\delta^*$, next $(m^{(1)})=(m^{*(i)})$ for some i, hence C_1 is birationally equivalent to C_1^*, so J and J^* are isomorphic. As the converse is evident, we see that, J and J^* have the same modules if and only if they are isomorphic. Moreover, the moduli of J^σ are clearly (M^σ), by definition. Thus (M) deserves certainly this name.

Lemma 33. (We do not assume in this lemma that R is the principal order, and that J is defined over k.) Under our assumption that $\mathcal{A}_0(J)$ is the field of degree $2g$, the moduli (M) of J are all algebraic numbers.

Proof. We first assume that R is the principal order. If (M) are not algebraic, then some $(m^{(i)})$ are not algebraic. Then the point $P=(1, m_1^{(i)},\cdots, m_n^{(i)})$ has a locus U over the algebraic closure $\bar{\bar{Q}}$ of Q in a projective space S^n. Now take as the set $\mathfrak{B}$ in Prop. 1, §1, the set of all discrete valuation of the field $\bar{\bar{Q}}(m^{(i)})$, which are trivial on $\bar{\bar{Q}}$. Let $\mathfrak{B}'$ be any proper open set in $\mathfrak{B}$. Then there is a non-void open set U' of U with the property that every valuation, with the centre in U' belongs to $\mathfrak{B}'$. Thus we can specialize $C_i, \varphi_i, J, \mathcal{A}(J)$ and a basis of differentials $\omega_1,\cdots,\omega_g$ in infinitely many different ways so that the statements of Prop. 1, 2 hold for $\tilde{C}_i, \tilde{\varphi}_i, \tilde{J}, \mathcal{A}(\tilde{J})$, and $\tilde{\omega}_1,\cdots,\tilde{\omega}_g$. Then, $\mathcal{A}(\tilde{J})$ contains a subring $\tilde{R}$, isomorphic to R, and we have $\tilde{S}(\tilde{\mu})=\widetilde{S(\mu)}=S(\mu)$, for μ in R as the coefficients of the diagonal representation $S(\mu)$ belong to $\bar{\bar{Q}}$. Thus the number of moduli $(\tilde{M})$ of these $\tilde{J}$'s must be finite, hence the number of moduli $(\tilde{m}^{(i)})$ must also be finite, which is a contradiction.

Next we consider the case where $\mathcal{A}(J^*)=R^*$ is not the principal order. We choose an ideal $\mathfrak{a}^*$ in R^*, having the principal order as its order in R_0^*, and put $J'=\lambda_{\mathfrak{a}^*}J^*$. Then $\mathcal{A}(J')=R'$ is the principal order of R_0', so, as was just proved, we can find some J, isomorphic to J' and defined over an algebraic number field $k^\#$. There is a homomorphism λ of J onto J^*. But we can find a homomorphism λ_1 of J onto an abelian variety A, having the same kernel as λ, such that A is defined over the same field $k^\#$ as J. (cf. [VA], Th. 17, §5 and Matsusaka [1], Th. 3). This A is isomorphic to J^* by [VA], Th. 17, and have algebraic moduli, which was to be proved.

Let σ be an automorphism of k, compatible with the representation $S_0(\mu)$, that is, $\mathcal{A}(J)$ and $\mathcal{A}(J^\sigma)$ admit the same $S(\mu)$, and let $J^\sigma \cong \lambda_{\mathfrak{a}(\sigma)}J$. Lemma 17 shows that $\mathfrak{a}(\sigma)\cdot\overline{\mathfrak{a}(\sigma)}=(\alpha)$ and α is a totally positive number in k_r. If we denote by h' the number of ideal classes $[\mathfrak{a}]$ of k_1, such that $\mathfrak{a}\bar{\mathfrak{a}}=(\alpha)$ belongs to the principal class in k_r in the narrower sense, then there are at most h' moduli $(M^{(1)}),\cdots,(M^{(i)})$ of jacobians having isomorphic R and admitting the same $S(\mu)$. In case $g=2$ or 3, there are just h' moduli (cf. Lemma 30). Let k^{***} be the subfield of $\bar{k}_1$, which corresponds to the subgroup of G, consisting of all σ's, compatible with $S_0(\mu)$. Then, every automorphism over k^{***} permutes $(M^{(1)}),\cdots,(M^{(i)})$ among themselves, that is, the "class equation" has all its coefficients in the field k^{***}. Here, "class equation" is considered to be formed of the "class-invariants" $(M^{(1)}),\cdots,(M^{(i)})$, of classes $[\mathfrak{a}_1],\cdots,[\mathfrak{a}_i]$, for which $\lambda_{\mathfrak{a}_1}J,\cdots,\lambda_{\mathfrak{a}_i}J$ are jacobians having $(M^{(1)}),\cdots,(M^{(i)})$ as their modules respectively.

It is clear that the field $\hat{k}$ in Prop. 5 is just the field $k^*(M)$ in this case, if J satisfies the condition $\Gamma(\mathfrak{p})^\sigma \neq \Gamma(\mathfrak{p})$ imposed in that proposition. If moreover, each of all the h' classes $[\mathfrak{a}]$ in k_1, having the norm $\mathfrak{a}\bar{\mathfrak{a}}$ to k_r in the principal class in the narrower sense, contains an ideal of the form $\Gamma(\mathfrak{b})$, then the isomorphism $\sigma\to[\sigma]$ of the Galois group of $k^*(M)$ over k^* into that ideal class group of h' classes, is onto. That is, there are just h' moduli $(M^{(1)}),\cdots,(M^{(h')})$ of jacobians corresponding to R and $S(\mu)$, and they are obtained from one of them by the automorphisms of $k^*(M)$ over k^*. This means the irreducibility of "class equation".

Thus we have proved the main theorem of this paper.

THEOREM 2. *Let k_1 be a totally imaginary algebraic number field of degree $2g$, which is an extension of a totally real field k_r of degree g, and let $\sigma_1,\cdots,\sigma_g$ be one half of all the $2g$ isomorphisms of k_1 into the complex number field, such that $\sigma_0\sigma_1,\cdots,\sigma_0\sigma_g$ give the another half. We assume that there exists a simple jacobin variety J, whose $\mathcal{A}(J)$ is isomorphic to the principal order of k_1 and admits the diagonal representation* $S(\mu)=\begin{pmatrix}\mu_1^{\sigma_1} & & \\ & \ddots & \\ & & \mu_1^{\sigma_g}\end{pmatrix}$. *Let k^*, k^{**}, k^{***} be the fields, defined as before. Let finally $[\mathfrak{a}_1],\cdots,[\mathfrak{a}_i]$ be the subset of the set of h' ideal classes having relative norms to k_r equal to the principal class in k_r in the narrower sense, for which $\lambda_{\mathfrak{a}_j}J$ are also jacobian varieties. We denote with $(M^{(j)})$ the moduli of $\lambda_{\mathfrak{a}_j}J$. Then* *the "class equation" of k_1, corresponding to $(\sigma_1, \cdots, \sigma_g)$, that is, composed of these $(M^{(1)}), \cdots, (M^{(i)})$ is rational over k^{***}. We assume further, that k_1 and k^* satisfy the condition: $\Gamma(\mathfrak{p})^\sigma \neq \Gamma(\mathfrak{p})$ for every isomorphism $\sigma \neq e$ and for almost all prime ideals of the first degree in k^*. Then, the field $k^{**}(M^{(j)})$ is abelian over k^{**}, that is, the irreducible components of "class equation" over k^{**} are abelian equations. Furthermore, the field $k^*(M^{(j)})$ is the class field over k^*, for the ideal group $H = \{\mathfrak{b} \mid \Gamma(\mathfrak{b}) \sim 1 \text{ in } k_1\}$, and the isomorphism $\sigma \leftrightarrow [\sigma]$ gives the explicit form of Artin's reciprocity law. If finally every one of all the h' ideal classes with the above property, contains an ideal of the form $\Gamma(\mathfrak{b})$, then the "class equation" is of degree h' and irreducible over k^*, consequently, $[k^*(M^{(j)}) : k^*] = h'$.*

§7. Decomposition law in the field of Teil-point

In this §, we assume that $\mathcal{A}(A) = R$ is the principal order of R_0, but we do not assume that φ is rational over k in case of jacobians.

In the following we identify μ and $\iota\mu$, as no confusion may occur. By Prop. 3 in §3, we have $(\pi_{\mathfrak{P}}) = \Gamma(\mathfrak{p})^f$, where $\mathfrak{P}$ is a prime ideal in $k_R \cup k^*$ and $\mathfrak{p}^f = N^*_{k_R \cup k^*/k^*}\mathfrak{P}$. We have also $|\pi^\rho_{\mathfrak{P}}| = N\mathfrak{P}^{\frac{1}{2}}$, for any isomorphism ρ of k_1, as recalled in §3. But we need more exact expression for $\pi_{\mathfrak{P}}$. For this purpose, we use a system of ideal numbers in k^*. Let us recall some basic properties thereof (cf. Hecke [3]). Let $\{\mathfrak{b}_1, \cdots, \mathfrak{b}_e\}$ be a basis of ideal class group in k^* and h_i be the order of $\mathfrak{b}_i : \mathfrak{b}_i^{h_i} = (b_i)$, with b_i in k^* $(i = 1, \cdots, e)$. We denote with $\widehat{\beta}_i$ a complex number $\sqrt[h_i]{b_i}$. Then, to any ideal $\mathfrak{c} = c\mathfrak{b}_1^{n_1} \cdots \mathfrak{b}_e^{n_e}$ in k^*, where c is in k^*, we attach a quantity $\hat{\gamma} = c\widehat{\beta}_1^{n_1} \cdots \widehat{\beta}_e^{n_e}$. σ being an isomorphism of k^*, we select one of the numbers $\sqrt[h_i]{b_i^\sigma}$ and denote it with $\widehat{\beta}_i^\sigma$, under one condition: $\widehat{\beta}_i^{\sigma\sigma_0} = \overline{\widehat{\beta}_i^\sigma}$ for any σ. Then, $\hat{\gamma}^\sigma$ is defined by $\hat{\gamma}^\sigma = c^\sigma \widehat{\beta}_1^{n_1\sigma} \cdots \widehat{\beta}_e^{n_e\sigma}$. The absolute norm of $\widehat{\beta}$ is defined by $N\widehat{\beta} = \prod_\sigma \widehat{\beta}^\sigma$, and relative norm is defined in the corresponding manner. If we choose these $\widehat{\beta}_i, \widehat{\beta}_i^\sigma$ arbitrarily but in a fixed way, then the system $\{\hat{\gamma}\}$ is determined by it, which is called a system of ideal numbers. For each ideal $\mathfrak{c}$ in k^*, there correspond ideal numbers $\hat{\gamma}$, which are different from each other by unit factors in k^*. The addition and subtraction, therefore the congruence relation, of ideal numbers can only be defined between those in the same class.

We define the correspondence Γ also for ideal numbers in k^* in the same way as for ideals: $\Gamma(\hat{\gamma}) = N_{\bar{k}_1/k_1}(\hat{\gamma}^{\tau_1^{-1}} \cdots \hat{\gamma}^{\tau_s^{-1}})$. Then we have $\Gamma(\hat{\gamma})^\sigma \cdot \overline{\Gamma(\hat{\gamma})^\sigma} = N\hat{\gamma}$ for any automorphism σ of $\bar{k}_1$ (cf. Lemma 27). But as we have $|N\hat{\gamma}| = N\mathfrak{c}$ clearly, we see that $|\Gamma(\hat{\gamma})^\sigma| = (N\mathfrak{c})^{\frac{1}{2}}$. Thus

$\Gamma(\hat{\gamma})$ is determined by $\mathfrak{c}$ up to a root of unity factor, (which is not necessarily in k_1), independent of the choice of the system $\{\hat{\gamma}\}$ and of the unit factor of $\hat{\gamma}$ in k^*. $\Gamma(\mathfrak{c})$ is a principal ideal in k_1 if and only if $\varepsilon'\Gamma(\hat{\gamma})$ belongs to k_1, with some root of unity ε'. If $k \supset k_R \cup k^*$, then we have,

$$\pi_{\mathfrak{P}} = \varepsilon_1 \Gamma(\hat{\varpi})^f,$$

where $(\hat{\varpi}) = \mathfrak{p}$, ε_1 is a root of unity and $N_{k/k^*}\mathfrak{P} = \mathfrak{p}^f$.

Denote with E the group of all the roots of unity in k_1. For any point b on A, we denote with $E(b)$ the zero cycle $\sum_{\varepsilon \in E} (\varepsilon b)$ on A, and $k(b)_0$ the smallest field containing k, over which the cycle $E(b)$ is rational.

We take $k = k_R \cup k^*$ and a proper $\mathfrak{a}$ Teil-point b on A, where $\mathfrak{a}$ is an ideal in R. Let $\mathfrak{p}$ be a prime ideal in k^*, prime to $\mathfrak{a}$, and $\mathfrak{P}$ be a prime divisor of $\mathfrak{p}$ in k, with relative degree f over k^*. We impose on $\mathfrak{P}$ one additional condition that the field $\widetilde{k(b)_0}$ is the smallest field containing $\tilde{k}$, over which $\tilde{E}(\tilde{b})$ is rational, which is certainly the case for almost all $\mathfrak{P}$. Let furthermore f_0 be the smallest exponent such that the congruence $\varepsilon'\Gamma(\hat{\varpi})^{f_0} \equiv \varepsilon$ (mod. $\mathfrak{a}$) holds with some roots of unity ε', ε, the latter being in E. Denote $F = l.c.m.[f, f_0]$. Then, as $\pi_{\mathfrak{P}}^{F/f} = \varepsilon_1^{F/f}\Gamma(\hat{\varpi})^F \equiv \varepsilon_2$ (mod $\mathfrak{a}$), with ε_2 in E, the cycle $E(b)$ is invariant by $\pi_{\mathfrak{P}}^{F/f}$. Hence $\tilde{E}(\tilde{b})$ is at most of degree F/f over $\tilde{k}$, that is, at most of degree F over $\tilde{k}^*$. Conversely, denote $F_0 = [\widetilde{k(b)_0} : \tilde{k}^*]$. It is clear that f divides F_0 and $F_0 \leq F$. As $\tilde{E}(\tilde{b})$ is rational over $\widetilde{k(b)_0}$, we have $\tilde{\pi}_{\mathfrak{P}}^{F_0/f}\tilde{b} = \tilde{\varepsilon}\tilde{b}$, where ε is in E, hence $\pi_{\mathfrak{P}}^{F_0/f} \equiv \varepsilon$ (mod $\mathfrak{a}$), as $\tilde{b}$ is a proper $\mathfrak{a}$ Teil-point on $\tilde{A}$. From this we see that $\varepsilon_1^{F_0/f}\Gamma(\hat{\varpi})^{F_0} \equiv \varepsilon$ (mod $\mathfrak{a}$), which means that f_0 divides F_0, hence $F = F_0$, as also f divides F_0. Thus we have proved that $k(b)_0$ is composite field of k and the class field over k^* for the ideal group $H_{\mathfrak{a}} = \{(\hat{\beta}) \mid \varepsilon'\hat{\beta} \equiv \varepsilon \text{ (mod } \mathfrak{a}),$ ε', ε are roots of unity, $\varepsilon \in E\}$.

The correspondence $\hat{\beta} \to \Gamma(\hat{\beta})$ induces clearly a homomorphism of "Strahlklassengruppe" mod. $N\mathfrak{a}$ in k^* into that mod. $N\mathfrak{a}$ in k_1. Especially, if $\beta \equiv$ a unit (mod. $N\mathfrak{a}$), then $\Gamma(\beta) \equiv \varepsilon$ (mod. $N\mathfrak{a}$) with some ε in E, as $|\Gamma(\eta)^\sigma| = 1$ for any unit η in k^*. Therefore, $H\mathfrak{a}$ contains the "Strahl" mod. $N\mathfrak{a}$.

If the condition of Prop. 5, §6, is satisfied, then we have $k_R = k \cup k^{**}$ by Lemmas 26 and 24. If furthermore, A is defined over the field $\hat{k}$, (the notation $\hat{k}$ having the same meaning as in Prop. 5,) this taking $k = \hat{k}$, we see from Prop. 5 that f_0 is a multiple of f, hence that $F = f_0$. We remark that if $A = J$ and $\gamma = [\eta : \bar{\zeta}\zeta] = 1$, then J is certainly defined over $\widehat{k}=k^*(M)$ by the result of Chow [1] and Deuring [1], as this condition means that there is just one curve C having J as its jacobian, hence $(M)=(m)$ is the modules of C. At any rate we have the following.

THEOREM 3. *Let A be a simple abelian variety, defined over an algebraic number field k, and let $\mathcal{A}(A)=R$ be the principal order of the field R_0 of degree $2g$. Then the field $(k^*\cup k_R)(b)_0$ is the composite field of k_R and the class field over k^* for the ideal group $H_{\mathfrak{a}}=\{(\widehat{\beta})\,|\,\varepsilon'\widehat{\beta}\equiv\varepsilon \pmod{\mathfrak{a}}\}$. This class field is contained in the "Strahlklassenkörper" mod $N\mathfrak{a}$ over k^*. Especially, if the condition of Proposition 5 is satisfied, and if A is defined over $\widehat{k}$, then $\widehat{k}(b)_0$ is just this class field over k^* for the group $H_{\mathfrak{a}}$.*

Appendix. Zeta function of an abelian variety

We first consider the congruence zeta function.

Let A be an abelian variety defined over the finite field k of $q=p^{f_0}$ elements. Denote with k^f the finite field of q^f elements, and with N_f the number of rational points of A over k^f. Denote also with π the endomorphism of A defined by $\pi x=x^q$ for a generic point x of A over k. Then, a point b is rational over k^f if and only if $\pi^f b=b$, namely, b is a (π^f-1) Teil-point. As the left ideal in $\mathcal{A}(A)$ generated by π, π^f-1 is $\mathcal{A}(A)$ itself, we see that $k(\pi x)\cup k((\pi^f-1)x)=k(x)$, that is, $k(x)$ is separable over $k((\pi^f-1)x)$. This means however that there are just $\nu(\pi^f-1)$ (π^f-1) Teil-points, that is, there are just $\nu(\pi^f-1)$ rational points over k^f. Denoting with $\pi_1,\cdots,\pi_{2g}$ the characteristic roots of the l-adic matrix $M_l(\pi)$, we have therefore $N_f=\prod_{i=1}^{2g}(\pi_i^f-1)$.

Congruence zeta function $\zeta(s)$ of a variety A over k is defined by A. Weil [7], by the equation

$$\frac{d}{du}\log Z(u)=\sum_{f=1}^{\infty}N_f u^{f-1},$$

where $u=q^{-s}$ and $Z(q^{-s})=\zeta(s)$. In our case of abelian variety A, we have therefore

$$\frac{d}{du}\log Z(u)=\sum_{f=1}^{\infty}\Big(\prod_{i=1}^{2g}(\pi_i^f-1)\Big)u^{f-1}=\sum_{\nu=0}^{2g}(-1)^{\nu}\cdot\Big[\sum_{i_1,\cdots,i_\nu}\sum_{f=1}^{\infty}(\pi_{i_1}\cdots\pi_{i_\nu})^f u^{f-1}\Big],$$

as the absolute convergence in some circle in u-plane is evident. But the right hand side is equal to

$$\sum_{\nu=0}^{2g}(-1)^{\nu}\Big[\sum_{i_1\cdots i_\nu}\pi_{i_1}\cdots\pi_{i_\nu}/(1-\pi_{i_1}\cdots\pi_{i_\nu}u)\Big]$$

hence we have

$$Z(u)=\prod_{\nu=0}^{2g}\Big[\prod_{i_1\cdots i_\nu}(1-\pi_{i1}\cdots\pi_{i_\nu}u)\Big]^{(-1)^{\nu+1}}.$$

Next, let A be an abelian variety over an algebraic number field k. Then, the zeta function $\zeta_A(s)$ of A is defined in the same way as that of a non-singular curve. Namely, $\zeta_A(s)=\prod'_{\mathfrak{P}}\zeta_{\mathfrak{P}}(s)$, where $\zeta_{\mathfrak{P}}(s)$ is the congruence zeta function of A mod. $\mathfrak{P}$, and $\prod'$ indicates the product of all non-exceptional $\mathfrak{P}$'s. From Prop. 4 in §4, we see the following theorem.

THEOREM. *Let A and k satisfy the condition of Proposition 4. Then, the zeta function $\zeta_A(s)$ of A over k has the following form:*

$$\zeta_A(s)=\Psi(s)\frac{\zeta(s)\zeta(s-g)\prod_{\nu=1}^{g-1}\Big[\prod_{i_1\cdots i_{2\nu}}(s-\nu,\chi_{i_1}\cdots\chi_{i_{2\nu}})\Big]\Big)}{\prod_{\mu=0}^{g-1}\Big[\prod_{j_1\cdots j_{2\mu+1}}L\Big(s-\mu-\frac{1}{2},\chi_{j_1}\cdots\chi_{j_{2\mu+1}}\Big)\Big]},$$

where $\zeta(s)$ is the zeta function of k, χ_i is "Grössencharakter" in k $(i=1,\cdots,2g)$.

$L(s,\chi_{i_1},\cdots,\chi_{i_\lambda})$ is the L-function of k with character $\chi_{i_1}\cdots\chi_{i_\lambda}$ and $\Psi(s)$ is a product of rational functions of q^{-s} for a finite number of $q=N\mathfrak{P}$.

Bibliography

C. Chevalley, [1] Theory of Lie groups. (Princeton Univ. Press, 1946.)

W. L. Chow, [1] The jacobian variety of an algebraic curve. (Amer. J. Math. 76 (1954), 453-476.)

H. Davenport und H. Hasse, [1] Die Nullstellen der Kongruenzzetafunktionen im gewissen zyklischen Fällen. (Crelles J. 172(1951), 151-182.)

M. Deuring, [1] Zur Theorie der Moduln algebraischen Funktionenkörper. (Math. Z. 47(1942) 34-46.)

[2] Algebraische Begründung der komplexen Multiplikation. (Abh. Math. Sem. Univ. Hamburg 16(1949) 32-47.)

[3] Die Struktur der elliptischen Funktionkörper und die Klassenkörper der imaginären quadratischen Zahlkörper. (Math. Ann. 124(1952), 393-426.)

[4] Die Zetafunktionen einer algebraischen Kurven vom Geschlecht Eins. (Nachr. Akad. Wiss. Göttingen, (1953), 85-94.)

H. Hasse, [1] Beweis des Analogons der Riemannschen Vermutung für die Artinschen und F. K. Schmidtschen Kongruenz Zetafunktionen in gewissen elliptischen Fälle. (Göttingen Nachr. (1933), 253-262.)

E. Hecke, [1] Höhere Modulfunktionen und ihre Anwendung auf die Zahlentheorie. (Math. Ann. 71(1912), 1-37.)

[2] Über die Konstruktion relativ Abelscher Zahlkörper durch Modulfunktionen von zwei Variabeln. (Math. Ann. 74(1913), 465-510.)

[3] Eine neue Art von Zetafunktionen und ihre Beziehung zur Verteilung der Primzahlen. II. (Math. Z. 6(1920), 11-51.)

S. Lefschetz, [1] On certain numerical invariants of algebraic varieties with applications to Abelian varieties. (Trans. Amer. Math. Soc. 22(1921), 327-482.)

T. Matsusaka, [1] Some theorems on abelian varieties. (Nat. Sci. Report Ochanomizu Univ. 4(1953), 22-35.)

A. Néron, [1] Problèmes arithmétiques et géométriques rattachés à la notion de rang d'une courbe algébrique dans un corps. (Bull. Soc. Math. Frances 80(1952), 101-161.)

G. Shimura, [1] Reduction of algebraic varieties with respect to a discrete valuation of the basic field. (Amer. J. Math. 77(1955), 134-176.)

T. Tamagawa, [1] On the differential forms on an algebraic variety. (Mathematical Report of Tôdai-Kyôyôgakubu, No. 4 (1951), 1-20.) (in Japanese.)

B. L. van der Waerden, [1] Zur algebraische Geometrie XI: Projective und birationale Äquivalenz und Moduln von evenen Kurven. (Math. Ann. 114(1937), 683-699.)

A. Weil, [1] Foundations of algebraic geometry. (New York, 1946.)

[2] Sur les courbes algébriques et les variétés qui s'en déduisent. (Paris, 1948.)

[3] Variétés abéliennes et courbes algébriques. (Paris, 1948.)

[4] Théorèmes fondamentaux de la théorie des fonctions thêta. (Seminaire Bourbaki, Paris, 1949.)

[5] Arithmetic on algebraic varieties. (Ann. of Math. (2), 53(1951), 412-44.)

[6] On Picard varieties. (Amer. J. Math. 74(1952), 865-894.)

[7] Number of solutions of equations in finite fields. (Bull. Amer. Math. Soc. 55(1949), 497-508.)

[8] Jacobi sums as "Grössencharaktere". (Trans. Amer. Math. Soc. 73(1952), 487-495.)

Jacobian Varieties and Number Fields[1)]

Proceedings of the International Symposium on Algebraic
Number theory, Tokyo-Nikko, 1955; pp. 31—45;
(October, 1956)

Introduction

The classical theory of complex multiplication solves the problem of construction of abelian extensions of imaginary quadratic fields. Apart from this classical theory, E. Hecke [2] [3] has treated successfully the problem of unramified abelian extensions of certain imaginary biquadratic fields by means of Hilbert modular functions.

The main purpose of this work is to develop a theory comprising both classical theory of complex multiplication and the theory of Hecke, by the method of algebraic geometry.

By the way, an arithmetic characterization of endomorphisms π of an abelian variety with sufficiently many complex multiplications is obtained. By means of this result we can prove in the affirmative the conjecture of Hasse on zeta functions of abelian varieties, and of curves, in certain singular cases. On this subject, we have some known results in special cases. A. Weil [14] proved namely that the zeta function of a curve defined by $ax^n+by^m+c=0$ over a certain algebraic number field k can be expressed by the zeta function of k and L-functions with "Grössencharaktere". M. Deuring [1] proved analogous result for singular elliptic curves. The result in this paper contains that of Deuring, and of Weil in case n, m are different prime numbers.

The same problem of construction of abelian extensions of algebraic number fields was also treated by G. Shimura and A Weil (cf. these proceedings pp. 23–30 and pp. 9–22). A. Weil pointed out moreover some important properties of characters of idèle class groups in connection with the zeta functions of abelian varieties (cf. these

1) The following exposition is somewhat different from the text presented to the symposium. The main differences are as follows.

i) The part in which special emphasis was made on jacobian varieties is omitted, as this part contained a mistake.

ii) The part concerning Galois theory of the field K' is revised and simplified.

iii) Some results in §3, especially Proposition 3, are generalized to contain the case where $[R_0:\boldsymbol{Q}]<g$.

iv) In §5, existence theorem of Lefschetz is added.

2) This study was done with the help of subsidy from the Ministry of Education (1954, n° 10429).

proceedings pp. 1–7). The complete exposition of the problem comprising the ideas and results of G. Shimura, A. Weil and myself will be published elsewhere in a joint paper of G. Shimura and myself.

Recently I have obtained a second proof of Hasse's conjecture in case of complex multiplications as a corollary of a theorem on characters of idèle class groups, which is in close connection with the properties pointed out by A. Weil. This will be exposed in a forthcoming paper of mine. Here I wish to express my hearty thanks to Professor A. Weil for his kind discussions and valuable suggestions on these subjects during and since the symposium and also to Professor S. Iyanaga for his constant encouragement.

Notations and terminologies.

The varieties considered in this paper are always supposed to be in some projective space or in a product of projective spaces. Except for this, we use mainly the terminologies as in Weil's books [8], [9], [10] and Shimura's paper [7].

Q denotes as usual the rational number field, C the complex number field. σ_0 denotes the complex conjugate automorphism of $C: \sigma_0\mu=\bar{\mu}$ for μ in C. Algebraic number fields are always supposed to be in C. By the *Galois closure* of an algebraic number field we understand the smallest absolutely normal field containing that field. N denotes always absolute norm. For any field k, $\bar{k}$ denotes the algebraic closure of k.

A, A′, B mean always abelian varieties. The dimensions of A, A′ are always supposed to be equal and are denoted by g. As in Weil's book [10], $\mathcal{A}(A)$ denotes the ring of endomorphisms of A, $\mathcal{H}(A, B)$ denotes the module of homomorphisms of A into B. $\mathcal{A}_0(A)$, $\mathcal{H}_0(A, B)$ denote the tensor products $\mathcal{A}(A)\otimes Q$, $\mathcal{H}(A, B)\otimes Q$ respectively. x denotes always a generic point of A. The field over which x is a generic point will be clear from the context, so that we shall need no reference to it. For any λ in $\mathcal{H}(A, B)$, $\nu(\lambda)$ denotes the degree $[k(x):k(\lambda x)]$ if this degree is finite, and otherwise we put $\nu(\lambda)=0$, where k is a common field of definition of A, B, λ. In case $\nu(\lambda)\neq 0$, $\nu_i(\lambda)$ denotes the inseparability degree of $k(x)$ over $k(\lambda x)$. Let A be defined over k and R be a subring of $\mathcal{A}(A)$. Then k_R denotes the smallest field containing k, over which all endomorphisms μ in R are rational.

Let a be any point on A. In case A is defined over a finite field k of $q=p^f$ elements, we denote by a^{p^h} the isomorphic image of a by the isomorphism $\xi\to\xi^{p^h}$ of the universal domain. Then the mapping $a\to a^q$ for all a on A determines an endomorphism of A, which is denoted by π_A, or $\pi_A(k)$.

Now, let k be a field and v a discrete valuation of k, and $\mathfrak{P}$ the maximal ideal of the valuation ring $\mathfrak{O}$ of v. Then we use the symbol $\sim$ to denote the object obtained from an object of the same kind by the reduction mod. $\mathfrak{P}$. For example, $\tilde{A}$ denotes the variety obtained from A mod. $\mathfrak{P}$. The use of symbol $\tilde{A}$ indicates implicitly that this variety is also an abelian variety. $\tilde{k}$ denotes therefore the residue field $\mathfrak{O}/\mathfrak{P}$. We do not use the symbol $\sim$ in "degenerate case".

Finally, let A be defined over an algebraic number field k of finite degree, and $\mathfrak{P}$ a "non-exceptional" prime ideal in k for A (cf. below Prop. 1, §1). If there is an endomorphism in $\mathcal{A}(A)$ from which $\pi_{\tilde{A}}(\tilde{k})$ is obtained by the reduction mod. $\mathfrak{P}$, then we denote this endomorphism by $\pi_{\mathfrak{P}}$.

§1. Preliminaries from reduction theory.

Let V be a variety defined over a field k and $\mathfrak{V}$ a set of normalized discrete valuations v on k. $(\alpha)=(\alpha_1,\cdots,\alpha_r)$ being a finite set of non-zero elements in k, we denote by $\mathfrak{V}(\alpha)$ the set of all v's in $\mathfrak{V}$ such that $v(\alpha_i)=0$ for $i=1,\cdots,r$. If an assertion holds for all v's in some union of $\mathfrak{V}(\alpha)$'s, we say that it holds for *almost all* v in $\mathfrak{V}$. Therefore, when k is an algebraic number field of finite degree and $\mathfrak{V}$ is the set of all normalized discrete valuations of k, "almost all" means "all but a finite number of".

Let V_1, V_2 be two non-singular varieties and Z be the graph of a mapping f of V_1 into V_2, everywhere defined on V_1. Let V_1, V_2, f be defined over k. Then, Shimura's theory [7] shows, together with the arithmetic on algebraic varieties (Weil, [13]), that, for almost all v, $\tilde{V}_1$, $\tilde{V}_2$ are non-singular varieties and $\tilde{f}$ is a mapping, everywhere defined over $\tilde{V}_1$ with graph $\tilde{Z}$. From this we see

PROPOSITION 1. Let A be an abelian variety defined over k, and $\mathfrak{V}$ be arbitrary. Then, for almost all v in $\mathfrak{V}$, $\tilde{A}$ is an abelian variety such that $a\to\tilde{a}$ is a homomorphism of A onto $\tilde{A}$. This homomorphism induces an isomorphism of the group of all points on A with finite orders prime to the characteristic of the residue field $\tilde{k}$. If, especially, A is a jacobian variety of a non-singular curve C of genus g with a canonical mapping φ, all defined over k, then, for almost all v in $\mathfrak{V}$, $\tilde{A}$ is a jacobian variety of the non-singular curve $\tilde{C}$ with the same genus g, and $\tilde{\varphi}$ is a canonical mapping of $\tilde{C}$ into $\tilde{A}$. Moreover, let A, B be two abelian varieties defined over k. Then, for almost all v in $\mathfrak{V}$, $\tilde{\lambda}$ is a homomorphism of $\tilde{A}$ onto $\tilde{B}$ for all λ in

$\mathcal{H}(A, B)$, and $\lambda \to \tilde{\lambda}$ is an isomorphism of $\mathcal{H}(A, B)$ into $\mathcal{H}(\tilde{A}, \tilde{B})$. The same hold especially for the ring $\mathcal{A}(A)$.

We call v *non-exceptional* for A, if $\tilde{A}$ is abelian variety and $\mu \to \tilde{\mu}$ is an isomorphism of $\mathcal{A}(A)$ into $\mathcal{A}(\tilde{A})$, and non-exceptional for (A, B) if v is non-exceptional for A and for B and $\lambda \to \tilde{\lambda}$ is an isomorphism of $\mathcal{H}(A, B)$ into $\mathcal{H}(\tilde{A}, \tilde{B})$. We call v *exceptional* if it is not non-exceptional.

The linear differentials of the first kind on A are just the linear differentials invariant under translations on A, and they form a vector space D(A) of dimension g over the universal domain. Let λ be in $\mathcal{H}(A, B)$. Then, λ induces a linear transformation $\delta\lambda$ of D(B) into D(A). We denote by $S(\lambda)$ the representation-matrix of $\delta\lambda$ with basis of D(B) and of D(A). Especially, the ring $\mathcal{A}(A)$ has an anti-representation $\mu \to S(\mu)$ as a linear transformation of D(A). Let now v be non-exceptional for A. Then, the invariant property of a differential ω in D(A) shows that if ω is defined at one point of $\tilde{A}$, then it is everywhere defined on $\tilde{A}$, i.e. $\tilde{\omega}$ belongs to $D(\tilde{A})$. Thus, for any $\omega \neq 0$ rational over k, we can find an α in k such that $\widetilde{\alpha\omega}$ belongs to D(A) and is not 0. This shows that there is a basis $(\omega) = (\omega_1, \cdots, \omega_g)$ of D(A) such that $(\tilde{\omega}) = (\tilde{\omega}_1, \cdots, \tilde{\omega}_g)$ forms a basis of $D(\tilde{A})$. Conversely, for any basis (ω) of D(A), $(\tilde{\omega})$ forms a basis of $D(\tilde{A})$ for almost all non-exceptional v. Moreover, we have clearly

PROPOSITION 2. Let A, B and all λ in $\mathcal{H}(A, B)$ be defined over k, and (ω), (η) be basis of D(A), D(B) respectively, rational over k. Let v be a non-exceptional valuation of k for (A, B) such that $(\tilde{\omega})$, $(\tilde{\eta})$ form basis of $D(\tilde{A})$, $D(\tilde{B})$ respectively. Then we have $\widetilde{S(\lambda)} = \tilde{S}(\tilde{\lambda})$ for any λ in $\mathcal{H}(A, B)$, where S, $\tilde{S}$ denote the representations of $\mathcal{H}(A, B)$, $\mathcal{H}(\tilde{A}, \tilde{B})$ with basis (ω), (η) and $(\tilde{\omega})$, $(\tilde{\eta})$ respectively. The same results hold especially for anti-representations S, $\tilde{S}$ of $\mathcal{A}(A)$, $\mathcal{A}(\tilde{A})$ with basis (ω), $(\tilde{\omega})$ respectively.

COROLLARY. In the same situation, $\widetilde{S(\lambda)} = 0$ (i.e. $S(\lambda) \equiv 0 \bmod. v$) if and only if $\tilde{k}(\tilde{x}^p) \supset \tilde{k}(\tilde{\lambda}\tilde{x})$, p being the characteristic of $\tilde{k}$.

§2. Isogenous abelian varieties.

We recall first some basic properties of abelian varieties (cf. Weil [10]). Two abelian varieties A, A′ (of the same dimension g) are called *isogenous* if there is a homomorphism of A onto A′. This is an equivalence relation, by which we classify all abelian varieties into categories. We call A *simple* if the category of A is simple. Let A, A′ be isogenous. Then, the relation $\mu^*\lambda = \lambda\mu$ determines an isomorphism $\mu\rightarrow\mu^*$ of the algebra $\mathcal{A}_0(A)$ (over Q) onto the algebra $\mathcal{A}_0(A')$. For a subring R of $\mathcal{A}(A)$, we denote by R^* the image of R by this isomorphism. l being a prime number different from the characteristic p of the universal domain, we denote by M_l the l-adic representation of $\mathcal{H}(A, B)$ or of $\mathcal{A}(A)$ (cf. Weil [10] n° 31). For $\mathcal{A}(A)$ this representation is of degree $2g$ and faithfull. For any μ in $\mathcal{A}(A)$, the characteristic equation of $M_l(\mu)$ is the same for all $l\neq p$, and has rational integral coefficients with the constant term $\nu(\mu)$. Thus, each element of $\mathcal{A}_0(A)$ is of degree at most $2g$ over Q. On the other hand, if A is simple, $\mathcal{A}_0(A)$ is a division algebra, and $\mathcal{A}_0(A\times\cdots\times A)$ is a matrix algebra over $\mathcal{A}_0(A)$. If A, B are simple and not isogenous, $\mathcal{H}_0(A, B)=0$. Hence in general if $\mathcal{A}_0(A)$ contains a field R_0 of degree $2g$, A must be isogenous to $B\times\cdots\times B$, B being simple, and the commutor of R_0 in $\mathcal{A}_0(A)$ is R_0 itself.

Now, let R be a subring of $\mathcal{A}(A)$ and k be a field of definition for A and for all μ in R. Then, for any left ideal $\mathfrak{a}$ of R, there exists an abelian variety B, and a $\lambda_{\mathfrak{a}}$ in $\mathcal{H}(A, B)$, both defined over k, with the property: $k(\lambda_{\mathfrak{a}}x)=\bigcup_{\mu\in\mathfrak{a}} k(\mu x)$. Indeed, $(\mu_1,\cdots,\mu_r)$ being a set of generators of R, we can take as B the locus of $\mu_1x\times\cdots\times\mu_rx$ in $A\times\cdots\times A$, and as $\lambda_{\mathfrak{a}}$ the homomorphism defined by $\lambda_{\mathfrak{a}}x=\mu_1x\times\cdots\times\mu_rx$. Clearly, B and $\lambda_{\mathfrak{a}}$ are determined by $\mathfrak{a}$ up to isomorphisms, and the kernel of $\lambda_{\mathfrak{a}}$ is the group $\mathfrak{g}_{\mathfrak{a}}$ of all points a such that $\mu a=0$ for all μ in $\mathfrak{a}$. Moreover, if R_1 is a subring of $\mathcal{A}(A)$ containing R, and $\mathfrak{a}_1$ is the left R_1-ideal $R_1\mathfrak{a}$, then we can take $\lambda_{\mathfrak{a}}$ as $\lambda_{\mathfrak{a}_1}$. It is clear that, if v is a non-exceptional discrete valuation of k for A, then $\tilde{B}$ is also an abelian variety and $\tilde{\lambda}_{\mathfrak{a}}$ is a homomorphism of $\tilde{A}$ onto $\tilde{B}$, which can be written as $\lambda_{\tilde{\mathfrak{a}}}$.

If μ is in the right order of the left ideal $\mathcal{A}(A)\mathfrak{a}$ in $\mathcal{A}_0(A)$, then the relation $\mu^*\lambda_{\mathfrak{a}}=\lambda_{\mathfrak{a}}\mu$ determines μ^* in $\mathcal{A}(B)$. If $R_0=R\otimes Q$ is semi-simple, $\mu\rightarrow\mu^*$ induces an isomorphism on R. Assume now that $\mathfrak{a}$ contains an element α such that $\nu(\alpha)\neq 0$. Then A and B are isogenous, $\mu\rightarrow\mu^*$ is an isomorphism of $\mathcal{A}_0(A)$ onto $\mathcal{A}_0(B)$ and $\mathcal{A}(B)$ contains the isomorphic image of the right order of $\mathcal{A}(A)\mathfrak{a}$. If then $\mathfrak{b}$ is a left ideal in $\mathcal{A}(A)$ with this right order of $\mathcal{A}(A)\mathfrak{a}$ as its left order, we have $\lambda_{\mathfrak{a}\mathfrak{b}}=\varepsilon\lambda^*_{\mathfrak{b}}\lambda_{\mathfrak{a}}$ with an isomorphism ε. Note that if $\mathfrak{a}=(\alpha)$ is principal, we have $\lambda_{\mathfrak{a}}=\eta\alpha$ with an isomorphism η.

If especially R is the principal order of a subfield R_0 (of $\mathcal{A}_0(A)$) of degree d, then we can prove easily

$$\nu(\lambda_{\mathfrak{a}})=(N\mathfrak{a})^{2g/d},$$

where $N\mathfrak{a}$ denotes the absolute norm of $\mathfrak{a}$ in the field R_0. Let furthermore R be the principal order of a field R_0 of degree $2g$, and B be isogenous to A. Assume that R^* is contained in $\mathcal{A}(B)$ for some λ

in $\mathcal{H}(\mathrm{A}, \mathrm{B})$, with $\nu_i(\lambda)=1$, and denote by $\mathfrak{h}$ the kernel of λ. Then our assumption implies that $\mu\mathfrak{h}\subset\mathfrak{h}$ for any μ in R. Denote by $\mathfrak{a}$ an ideal in R consisting of all μ in R such that $\mu(\mathfrak{h})=0$, then the kernel of $\lambda_{\mathfrak{a}}$ is $\bigcup_{\mu\in R}\mu\mathfrak{h}$, hence it is equal to $\mathfrak{h}$. Thus we see that λ can be written as $\lambda_{\mathfrak{a}}$. On the other hand, if A is simple and $\mathrm{R}=\mathcal{A}(\mathrm{A})$ is the principal order of the field $\mathcal{A}_0(\mathrm{A})$ of degree $2g$, then we have clearly $\mathcal{A}(\mathrm{B})\subset\mathrm{R}^*$ for any λ. If $\mathcal{A}(\mathrm{B})$ is also the principal order of $\mathcal{A}_0(\mathrm{B})$, this shows that $\mu\to\mu^*$ induces an isomorphism of $\mathcal{A}(\mathrm{A})$ onto $\mathcal{A}(\mathrm{B})$. Hence in this case, every λ in $\mathcal{H}(\mathrm{A}, \mathrm{B})$ with $\nu_i(\lambda)=1$ must be of the form $\lambda_{\mathfrak{a}}$.

§3. Ideal decomposition of the endomorphism π.

Let A be defined over k, and put $k_1=k_{\mathcal{A}(\mathrm{A})}$. For any automorphism σ of $\bar{k}_1$, A^σ is also an abelian variety and $\mu\to\mu^\sigma$ gives an isomorphism of $\mathcal{A}(\mathrm{A})$ onto $\mathcal{A}(\mathrm{A}^\sigma)$, which is an automorphism of $\mathcal{A}(\mathrm{A})$ if σ fixes all elements of k. From this we see that k_1 is a finite algebraic normal extension of k, and its Galois group over k operates faithfully on $\mathcal{A}(\mathrm{A})$.

Assume now the characteristic of k to be zero. Let (ω) be a basis of D(A), rational over k. Denote by $\mathrm{S}(\mu)$, $\mathrm{S}^\sigma(\mu^\sigma)$ the (faithful) anti-representation of $\mathcal{A}_0(\mathrm{A})$, $\mathcal{A}_0(\mathrm{A}^\sigma)$ with basis (ω), (ω^σ) respectively. Given a commutative semi-simple subalgebra R_0 of $\mathcal{A}_0(\mathrm{A})$, $\mathrm{S}(\mu)$ can clearly be transformed in $\bar{k}$ into diagonal forms $\mathrm{S}_1(\mu)$ simultaneously for all μ in R_0. If especially R_0 is a field, diagonal elements of $S_1(\mu)$ must be of the form $\sigma_1\iota\mu,\cdots,\sigma_j\iota\mu, 0,\cdots,0$, where ι is an isomorphism of R_0 into C and $\sigma_1,\cdots,\sigma_j$ are isomorphisms of the field $\iota\mathrm{R}_0$ into C, determined uniquely by A, R_0 and ι. Denote by K' the Galois closure of $\iota\mathrm{R}_0$, by G the Galois group of K' over Q and by H the subgroup of G corresponding to ιR_0. We consider σ_i as an element in G. Then the set H^* of all σ in G such that $\sum_{i=1}^{j}\sigma\sigma_iH=\sum_{i=1}^{j}\sigma_iH$ is a subgroup of G (uniquely determined by A and R_0). Denote by K^* the subfield of K' corresponding to H^*. Then we can find $\tau_1,\cdots,\tau_s$ in G for which $\sum_{i=1}^{j}\sigma_iH=\sum_{i=1}^{s}H^*\tau_i$ holds. Put $\mathrm{R}=\mathrm{R}_0\cap\mathcal{A}(\mathrm{A})$, then we see that k_{R} always contains K^*, as we have clearly $S^\sigma(\mu^\sigma)=\sigma[S(\mu)]$ for any μ in $\mathcal{A}_0(\mathrm{A})$.

Here we shall recall some properties of endomorphisms π (cf. Weil [9], [10]). Let C be a non-singular curve of genus g, J a jacobian variety of C and φ a canonical mapping of C into J, all defined over the finite field k of $q=p^f$ elements. Then, the zeta-function $\mathrm{Z}(u)$ of C over k is of the form $[(1-u)(1-qu)]^{-1}\prod_{i=1}^{2g}(1-\varpi_iu)$, where $\varpi_1,\cdots,\varpi_{2g}$ are characteristic roots of $\mathrm{M}_l(\pi_{\mathrm{J}})$ for $l\neq p$. The

"Riemann hypothesis" shows now that $|\varpi_i|=q^{1/2}$ for $i=1,\cdots,2g$. The same holds for general abelian variety A and π_A if k is large enough, as, in case A is simple, there is a homomorphism of a jacobian onto A, and in general case, as π_A belongs to the centre of $\mathcal{A}(A)$, which is isomorphic to the direct sum of centres of $\mathcal{A}(B)$'s, B being simple. However, as $\pi_A(k')=\pi_A(k)^i$ if $[k':k]=i$, we have $|\varpi_i|=q^{1/2}$ for any field of definition k of A.

Now, A and k being as above, let R_0 be a subfield of $\mathcal{A}_0(A)$. Assume that $R=R_0\cap\mathcal{A}(A)$ is the principal order of R_0. By the definition of $\lambda_{\mathfrak{a}}$, $S(\lambda_{\mathfrak{a}})=0$ if and only if $k_R(x^p)\supset k_R(\mu x)$ for all μ in $\mathfrak{a}$, where $\mathfrak{a}$ is an ideal of R and x denotes a generic point of A over k_R. Hence there is the largest ideal $\mathfrak{Q}$ in R such that $S(\lambda_{\mathfrak{Q}^r})=0$ with some natural number r. Clearly, $\mathfrak{Q}$ contains no multiple factor. Assume now some power π_A^h of π_A belongs to R. Then $\mathfrak{Q}$ divides π_A^h by definition. Conversely, for any ν in $\mathfrak{Q}$ we have $S(\nu^r)=0$, that is, $k_R(x^p)\supset k_R(\nu^r x)$, and by induction, $k_R(\pi_A^h x)=k_R(x^{p^{fh}})\supset k_R(\nu^{rfh}x)$, which shows that π_A^h divides $\mathfrak{Q}^{rfh}$. Hence, if we denote by $\mathfrak{p}_1,\cdots,\mathfrak{p}_s$ all the different prime factors of π_A^h in R, we have $\mathfrak{Q}=\mathfrak{p}_1\cdots\mathfrak{p}_s$. Remark that if $S(\mu)$ is diagonal for all μ in R, this $\mathfrak{Q}$ can be characterized as the largest ideal such that $S(\lambda_{\mathfrak{Q}})=0$.

On the other hand, for arbitrary ground field k, the algebra $\mathcal{A}_0(A)$ has an involutorial anti-automorphism $\mu\rightarrow\mu'$ such that the trace of $M_l(\mu\mu')$ is positive for any $\mu\neq 0$ in $\mathcal{A}_0(A)$ (and for any $l\neq$characteristic of k) (cf. Weil [10]). Then, if a subfield R_0 of $\mathcal{A}_0(A)$ is invariant by $\mu\rightarrow\mu'$ as a whole, we have $\iota\mu'=\overline{\iota\mu}$ for each isomorphism ι of R_0 into C and for $\mu\in R_0$, (cf. Morikawa [6]). Thus, ιR_0 is either a totally real field or a totally imaginary quadratic extension of a totally real field.

We now prove the following

PROPOSITION 3. Let A be defined over an algebraic number field k' of finite degree. Let R_0 be a subfield of $\mathcal{A}_0(A)$, and put $R=R_0\cap\mathcal{A}(A)$, $K=\iota R_0$. Assume that $k'=k'_R$, that k' is absolutely normal, and contains K'. Assume furthermore that, for any non-exceptional $\mathfrak{P}'$ in k', the endomorphism $\pi_{\mathfrak{P}'}$ exists and belongs to R. Denote by $\sigma_1,\cdots,\sigma_r$ all the different isomorphisms among $\sigma_1,\cdots,\sigma_j$, determined by A, R_0 and ι. Then, we have the following ideal decomposition:

$$(\iota\pi_{\mathfrak{P}'})=N_{k'/K}(\sigma_1^{-1}\mathfrak{P}'\cdots\sigma_r^{-1}\mathfrak{P}')$$

in ιR_0, where σ_i are supposed to be extended to k'. Moreover $[K:Q]=2r$ and $\sigma_1,\cdots,\sigma_r$, $\sigma_0\sigma_1,\cdots,\sigma_0\sigma_r$ give all the $2r$ isomorphisms of K into C.

PROOF. (Note that $S(\mu)$ can be transformed simultaneously into diagonal forms $S_1(\mu)$ in k' for all μ in R, as $k'\supset K'$.)

At first we assume that R is the principal order. The condition $S_1(\nu)\equiv 0$ mod. $\mathfrak{P}'$ for ν in R is equivalent to $\iota\nu\equiv 0$ mod. $\sigma_i^{-1}(\mathfrak{P}')$ $i=$

$1,\cdots,r$. Then, as was seen above, the prime ideal $\mathfrak{p}_i$ in K divisible by $\sigma_i^{-1}(\mathfrak{P}')$ are all the prime divisors of $\iota\pi_{\mathfrak{P}'}$ in ιR_0. If $\mathfrak{P}'$ is of the first degree, then $\iota\pi_{\mathfrak{P}'}\cdot\overline{\iota\pi_{\mathfrak{P}'}}=p$. If moreover $\mathfrak{P}'$ is unramified over $\mathbf{Q}$, then $\mathfrak{p}_1,\cdots,\mathfrak{p}_r$ are all different, hence $(\iota\pi_{\mathfrak{P}'})=\mathfrak{p}_1\cdots\mathfrak{p}_r$ and $p=\mathfrak{p}_1\cdots\mathfrak{p}_r\cdot\overline{\mathfrak{p}_1\cdots\mathfrak{p}_r}$. This shows that $[K:\mathbf{Q}]=2r$ and $\sigma_1,\cdots,\sigma_r$, $\sigma_0\sigma_1,\cdots,\sigma_0\sigma_r$ are all the isomorphisms of K. Then, for general $\mathfrak{P}'$, the relation $\prod_{i=1}^{r}[N_{k'/K}(\sigma_i^{-1}\mathfrak{P}')\cdot\overline{N_{k'/K}(\sigma_i^{-1}\mathfrak{P}')}]=N\mathfrak{P}'=(\iota\pi_{\mathfrak{P}'})(\overline{\iota\pi_{\mathfrak{P}'}})$ proves our proposition in this case.

In case of general order R, we take an ideal $\mathfrak{a}$ of the principal order, contained in R, and put $B=\lambda_{\mathfrak{a}}A$. Then $\mu\rightarrow\mu^*$ induces an isomorphism on R_0 and maps the principal order of R_0 into $\mathcal{A}(B)$. B, R_0^*, ι' have the same system $\sigma_1,\cdots,\sigma_j$ as A, R_0, ι for ι' defined by $\iota'\mu^*=\iota\mu$, because non-zero characteristic roots of $S(\mu)$ and of $S(\mu^*)$ are equal. As $\lambda_{\mathfrak{a}}$ is defined over k, $\pi_{\mathfrak{P}'}^*$ is just the $\pi_{\mathfrak{P}'}$ for B. Then the above result, applied to $\pi_{\mathfrak{P}'}^*$, completes the proof.

Now assume that R_0 is of degree $2g$. Then $r=j=g$. Let k be any field of definition of A and k' be an overfield of k satisfying the condition of Prop. 3. For non-exceptional $\mathfrak{P}$ in k_R, $\pi_{\mathfrak{P}}$ belongs to the commutor of R_0, hence to R_0 itself. Let $\mathfrak{P}'$ be a prime divisor of $\mathfrak{P}$ in k', and put $N_{k'/k_R}\mathfrak{P}'=\mathfrak{P}^d$, $N_{k_R/K^*}\mathfrak{P}=\mathfrak{p}^f$. Then, by Prop. 3, we have $(\iota\pi_{\mathfrak{P}})^d=(\iota\pi_{\mathfrak{P}'})=N_{k'/K}(\sigma_1^{-1}\mathfrak{P}'\cdots\sigma_g^{-1}\mathfrak{P}')=(\tau_1^{-1}\mathfrak{p}\cdots\tau_s^{-1}\mathfrak{p})^{df}$, as $\sum\sigma_iH=\sum H^*\tau_j$. Hence we have

COROLLARY 1. Assume that R_0 is of degree $2g$. Then, for non-exceptional $\mathfrak{P}$ in k_R, with relative degree f over K^*, we have

$$(\iota\pi_{\mathfrak{P}})=(\tau_1^{-1}\mathfrak{p}\cdots\tau_s^{-1}\mathfrak{p})^f,$$

where prime ideal $\mathfrak{p}$ in K^* is considered as an ideal in K'.

COROLLARY 2. Under the same assumption as in Cor. 1, all conjugates of $N_{k'/K}(\sigma_1^{-1}\beta\cdots\sigma_g^{-1}\beta)$ have the same absolute values $|N\beta|^{1/2}$ for any β in k'.

(This Cor. is evident, as we have $(\rho\iota\pi_{\mathfrak{P}})\cdot(\sigma_0\rho\iota\pi_{\mathfrak{P}})=N\mathfrak{P}$ for any isomorphism ρ of K.)

Notations being as above, we make furthermore the assumption:

(A) R_0 is a field of degree $2g$, and there is no not-identical isomorphism σ of K over some imaginary subfield of it such that $\sum_{i=1}^{g}\sigma_iH\sigma=\sum_{i=1}^{g}\sigma_iH$.

Now, let $\mathfrak{p}$ be a prime ideal of the first degree in K^* such that $p=N\mathfrak{p}$ is unramified in K' and that a prime divisor $\mathfrak{P}$ of $\mathfrak{p}$ in k_R is non-exceptional. Then, as $p=\prod\tau_i^{-1}\mathfrak{p}\cdot\overline{\prod\tau_i^{-1}\mathfrak{p}}$, $\prod\tau_i^{-1}\mathfrak{p}$ is not a real ideal. Moreover, for any σ in G, not belonging to H, we have $\sigma(\prod\tau_i^{-1}\mathfrak{p})\neq\prod\tau_i^{-1}\mathfrak{p}$. Indeed, if this is not the case, we should have $\sum\sigma\tau_i^{-1}H^*=\sum\tau_i^{-1}H^*$, hence $\sum\sigma_iH\sigma=\sum\sigma_iH$, and $\prod\tau_i^{-1}\mathfrak{p}$ should be real, a contradiction. This shows that $\iota\pi_{\mathfrak{P}}^h$ generates K for any

$h \neq 0$. As $\tilde{\pi}_{\mathfrak{P}}^h$ belongs to the centre of $\mathcal{A}_0(\tilde{\mathrm{A}})$ for some h, we see that $\mathcal{A}_0(\tilde{\mathrm{A}})=\tilde{\mathrm{R}}_0$. Thus, $\tilde{\mathrm{A}}$, and a fortiori A, must be simple. Moreover, for any not-identical automorphism σ of K, we see that $\sum \sigma_i\sigma H \neq \sum \sigma_i H$. This last assertion shows that, any automorphism $\mu \rightarrow \mu^\alpha$ of R_0 is the identity if and only if $\mathrm{S}(\mu)$ and $\mathrm{S}(\mu^\alpha)$ have the same characteristic roots. From this we see especially that $k_{\mathrm{R}}=k \cup K^*$, as $k_{\mathrm{R}} \supset k \cup K^*$ in general.

§4. Zeta functions.

In this § we assume that A is defined over an algebraic number field k of finite degree, and that $\mathcal{A}_0(\mathrm{A})$ contains a subfield R_0 of degree $2g$. We assume furthermore that k contains K' and $k=k_{\mathrm{R}}$, where $\mathrm{R}=\mathrm{R}_0 \cap \mathcal{A}(\mathrm{A})$.

At first, we assume that R is the principal order of R_0. For an ideal $\mathfrak{a}$ in R, we call "$\mathfrak{a}$ *division-point*" on A a point a such that $\mu a=0$ for all μ in $\mathfrak{a}$.

The number of such points is just $\mathrm{N}\mathfrak{a}$. We call an $\mathfrak{a}$ division-point b *proper* if it is no $\mathfrak{b}$ division-point for any $\mathfrak{b} \supsetneqq \mathfrak{a}$, that is, $\mu b=\nu b$ implies $\mu \equiv \nu \bmod \mathfrak{a}$. Let b be fixed one proper $\mathfrak{a}$ division-point. Then all proper $\mathfrak{a}$ division-points can be written as μb with μ in R prime to $\mathfrak{a}$, and conversely. Especially, for any isomorphism σ of $k(b)$ over k, $b^\sigma=\mu_\sigma b$ with some μ_σ in R, as $k=k_{\mathrm{R}}$. This shows that b^σ is rational over $k(b)$, that is, $k(b)$ is normal over k. Denote then by $\mathfrak{G}$ the Galois group of $k(b)$ over k. Moreover, for any σ in $\mathfrak{G}$, the class of $\mu_\sigma \bmod \mathfrak{a}$ is determined uniquely by σ, hence we denote it by $[\sigma]$. Notice that $[\sigma]=1$ if and only if σ is identity. As μ_σ is rational over k, $\sigma \rightarrow [\sigma]$ is an isomorphism of $\mathfrak{G}$ into the prime residue class group $\bmod \mathfrak{a}$ in R, which shows that $k(b)$ is abelian over k. Therefore, $k(b)$ is contained in some "Strahlklassenkörper" $\bmod F$ over k, here F can be assumed to be a natural number divisible by $\mathfrak{a}$.

Let now $\mathfrak{P}$ be a non-exceptional prime ideal in k, prime to F, and denote by $\sigma_{\mathfrak{P}}$ the Frobenius automorphism of $\mathfrak{P}$ in $\mathfrak{G}$. Then we have $\widetilde{b^{\sigma_{\mathfrak{P}}}}=\widetilde{\pi_{\mathfrak{P}} b}$. Then Prop. 1, §1 shows that $b^{\sigma_{\mathfrak{P}}}=\pi_{\mathfrak{P}} b$, as the order of b is prime to $\mathrm{N}\mathfrak{P}$. Put $\pi_{\mathfrak{B}}=\pi_{\mathfrak{P}_1}\cdots\pi_{\mathfrak{P}_r}$ for an ideal $\mathfrak{B}=\mathfrak{P}_1\cdots\mathfrak{P}_r$ prime to any exceptional $\mathfrak{P}$. If moreover $\mathfrak{B}$ is prime to F, denote by $\sigma_{\mathfrak{B}}$ the Artin-symbol of $\mathfrak{B}$ in $\mathfrak{G}$. Then the above result shows $b^{\sigma_{\mathfrak{B}}}=\pi_{\mathfrak{B}} b$, that is, $[\sigma_{\mathfrak{B}}] \ni \pi_{\mathfrak{B}}$. Hence, if β in k belongs to the "Strahl" $\bmod F$, we have $\pi_{(\beta)} \equiv 1 \bmod \mathfrak{a}$. But we have in general $(\iota\pi_{\mathfrak{B}})=\mathrm{N}_{k/K}(\sigma_1^{-1}\mathfrak{B}\cdots\sigma_g^{-1}\mathfrak{B})$, so especially from β in k,

$$(\iota\pi_{(\beta)})=\varepsilon(\beta)\mathrm{N}_{k/K}(\sigma_1^{-1}\beta\cdots\sigma_g^{-1}\beta),$$

where $\varepsilon(\beta)$ is a unit in K. As $|\sigma\varepsilon(\beta)|=1$ for any σ in G by Cor. 2

of Prop. 3, $\varepsilon(\beta)$ must be a root of unity in K. Now, if $\beta \equiv 1$ mod. F, then $\pi_{(\beta)} \equiv 1$ mod. $\mathfrak{a}$, so $\varepsilon(\beta) \equiv 1$ mod. $\mathfrak{a}$. This being true for any $\mathfrak{a}$, we assume here that $\mathfrak{a}$ is prime to twice the discriminant of K. Then we have $\varepsilon(\beta)=1$ for $\beta \equiv 1$ mod. F. Thus *the symbol* $\chi(\mathfrak{B})=\iota\pi_{\mathfrak{B}}/|\iota\pi_{\mathfrak{B}}|$ *is a "Grössencharakter"* in k (cf. Hecke [4]). Similarly we see that $\chi^{\sigma}(\mathfrak{B})=\sigma\iota\pi_{\mathfrak{B}}/|\sigma\iota\pi_{\mathfrak{B}}|$ is also a "Grössencharakter" for any σ in G.

The case where R is not the principal order can be treated just as in the second half of the proof of Prop. 3. Then, as the characteristic roots of $M_l(\pi_{\mathfrak{P}})$ and of $M_l(\pi^*_{\mathfrak{P}})$ are equal, we have the same conclusion in this case also.

Now, let B^g be an abelian variety defined over the finite field κ of $q=p^f$ elements. Denote by κ_n the finite field of q^n elements, and by N_n the number of rational points of A over κ_n. Then the zeta function $Z(u)$ of B is defined by $\frac{d}{du}\log Z(u)=\sum_{n=1}^{\infty} N_n u^{n-1}$ (cf. Weil [12]). Clearly, N_n is the number of (π_B^n-1) division-points. As π_B^n-1 is prime to π_B, we have $\nu_i(\pi_B^n-1)=1$ for any n. Thus $N_n=\nu(\pi_B^n-1)=\prod_{i=1}^{2g}(\varpi_i^n-1)$, where $\varpi_1,\cdots,\varpi_{2g}$ are characteristic roots of $M_l(\pi_B)$ for $l\neq p$. From this we see, by a simple calculation, that

$$Z(u)=\prod_{\nu=0}^{2g}\Big[\prod_{i_1\cdots i_\nu}(1-\varpi_{i_1}\cdots\varpi_{i_\nu}u)\Big]^{(-1)^{\nu+1}},$$

where $i_1,\cdots,i_\nu$ run over all combinations of $1,\cdots,2g$.

Coming back to original A defined over k, let $Z_{\mathfrak{P}}(s)$ be the zeta function of $\tilde{A}$ (mod. $\mathfrak{P}$) over $\tilde{k}$, with $(N\mathfrak{P})^{-s}=u$. We define as usual the zeta function of A over k by

$$\zeta_A(s)=\prod_{\mathfrak{P}} Z_{\mathfrak{P}}(s),$$

where $\mathfrak{P}$ runs over all non-exceptional $\mathfrak{P}$'s for A. Then we have proved

THEOREM 1. *The zeta function $\zeta_A(s)$ of A has the form:*

$$\zeta_A(s)=\Psi(s)\prod_{\nu=0}^{2g}\Big[\prod_{i_1\cdots i_\nu} L\big(s-\frac{\nu}{2},\chi_{i_1\cdots i_\nu}\big)\Big]^{(-1)^\nu}.$$

where $L(s,\chi_{i_1\cdots i_\nu})$ are L-functions of k with "Grössencharaktere" $\chi_{i_1\cdots i_\nu}=\chi^{\sigma_{i_1}}\cdots\chi^{\sigma_{i_\nu}}$, Ψ is a product of rational functions of q^{-s} for a finite number of $q=N\mathfrak{P}$, and $\sigma_{i_1},\cdots,\sigma_{i_\nu}$ run over all combinations of isomorphisms of K.

We have similarly,

THEOREM 1$'$. *Let* C *be a non-singular curve. Assume that* C *and the jacobian variety* J *of* C *and the canonical mapping φ of* C *into* J *are defined over k, and that* J, k *satisfy the condition of this* §. *Then the zeta function ζ_C of* C *(defined similarly as ζ_A) has the form:*

$$\zeta_C(s)=\Psi(s)\zeta(s)\zeta(s-1)\prod_{i=1}^{2g} L(s-\tfrac{1}{2},\chi_i)^{-1},$$

where L and Ψ are as in Theorem 1, and ζ(s) is the zeta function of k.

Remark finally that, if in general there is an abelian variety B^g of characteristic zero, whose ring $\mathcal{A}_0(B)$ contains a field R_0 of degree $2g$ with a system $\iota, \sigma_1, \cdots, \sigma_g$, then there is an A defined over an algebraic number field of finite degree, whose ring $\mathcal{A}_0(A)$ contains a field isomorphic to R_0 with the same system $\sigma_1, \cdots, \sigma_g$. Indeed, B can be defined over a finitely generated field $\kappa = \mathbf{Q}(y_1, \cdots, y_r)$ with transcendency degree d. Then, denote by V the locus of $(y_1, \cdots, y_r)$ over $\overline{\mathbf{Q}}$. Taking $\mathfrak{V}$ as the set of all divisorial valuations of κ, Prop. 1 shows that there is an abelian variety B′, with the same property as B and defined over a field κ' of transcendency degree $d-1$. Repeating this process d times, we arrive at a desired A.

§5. Lemmas from analytic theory.

In this §, we consider only the case of universal domain C.

Let K be a totally imaginary quadratic extension of a totally real field K_0 of degree g, and let $\sigma_1, \cdots, \sigma_g$ be a system of isomorphisms of K, inducing all the g isomorphisms of K_0. Then, σ_0 commutes with all automorphisms of the Galois closure of K, hence we see from Lefschetz's criterion (cf. Lefschetz's [5]) that there is an abelian variety A, whose ring $\mathcal{A}_0(A)$ contains a subfield R_0 isomorphic to K with an isomorphism ι, and $\sigma_1, \cdots, \sigma_g$ are exactly the system of isomorphisms of K determined by A, R_0, ι.

Now, let A have this property, and let Ω be a period matrix of A. Then we have $S(\mu)\Omega = \Omega C(\mu)$ for μ in $\mathcal{A}(A)$, where $C(\mu)$ is a rational integral matrix of degree $2g$. Transforming $S(\mu)$ into diagonal form for μ in R, we see that Ω can be isomorphically transformed into the form (ω_{ij}), where $\omega_{ij} = \sigma_i \omega_j$ and $\omega_1, \cdots, \omega_{2g}$ form a basis of an ideal of $\iota(R)$. Therefore, if A, A′ have the above property with the same system $\sigma_1, \cdots, \sigma_g$, they must be isogenous.

Finally, we remark the following facts: Let A be simple and $\mathcal{A}(A) = R$ the principal order of the field $\mathcal{A}_0(A)$ of degree $2g$. Assume that there is a positive divisor X whose all elementary divisors are 1, i.e. $l(X) = 1$ (cf. Weil [11]). Then, for an ideal $\mathfrak{a}$ of R, $\lambda_{\mathfrak{a}} A$ has also a positive divior Y with $l(Y) = 1$ if and only if $\mathfrak{a}\mathfrak{a}' = (\alpha)$ and $\iota\alpha$ is a totally positive number in K_0. The proof is omitted here.

§6. Unramified extension k_0.

In the following §§6, 7, we assume that A is defined over an algebraic number field k of finite degree, and the condition (A) in §3 is satisfied. Hence A is simple. We assume furthermore that $R = \mathcal{A}(A)$ is the principal order of $\mathcal{A}_0(A)$ and that k contains K^*.

This implies $k_{\mathrm{R}}=k$. Let finally k' be the Galois closure of k and $\mathfrak{G}$ the Galois group of k' over K^*.

Let A, A$'$ satisfy these conditions with the same $K=\iota\mathrm{R}=\iota'\mathrm{R}'$ and the same system $\sigma_1,\cdots,\sigma_g$ for these ι, ι'. Then they are isogenous (§5), and we can write $\mathrm{A}'=\lambda_{\mathfrak{a}}\mathrm{A}$ with an ideal $\mathfrak{a}$ in R (§2). Conversely, for any ideal $\mathfrak{b}$ in R, $\lambda_{\mathfrak{b}}\mathrm{A}$ has the same property as A with the same K and $\sigma_1,\cdots,\sigma_g$. Thus, for a given K and $\sigma_1,\cdots,\sigma_g$, there are just h non-isomorphic $\mathrm{A}_1,\cdots,\mathrm{A}_h$, h being the class number of K.

Let σ be an automorphism of k'. If there is an automorphism τ of K such that $\sum\sigma\sigma_i H=\sum\sigma_i H\tau$, then $\mathrm{S}^\sigma(\mu^\sigma)=\sigma[\mathrm{S}(\mu)]$ has characteristic roots $\{\sigma\sigma_1\iota\mu,\cdots,\sigma\sigma_g\iota\mu\}=\{\sigma_1\tau\iota\mu,\cdots,\sigma_g\tau\iota\mu\}$. Thus, if we define ι' by $\tau\iota\mu=\iota'\mu^\sigma$, A, R, ι and A^σ, R^σ, ι' have the same system $\sigma_1,\cdots,\sigma_g$. Hence σ permutes the isomorphism-classes of $\mathrm{A}_1,\cdots,\mathrm{A}_h$ among themselves. The relation $\mathrm{A}^\sigma\cong\lambda_{\mathfrak{a}}\mathrm{A}$ ($\cong$ denoting the isomorphism of abelian varieties) determines the class of $\mathfrak{a}$ uniquely, so we can write this class as $\langle\sigma\rangle$. Consider now σ in $\mathfrak{G}$, then we have $\sum\sigma\sigma_i H=\sum\sigma_i H$, hence $\langle\sigma\rangle$ can be defined. Moreover, as $\mathrm{S}^\sigma(\mu^\sigma)=\sigma[\mathrm{S}(\mu)]$ has the same characteristic roots as $\mathrm{S}(\mu)$, hence as $\mathrm{S}(\mu^*)$, and as $\mu^\sigma\to\mu^*$ is an automorphism of $\mathrm{R}_0=\mathcal{A}_0(\mathrm{A})$, we see $\mu^*=\mu^\sigma$. Let τ be also in $\mathfrak{G}$ and put $\mathrm{A}^\tau\cong\lambda_{\mathfrak{b}}\mathrm{A}$. Then we have $(\mathrm{A}^\tau)^\sigma\cong\lambda_{\mathfrak{b}}^\sigma\mathrm{A}^\sigma\cong\lambda_{\mathfrak{b}^\sigma}\lambda_{\mathfrak{a}}\mathrm{A}\cong\lambda_{\mathfrak{ab}}\mathrm{A}$ as $\mathfrak{b}^\sigma=\mathfrak{b}^*$. This shows that $\sigma\to\langle\sigma\rangle$ is a homomorphism of $\mathfrak{G}$ into the ideal class group of K. Denote by k_0 the subfield of k' corresponding to the kernel of this homomorphism. Then, any automorphism σ of $\bar{k}$ over K^* fixes all elements of k_0 if and only if $\mathrm{A}^\sigma\cong\mathrm{A}$. Clearly k_0 is contained in k, and is abelian over K^*. If finally there is a positive divisor X on A with $l(\mathrm{X})=1$, then there are just h' abelian varieties among $\mathrm{A}_1,\cdots,\mathrm{A}_h$, having X_i with $l(\mathrm{X}_i)=1$, where h' denotes the number of classes in K, whose norm to K_0 are the principal class in the narrower sense. Remark that $l(\mathrm{X}^\sigma)=1$ if $l(\mathrm{X})=1$.

Let now $\mathfrak{p}$ be a prime ideal of the first degree in K^* such that a prime divisor $\mathfrak{P}$ of $\mathfrak{p}$ in k is non-exceptional for A and $\mathrm{N}\mathfrak{p}$ is unramified in k_0. Denote by $\sigma_{\mathfrak{p}}$ the Frobenius automorphism of $\mathfrak{p}$ in k_0/K^*, and put $\mathfrak{c}=\tau_1^{-1}\mathfrak{p}\cdots\tau_s^{-1}\mathfrak{p}$. Considering mod. $\mathfrak{P}$, we have $\tilde{k}(\tilde{x}^p)\supset\tilde{k}(\lambda_{\mathfrak{c}}\tilde{x})$, as all prime divisors of $(\iota\pi_{\mathfrak{p}})$ in K divide $\mathfrak{c}$. As $\mathrm{N}\mathfrak{c}=p^g$, we have moreover $\tilde{k}(\tilde{x}^p)=\tilde{k}(\lambda_{\mathfrak{c}}\tilde{x})$, that is $\widetilde{\mathrm{A}^{\sigma_{\mathfrak{p}}}}\cong\lambda_{\mathfrak{c}}\tilde{\mathrm{A}}$. Now, $\tilde{\mathrm{A}}$ being simple, we have $\mathcal{A}(\tilde{\mathrm{A}})=\mathrm{R}$. As we have $\mathrm{A}^{\sigma_{\mathfrak{p}}}\cong\lambda_{\mathfrak{a}}\mathrm{A}$ with $\mathfrak{a}\in\langle\sigma_{\mathfrak{p}}\rangle$, this shows that $\mathfrak{c}\in\langle\sigma_{\mathfrak{p}}\rangle$. We have seen therefore that $(\tau_1^{-1}\mathfrak{p}\cdots\tau_s^{-1}\mathfrak{p})^f$ is a principal ideal in K if and only if $\sigma_{\mathfrak{p}}^f$ is the identity. Summing up, we have

THEOREM 2. *Let K be a totally imaginary quadratic extension of a totally real field K_0 of degree g, and $\sigma_1,\cdots,\sigma_g$ be a set of isomorphisms of K into* C, *inducing all the g isomorphisms of K_0.*

Assume that $\sigma_1,\cdots,\sigma_g$ *satisfy the condition in* (A). *Then there are just* h *non-isomorphic abelian varieties* $A_1,\cdots,A_h$, *defined over an algebraic number field* k *of finite degree such that* $\iota_i\mathcal{A}(A_i)$ *is the principal order of* K, *and* A_i, $\mathcal{A}(A_i)$, ι_i *have the system* $\sigma_1,\cdots,\sigma_g$, *where* ι_i *are isomorphisms:* $\iota_i\mathcal{A}_0(A_i)=K$. *Then, the field* k_0 *defined as above for these* A *is the class field of* K^* *for the ideal group* $I=\{\mathfrak{b}\mid(\tau_1^{-1}\mathfrak{b}\cdots\tau_s^{-1}\mathfrak{b})\sim 1$ *in* $K\}$, *and* $\sigma\leftrightarrow\langle\sigma\rangle$ *gives an explicit form of Artin's reciprocity law.*

COROLLARY. *Notations and assumptions being the same as in the theorem 2, if some* A_i *has a positive divisor* X_i *with* $l(X_i)=1$, *then just* h' A_i*'s have* X_i *with* $l(X_i)=1$. *If moreover each one of* h' *classes in* K, *whose norms to* K *are the principal class in the narrower sense, contains an ideal of the form* $\tau_1^{-1}\mathfrak{b}\cdots\tau_s^{-1}\mathfrak{b}$ *with* $\mathfrak{b}$ *in* K^*, *then isomorphism-classes of these* h' A_i*'s are conjugate to each other over* K^*.

§7. Field of division-points.

Let $\mathfrak{P}$ be a non-exceptional prime ideal in $k=k_R$ for A, and put $N_{k/K^*}\mathfrak{P}=\mathfrak{p}^f$. By the Cor. 1 of Prop. 3, §3, we have $(\iota\pi_{\mathfrak{P}})=(\tau_1^{-1}\mathfrak{p}\cdots\tau_s^{-1}\mathfrak{p})^f$. Moreover any conjugate of $\iota\pi_{\mathfrak{P}}$ has absolute value $(N\mathfrak{p})^{1/2}$. Now, take a system of ideal numbers in K^* (cf. Hecke [4]), and denote by $\hat{\alpha},\hat{\beta},\hat{\varpi},\cdots$, the ideal numbers representing ideals $\mathfrak{a},\mathfrak{b},\mathfrak{p},\cdots$ in K^*. Then, as in §3, we see that any conjugate of $\prod_{i=1}^{s}\tau_i^{-1}\hat{\alpha}$ has the absolute value $|N\hat{\alpha}|^{1/2}=|N\mathfrak{a}|^{1/2}$. Thus, $\prod_{i=1}^{s}\tau_i^{-1}\hat{\alpha}$ is determined up to a root of unity factor (not necessarily in K) by $\mathfrak{a}$ only. This shows especially:

$$\iota\pi_{\mathfrak{P}}=\eta\prod_{i=1}^{s}\tau_i^{-1}\hat{\varpi}^f,$$

where η is a root of unity.

Denote by E the group of all roots of unity in R. For any point b on A, we denote by Eb the 0-cycle $\sum_{\varepsilon\in E}(\varepsilon b)$ on A, and by $k(Eb)$ the smallest field containing k, over which the cycle Eb is rational.

$\mathfrak{a}$ being an ideal in R, let b be a proper $\mathfrak{a}$ division-point on A. Assume that $N\mathfrak{p}$ is prime to $\iota\mathfrak{a}$, and that $k(Eb)$ (mod. $\mathfrak{P}$) is equal to $\tilde{k}(E\tilde{b})$, which is certainly the case for almost all $\mathfrak{P}$. Let now f_0 be the smallest exponent such that the congruence $\eta[\prod_{i=1}^{s}\tau_i^{-1}\hat{\varpi}]^{f_0}\equiv\iota\varepsilon \bmod. \iota\mathfrak{a}$ holds, with some roots of unity η, $\iota\varepsilon$, the latter being in K. Denote then $F=l.c.m.[f,f_0]$. As $\pi_{\mathfrak{P}}^e$ leaves Eb invariant if and only if $\pi_{\mathfrak{P}}^e\equiv\varepsilon \bmod. \mathfrak{a}$, $\varepsilon\in E$, and as $\tilde{b}$ is a proper $\tilde{\mathfrak{a}}$ division-point on $\tilde{A}$, the above expression for $\iota\pi_{\mathfrak{P}}$ shows that F is equal to the relative degree of any prime divisor of $\mathfrak{P}$ in $k(Eb)$ over K^*. This implies that $k(Eb)$

is equal to the composite field of k and the class field $k_{\mathfrak{a}}$ over K^* for the ideal group $I_{\mathfrak{a}}=\{(\hat{\beta})\mid \eta\prod \tau_i^{-1}\hat{\beta}\equiv \iota\varepsilon \text{ mod. } \iota\mathfrak{a};\ \eta$ a root of unity and $\varepsilon\in \mathrm{E}\}$. Evidently, this class field contains k_0. Hence we have proved

THEOREM 3. *For proper* $\mathfrak{a}$ *division-point* b, *we have* $k(\mathrm{E}b)=k\cup k_{\mathfrak{a}}$. *If especially* A *is defined over* k, *then we have* $k(\mathrm{E}b)=k_{\mathfrak{a}}$.

Bibliography

[1] M. Deuring, Die Zetafunktionen einer algebraischen Kurven von Geschlecht Eins, Nachr. Akad. Wiss. Göttingen, 1953, 85-94.

[2] E. Hecke, Höhere Modulfunktionen und ihre Anwendung auf die Zahlentheorie, Math. Ann., **71** (1912), 1-37.

[3] E. Hecke, Über die Konstruktion relative Abelscher Zahlkörper durch Modulfunktionen von zwei Variabeln, Math. Ann., **74** (1913), 465-510.

[4] E. Hecke, Eine neue Art von Zetafunktionen und ihre Beziehung zur Verteilung der Primzahlen, II, Math. Z., **6** (1920), 11-51.

[5] S. Lefschetz, On certain numerical invariants of algebraic varieties with application to Abelian varieties, Trans. Amer. Math. Soc., **22** (1921), 327-482.

[6] H. Morikawa, On abelian varieties, Nagoya Math. J., **6** (1953), 151-170.

[7] G. Shimura, Reduction of algebraic varieties with respect to a discrete valuation of the basic field, Amer. J. Math., **77** (1955), 134-176.

[8] A. Weil, Foundations of algebraic geometry, New York, 1946.

[9] A. Weil, Sur les courbes algébriques et les variétés qui s'en déduisent, Paris, 1948.

[10] A. Weil, Variétés abéliennes et courbes algébriques, Paris, 1948.

[11] A. Weil, Théorèmes fondamentaux de la théorie des fonctions thêta, Séminaire Bourbaki, Paris, 1949.

[12] A. Weil, Number of solutions of equations in finite fields, Bull. Amer. Math. Soc., **55** (1949), 497-508.

[13] A. Weil, Arithmetic on algebraic varieties, Ann. of Math. (2), **53** (1951), 412-444.

[14] A. Weil, Jacobi sums as "Grössencharaktere", Trans. Amer. Math. Soc., **73** (1952), 487-495.

On a Certain Relation between L-functions

An unpublished manuscript dated February 4, 1956

Introduction

It has been proved recently by A. Weil, M. Deuring and myself, that Hasse's ζ-functions of algebraic curves and of abelian varieties having sufficiently many complex multiplications, can be expressed by Hecke's L-functions with "Grössencharaktere" as a product. On the other hand, M. Eichler showed that Hasse's ζ-functions of certain modular function fields are related to the theory of Hecke's operators. But those results might still appear somewhat accidental.

On the other hand, A. Weil remarked[1] that to an algebraic-valued "Grössencharakter" we can attach infinitely many characters of finite order, and suggested the possibility of finding some connections of Hasse's ζ-functions with Hecke's L-functions with "Grössencharaktere" from this standpoint.

In this paper, I shall show that there is an actual relation between these two kinds of functions, giving a new relation between Hecke's L-functions. Namely, in §1, we shall show that an infinite product of L-functions with characters attached to a given "regularly algebraic-valued" "Grössencharakter" χ can be expressed as a finite product of L-functions with conjugate-product characters of original χ. In §2, we shall introduce a notion of the local system, and attach to it infinitely many "characters". Then, a similar type of relation is shown to hold between L-functions $L_b(s)$ with these "characters" and other well defined L-function. Here, the relation of this $L_b(s)$ and Artin's L-function $L(s, \chi_0^{(b)})$, is something like the well-known relation between Hecke's L-functions with imprimitive and primitive characters, as may be seen in §3. At any rate, $L_b(s)$ is obtained from $L(s, \chi_0^{(b)})$ by leaving out a finite number of factors. This "degeneration" of $L(s, \chi_0^{(b)})$ to $L_b(s)$ brings about in particular the absolute convergence of infinite product $\prod_b L_b(s)$ in some domain. Finally, in §3, we shall give two examples of local system, showing that the result in §1 can be regarded as a special case of §2, and also that Hasse's ζ-functions for general abelian varieties can also be regarded as an L-function of local system. These two examples coincide in case where the abelian variety have sufficiently many

1) Symposium on algebraic number theory, 1955, Tokyo.

complex multiplications. Thus the results mentioned at the beginning seem to be non-accidental: our results seem to suggest the possibility of treating all these matters from a common point of view.

§1. L-function with "Grössencharaktere"

Let k be an algebraic number field of finite degree, and χ a "Grössencharakter" of k with conductor $\mathfrak{f}$, satisfying the following conditions:

(i) There is a real number ν and an algebraic number field K of finite degree, such that $\chi(\mathfrak{p})[N(\mathfrak{p})]^{\nu}$ is an integer in K for any prime ideal $\mathfrak{p}$ in k not dividing $\mathfrak{f}$. Here $N(\mathfrak{p})$ means the absolute norm of $\mathfrak{p}$.

(ii) $\chi(\alpha)[N(\alpha)]^{\nu}$ is of the form $\prod_{\sigma}(\alpha^{\sigma})^{n_{\sigma}}$ for any number $\alpha\equiv 1$ mod. $\mathfrak{f}$, where σ runs over all isomorphisms of k (into the complex number field) and n_{σ} are non-negative integers independent of α.

We call χ's satisfying (i), (ii) *regularly algebraic-valued "Grössencharaktere"*. Such is for example a "Grössencharakter" χ of the form

$$\chi((\alpha))=\prod_{i}\left(\frac{\alpha^{(i)}}{\overline{\alpha^{(i)}}}\right)^{\nu_i} \qquad \text{for } \alpha\equiv 1 \text{ mod. } \mathfrak{f},$$

where $\alpha^{(i)}$ are imaginary conjugates of α with complex conjugates $\overline{\alpha^{(i)}}$, and ν_i are rational integers. Another example of the regulary algebraic-valued character is given as follows. Let k be a totally imaginary quadratic extension of a totally real field and let

$$\chi((\alpha))=\frac{\alpha^{(1)}\cdots\alpha^{(r)}}{|N(\alpha)|^{1/2}} \qquad \text{for } \alpha\equiv 1 \text{ mod. } \mathfrak{f},$$

where $\alpha^{(1)},\cdots,\alpha^{(r)}$ are conjugates of α such that, together with $\overline{\alpha^{(1)}},\cdots,\overline{\alpha^{(r)}}$, give all the conjugates of α in k over the rational number field $\boldsymbol{Q}$. Then, χ is a regularly algebraic-valued "Grössencharakter".

We denote by J the idèle group of k. Let $\tilde{\mathfrak{a}}\in J$ be an idèle of k. Then, we denote by $\tilde{\mathfrak{a}}_0$, $\tilde{\mathfrak{a}}_{\infty}$ the finite and the infinite parts of $\tilde{\mathfrak{a}}$ respectively: $\tilde{\mathfrak{a}}=\tilde{\mathfrak{a}}_0\cdot\tilde{\mathfrak{a}}_{\infty}$. Let $\mathfrak{p}$ be a prime ideal in k and $k_{\mathfrak{p}}$ be the $\mathfrak{p}$-adic completion of k. We denote by $\pi_{\mathfrak{p}}$, $u_{\mathfrak{p}}$ a $\mathfrak{p}$-prime element and a $\mathfrak{p}$-unit in $k_{\mathfrak{p}}$ respectively, and by $\tilde{\pi}_{\mathfrak{p}}$, $\tilde{u}_{\mathfrak{p}}$ the idèles with $\mathfrak{p}$-components $\pi_{\mathfrak{p}}$, $u_{\mathfrak{p}}$ respectively and all $\mathfrak{p}'$-components 1 for $\mathfrak{p}'\neq\mathfrak{p}$. Furthermore we denote by $\tilde{\alpha}$ the principal idèle corresponding to a number α in the multiplicative group k^*.

As is well known, for any "Grössencharakter" χ, there is a character $\tilde{\chi}$ of J, satisfying:

(iii) $\qquad \tilde{\chi}(\tilde{\alpha})=1 \qquad$ for all α in k^*;

(iv) $\tilde{\chi}(\tilde{\pi}_{\mathfrak{p}})=\chi(\mathfrak{p})$, $\tilde{\chi}(\tilde{u}_{\mathfrak{p}})=1$, for every prime ideal $\mathfrak{p}$ prime to $\mathfrak{f}$.

By the approximation theorem for valuations, $\tilde{k}^*\cdot\prod_{\mathfrak{p}}'\tilde{k}_{\mathfrak{p}}^*$ is dense in J, provided that in the restricted direct product $\prod'$, $\mathfrak{p}$ runs over all but a finite number of prime divisors of k. Thus $\tilde{\chi}$ is determined uniquely by these conditions (iii), (iv). Then, for any idèle $\tilde{\mathfrak{a}}$, having components 1 at all prime divisors of the conductor $\mathfrak{f}$ and all infinite primes, we have $\tilde{\chi}(\tilde{\mathfrak{a}})=\chi(\mathfrak{a})$, $\mathfrak{a}$ denoting the divisor of $\tilde{\mathfrak{a}}$.

Now, let χ be a regularly algebraic-valued "Grössencharakter". Then, we define a representation (i.e. algebraic homomorphism which is continuous) $\tilde{\Psi}$ of J into the multiplicative group of complex number field by the relation:

$$\tilde{\Psi}(\tilde{\mathfrak{a}})=\tilde{\chi}(\tilde{\mathfrak{a}})\cdot[V(\tilde{\mathfrak{a}})]^{-\nu},$$

where $V(\tilde{\mathfrak{a}})$ denotes the volume of $\tilde{\mathfrak{a}}$ and ν is the real number in condition (i). Then, for any α in k^*, we have, by condition (ii),

$$\tilde{\Psi}(\tilde{\alpha}_\infty)=\prod_\sigma\,[\varepsilon_\sigma(\alpha)(\alpha^\sigma)^{-n_\sigma}],$$

where $\varepsilon_\sigma(\alpha)$ is the "Vorzeichencharakter" of α^σ due to the infinite real prime (∞,σ) contained in $\mathfrak{f}$.

It is clear that all values $\tilde{\Psi}(\tilde{\mathfrak{a}}_0)$ are contained in K. Here we may assume that K contains k and is absolutely normal. Then we have,

LMMMA 1. (Weil) *Let $\mathfrak{Q}$ be any prime ideal in K. Then, there is a uniquely determined representation $\psi_\mathfrak{Q}$ of J into the multiplicative group $K_\mathfrak{Q}^*$ of the $\mathfrak{Q}$-adic completion of K, satisfying the following conditions*:

(i) $\quad\psi_\mathfrak{Q}(\tilde{\alpha})=1\qquad$ *for all α in k.*

(ii) *For any prime ideal $\mathfrak{p}$ in k, not dividing $\mathfrak{f}\cdot N\mathfrak{Q}$,*

$$\psi_\mathfrak{Q}(\tilde{\pi}_\mathfrak{p})=\tilde{\Psi}(\tilde{\pi}_\mathfrak{p})\quad(=\chi(\mathfrak{p})[N\mathfrak{p}]^\nu),\quad \psi_\mathfrak{Q}(\tilde{u}_\mathfrak{p})=1.$$

Proof. Uniqueness is evident. We must prove the existence. Let $\mathfrak{p}_1,\cdots,\mathfrak{p}_g$ be all the prime divisors of $N\mathfrak{Q}$ in k. Let $S^{(i)}=\{\sigma_1^{(i)},\cdots,\sigma_{\mu_i}^{(i)}\}$ be the system of all isomorphisms of k which can be extended to a continuous isomorphism of $k_{\mathfrak{p}_i}$ into $K_\mathfrak{Q}$ in the $\mathfrak{p}_i$-adic and $\mathfrak{Q}$-adic topologies. We have clearly $\mu_i=e_if_i$, where e_i, f_i are ramification order and degree of $\mathfrak{p}_i$ respectively. It is also clear that σ can be extended to a continuous isomorphism of $k_{\mathfrak{p}_i}$ into $K_\mathfrak{Q}$ if and only if $\mathfrak{Q}$ divides $\mathfrak{p}_i^\sigma$. Thus the set $\{\sigma_1^{(1)},\cdots,\sigma_{\mu_1}^{(1)},\cdots,\sigma_1^{(g)},\cdots,\sigma_{\mu_g}^{(g)}\}$ coincides with the set of all the isomorphisms of k.

Now, the mapping $k^*\ni\alpha\to\prod_{\sigma\in S^{(i)}}(\alpha^\sigma)^{-n_\sigma}$ can be extended (uniquely) to a representation $\varphi_{\mathfrak{p}_i}$ of $k_{\mathfrak{p}_i}^*$ into $K_\mathfrak{Q}^*$. We have clearly

$$\prod_{\mathfrak{p}_i|N\mathfrak{Q}}\varphi_{\mathfrak{p}_i}(\alpha)=\pm\tilde{\Psi}(\tilde{\alpha}_\infty)\qquad\text{for } \alpha \text{ in } k^*.$$

On the other hand, the mappings $\tilde{\mathfrak{a}}\to\tilde{\Psi}(\tilde{\mathfrak{a}}_0)$ and $\tilde{\mathfrak{a}}\to\prod_\sigma\varepsilon_\sigma(\mathfrak{a}_\sigma)=\pm1$ are clearly continuous also in the topology of $K_\mathfrak{Q}^*$. Thus

$$\psi_{\mathfrak{Q}}(\tilde{\mathfrak{a}})=\tilde{\psi}(\tilde{\mathfrak{a}}_0)[\prod_{\sigma}\varepsilon_\sigma(\mathfrak{a}_\sigma)]\prod_{\mathfrak{p}|N\mathfrak{Q}}\varphi_{\mathfrak{p}}(\mathfrak{a}_{\mathfrak{p}})$$

satisfies our condition, where $\mathfrak{a}_{\mathfrak{p}}$ denotes the $\mathfrak{p}$-components of $\tilde{\mathfrak{a}}$ and $\mathfrak{a}_\sigma$ denotes the component of $\tilde{\mathfrak{a}}$ at infinite real prime (∞, σ).

Lemma 2. *All values $\psi_{\mathfrak{Q}}(\tilde{\mathfrak{a}})$ are $\mathfrak{Q}$-units.*

Proof. By definition and by condition (ii), $\tilde{\psi}(\mathfrak{a}_{\mathfrak{p}})$ is prime to $\mathfrak{Q}$ if $\mathfrak{p}$ is prime to $N\mathfrak{Q}$. If $\mathfrak{p}_i$ divides $N\mathfrak{Q}$, the $\mathfrak{Q}$-order of $\tilde{\psi}(\tilde{\pi}_{\mathfrak{p}_i})$ is just $\sum_{\sigma\in S^{(i)}} n_\sigma E_\sigma$, where E_σ denotes the $\mathfrak{Q}$-order of $\mathfrak{p}_i^\sigma$, and the $\mathfrak{Q}$-order of $\varphi_{\mathfrak{p}_i}(\pi_{\mathfrak{p}_i})$ is $-\sum_{\sigma\in S^{(i)}} n_\sigma E_\sigma$. As $\tilde{\psi}(\tilde{u}_{\mathfrak{p}_i})$, $\varphi_{\mathfrak{p}_i}(u_{\mathfrak{p}_i})$ are $\mathfrak{Q}$-units, this proves our lemma.

Lemma 3. *If $\mathfrak{p}$ divides $N\mathfrak{Q}$ and $\tilde{\psi}(\tilde{\pi}_{\mathfrak{p}})$ is prime to $\mathfrak{Q}$, then* $\varphi_{\mathfrak{p}}=1$.

Proof. We first remark that the second condition does not depend on the choice of $\pi_{\mathfrak{p}}$ even if $\mathfrak{p}$ divides $\mathfrak{f}$. Now, as $\mathfrak{Q}$ is prime to $(\tilde{\psi}(\tilde{\pi}_{\mathfrak{p}}))=\prod_{\sigma}(\mathfrak{p}^\sigma)^{n_\sigma}$, and as all n_σ are non-negative, so $\mathfrak{Q}$ must be prime to $\mathfrak{p}^\sigma$ for which $n_\sigma\neq 0$. Thus we see $\varphi_{\mathfrak{p}}(\alpha)=\prod_{\mathfrak{Q}|\mathfrak{p}^\sigma}(\alpha^\sigma)^{-n_\sigma}=1$.

Lemma 4. *If $\mathfrak{p}$ is prime to $\mathfrak{f}$ and $\tilde{\psi}(\tilde{\pi}_{\mathfrak{p}})$ is prime to $\mathfrak{Q}$, then*

$$\psi_{\mathfrak{Q}}(\tilde{\pi}_{\mathfrak{p}})=\tilde{\psi}(\tilde{\pi}_{\mathfrak{p}})=\chi(\mathfrak{p})[N(\mathfrak{p})]^{\nu} \quad\text{and}\quad \psi_{\mathfrak{Q}}(\tilde{u}_{\mathfrak{p}})=1.$$

Proof is evident by definition and by Lemma 3.

It may happen that all values $\tilde{\psi}(\tilde{\mathfrak{a}}_0)$ and $\prod_{\mathfrak{p}|N\mathfrak{Q}}\varphi_{\mathfrak{p}}(\alpha)$ $(\alpha\in k^*)$ belong to some subfield Σ of K. Then, $\psi_{\mathfrak{Q}}$ is also a representation of J into $\Sigma_{\mathfrak{q}}^*$ for a prime ideal $\mathfrak{q}$ in Σ divisible by $\mathfrak{Q}$. We can denote this representation by $\psi_{\mathfrak{q}}$, as this does not depend on the choice of prime divisor $\mathfrak{Q}$ of $\mathfrak{q}$ in K. Then, for any ideal $\mathfrak{b}=\mathfrak{q}_1^{m_1}\cdots\mathfrak{q}_s^{m_s}$ in Σ, we form the representation $\psi_{\mathfrak{q}_1}\times\cdots\times\psi_{\mathfrak{q}_s}$ of J into $\Sigma_{\mathfrak{q}_1}^*\times\cdots\times\Sigma_{\mathfrak{q}_s}^*$, and reduce it modulo $\mathfrak{b}$, namely, reduce each component $\psi_{\mathfrak{q}_i}(\tilde{\mathfrak{a}})$ modulo $\mathfrak{q}_i^{m_i}$. Then we obtain a representation $\chi_{\mathfrak{b}}$ of J into "Strahlklassengruppe" $G_{\mathfrak{b}}$ mod. $\mathfrak{b}$ in Σ:

$$\chi_{\mathfrak{b}}(\tilde{\mathfrak{a}})=[\psi_{\mathfrak{q}}(\tilde{\mathfrak{a}})\text{ mod. }\mathfrak{q}_1^{m_1}]\times\cdots\times[\psi_{\mathfrak{q}_s}(\tilde{\mathfrak{a}})\text{ mod. }\mathfrak{q}_s^{m_s}].$$

Finally, denoting by μ a character of the finite abelian group $G_{\mathfrak{b}}$, we obtain a character $\chi_{\mathfrak{b},\mu}$ of J of finite order by $\chi_{\mathfrak{b},\mu}=\mu\cdot\chi_{\mathfrak{b}}$. Then, by Lemma 4, we see

Lemma 5. *$\chi_{\mathfrak{b},\mu}$ is unramified at each prime $\mathfrak{p}$, which is relatively prime to $\mathfrak{f}$, such that $\tilde{\psi}(\tilde{\pi}_{\mathfrak{p}})$ is also prime to $\mathfrak{b}$.*

We denote by $\mathfrak{F}(\mathfrak{b})$ the product of all prime ideals $\mathfrak{p}$ in k such that either $\mathfrak{p}$ divides $\mathfrak{f}$ or $\tilde{\psi}(\tilde{\pi}_{\mathfrak{p}})$ is not prime to $\mathfrak{b}$. Then, by Lemma 5, we may take a suitable power $\mathfrak{F}(\mathfrak{b})^n$ of $\mathfrak{F}(\mathfrak{b})$ as a defining module of the character $\chi_{\mathfrak{b},\mu}$. In the following, if we speak of $\chi_{\mathfrak{b},\mu}$, we always suppose that $\chi_{\mathfrak{b},\mu}$ is defined modulo $\mathfrak{F}(\mathfrak{b})^n$. Thus, $\chi_{\mathfrak{b},\mu}$ *are in general not primitive.*

Clearly, $\chi_{\mathfrak{b},\mu}$ is also a character of the idèle class group C of k. We denote by $C_{\mathfrak{b}}$ the subgroup of C defined by $C_{\mathfrak{b}}=\{\tilde{\mathfrak{a}}' \mid \chi_{\mathfrak{b},\mu}(\tilde{\mathfrak{a}})=1$ for all characters μ of $G_{\mathfrak{b}}\}$, where $\tilde{\mathfrak{a}}'$ denotes the class of $\tilde{\mathfrak{a}}$. As $C_{\mathfrak{b}}$ is a closed subgroup of C with finite index, there is a finite abelian extension $k_{\mathfrak{b}}$ of k corresponding to it in the sense of class field theory. For any $\mathfrak{p}$ in k, we denote by $f(\mathfrak{p}, \mathfrak{b})$ the relative degree of $\mathfrak{p}$ in $k_{\mathfrak{b}}$ over k. Then we have

LEMMA 6. *For any $\mathfrak{p}$ prime to $\mathfrak{f}$, and for any natural number n, we have*

$$\sum_{\substack{\mathfrak{b} \\ \mathfrak{p}\nmid\mathfrak{F}(\mathfrak{b});\, f(\mathfrak{p},\mathfrak{b})\mid n}} \varphi(\mathfrak{b}) = N(\tilde{\psi}(\tilde{\pi}_{\mathfrak{p}})^n-1),$$

where φ denotes Euler function and N denotes always absolute norm.

Proof. By Artin's reciprocity law, $f(\mathfrak{p}, \mathfrak{b})$ is the smallest natural number f such that $\chi_{\mathfrak{b},\mu}(\tilde{\pi}_{\mathfrak{p}})^f=1$ for all characters μ of $G_{\mathfrak{b}}$. By definition, this last condition is equivalent to $\psi_{\mathfrak{a}_i}(\tilde{\pi}_{\mathfrak{p}})^f=1 \pmod{\mathfrak{q}_i^{m_i}}$ $(i=1,\cdots,s)$. The latter is also equivalent to $\tilde{\psi}(\tilde{\pi}_{\mathfrak{p}})^f\equiv 1 \pmod{\mathfrak{b}}$ by Lemma 4. Thus, $\tilde{\psi}(\tilde{\pi}_{\mathfrak{p}})$ being integral,

$$\sum_{\substack{\mathfrak{b} \\ \mathfrak{p}\nmid\mathfrak{F}(\mathfrak{b}),\, f(\mathfrak{p},\mathfrak{b})\mid n}} \varphi(\mathfrak{b}) = \sum_{\substack{\mathfrak{b} \\ \tilde{\psi}(\tilde{\pi}_{\mathfrak{p}})^n\equiv 1(\mathfrak{b})}} \varphi(\mathfrak{b}) = N(\tilde{\psi}(\tilde{\pi}_{\mathfrak{p}})^n-1). \qquad \text{q.e.d.}$$

We now form Hecke's L-function $L(s\,;\chi_{\mathfrak{b},\mu})$ of k with character $\chi_{\mathfrak{b},\mu}$, and define

$$L_{\mathfrak{b}}(s)=\prod_{\mu} L(s\,;\chi_{\mathfrak{b},\mu}),$$

where μ runs over all the $\varphi(\mathfrak{b})$ characters of $G_{\mathfrak{b}}$. Then, $L_{\mathfrak{b}}(s)$ is a ζ-function of $k_{\mathfrak{b}}$, raised to the power $\varphi(\mathfrak{b})/[k_{\mathfrak{b}}:k]$ and from which a finite number of factors $(1-N\mathfrak{p}^{-s})^{-1}$ due to prime factors of $\mathfrak{F}(\mathfrak{b})$ are omitted. More explicitly,

$$L_{\mathfrak{b}}(s)=\prod_{\mathfrak{p}\nmid\mathfrak{F}(\mathfrak{b})}\left(1-\frac{1}{(N\mathfrak{p})^{f(\mathfrak{p},\mathfrak{b})s}}\right)^{-\varphi(\mathfrak{b})/f(\mathfrak{b},\mathfrak{p})}.$$

We consider the infinite product

$$L(s)=\prod_{\mathfrak{b}} L_{\mathfrak{b}}(s),$$

where $\mathfrak{b}$ runs over all the integral ideals in Σ.

LEMMA 7. *The infinite product*

$$\prod_{\mathfrak{b}}\prod_{\mathfrak{p}\nmid\mathfrak{F}(\mathfrak{b})}\left(1-\frac{1}{(N\mathfrak{p})^{f(\mathfrak{p},\mathfrak{b})s}}\right)^{-\varphi(\mathfrak{b})/f(\mathfrak{b},\mathfrak{p})}$$

converges absolutely in the half plane $\operatorname{Re} s>c$, *where c is some finite real constant.*

Proof. We denote $\sigma=\operatorname{Re} s$. Then, denoting for a fixed $\mathfrak{p}$, $q=N(\mathfrak{p})$, $f(\mathfrak{b})=f(\mathfrak{p}, \mathfrak{b})$ and

$$S_{\mathfrak{p}}=\sum_{\substack{\mathfrak{b} \\ \mathfrak{p}\nmid\mathfrak{F}(\mathfrak{b})}} \frac{\varphi(\mathfrak{b})}{f(\mathfrak{b})\cdot q^{f(\mathfrak{b})\sigma}},$$

we have by Lemma 6,

$$S_{\mathfrak{p}} < \sum_{n=1}^{\infty}\Big(\sum_{\substack{\mathfrak{b}\\ f(\mathfrak{b})|n}}\varphi(\mathfrak{b})\Big)\frac{1}{q^{n\sigma}} = \sum_{n=1}^{\infty}\frac{N(\widetilde{\Psi}(\widetilde{\pi}_{\mathfrak{p}})^n-1)}{q^{n\sigma}}.$$

But by definition there is a constant A, independent of $\mathfrak{p}$, such that $N(\widetilde{\Psi}(\widetilde{\pi}_{\mathfrak{p}})^n-1)<(N\mathfrak{p})^{An}$. Hence $\sum_{\mathfrak{p}} S_{\mathfrak{p}}$ converges for $\sigma>A+1$, which was to be proved.

By this Lemma 7, we can change the order of product of $L(s)$.

$$L(s)=\prod_{\mathfrak{p}\nmid\mathfrak{f}} L^{(\mathfrak{p})}(s),$$

where

$$L^{(\mathfrak{p})}(s)=\prod_{\substack{\mathfrak{b}\\ \mathfrak{p}\nmid\mathfrak{F}(\mathfrak{b})}}\Big(1-\frac{1}{(N\mathfrak{p})^{f(\mathfrak{p},\mathfrak{b})s}}\Big)^{-\varphi(\mathfrak{b})/f(\mathfrak{p},\mathfrak{b})}.$$

Then, for $\mathfrak{p}$ prime to $\mathfrak{f}$, denoting $u=(N\mathfrak{p})^{-s}$, $f(\mathfrak{p},\mathfrak{b})=f(\mathfrak{b})$, we have again by Lemma 6,

$$\frac{d}{du}\log L^{(\mathfrak{p})}(u)=\sum_{\mu=1}^{\infty}\sum_{\substack{\mathfrak{b}\\ \mathfrak{p}\nmid\mathfrak{F}(\mathfrak{b})}}\varphi(\mathfrak{b})u^{\mu f(\mathfrak{b})-1}=\sum_{n=1}^{\infty}\Big(\sum_{\substack{\mathfrak{b}\\ \mathfrak{p}\nmid\mathfrak{F}(\mathfrak{b}),\,f(\mathfrak{b})|n}}\varphi(\mathfrak{b})\Big)u^{n-1}$$

$$=\sum_{n} N(\widetilde{\Psi}(\widetilde{\pi}_{\mathfrak{p}})^n-1)u^{n-1}.$$

But $N(\widetilde{\Psi}(\widetilde{\pi}_{\mathfrak{p}})^n-1)=\pm\prod_{\tau}(\widetilde{\Psi}(\widetilde{\pi}_{\mathfrak{p}})^{n\tau}-1)$, where τ runs over all the $D=[\Sigma:\boldsymbol{Q}]$ isomorphism of Σ. Replacing Σ by a quadratic extension, if necessary, we can always assume that this last formula holds with the sign $+$. Thus

$$\frac{d}{du}\log L^{(\mathfrak{p})}(u)=\sum_{t=0}^{D}(-1)^t\sum_{i_1,\cdots,i_t}\sum_{n}(\widetilde{\Psi}(\widetilde{\pi}_{\mathfrak{p}})^{\tau_{i_1}+\cdots+\tau_{i_t}})^n u^{n-1},$$

hence

$$L^{(\mathfrak{p})}(u)=\prod_{t=0}^{D}\prod_{i_1,\cdots,i_t}(1-\widetilde{\Psi}(\widetilde{\pi}_{\mathfrak{p}})^{\tau_{i_1}+\cdots+\tau_{i_t}}u)^{(-1)^{t+1}}.$$

Here $(\tau_{i_1},\cdots,\tau_{i_t})$ runs over all combinations of isomorphisms τ of Σ. Recalling the definition $\widetilde{\Psi}(\widetilde{\pi}_{\mathfrak{p}})=\chi(\mathfrak{p})(N\mathfrak{p})^{\nu}$ for $\mathfrak{p}$ prime to $\mathfrak{f}$, we have thus proved our main theorem:

THEOREM 1. *For a regularly algebraic-valued "Grössencharakter" χ, we have the following relation*:

$$\prod_{\mathfrak{b}} L_{\mathfrak{b}}(s)=\prod_{t=0}^{D}\prod_{i_1,\cdots,i_t} L(s-\nu t,\chi^{\tau_{i_1}+\cdots+\tau_{i_t}})^{(-1)^t},$$

where $\mathfrak{b}$ runs over all integral ideals in Σ.

Here, factors in right hand side:

$$L(s-\nu t,\chi^{\tau_{i_1}+\cdots+\tau_{i_t}})=\prod_{\substack{\mathfrak{p}\\ \mathfrak{p}\nmid\mathfrak{f}}}\Big(1-\frac{\chi(\mathfrak{p})^{\tau_{i_1}+\cdots+\tau_{i_t}}}{(N\mathfrak{p})^{s-\nu t}}\Big)^{-1}$$

are L-functions with "Grössencharaktere" $\chi^{\tau_{i_1}+\cdots+\tau_{i_t}}$.

§2. Local system for normal extension.

Let k be as usual an algebraic number field of finite degree, $\bar{k}$ an infinite normal extension of k and $\mathfrak{G}$ the Galois group of $\bar{k}$ over k. Let q be a rational prime. By a local representation M_q of $\mathfrak{G}$ at q (or in $\boldsymbol{Q}_q$), we understand a representation $\sigma \rightarrow M_q(\sigma)$ of $\mathfrak{G}$ into the multiplicative group of matrix algebra over the ring of q-adic integers. (This means especially that the determinants of $M_q(\sigma)$ are q-units.) Given a local representation M_q of $\mathfrak{G}$, we can reduce it modulo q^n and obtain a representation $M_q^{(n)}$ of $\mathfrak{G}$ with matrices over residue class ring mod. q^n. The kernel of $M_q^{(n)}$ is a closed subgroup of $\mathfrak{G}$ with finite index. We denote by k_{q^n} the subfield of $\bar{k}$ corresponding to this kernel, and put $k^{(q)} = \bigcup_{n=1}^{\infty} k_{q^n}$. Clearly $k^{(q)}$ is the subfield of $\bar{k}$ corresponding to the kernel of M_q. The Galois group of $k^{(q)}$ over k is denoted by $\mathfrak{G}^{(q)}$. Finally we denote by $\sigma(\mathfrak{p})$ the Frobenius automorphism of a prime divisor in $\bar{k}$ of a prime ideal $\mathfrak{p}$ in k, and also its image in $\mathfrak{G}^{(q)}$. The latter is determined up to inner automorphisms in $\mathfrak{G}^{(q)}$ if $\mathfrak{p}$ is unramified in $k^{(q)}$.

DEFINITION. A *local system* (of representations) for a normal extention $\bar{k}$ over k is a system $\{M_q\}$ of local representations of $\mathfrak{G}$ of the same degree d for all rational primes q, satisfying the following conditions:

(i) There exists a fixed integral ideal $\mathfrak{f}$ in k, such that every $\mathfrak{p}$ prime to $\mathfrak{f}$ is unramified in $k^{(q)}$ for any rational prime q prime to $\mathfrak{p}$.

(ii) The characteristic polynomials of $M_q(\sigma(\mathfrak{p}))$ for $\mathfrak{p}$ prime to $\mathfrak{f}$ are independent of q for all q prime to $\mathfrak{p}$. We denote this polynomial by $F_{\mathfrak{p}}(X)$.

(iii) $F_{\mathfrak{p}}(X)$ has the form:

$$F_{\mathfrak{p}}(X) = X^d + a_1 X^{d-1} + \cdots + a_{d-1} X \pm p^n,$$

where $a_1, \cdots, a_{d-1}$ are rational integers, p is the rational prime divisible by $\mathfrak{p}$ and $n = n(\mathfrak{p})$ are natural numbers bounded for all $\mathfrak{p}$ prime to $\mathfrak{f}$. We denote by $\pi_1(\mathfrak{p}), \cdots, \pi_d(\mathfrak{p})$ the roots of $F_{\mathfrak{p}}(X) = 0$.

(iv) Let $\mathfrak{K}$ be an absolutely normal field containing $k^{(p)}$ and all $\pi_1(\mathfrak{p}), \cdots, \pi_d(\mathfrak{p})$. Let $\tilde{\mathfrak{P}}$ be a prime divisor of $\mathfrak{p}$ in $\mathfrak{K}$, and $\mathfrak{P}$ be the one in $k^{(p)}$, divisible by $\tilde{\mathfrak{P}}$. We imbed the coefficient field $\boldsymbol{Q}_p$ of M_p into the $\tilde{\mathfrak{P}}$-adic completion $\mathfrak{K}_{\tilde{\mathfrak{P}}}$ of $\mathfrak{K}$ by a continuous isomorphism ι. Let $\iota\pi_1(\mathfrak{p}), \cdots, \iota\pi_d(\mathfrak{p})$ be images of $\pi_1(\mathfrak{p}), \cdots, \pi_d(\mathfrak{p})$ by a fixed extension of ι to $\boldsymbol{Q}_p(\pi_1(\mathfrak{p}), \cdots, \pi_d(\mathfrak{p}))$. Then, let $\iota\pi_1(\mathfrak{p}), \cdots, \iota\pi_\lambda(\mathfrak{p})$ be prime to $\tilde{\mathfrak{P}}$ while $\iota\pi_{\lambda+1}(\mathfrak{p}), \cdots, \iota\pi_d(\mathfrak{p})$ be divisible by $\tilde{\mathfrak{P}}$. (We change hereby the index of $\pi_i(\mathfrak{p})$ if necessary). Then the representation ιM_p, restricted to the decomposition group $D(\mathfrak{P})$ of $\mathfrak{P}$ in $\mathfrak{G}^{(p)}$, can be transformed in the form,

$$\iota M_p(\sigma)=\begin{pmatrix} {}^uM_p(\sigma) & 0 \\ * & {}^rM_p(\sigma)\end{pmatrix}.$$

where ${}^uM_p(\sigma)$, "unramified part" of $\iota M_p(\sigma)$, has the following properties: ${}^uM_p(\sigma)$ is of degree λ, and denoting by ${}^uk^{(\mathfrak{P})}$ the extension of the decomposition field $\mathfrak{H}(\mathfrak{P})$ of $\mathfrak{P}$, corresponding to the kernel of ${}^uM_p$, $\mathfrak{P}$ is unramified in ${}^uk^{(\mathfrak{P})}$ over k, and the characteristic roots of ${}^uM_p(\sigma(\mathfrak{P}))$, where $\sigma(\mathfrak{P})$ is the Frobenius automorphism of $\mathfrak{P}$ over $\mathfrak{H}(\mathfrak{P})$, are just $\iota\pi_1(\mathfrak{p}),\cdots,\iota\pi_\lambda(\mathfrak{p})$.

Clearly, this condition (iv) does not depend on the choices of $\tilde{\mathfrak{P}}$ and of ι.

We call $\mathfrak{f}$ in (i) the *defining ideal*, the set $\{F_\mathfrak{p}(X)\}$ the *characteristic set* of the local system $\{M_q\}$. A prime ideal $\mathfrak{p}$ is called *proper* for $\{M_q\}$ if $\mathfrak{p}$ does not divide $\mathfrak{f}$.

Now, for any natural number $b=q_1^{m_1}\cdots q_s^{m_s}$, we form a representation $D_b=M_{q_1}^{(m_1)}\times\cdots\times M_{q_s}^{(m_s)}$ of $\mathfrak{G}$, with matrices over residue class ring $R(b)$ of the ring $\boldsymbol{Z}$ of rational integers mod. b. Also, for a proper prime ideal $\mathfrak{p}$ in k, we take a prime divisor $\mathfrak{P}$ in $k^{(p)}$ as in (iv), and transform M_p in the form $\begin{pmatrix} {}^uM_p & 0 \\ * & {}^rM_p\end{pmatrix}$. Then we put ${}^uD_b(\sigma)=D_b(\sigma)$ if b is prime to p, and ${}^uD_b={}^uM_p^{(m_1)}\times M_{q_2}^{(m_2)}\times\cdots\times M_{q_s}^{(m_s)}$ if some q_i, say q_1, is p. This ${}^uD_b$ is a representation of the decomposition group $D(\mathfrak{P})$ of $\mathfrak{P}$.

By b-vectors of degree d we understand vectors $\mathfrak{x}, \mathfrak{y},\cdots$, with d components in the ring $R(b)$. We call $\mathfrak{x}$ a proper b-vector if at least one component of $\mathfrak{x}$ is regular in $R(b)$. The number of b-vectors is b^d, that of proper b-vectors is $b^d\prod_i(1-q_i^{-d})$, which we denote as $\Phi(b)$. Also by reduced b-vectors for $\mathfrak{p}$ we understand b-vectors if b is prime to $\mathfrak{p}$, and the product $\bar{\mathfrak{x}}_p\times\mathfrak{x}_{q_2}\times\cdots\times\mathfrak{x}_{q_s}$ if say, $q_1=p$, where $\mathfrak{x}_{q_i}$ are $q_i^{m_i}$-vectors of degree d $(i=2,\cdots,s)$ and $\bar{\mathfrak{x}}_p$ is a p^{m_1}-vectors of degree λ. If each $\mathfrak{x}_{q_i}$, $\bar{\mathfrak{x}}_p$ are proper, we speak of reduced proper b-vectors. We denote by $\Phi_0(b;\mathfrak{p})$ the number of reduced proper b-vectors for $\mathfrak{p}$.

We denote by k_b the subfield of $\bar{k}$ corresponding to the kernel of D_b, and by G_b the Galois group of k_b over k. Then, D_b may be considered as a representation of G_b.

As determinants of $M_q(\sigma)$, ${}^uM_q(\sigma)$ are q-units, $D_b(\sigma)$, ${}^uD_b(\sigma)$ permute proper b-vectors, reduced proper b-vectors for $\mathfrak{p}$ among themselves respectively. Thus we obtain a representation $\mathfrak{D}_b$ of $\mathfrak{G}$, of degree $\Phi(b)$, as permutations of $\Phi(b)$ proper b-vectors, and also we obtain a representation ${}^u\mathfrak{D}_b$ of the decomposition group $D(\mathfrak{P})$ of $\mathfrak{P}$ as permutations of $\Phi_0(b,\mathfrak{p})$ reduced proper b-vectors for $\mathfrak{p}$. We denote by $\chi_0^{(b)}(\sigma)$ the character of $\mathfrak{D}_b(\sigma)$ and by $\chi_b(\mathfrak{p}^n)$ the character of

${}^{u}\mathfrak{D}_b(\sigma(\mathfrak{P})^n)$ for the Frobenius automorphism $\sigma(\mathfrak{P})$ of $\mathfrak{P}$. This $\chi_b(\mathfrak{p}^n)$ is independent of the choice of $\mathfrak{P}$, as is seen from the condition (iv), even if $\mathfrak{p}$ divides b. If $\mathfrak{p}$ is prime to b, we have clearly $\chi_b(\mathfrak{p}^n)=\chi_0^{(b)}(\sigma(\mathfrak{p})^n)$. We remark that in any case $\chi_0^{(b)}(\sigma)$, $\chi_b(\mathfrak{p}^n)$ are non-negative rational integers.

With this "character" $\chi(\mathfrak{p}^n)$, we define an "L-function" $L_b(s)$ as follows:

$$\log L_b(s)=\sum_{\substack{\mathfrak{p}\\ \mathfrak{p}\nmid\mathfrak{f}}}\sum_{n=1}^{\infty}\frac{\chi_b(\mathfrak{p}^n)}{n(N\mathfrak{p})^{ns}}.$$

This $L_b(s)$ is different from Artin's L-function $L(s,\chi_0^{(b)})$ only by some factors due to prime divisors of $\mathfrak{f}(b)$. Thus, $L_b(s)$ is a product of ζ-functions of some intermediate fields between k_b and k, except for a finite number of the form $(1-N\mathfrak{q}^{-s})^{-1}$.

We form here again an infinite product

$$L(s)=\prod_b L_b(s),$$

where b runs over all natural numbers. Then, using Lemma 3 below, we can easily see as in §1 that the infinite series

$$\log L(s)=\sum_b\sum_{\substack{\mathfrak{p}\\ \mathfrak{p}\nmid\mathfrak{f}}}\sum_n\frac{\chi_b(\mathfrak{p}^n)}{n(N\mathfrak{p})^{ns}}$$

converges absolutely in some half plane $\mathrm{Re}\, s>c$, c being a finite real constant.

Thus we can change the order of summation of this series as follows:

$$\log L(s)=\sum_{\substack{\mathfrak{p}\\ \mathfrak{p}\nmid\mathfrak{f}}}\log L^{(\mathfrak{p})}(s),$$

where

$$\log L^{(\mathfrak{p})}(s)=\sum_{n=1}^{\infty}\sum_{b=1}^{\infty}\frac{\chi_b(\mathfrak{p}^n)}{n(N\mathfrak{p})^{ns}}.$$

Denoting again $(N\mathfrak{p})^{-s}=u$, we see,

$$\frac{d}{du}\log L^{(\mathfrak{p})}(s)=\sum_{n=1}^{\infty}\Bigl(\sum_{b=1}^{\infty}\chi_b(\mathfrak{p}^n)\Bigr)u^{n-1}.$$

Here we must calculate $\sum_b\chi_b(\mathfrak{p}^n)$.

LEMMA 1. *If $b=a\cdot c$ and a, c are coprime, then*

$$\chi_b(\mathfrak{p}^n)=\chi_a(\mathfrak{p}^n)\cdot\chi_c(\mathfrak{p}^n).$$

Proof. By definition, we have ${}^{u}D_b(\sigma)={}^{u}D_a(\sigma)\times{}^{u}D_c(\sigma)$, hence the representation ${}^{u}\mathfrak{D}_b(\sigma)$ is the tensor product of ${}^{u}\mathfrak{D}_a(\sigma)$ and ${}^{u}\mathfrak{D}_c(\sigma)$.

LEMMA 2. $\sum_{i=1}^{\infty}\chi_{q^i}(\mathfrak{p}^n)$ *is the highest exponent of q dividing* $\det|{}^{u}M_q(\sigma(\mathfrak{P})^n)-E|$ *considered as a q-adic number, where E is the unit matrix.*

Proof. By definition, $\chi_{q^i}(\mathfrak{p}^n)$ is the number of reduced proper q^i-vectors $\mathfrak{x}$ for $\mathfrak{p}$ invariant by ${}^{u}D_{q^i}(\sigma(\mathfrak{P})^n)$. Thus there is a sufficiently large number m such that for $i>m$ $\chi_{q^i}(\mathfrak{p}^n)=0$. Also the finite sum $\sum_{i=0}^{m}\chi_{q^i}(\mathfrak{P}^n)$ is the number of reduced q^m-vectors for $\mathfrak{p}$ invariant by ${}^uD_{q^m}(\sigma(\mathfrak{P})^n)$. This number is equal to q^ν if $\det|{}^uM_q(\sigma)-E|$ is of order ν in the q-adic valuation, as ${}^uD_{q^m}(\sigma)\equiv{}^uM_q(\sigma)$ (mod. q^m).

REMARK. As is easily seen, even if $q=p$, $\det|{}^uM_q(\sigma(\mathfrak{P}))-E|$ belongs to the q-adic number field $\boldsymbol{Q}_q$ (by our condition (iv)).

LEMMA 3. *For every proper* $\mathfrak{p}$, *we have,*

$$\sum_{b=1}^{\infty}\chi_b(\mathfrak{p}^n)=|\prod_{i=1}^{d}(\pi_i(\mathfrak{p})^n-1)|.$$

Proof. By Lemma 1, 2,

$$\sum_b\chi_b(\mathfrak{p}^n)=\prod_q\left(\sum_{i=1}^{\infty}\chi_{q^i}(\mathfrak{p}^n)\right)=\prod_q q^{\nu(q)},$$

where $\nu(q)$ is the order of $\prod_{i=1}^{\lambda}(\iota\pi_i(\mathfrak{p})^n-1)$ in q-adic valuation. Here, if $\lambda\neq d$, then $\iota\pi_{\lambda+1}(\mathfrak{p}),\cdots,\iota\pi_d(\mathfrak{p})$ are divisible by $\tilde{\mathfrak{P}}$ in the notation of condition (iv), hence the p-adic order of $\prod_{i=1}^{d}(\iota\pi_i(\mathfrak{p})^n-1)$ is equal to that of $\prod_{i=1}^{d}(\iota\pi_i(\mathfrak{p})^n-1)$, the latter is clearly equal to the p-adic order of $\prod_{i=1}^{d}(\pi_i(\mathfrak{p})^n-1)$. Thus, for all q, $q^{\nu(q)}$ is the q-component of rational integer $\prod_{i=1}^{d}(\pi_i(\mathfrak{p})^n-1)$, which gives our lemma.

Replacing $\{M_q\}$ by $\left\{\begin{pmatrix}M_q & 0\\ 0 & M_q\end{pmatrix}\right\}$ if necessary, we can always suppose that $\prod_{i=1}^{d}(\pi_i(\mathfrak{p})^n-1)$ are all positive. Then we have,

$$\frac{d}{du}\log L^{(\mathfrak{p})}(u)=\sum_{i=1}^{\infty}\left(\prod_{i=1}^{d}(\pi_i(\mathfrak{p})^n-1)\right)u^{n-1}$$

$$=\sum_{t=0}^{d}(-1)^t\sum_{i_1,\cdots,i_t}\sum_{n=1}^{\infty}(\pi_{i_1}(\mathfrak{p})\cdots\pi_{i_t}(\mathfrak{p}))^n u^{n-1}.$$

Thus, as in §1, we see

THEOREM 2. *For a local system* $\{M_q\}$, *we have*

$$\prod_{b=1}^{\infty}L_b(s)=\prod_{t=0}^{d}\prod_{i_1,\cdots,i_t}L(s,\chi_{i_1\cdots i_t})^{(-1)^t}.$$

Here, $\chi_{i_1\cdots i_t}(\mathfrak{p})=\pi_{i_1}(\mathfrak{p})\cdots\pi_{i_t}(\mathfrak{p})$, and

$$L(s,\chi_{i_1\cdots i_t})=\prod_{\substack{\mathfrak{p}\\ \mathfrak{p}\nmid\mathfrak{f}}}\left(1-\frac{\chi_{i_1\cdots i_t}(\mathfrak{p})}{(N\mathfrak{p})^s}\right)^{-1}$$

is an "L-function" with "character" $\chi_{i_1\cdots i_t}$. We call $L(s)=\prod_b L_b(s)$ *ζ-function of the local system* $\{M_q\}$.

§3. Examples

1°. Theorem 1 may be subsumed in Theorem 2 as we shall now explain. Here we use the notations in §1 and §2.

Let q be a rational prime and $(q)=\mathfrak{q}_1^{E_1}\cdots\mathfrak{q}_s^{E_s}$ be its prime decomposition in Σ. It is clear that the value $\psi_{\mathfrak{q}_i}(\tilde{\mathfrak{a}})$ for any $\tilde{\mathfrak{a}}\in J$ belongs to $\sum_{\mathfrak{q}_i}^*$. Now, the ring $\mathfrak{O}_{\mathfrak{q}_i}$ of $\mathfrak{q}_i$-adic integers can be considered as an algebra of dimension E_iF_i over the ring $\mathfrak{O}_{\mathfrak{q}}$ of q-adic integers, F_i denoting the degree of $\mathfrak{q}_i$. Let $\Lambda''_{\mathfrak{q}_i}$ be the regular representation of this algebra $\mathfrak{O}_{\mathfrak{q}_i}$. Then we obtain by $M''_{\mathfrak{q}_i}(\tilde{\mathfrak{a}})=\Lambda''(\psi_{\mathfrak{q}_i}(\tilde{\mathfrak{a}}))$ a representation of J with q-adic integral matrices of degree E_iF_i, with q-adic unit determinants.

Let K be as in §1, Lemma 1. For each q, we fix a prime divisor $\mathfrak{Q}$ of q in K. Let $\{\tau_1^{(i)},\cdots,\tau_{\nu_i}^{(i)}\}$ be as in §1, the system of isomorphisms $\tau_\rho^{(i)}$ of Σ which can be extended to a continuous isomorphism of $\Sigma_{\mathfrak{q}_i}$ into $K_{\mathfrak{Q}}$, that is, such that $\mathfrak{q}_i^{\tau_\rho^{(i)}}$ is divisible by $\mathfrak{Q}$. (Then $\nu_i=E_i\cdot F_i$). By one of these isomorphisms $\tau^{(i)}$, we inbed $\mathfrak{O}_{\mathfrak{q}_i}$ into $K_{\mathfrak{Q}}$, and denote by $M'_{\mathfrak{q}_i}(\tilde{\mathfrak{a}})$ the image of $M''_{\mathfrak{q}_i}(\tilde{\mathfrak{a}})$ by this isomorphism. With these $M'_{\mathfrak{q}_i}$, we obtain a representation

$$M'_q=\begin{pmatrix} M'_{\mathfrak{q}_1} & & 0 \\ & \ddots & \\ 0 & & M'_{\mathfrak{q}_s}\end{pmatrix}$$

of degree $D=[\Sigma:\boldsymbol{Q}]$. Here we remark that, if we map $\Lambda''_{\mathfrak{q}_i}(\alpha)$ for $\alpha\in\Sigma$ to $\Lambda'_{\mathfrak{q}_i}(\alpha)$ by $\tau^{(i)}$, then the characteristic roots of $\Lambda'_{\mathfrak{q}_i}(\alpha)$ are $\alpha^{\tau_1^{(i)}},\cdots,\alpha^{\tau_{\nu_i}^{(i)}}$. Hence the characteristic equation of

$$\begin{pmatrix} \Lambda'_{\mathfrak{q}_1}(\alpha) & & 0 \\ & \ddots & \\ 0 & & \Lambda'_{\mathfrak{q}_s}(\alpha)\end{pmatrix}$$

must have rational coefficients.

As $\mathfrak{O}_{\mathfrak{q}_i}$ are totally disconnected, $\psi_{\mathfrak{q}_i}$ is trivial on the connected component of 1 in the idèle class group of k. Hence M'_q induces, by Artin's reciprocity law, a local representation M_q of the Galois group $\mathfrak{G}$ of maximal abelian extension $\bar{k}$ of k. We shall show that the system $\{M_q\}$ is a local system for $\bar{k}/k$.

We take as $\mathfrak{f}$ in §2 condition (i), the conductor of χ from which our $\psi_{\mathfrak{q}}$ are defined. If $\mathfrak{p}$ is prime to $\mathfrak{f}(q)$, then, by Lemma 4 in §1, $\psi_{\mathfrak{q}_i}(\tilde{\pi}_{\mathfrak{p}})=\tilde{\psi}(\tilde{\pi}_{\mathfrak{p}})=\chi(\mathfrak{p})(N\mathfrak{p})^\nu$. Thus, as is remarked above, the characteristic roots of $M'_q(\tilde{\pi}_{\mathfrak{p}})$ are all the D conjugates of $\chi(\mathfrak{p})(N\mathfrak{p})^\nu$. Hence, the characteristic equation $F_{\mathfrak{p}}(X)$ of $M'_q(\tilde{\pi}_{\mathfrak{p}})$ have rational integral coefficients, and its constant term is a power of p. Clearly $F_{\mathfrak{p}}(X)$ does not depend on q. But as $k^{(q)}=\bigcup_n k_{q^n}$, Lemma 5 in §1, together with class field theory, shows that $\mathfrak{p}$ is unramified in $k^{(q)}$. Then, as we have $M_q(\sigma(\mathfrak{p}))=M'_q(\tilde{\pi}_{\mathfrak{p}})$, again by class field theory, our conditions (i), (ii), (iii) for local system are satisfied.

Next, let $q=p$ be divisible by $\mathfrak{p}$. For simplicity, we write $\pi_1=\chi(\mathfrak{p})(N\mathfrak{p})^\nu$, and denote by $\{\pi_1,\cdots,\pi_d\}$ all the roots of $F_{\mathfrak{p}}(X)=0$.

Denoting also $(p)=\mathfrak{q}_1^{l_1}\cdots\mathfrak{q}_s^{l_s}$, $\mathfrak{q}_i$ is prime to π_1 if and only if $\psi_{\mathfrak{q}_i}(\tilde{\pi}_{\mathfrak{p}})=\pi_1$. "Only if" part is proved in Lemma 4, §1, "if" part follows from the fact that $\psi_{\mathfrak{q}_i}(\tilde{\pi}_{\mathfrak{p}})$ is a $\mathfrak{q}_i$-unit. But $\mathfrak{q}_i$ is prime to π_1 if and only if $\mathfrak{Q}$ is prime to $\pi_1^{\tau_p^{(i)}}$. We denote by $\mathfrak{q}_1,\cdots,\mathfrak{q}_r$ all $\mathfrak{q}_i$'s prime to π_1 changing the index if necessary. In condition (iv), §2, we take $\tilde{\mathfrak{P}}$ arbitrarily and map $K_{\mathfrak{Q}}$ into $\mathfrak{K}_{\tilde{\mathfrak{P}}}$ by a continuous isomorphism ι. Let $\{\iota\pi_1,\cdots,\iota\pi_\lambda\}$ be the set of all $\tilde{\mathfrak{P}}$-units among $\{\iota\pi_i\}$. Then $\{\pi_1,\cdots,\pi_\lambda\}$ is just the set of all $\mathfrak{Q}$-units among $\{\pi_i\}$, hence the set $\{\pi_1^{\tau_1^{(1)}},\cdots,\pi_1^{\tau_{\nu_1}^{(1)}},\cdots,\pi_1^{\tau_1^{(r)}},\cdots,\pi_1^{\tau_{\nu_r}^{(r)}}\}$ has a similar property. As we have $\psi_{\mathfrak{q}_i}(\tilde{\pi}_{\mathfrak{p}})=\pi_1$ for $i=1,\cdots,r$, $\pi_1^{\tau_1^{(i)}},\cdots,\pi_1^{\tau_{\nu_i}^{(i)}}$ are just the characteristic roots of $M'_{\mathfrak{q}_i}(\tilde{\pi}_{\mathfrak{p}})$. Thus, if we take ${}^uM_p$ in (iv), §2 as the part of ιM_p induced by $\iota\begin{pmatrix} M'_{\mathfrak{q}_1} & & 0 \\ & \ddots & \\ 0 & & M'_{\mathfrak{q}_r}\end{pmatrix}$, then, Lemma 5, §1 shows that $\mathfrak{p}$ is unramified in ${}^uk^{(p)}$. Thus all the conditions in (iv) are also satisfied, and $\{M_q\}$ is shown to be a local system.

Finally, the following relation holds, as is easily verified, between $L_{\mathfrak{b}}(s)$ in §1 and $L_b(s)$ in §2,

$$\prod_{\mathfrak{b}\mid B} L_{\mathfrak{b}}(s)=\prod_{b\mid B} L_b(s)$$

where B is a natural number. In this case $L_b(s)$ is a product of a finite number of $L_{\mathfrak{b}}(s)$. Thus Theorem 1 can be regarded as a special case of Theorem 2.

2°. Another example of a local system is found in the theory of abelian varieties. Let A be an abelian variety of dimension g defined over an algebraic number field k of finite degree. Let $\bar{k}$ be the extension of k obtained from k by "adjunction" of all points of finite order of A. Then $\bar{k}$ is an infinite normal extension of k; infiniteness of the degree $[\bar{k}:k]$ is indeed assured by Weil's finiteness theorem. Let q be a rational prime. Then, all points of A, whose orders are some powers of q, form an abelian group $\mathfrak{g}(q;A)$ isomorphic to the direct product of $2g$ additive groups of q-adic numbers modulo 1. As any automorphism σ of $\bar{k}$ over k permutes the points of A with the same order q^n among themselves, the Galois group $\mathfrak{G}$ of $\bar{k}$ over k can be represented by q-adic integral matrices $M_q(\sigma)$ of degree $2g$, whose determinants are q-units. Here, continuity of this representation M_q is assured by the definition of the topology of Galois group. Then, $\{M_q\}$ forms a local system for $\bar{k}/k$.

The proof is based on the theory of reduction modulo $\mathfrak{p}$ of abelian varieties, as is used in the theory of complex multiplication of abelian varieties by G. Shimura and the author.[2] Here we shall limit ourselves to a sketch of the proof.

Let $\mathfrak{p}$ be a prime ideal in k. If we reduce A modulo $\mathfrak{p}$ and obtain also an abelian variety $\tilde{A}$ (over a finite field $\tilde{k}$), we call $\mathfrak{p}$ proper. Then, let $\mathfrak{f}$ be the product of all non-proper prime ideals in k. As non-proper $\mathfrak{p}$'s are finite in number, $\mathfrak{f}$ is certainly an ideal in k. Let $\mathfrak{p}$ be proper and x be a generic point of A over k and $\tilde{x}$ be that of $\tilde{A}$ over $\tilde{k}$, obtained from x mod. $\mathfrak{p}$. For any point $\tilde{y}$ on $\tilde{A}$, we denote by $\tilde{y}^{N\mathfrak{p}}$ the point on $\tilde{A}$ with coordinates obtained from those of $\tilde{y}$ by raising to the power $(N\mathfrak{p})$. Then, $\tilde{x}\rightarrow\tilde{x}^{N\mathfrak{p}}$ is an endomorphism $\tilde{\pi}_{\mathfrak{p}}$ of $\tilde{A}$. Then, denoting always by $\sigma(\mathfrak{p})$ a Frobenius automorphism of $\mathfrak{p}$ in $\bar{k}/k$, we have

$$\widetilde{x^{\sigma(\mathfrak{p})}}=\tilde{x}^{N\mathfrak{p}}=\tilde{\pi}_{\mathfrak{p}}\tilde{x}.$$

Now, if $\mathfrak{p}$ is prime to q, the group $\mathfrak{g}(q;A)$ is mapped isomorphically onto $\mathfrak{g}(q,\tilde{A})$ by reduction modulo $\mathfrak{p}$. Hence $\mathfrak{p}$ is unramified in $k^{(q)}$, and $M_q(\sigma(\mathfrak{p}))$ is equal to the q-adic representation matrix of the endomorphism $\tilde{\pi}_{\mathfrak{p}}$ by $\mathfrak{g}(q;\tilde{A})$. The characteristic polynomial $F_{\mathfrak{p}}(X)$ of the latter has rational integral coefficients, with constant term $(N\mathfrak{p})^g$, and is independent of q. Hence our conditions (i), (ii), (iii) for local system are satisfied.

Now, let $\mathfrak{p}$ divide $q=p$. Let $\mathfrak{P}$ be a prime divisor of $\mathfrak{p}$ in $\bar{k}$, and $D(\mathfrak{P})$ the decomposition group of $\mathfrak{P}$. Then the points in $\mathfrak{g}(p;A)$ which are mapped to $\tilde{0}$ by the reduction modulo $\mathfrak{P}$, form a subgroup $\mathfrak{g}_0(p;A)$ of $\mathfrak{g}(p;A)$, which is invariant by all automorphisms σ in $D(\mathfrak{P})$. Hence, restricted to $D(\mathfrak{P})$, the representation M_p can be transformed into the form

$$\begin{pmatrix} {}^uM_p & 0 \\ * & {}^rM_p \end{pmatrix}$$

where ${}^rM_p$ corresponds to $\mathfrak{g}_0(p;A)$. Then, ${}^uM_p(\sigma)$ is just the representation of $\tilde{\sigma}$ by p-adic matrices as the transformation of $\mathfrak{g}(p;\tilde{A})$, where $\tilde{\sigma}$ is the automorphism in residue class field mod. $\mathfrak{P}$ induced by σ. Such a representation as transformations of $\mathfrak{g}(p;\tilde{A})$ will be called a p-adic representation in $\tilde{A}$. From this decomposition $\begin{pmatrix} {}^uM_p & 0 \\ * & {}^rM_p \end{pmatrix}$ we can see also, by the consideration modulo $\mathfrak{P}$, that $\mathfrak{P}$ is unramified in ${}^uk^{(p)}$. Especially, ${}^uM_p(\sigma(\mathfrak{p}))$ is equal to the matrix of $\tilde{\pi}_{\mathfrak{p}}$ in a p-adic representation in $\tilde{A}$.

2) Cf. the forthcoming papers in the Proceedings of the Symposium on algebraic numbers, 1955, Tokyo.

We remark here, as the ring $\mathcal{A}(\tilde{A})$ of endomorphisms is an order of a semi-simple algebra $\mathcal{A}_0(\tilde{A})$, $\tilde{\pi}_{\mathfrak{p}}$ is expressed as the sum of simple components: $\tilde{\pi}_{\mathfrak{p}}=\tilde{\pi}^{(1)}+\cdots+\tilde{\pi}^{(v)}$. But, as a suitable power $\tilde{\pi}_{\mathfrak{p}}^M$ belongs to the centre of $\mathcal{A}(\tilde{A})$, $\tilde{\pi}^{(i)M}$ generates a field C_i in $\mathcal{A}_0(\tilde{A})$. We also remark that the group $\mathfrak{g}(q;\tilde{A})$ is also decomposed into $\mathfrak{g}(q;\tilde{A})=\mathfrak{g}^{(1)}+\cdots+\mathfrak{g}^{(v)}$ according to the decomposition of endomorphisms.

Let $(p)=\tilde{\mathfrak{Q}}_1^{e_1}\cdots\tilde{\mathfrak{Q}}_s^{e_s}$ be a decomposition of p in the field C_i. Let $\tilde{\mathfrak{Q}}_1\cdots\tilde{\mathfrak{Q}}_r$ be prime to $\pi^{(i)M}$ and let $\tilde{\mathfrak{Q}}_{r+1},\cdots,\tilde{\mathfrak{Q}}_s$ divide $\pi^{(i)M}$. Then, $\mathfrak{g}^{(i)}$ is a direct product of groups of the points, corresponding to the i-th factor of $\mathcal{A}(\tilde{A})$ and with "order" $\tilde{\mathfrak{Q}}_j^n$ $(n=1,2,\cdots)$ for $j=1,\cdots,r$. Hence $\mathfrak{g}^{(i)}$ is isomorphic to the direct product of $e_1f_1+\cdots+e_rf_r$ additive groups of p-adic numbers modulo 1, where e_i, f_i are ramification order and degree of $\tilde{\mathfrak{Q}}_i$ respectively.

From this we can see, just as in the case $1°$, that the characteristic roots of the matrix of $\tilde{\pi}_{\mathfrak{p}}$ in a p-adic representation in $\tilde{A}$ are just the images of roots of $F_{\mathfrak{p}}(x)=0$, which are prime to some prime divisor $\mathfrak{P}_0$ in a suitable large field k containing all roots of $F_{\mathfrak{p}}(x)=0$. But as this representation matrix of $\tilde{\pi}_{\mathfrak{p}}$ is equal to ${}^uM_p$, our condition (iv) for local system is also satisfied.

In this case $L(s)$ is equal to Hasse's ζ-function of A, as is immediately seen by definition.

If moreover, the ring $\mathcal{A}(A)$ of endomorphisms of A contains a subring isomorphic to an order of an algebraic number field of degree $2g$, then, $\chi(\mathfrak{p})=\tilde{\pi}_{\mathfrak{p}}/|\tilde{\pi}_{\mathfrak{p}}|$ is a "Grössencharakter" of k, as has been proved by the theory of complex multiplication. But we can also prove this fact directly, using our notion of local system, as we shall expose in a forthcoming paper.

Regular Local System for an Abelian Extension

An unpublished manuscript written on February 17, 1956.

Introduction

In a former paper,[1] I have introduced the notion of local systems for normal extensions of algebraic number fields, and pointed out a property of ζ-functions of local systems. I have also remarked that Hasse's ζ-function of an abelian variety is the ζ-function of a certain local system. Now, the object of this paper is this: To prove the conjecture of Hasse in case of complex multiplication more arithmetically, using the notion of local system, and to find some inner relation of this conjecture with the nature of the class field obtained by the general complex multiplication.[2]

For this purpose, we shall introduce in §1 the notion of the regular local system, extracting certain properties of local systems obtained from abelian varieties with sufficiently many complex multiplications, and prove that any regular local system can be obtained from certain "Grössencharaktere" (Theorem 1). This gives especially a proof of the conjecture of Hasse for abelian varieties with sufficiently many complex multiplications under somewhat weaker condition on the ground field k (Theorem 3, §3). Our Theorem 1 signifies however more than this, as there are regular local systems which are not obtained from abelian varieties; it gives rather a characterization of regularly algebraic-valued "Grössencharacter". In §2, we shall discuss the nature of the infinite abelian extension "attached" to a regular local system, and point out a relation of the Galois group of this extension with that of the maximally abelian extension of a certain number field, which is in general different from the original ground field (Theorem 2). This Theorem 2 gives an interpretation of regularly algebraic-valued "Grössencharakter" in a certain sense, and shows, together with Theorem 1 and the proof of Theorem 3, some relation between Hasse's conjecture and the class field obtained by the general complex multiplication. In §3, we shall give, besides a proof of Hasse's conjecture, two examples of regular local systems.

A. Weil, who has systematically used the method of local repre-

1) Y. Taniyama, "On a certain relation of L-functions", to appear. This paper is quoted by [L] in the following.

2) Cf. Contributions by G. Shimura, Y. Taniyama and A. Weil to the symposium on algebraic number theory, 1955, Tokyo.

sentations in the theory of abelian varieties, has suggested[3] that this method might be also effective in algebraic number theory. Our theorems show indeed that this is the case. In fact, our proof is essentially based on the continuity of local representations, from which the existence of global representation of the idèle group is concluded.

In the general case of normal extensions, we have now a new problem: Is it possible to find some structure, which is to be regarded as a generalization of the idèle group in case of abelian extensions, and from whose global representation our local system can be obtained in a natural way? Clearly, this problem is closely connected with the conjecture of Hasse for general abelian varieties. It is true that our method used in this paper is too "abelian" and might have but little meaning on this problem. But our result still suggests a possibility of attacking this conjecture from the side of arithmetic.

In the following, we use the notations and terminologies in [L]. Especially, "representation" means as usual an algebraic homomorphism which is continuous.

§1. Fundamental theorem.

Let k be an algebraic number field of finite degree, $\bar{k}$ an infinite abelian extension of k and $\mathfrak{G}$ be the Galois group of $\bar{k}$ over k. Let $\{M_q\}$ be a system of local representations (i.e. q-adic unimodular representations) of $\mathfrak{G}$ with the same degree d, at all rational primes q. Then, for every q, $M_q(\sigma)$ can be transformed simultaneously in the form

$$(*)\qquad \begin{pmatrix} \mu_{q1}(\sigma) & & * \\ & \ddots & \\ 0 & & \mu_{qd}(\sigma) \end{pmatrix}$$

for all σ in $\mathfrak{G}$, in a suitable extension of the q-adic number field $\boldsymbol{Q}_q$. We denote by $k^{(q)}$ the subfield of $\bar{k}$ corresponding to the kernel of M_q.

Definition. A system $\{M_q\}$ of local representations M_q of $\mathfrak{G}$ with the same degree d for all rational primes q is called a *simple regular local system* for the abelian extension $\bar{k}$ over k, if it satisfies the following conditions.

(a) There exists a fixed integral ideal $\mathfrak{f}$ in k, such that every $\mathfrak{p}$ prime to $\mathfrak{f}$ is unramified in $k^{(q)}$ for any rational prime q prime to $\mathfrak{p}$.

We shall call $\mathfrak{p}$ *proper* if $\mathfrak{p}$ is prime to $\mathfrak{f}$.

(b) There is an algebraic number field Σ of degree d, and for each proper $\mathfrak{p}$, there is an integer $\pi(\mathfrak{p})$ in Σ, with the following properties:

3) Symposium on algebraic number theory, 1955, Tokyo.

1) We denote by ι_{qi} $(1=1,\cdots,d)$ all the isomorphisms of Σ into the algebraic closure $\bar{\boldsymbol{Q}}_q$ of $\boldsymbol{Q}_q$ for each q and i, and we denote by $\sigma(\mathfrak{p})$ a Frobenius automorphism of $\mathfrak{p}$ in $\bar{k}$. Then, if we transform all M_q into the form (*) and determine thereby the index i in a suitable way, we have

$$\mu_{qi}(\sigma(\mathfrak{p}))=\iota_{qi}(\pi(\mathfrak{p}))$$

for all i, for all proper $\mathfrak{p}$ and for all q prime to $\mathfrak{p}$.

2) We have

$$|\pi(\mathfrak{p})^{\tau}|=(N\mathfrak{p})^{\nu}$$

for all proper $\mathfrak{p}$ and for all isomorphisms τ of Σ into the complex number field. Here N denotes absolute norm, and ν is a rational number independent of $\mathfrak{p}$ and τ.

(c) The ideal $(\pi(\mathfrak{p}))$ in Σ can be expressed by $\mathfrak{p}$ in the form:

$$(**)\qquad (\pi(\mathfrak{p}))=\prod_{\sigma}(\mathfrak{p}^{\sigma})^{n_{\sigma}(\mathfrak{p})}$$

for all proper $\mathfrak{p}$. Here σ runs over all isomorphisms of k, and $n_{\sigma}(\mathfrak{p})$ denotes a non-negative rational integer, which *may depend* on $\mathfrak{p}$.

(d) Every representation M_q is completely reducible.

A direct sum of a finite number of simple regular local systems for the same abelian extension $\bar{k}$ over k is called a *regular local system* for $\bar{k}$ over k.

The condition (a) is the same as the condition (i) for local systems in [L], and the conditions (a), (b) contain also the conditions (ii), (iii) for local systems. Moreover, as we shall show below, (a), (b), (c), (d) contain also the condition (iv) for local systems, so our $\{M_q\}$ is indeed a local system.

Remark on the condition (c). If k is absolutely normal and contains Σ, this condition is trivially satisfied. But in general, k can be smaller and this may not be the case. In any case, however, if (a), (b) are satisfied we can replace k by some overfield, and restricting M_q to the corresponding subgroup of $\mathfrak{G}$, get a system $\{M_q'\}$ satisfying (c). Thus the condition (c) is not too restrictive. Moreover, we can find the smallest overfield of k for which (c) is satisfied, if we determine the exponents n_{σ} in (**) for $\{M_q'\}$ explicitly.

We remark here also that we do not use the condition (d) for a while, and deal with the triangular form (*) of M_q, until we indicate explicitly the use of (d).

Any simple regular local system for an abelian extension can be considered as the one for the maximal abelian extension of k. Thus, from any representation μ_{qi} of $\mathfrak{G}$ we get a representation ψ_{qi} of the idèle group J of k into the multiplicative group $\Sigma_{\mathfrak{q}}^{*}$ of $\mathfrak{q}$-adic completion of Σ, with the following properties, where $\mathfrak{q}$ is a prime divisor of q

in Σ such that ι_{qi} can be extended to an isomorphism of $\Sigma_{\mathfrak{q}}$ into $\bar{\boldsymbol{Q}}_q$:

α) $\psi_{qi}(\tilde{\alpha})=1$ for all principal idèles $\tilde{\alpha}$.

β) $\psi_{qi}(\tilde{\pi}_{\mathfrak{p}})=\iota_{qi}^{-1}(\mu_{qi}(\sigma(\mathfrak{p})))=\pi(\mathfrak{p})$, $\psi_{qi}(\tilde{u}_{\mathfrak{p}})=1$

for all $\mathfrak{p}$ prime to $\mathfrak{f}\cdot q$. Here $\pi_{\mathfrak{p}}$, $u_{\mathfrak{p}}$ denote, as in [L], a $\mathfrak{p}$-prime element, a $\mathfrak{p}$-unit in $k_{\mathfrak{p}}^*$, respectively.

Let q be any rational prime. For any idèle $\tilde{\mathfrak{a}}$ of k, we denote by $\tilde{\mathfrak{a}}_q$ the "q-part" of $\tilde{\mathfrak{a}}$, namely, $\tilde{\mathfrak{a}}_q$ is the idèle with component 1 at each $\mathfrak{p}$ prime to q, and with the same component as $\tilde{\mathfrak{a}}$ at each $\mathfrak{p}$ dividing q. Then, we put $\tilde{\mathfrak{a}}^{(q)}=\tilde{\mathfrak{a}}\cdot\tilde{\mathfrak{a}}_q^{-1}$, so that $\tilde{\mathfrak{a}}^{(q)}$ has component 1 at each $\mathfrak{p}$ dividing q. We also denote by $\tilde{k}_q^*$ the subgroup of J composed of idèles $\tilde{\alpha}_q$ for all α in k^*. Thus, $\tilde{\mathfrak{a}}_\infty$ denotes the infinite part of $\tilde{\mathfrak{a}}$, and the finite part $\tilde{\mathfrak{a}}_0$ of $\tilde{\mathfrak{a}}$ is denoted by $\tilde{\mathfrak{a}}^{(\infty)}$ in this notation. Then, we call *q-topology* of k^* the topology of k^* induced by the topology of J by the imbedding $k^*\to\tilde{k}_q^*$.

Hereafter we consider only μ_{qi}, ψ_{qi} for a fixed i, so we omit the index i in ι_{qi}, μ_{qi} and ψ_{qi}.

By our condition (a), ψ_q is unramified at every $\mathfrak{p}$ prime to $\mathfrak{f}\cdot q$. Let $\mathfrak{f}=\mathfrak{h}_1^{e_1}\cdots\mathfrak{h}_r^{e_r}$ be the prime decomposition of $\mathfrak{f}$ in k. Let $U_{\mathfrak{h}_i}$ be the $\mathfrak{h}_i$-unit group in $k_{\mathfrak{h}_i}^*$. Then, comparing topologies of $U_{\mathfrak{h}_i}$ and of $\Sigma_{\mathfrak{q}}^*$, we see that the image of $\tilde{U}_{\mathfrak{h}_i}=\{\tilde{u}_{\mathfrak{h}_i}\mid u_{\mathfrak{h}_i}\in U_{\mathfrak{h}_i}\}$ by ψ_q must be a finite group, hence a group of roots of unity, for each i and for each q prime to $\mathfrak{f}$. Hence there is the smallest power $\mathfrak{h}_i^{c_i}$ of $\mathfrak{h}_i$ such that $u_{\mathfrak{h}_i}\equiv 1$ mod. $\mathfrak{h}_i^{c_i}$ implies $\psi_q(\tilde{u}_{\mathfrak{h}_i})=1$ for any $u_{\mathfrak{h}_i}$ in $U_{\mathfrak{h}_i}$. We call the product $\mathfrak{f}_q=\prod_{i=1}^{r}\mathfrak{h}_i^{c_i}$ the $\mathfrak{f}$-part of the conductor of ψ_q. Then, for any α in k^* prime to $\mathfrak{f}$, and for any q prime to $\mathfrak{f}$, we have

$$\psi_q(\tilde{\alpha}_q)=\psi_q(\tilde{\alpha}^{(q)})^{-1}=\varepsilon_q'(\alpha)\cdot\prod_{\mathfrak{p}\nmid q}\pi(\mathfrak{p})^{-\nu_{\mathfrak{p}}(\alpha)},$$

where $\nu_{\mathfrak{p}}(\alpha)$ denotes the order of α at $\mathfrak{p}$ and $\varepsilon_q'(\alpha)$ is a root of unity, clearly in $\Sigma_{\mathfrak{q}}^*$, depending only on the class of α modulo $\mathfrak{f}_q$. If we put

$$\varphi_q(\alpha)=\psi_q(\tilde{\alpha}_q)\cdot\prod_{\mathfrak{p}/q}\pi(\mathfrak{p})^{-\nu_{\mathfrak{p}}(\alpha)}$$

for all α in k^*, we have therefore

$$(***)\qquad \varphi_q(\alpha)=\varepsilon_q'(\alpha)\prod_{\mathfrak{p}}\pi(\mathfrak{p})^{-\nu_{\mathfrak{p}}(\alpha)}$$

for any α prime to $\mathfrak{f}$ and for any q prime to $\mathfrak{f}$. At any rate, as ψ_q and $\alpha\to\prod_{\mathfrak{p}/q}\pi(\mathfrak{p})^{\nu_{\mathfrak{p}}(\alpha)}$ is continuous, φ_q is a representation of k^* with q-topology into $\Sigma_{\mathfrak{q}}^*$.

LEMMA 1. *Let W_q be the number of roots of unity which appear in (***) as $\varepsilon_q'(\alpha)$, where α runs over all numbers in k^* prime to $\mathfrak{f}$. Then, there is a certain infinite set Λ of rational primes q such that W_q are bounded for all q in Λ.*

Proof. (In this proof we assume that q is prime to $\mathfrak{f}$.) We recall that $\mathfrak{f}_q$ has no other prime factors than those of $\mathfrak{f}$, namely, than $\mathfrak{h}_1, \cdots, \mathfrak{h}_r$. We denote by h_i the rational prime divisible by $\mathfrak{h}_i$. If $\alpha \equiv 1$ mod. $\mathfrak{h}_1 \cdots \mathfrak{h}_r$, then, as we have $\alpha^{(h_1 \cdots h_r)^t} \equiv 1$ mod. $(\mathfrak{h}_1 \cdots \mathfrak{h}_r)^t$ for any natural number t, the order of $\varepsilon'_q(\alpha)$ must divide some power $(h_1 \cdots h_r)^t$. On the other hand, this order of $\varepsilon'_q(\alpha)$ must divide $\Phi(\mathfrak{q})$, as $\varepsilon'_q(\alpha)$ belong to $\Sigma^*_{\mathfrak{q}}$ and q is prime to $h_1 \cdots h_r$, where Φ denotes the Euler function in Σ. Hence, if $\Phi(\mathfrak{q}) \not\equiv 0$ mod. h_i^t, $i=1, \cdots, r$, then the number of $\varepsilon'_q(\alpha)$ for $\alpha \equiv 1$ mod. $(\mathfrak{h}_1 \cdots \mathfrak{h}_r)$ is not greater than $(h_1 \cdots h_r)^t$. As the number of classes of α mod. $(\mathfrak{h}_1 \cdots \mathfrak{h}_r)$ is finite and independent of q, this shows that W_q is bounded for all q for which $\Phi(\mathfrak{q}) \not\equiv 0$ mod. h_i^t $(i=1, \cdots, r)$ holds. But this condition on $\mathfrak{q}$ is equivalent to another condition on $\mathfrak{q}$ that it does not decompose completely in the extension $\Sigma(\zeta_i)$ of Σ for each i, where ζ_i means a primitive h_i^t-th root of unity. Therefore, if we take t so large that any one of ζ_i's does not belong to the smallest absolutely normal field containing Σ, then there are certainly infinitely many rational primes q, with the property that any prime divisor $\mathfrak{q}$ of q in Σ satisfies this last condition. This completes the proof.

We denote by E the finite group of roots of unity generated by roots of unity in the smallest absolutely normal field K containing k and Σ, and also by all the roots of unity appearing in (***) for some q in Λ and for some α in k^* prime to $\mathfrak{f}$.

Now, we denote by σ_0 the complex conjugate automorphism of the complex number field. Then, as any $\pi(\mathfrak{p})^\tau$ has absolute value $N\mathfrak{p}^\nu$ (Condition (b), 2)), we have

$$n_\sigma(\mathfrak{p}) + n_{\sigma\tau\sigma_0\tau^{-1}}(\mathfrak{p}) = 2\nu$$

for any proper prime ideal $\mathfrak{p}$ of the first degree in k, unramified over $\boldsymbol{Q}$, and for any pair (σ, τ) of isomorphisms of k. Then, we denote by $\{n_\sigma^{(\lambda)}\}$ the set of non-negative rational integers $n_\sigma^{(\lambda)}$ satisfying

$$n_\sigma^{(\lambda)} + n_{\sigma\tau\sigma_0\tau^{-1}}^{(\lambda)} = 2\nu$$

for all pairs (σ, τ). Clearly the number of these sets $\{n_\sigma^{(\lambda)}\}$ is finite; we denote this number by v. The set of proper prime ideals $\mathfrak{p}$ of the first degree in k, unramified over $\boldsymbol{Q}$, is divided into mutually disjoint subsets T_λ $(\lambda=1, \cdots, v)$ so that $\mathfrak{p}$ belongs to T_λ if and only if $n_\sigma(\mathfrak{p}) = n_\sigma^{(\lambda)}$ for all σ. We denote by T'_λ the group of ideals generated by all prime ideals in T_λ.

For any λ, we put

$$\gamma_\lambda(\alpha) = \prod_\sigma \alpha^{\sigma n_\sigma^{(\lambda)}}$$

for all α in k^*, where σ runs over all isomorphisms of k. Then, any conjugate of $\gamma_\lambda(\alpha)$ has the same absolute value $|N\alpha|^\nu$. We have therefore

$$\prod_{\mathfrak{p}} \pi(\mathfrak{p})^{\nu_{\mathfrak{p}}(\alpha)} = \varepsilon_\lambda(\alpha)\, \gamma_\lambda(\alpha)$$

for any α such that the ideal (α) belongs to T'_λ, where $\varepsilon_\lambda(\alpha)$ is a root of unity contained in K. Especially, for any unit η in k, $\gamma_\lambda(\eta)$ is a root of unity in K. On the other hand, it is clear that γ_λ is a representation of k^* with q-topology into $K^*_{\mathfrak{Q}}$ for any q and for any prime divisor $\mathfrak{Q}$ of $\mathfrak{q}$ in K.

LEMMA 2. *Let $R_{\lambda q}$ be the image of T'_λ in the "Strahlklassengruppe" S_q* mod. q *in k by natural homomorphism. Then, there is at least one λ such that $R_{\lambda q}$ covers the whole group S_q for all but a finite number of q.*

Proof. We assume that, for each λ, there are infinitely many different rational primes $q_{\lambda i}$ $(i=1,2,\cdots)$ such that $R_{\lambda q}$ is a proper subgroup of S_q for $q=q_{\lambda i}$. We denote by $\overline{T}_{\lambda i}$ the set of all prime ideals, whose image in $S_{q_{\lambda i}}$ belong to $R_{\lambda q_{\lambda i}}$. Then T_λ is contained in $\bigcap_{i=1}^{m} \overline{T}_{\lambda i}$ for any natural number m. The set $\bigcap_{i=1}^{m} \overline{T}_{\lambda i}$ has clearly a definite density (in the sense of Kronecker), and this density is not greater than 2^{-m}, as $R_{\lambda q_{\lambda 1}} \times \cdots \times R_{\lambda q_{\lambda m}}$ has the index not smaller than 2^m in $S_{q_{\lambda 1}} \times \cdots \times S_{q_{\lambda m}}$. Thus, if we take m large enough, the density of the set $\bigcup_{\lambda=1}^{v} \bigcap_{i=1}^{m} \overline{T}_{\lambda i}$ becomes arbitrarily small, which is impossible as this set $\bigcup_{\lambda=1}^{v} \bigcap_{i=1}^{m} \overline{T}_{\lambda i}$ must contain all T_λ, hence all but a finite number of prime ideals of the first degree in k. We have arrived therefore at a contradiction.

We denote by T_0 any fixed one of T_λ, for which $R_{\lambda q}=S_q$ for all but a finite number q, and we denote the corresponding $n_\sigma^{(\lambda)}$, T'_λ by $n_\sigma^{(0)}$, T'_0 respectively.

LEMMA 3. *We have*

$$(\pi(\mathfrak{p})) = \prod_{\sigma} \mathfrak{p}^{\sigma n_\sigma^{(0)}}$$

for all proper $\mathfrak{p}$.

Proof. We denote by h the class number of k. Then, for any prime ideal $\mathfrak{p}$ in k, $\mathfrak{p}^h=(\varpi)$ is a principal ideal, and we have only to prove $(\pi(\mathfrak{p}))^h = \prod_{\sigma} (\varpi)^{\sigma n_\sigma^{(0)}}$. At any rate, we have

$$(\pi(\mathfrak{p}))^h = \prod_{\sigma} (\varpi)^{\sigma n_\sigma^{(\lambda)}}$$

for some λ. We put $\varpi_\mu = \prod_{\sigma} \varpi^{\sigma(n_\sigma^{(0)} - n_\sigma^{(\mu)})}$ for $\mu=1,\cdots,v$, and further

$$\Pi = \prod_{\varepsilon} \prod_{\mu} (\varpi_\mu - \varepsilon),$$

where μ runs over all index μ, for which ϖ_μ is not a unit, and ε runs over the finite group E. By Lemma 2, there is a rational prime q in Λ, prime to Π and to $\mathfrak{p}$, with the property that the image of T'_0 into S_q covers the whole S_q. We can find therefore a principal ideal

(α) in T_0' such that $(\alpha)(\varpi)\equiv 1$ (mod. q), hence also $\eta\alpha\varpi\equiv 1$ (mod. q) with a unit η in k. But we have

$$\begin{aligned}\varphi_q(\eta\alpha\varpi)\gamma_0(\eta\alpha\varpi)&=\varepsilon_q'(\eta\alpha\varpi)\prod_{\mathfrak{p}}\pi_{\mathfrak{p}}^{-\nu_{\mathfrak{p}}(\alpha\varpi)}\prod_{\sigma}(\eta\alpha\varpi)^{\sigma n_\sigma^{(0)}}\\&=\varepsilon_q'(\eta\alpha\varpi)\varepsilon_0^{-1}(\eta\alpha)\varepsilon_\lambda^{-1}(\varpi)\prod_{\sigma}\varpi^{\sigma(n_\sigma^{(0)}-n_\sigma^{(\lambda)})}\\&=\varepsilon\varpi_\lambda,\end{aligned}$$

and ε belongs to E, as $\eta\alpha\varpi$ is prime to $\mathfrak{f}$. Now, $\varphi_q\cdot\gamma_0$ is a continuous mapping from k^* with q-topology into $K_{\mathfrak{Q}}^*$, where $\mathfrak{Q}$ is a prime divisor of $\mathfrak{q}$ in K. But from $\eta\alpha\varpi\equiv 1$ mod. q follows $[\eta\alpha\varpi]^{q^t}\equiv 1$ mod. q^t for any natural number t, so the continuity of $\varphi_q\cdot\gamma_0$ implies that $[\varepsilon\varpi_\lambda]^{q^t}$ converges to 1 in $K_{\mathfrak{Q}}^*$ as $t\to\infty$, which is impossible unless $\varepsilon\varpi_\lambda\equiv 1$ mod. $\mathfrak{Q}$, that is, $\mathfrak{Q}$ divides $\varpi_\lambda-\varepsilon^{-1}$. As q is prime to II, this means that ϖ_λ must be a unit, which was to be proved.

From this Lemma 3, we see that, for any α in k^* prime to $\mathfrak{f}$, we have

$$\prod_{\mathfrak{p}}\pi(\mathfrak{p})^{\nu_{\mathfrak{p}}(\alpha)}=\varepsilon(\alpha)\prod_{\sigma}\alpha^{\sigma n_\sigma^{(0)}},$$

where $\varepsilon(\alpha)$ is a root of unity in K. Hence we have, by (***),

$$\varphi_q(\alpha)\gamma_0(\alpha)=\varepsilon_q'(\alpha)\varepsilon(\alpha)^{-1}$$

for any α in k^* prime to $\mathfrak{f}$ and for any q prime to $\mathfrak{f}$.

LEMMA 4. *We have, for any q prime to $\mathfrak{f}$,*

$$\varphi_q(\alpha)\gamma_0(\alpha)=1$$

for all α in k^*.

Proof. By the continuity of $\varphi_q\cdot\gamma_0$, we see that $\varepsilon_q'(\alpha)\varepsilon(\alpha)^{-1}=1$ for any α prime to $\mathfrak{f}$ satisfying $\alpha\equiv 1$ mod. q^n with a suitable n. Thus, $\varepsilon(\alpha)$ depends only on the class of α mod. $\mathfrak{f}_q\cdot q^n$ for α prime to $\mathfrak{f}$. But clearly $\varepsilon(\alpha)$ does not depend on the choice of q. Hence, $\varepsilon(\alpha)$ depends only on the class of α mod. $\mathfrak{f}_l\cdot l^m$ on the other hand, where l is a rational prime which is prime to $\mathfrak{f}q$. Consequently, $\varepsilon(\alpha)$ depends only on the class of α mod $\mathfrak{f}_q\mathfrak{f}_l$. Thus, for any α prime to $\mathfrak{f}$, $\varepsilon_q'(\alpha)\varepsilon(\alpha)^{-1}$ depends only on the class of α mod. $\mathfrak{f}_q\mathfrak{f}_l$ on the one hand, but on the other hand only on the class of α mod. q^n. We see therefore $\varepsilon_q'(\alpha)\varepsilon(\alpha)^{-1}$ must be 1 for any α prime to $\mathfrak{f}$. But as the set of α prime to $\mathfrak{f}$ is dense in k^* with q-topology, the continuity of $\varphi_q\cdot\gamma_0$ shows that $\varepsilon_q'(\alpha)\varepsilon(\alpha)^{-1}=1$ for all α in k^*, and our lemma is proved.

This lemma shows especially that the $\mathfrak{f}$-part $\mathfrak{f}_q$ of the conductor of ψ_q does not depend on q prime to $\mathfrak{f}$.

Now, as $\tilde{\alpha}_\infty$ is dense in the infinite part $\prod_{\mathfrak{p}_\infty}\tilde{k}_{\mathfrak{p}_\infty}^*$ of J, and as γ_0 is a continuous mapping from k^* with ∞-topology into the multiplicative group $\boldsymbol{C}^*$ of the complex number field $\boldsymbol{C}$, γ_0 can be extended uniquely to a representation γ of $\prod_{\mathfrak{p}_\infty}\tilde{k}_{\mathfrak{p}_\infty}^*$ into $\boldsymbol{C}^*$. Then we define, for any q prime to $\mathfrak{f}$,

$$\tilde{\psi}(\tilde{\mathfrak{a}})=\psi_q(\tilde{\mathfrak{a}}^{(q)})\prod_{\mathfrak{p}/q}\pi(\mathfrak{p})^{\nu_{\mathfrak{p}}(\tilde{\mathfrak{a}})}\cdot\gamma(\tilde{\mathfrak{a}}_\infty)^{-1}.$$

This $\tilde{\psi}$ is clearly a representation of J into $\boldsymbol{C}^*$. Moreover we see immediately,

$$\gamma)\qquad \tilde{\psi}(\tilde{\alpha})=\psi_q(\tilde{\alpha}^{(q)})\prod_{\mathfrak{p}/q}\pi(\mathfrak{p})^{\nu_{\mathfrak{p}}(\alpha)}\cdot\varphi_q(\alpha)=\psi_q(\tilde{\alpha}^{(q)})\psi_q(\tilde{\alpha}_q)=1$$

for any α in k^*.

$$\delta)\qquad \tilde{\psi}(\tilde{\pi}_{\mathfrak{p}})=\psi_q(\tilde{\pi}_{\mathfrak{p}})=\pi(\mathfrak{p}),\quad \psi(\tilde{u}_{\mathfrak{p}})=\psi_q(\tilde{u}_{\mathfrak{p}})=1$$

for any $\mathfrak{p}$ prime to $\mathfrak{f}q$.

This shows especially that $\tilde{\psi}$ is determined uniquely, independent of q, and that δ) holds for any proper $\mathfrak{p}$. We see also that there is a character $\tilde{\chi}$ of J such that $\tilde{\psi}(\tilde{\mathfrak{a}})=\tilde{\chi}(\tilde{\mathfrak{a}})V(\tilde{\mathfrak{a}})^{-\nu}$, where $V(\tilde{\mathfrak{a}})$ denotes the volume of $\tilde{\mathfrak{a}}$.

We recall that, for *all* rational primes q, ψ_q has always the properties α), β). Thus, this ψ_q can be obtained from our $\tilde{\psi}$ as $\psi_{\mathfrak{q}}$ by the method used in [L], §1, Lemma 1.

Hitherto we have discussed only ψ_{qi} with a fixed i. Now, let $\mathfrak{q}$ be a prime divisor of q in Σ, and let $\iota_{qi_1},\cdots,\iota_{qi_u}$ be all the isomorphisms of $\Sigma_{\mathfrak{q}}$ into $\overline{\boldsymbol{Q}}_q$, where u is the degree $[\Sigma_{\mathfrak{q}}:\boldsymbol{Q}_q]$. Then, it is clear that $\{\mu_{qi_1},\cdots,\mu_{qi_u}\}$ is *a* complete set of conjugates over $\boldsymbol{Q}_q$. Thus the representations $\mu_{q1},\cdots,\mu_{qd}$ can be obtained from the representation M_q' obtained by $\tilde{\psi}$ by the method used in [L], §3, 1°. But $\tilde{\chi}$ corresponds to a "Grössencharakter" χ of k, which is clearly regularly algebraic-valued.

Now, we use the assumption (d) of complete reducibility of M_q. Then, $\{M_q\}$ is just the local system obtained from this χ. This shows especially that our conditions (a), (b), (c), (d) contain the condition (iv) for local system in [L]. Conversely, it is evident that the local system obtained from any regularly algebraic-valued "Grössencharakter" is a simple regular local system. Hence we have obtained the *fundamental theorem* on regular local system for abelian extension:

THEOREM 1. *Let $\{M_q\}$ be a regular local system for an infinite abelian extension $\bar{k}$ over k. Then there is a finite number of regularly algebraic-valued "Grössencharaktere" $\chi_1,\cdots,\chi_s$ in k, from which $\{M_q\}$ can be generated, in the sense that $\{M_q\}$ is a direct sum of local systems obtained from these $\chi_1,\cdots,\chi_s$. Conversely, the local system obtained from any regularly algebraic-valued "Grössencharakter" in k is a simple regular local system for some infinite abelian extension of k.*

Especially, the zeta function of any regular local system can be expressed by L-functions of regularly algebraic-valued "Grössen-

charaktere".

§2. Infinite abelian extension attached to regular local system.

Let $\{M_q\}$ be a simple regular local system for an infinite abelian extension $\bar{k}$ over k. By Theorem 1, $\{M_q\}$ can be obtained from a regularly algebraic-valued "Grössencharakter" χ. We denote as above by $k^{(q)}$ the subfied of $\bar{k}$ corresponding to the kernel of M_q. Then, the composite field $k_\chi = \bigcup_q k^{(q)}$ for all q will be called to be *attached to* $\{M_q\}$.

For any prime ideal $\mathfrak{q}$ in Σ, the representation $\psi_\mathfrak{q}$ of the idèle group J_k of k into $\Sigma_\mathfrak{q}^*$ obtained from this χ is one of the ψ_{qi} obtained from μ_{qi} in the diagonal form of M_q. For any infinite prime $\mathfrak{q}_\sigma$ of Σ, corresponding to an isomorphism σ of Σ, we define $\psi_{\mathfrak{q}_\sigma} = (\tilde{\chi}\cdot V^{-\nu})^\sigma$. Then, we denote by $\Psi(\tilde{\mathfrak{a}})$ the element of the idèle group J_Σ of Σ, whose $\mathfrak{q}$-component is $\psi_\mathfrak{q}(\tilde{\mathfrak{a}})$ for any $\mathfrak{q}$. As any $\psi_\mathfrak{q}(\tilde{\mathfrak{a}})$ is a $\mathfrak{q}$-unit, $\Psi(\tilde{\mathfrak{a}})$ is indeed an idèle. Evidently, $\tilde{\mathfrak{a}} \to \Psi(\tilde{\mathfrak{a}})$ is a representation of J_k into J_Σ, which may also be regarded as a representation of the idèle class group C_k of k into J_Σ.

Now, if $\Psi(\tilde{\mathfrak{a}})$ is a principal idèle: $\Psi(\tilde{\mathfrak{a}}) = \tilde{\beta}$, $\beta \in \Sigma^*$, then β must be a unit in Σ, as each $\psi_\mathfrak{q}(\tilde{\mathfrak{a}})$ is a $\mathfrak{q}$-unit. This means especially $\prod_\sigma |\beta^\sigma| = 1$. But as $|\beta^\sigma| = V(\tilde{\mathfrak{a}})^{-\nu}$ is independent of σ, we have $|\beta^\sigma| = 1$ for all σ, hence β is a root of unity in Σ.

Considering $\Psi(\tilde{\mathfrak{a}})$ modulo $\tilde{\Sigma}^*$, we obtain a mapping Ψ_1 of J_k into C_Σ, which is clearly a representation. Especially, for any $\mathfrak{p}$ prime to $\mathfrak{f}$, $\Psi_1(\tilde{\pi}_\mathfrak{p}), \Psi_1(\tilde{u}_\mathfrak{p})$ are equal to images of idèles in Σ, with components 1 at any $\mathfrak{q}$ prime to $N\mathfrak{p}$, and with components $\psi_\mathfrak{q}(\tilde{\pi}_\mathfrak{p})/\tilde{\psi}(\tilde{\pi}_\mathfrak{p})$, $\psi_\mathfrak{q}(\tilde{u}_\mathfrak{p})$ respectively, at each $\mathfrak{q}$ dividing $N\mathfrak{p}$.

This Ψ_1 can also be regarded as a representation of C_k into C_Σ. Let D_k, D_Σ denote the connected component of identities in C_k, C_Σ respectively. Then, as $\Psi_1(D_k)$ is clearly contained in D_Σ, Ψ_1 induces a representation Ψ_2 of $C'_k = C_k/D_k$ into $C'_\Sigma = C_\Sigma/D_\Sigma$. If we denote by C_0 the kernel of Ψ_2, C'_k/C_0 is isomorphic to the subgroup $\Psi_2(C'_k)$ of C'_Σ (as topological groups), because C'_k is compact. We denote furthermore by B the kernel of Ψ in J_k, and by B' the natural image of B in C'_k.

Now, as $\Sigma_\mathfrak{q}^*$ is totally disconnected, $\psi_\mathfrak{q}$ can be regarded as a representation of C'_k into $\Sigma_\mathfrak{q}^*$ for any finite prime $\mathfrak{q}$ in Σ. Thus the finite part of $\Psi(\tilde{\mathfrak{a}})$ and the signatures of $\Psi(\tilde{\mathfrak{a}})$ at infinite real primes $\mathfrak{q}_{\sigma_r}$ are determined uniquely by the image a' of $\tilde{\mathfrak{a}}$ in C'_k. We denote by $\Psi_0(a')$ the idèles with the same component as $\Psi(\tilde{\mathfrak{a}})$ at each finite $\mathfrak{q}$, and the component ± 1 at each infinite real prime $\mathfrak{q}_{\sigma_r}$ determined by the signature of $\Psi(\tilde{\mathfrak{a}})$ at $\mathfrak{q}_{\sigma_r}$, and the component 1 at each infinite imaginary prime. Then, a' is in C_0 if and only if $\Psi_0(a')=\tilde{\eta}^{(\infty)}\prod_{\sigma_r}\widetilde{(\operatorname{sgn}\eta^{\sigma_r})}$, where η is a unit in Σ and $\widetilde{(\operatorname{sgn}\eta^{\sigma_r})}$ denotes the idèle with component $\operatorname{sgn}\eta^{\sigma_r}$ at the real prime $\mathfrak{q}_{\sigma_r}$, and with component 1 at any other primes. The mapping $a'\to\eta=\eta(a')$ is clearly a representation of C_0 into the unit group H of Σ^* with discrete topology. As C_0 is compact, the image $\{\eta(a')\}$ of C_0 must be a finite group, hence must be contained in the group Z of roots of unity in Σ. From this we see that B' has a finite index i in C_0', dividing the order of Z.

Let k_{χ_0} be the subfield of the maximal abelian extension of k corresponding to C_0, then k_{χ_0} has the index i in k_χ. Let $\mathfrak{G}_0$ be the Galois group of k_{χ_0} over k. Then $\mathfrak{G}_0$ is isomorphic to C_k'/C_0, hence to the subgroup $\Psi(C_k')$ of C_Σ' (as topological groups).

Thus we have obtained

THEOREM 2. *The field k_χ attached to a simple regular local system $\{M_q\}$ for an abelian extension $\bar{k}$ over k contains a subfield k_{χ_0} with the following properties:*

(ε) *k_{χ_0} has a finite index in k_χ, which divides the number of roots of unity in Σ.*

(ζ) *The Galois group of k_{χ_0} over k is isomorphic to a subgroup of the Galois group of the maximal abelian extension of Σ over Σ.*

(η) *This isomorphism is induced by the representation Ψ_0 of C_k' into J_Σ, which is obtained from irreducible representations ψ_{qi} of J_k given by $\{M_q\}$.*

§3. Application to the conjecture of Hasse. Examples.

In this §, special reference is made to §3, 2° of [L].

Let A be an abelian variety of dimension g defined over an algebraic number field k of finite degree. We denote by $\mathcal{A}(A)$ the ring of endomorphisms of A, which is an order of a semi-simple algebra $\mathcal{A}_0(A)$ over $\boldsymbol{Q}$. Let $k^{(q)}$ be the field obtained from k by "adjunction" of all points in $\mathfrak{g}(q;A)$ (group of all points on A with order q^n for some n), and $\bar{k}$ be the union $\bigcup_q k^{(q)}$, where q runs over all rational primes. We denote by M_q the q-adic representation in A, (representation as a transformation of $\mathfrak{g}(q;A)$) of the Galois group $\mathfrak{G}$ of $\bar{k}$ over k, and by L_q the q-adic representation in A of the ring $\mathcal{A}(A)$. Then, it is clear that $k^{(q)}$ is just the subfield of $\bar{k}$ corresponding to the kernel of M_q.

If we reduce A modulo a prime ideal $\mathfrak{p}$ in k and obtain also an abelian variety $\tilde{A}$ (over the finite field $\tilde{k}$ of $N\mathfrak{p}$ element), then we call $\mathfrak{p}$ *proper* as in [L], §3, 2°. We recall here that there are but a finite number of non-proper $\mathfrak{p}$'s and that any proper $\mathfrak{p}$ prime to q is unramified in $k^{(q)}$. We denote by $\mathfrak{f}$ the product of all non-proper $\mathfrak{p}$'s. For any proper $\mathfrak{p}$, there is an endomorphism $\widetilde{\pi(\mathfrak{p})}$ of $\tilde{A}$ with the property: $\widetilde{\pi(\mathfrak{p})}\tilde{x}=\tilde{x}^{N\mathfrak{p}}$ for any point $\tilde{x}$ on $\tilde{A}$.

Now, we assume that A, k satisfy the following conditions:

(A). For any proper $\mathfrak{p}$, the ring $\mathcal{A}(A)$ contains an element $\pi(\mathfrak{p})$, from which $\widetilde{\pi(\mathfrak{p})}$ is obtained by the reduction mod. $\mathfrak{p}$. Moreover, all $\pi(\mathfrak{p})$ belong to a commutative semi-simple subalgebra R_0 of $\mathcal{A}_0(A)$.

(B). Any endomorphism in $R=R_0\frown\mathcal{A}(A)$ is defined over k.

Then, we have

LEMMA. *All representations M_q are completely reducible, and $\bar{k}$ is abelian over k.*

Proof. Let $\sigma(\mathfrak{P})$ be the Frobenius automorphism of a prime divisor $\mathfrak{P}$ of $\mathfrak{p}$ in $k^{(q)}$, for any $\mathfrak{p}$ prime to $\mathfrak{f}q$. Then we have, considering mod. $\mathfrak{P}$, $\widetilde{x^{\sigma(\mathfrak{P})}}=\tilde{x}^{N\mathfrak{p}}=\widetilde{\pi(\mathfrak{p})}\tilde{x}=\widetilde{\pi(\mathfrak{p})x}$ for any point x in $\mathfrak{g}(q; A)$. As $x\to\tilde{x}$ is one to one for x in $\mathfrak{g}(q; A)$, we see that $x^{\sigma(\mathfrak{P})}=\pi(\mathfrak{p})x$, consequently that $M_q(\sigma(\mathfrak{P}))=L_q(\pi(\mathfrak{p}))$. But L_q induces a faithful representation of the commutative semi-simple algebra R_0, and the representation of R_0 thus obtained can be transformed into diagonal form (in the algebraic closure $\overline{\boldsymbol{Q}}_q$ of $\boldsymbol{Q}_q$). Then, by the same transformation, $M_q(\sigma(\mathfrak{P}))$ are clearly transformed into diagonal form for any $\mathfrak{P}$ dividing proper $\mathfrak{p}$ prime to q. As the set of Frobenius automorphisms $\sigma(\mathfrak{P})$ for such $\mathfrak{P}$'s is dense in the Galois group $\mathfrak{G}^{(q)}$ of $k^{(q)}$ over k, this shows that M_q can be transformed into diagonal form. Then, the faithfulness of M_q for $\mathfrak{G}^{(q)}$ shows that $\mathfrak{G}^{(q)}$ is abelian, which was to be proved.

Let $R_0=\Sigma_1+\cdots+\Sigma_r$, where Σ_i is an algebraic number field of finite degree d_i. Then, $\pi(\mathfrak{p})$ can be expressed as $\pi_1(\mathfrak{p})+\cdots+\pi_r(\mathfrak{p})$, where $\pi_i(\mathfrak{p})$ belongs to Σ_i. As the degree of M_q is $2g$, we have $\sum n_i d_i=2g$, with some natural numbers $n_1,\cdots, n_r$. Each L_q can be transformed correspondingly into a direct sum of representation L_{qi} of degree d_i, each repeated n_i-times, in the field $\boldsymbol{Q}_q$. Then, M_q is also transformed (in $\boldsymbol{Q}_q$) into a direct sum of M_{qi}, determined by $M_{qi}(\sigma(\mathfrak{p}))=L_{qi}(\pi_i(\mathfrak{p}))$ for proper $\mathfrak{p}$ prime to q.

It is clear that, for each i, the system $\{M_{qi}\}$ satisfies the conditions (a), (b), (d) for simple regular local systems for $\bar{k}$ over k, with $\Sigma=\Sigma_i$, $\pi(\mathfrak{p})=\pi_i(\mathfrak{p})$ and $\nu=1/2$. If we use the representation of R with invariant differentials on A, it is not difficult to prove that from (B) follows also (c).[4] Without assuming this, however, we can replace k by some finite algebraic entension of it, for which (c) holds. We

4) Cf. The contribution of Y. Taniyama to the symposium of algebraic number theory, 1955, Tokyo.

see therefore that our system $\{M_q\}$ is a regular local system for $\bar{k}$ over k, and also that $\bar{k}$ is just the field attached to $\{M_q\}$. As the zeta-function of A is the zeta-function of $\{M_q\}$, we have obtained

THEOREM 3. *The conjecture of Hasse holds for the zeta-function of an abelian variety A defined over k, provided the conditions* (A), (B) *are satisfied.*

COROLLARY. *Let C be an algebraic curve defined over an algebraic number field k of a finite degree, and J be a jacobian variety of C defined over the same field k. Then, the conjecture of Hasse holds for zeta-functions of C provided a canonical mapping from C into J is defined over k, and J, k satisfy the conditions* (A), (B).[5]

The author once obtained the same result by another method,[6] under the condition however that the ground field k satisfies a more restrictive condition. It was required, namely, that k contains all the conjugates of Σ, even in the case where $R_0=\Sigma$ is a field of degree $2g$. This condition is, in general, stronger than our condition (B), or even than (c). If moreover R_0 is not simple, the simple components of A, and homomorphisms of A onto them, are not always defined over k. This impose another condition on k in the former proof.[6]

EXAMPLE 1. We remark that, if $\mathcal{A}(A)$ contains an element μ, whose minimal polynomial over $\boldsymbol{Q}$ has only simple factors and of degree $2g$, then our conditions are satisfied. We can take namely, $R_0=\boldsymbol{Q}[\mu]_0$, which is clearly commutative and semi-simple. Next, take k satisfying (B). Then, as $\widetilde{\pi(\mathfrak{p})}$ commutes with any element of the algebra $\tilde{R}_0$, which is of degree $2g$, $\widetilde{\pi(\mathfrak{p})}$ must belong to $\tilde{R}$. Hence the condition (A) is also satisfied.

One of the examples of such A is the jacobian variety J of the plane algebraic curve C defined by

$$ax^n+by^m=1,$$

where n, m are coprime natural numbers and a, b are rational integers.[7] Let ζ, η be primitive n-th and m-th roots of unity respectively. Then the correspondence $H:(x, y)\rightarrow(\zeta x, \eta y)$ of C with itself induces an endomorphism μ of A, whose minimal polynomial is the product of irreducible polynomials (over $\boldsymbol{Q}$) for primitive $n_i\cdot m_j$ -th roots of unity for all divisors $n_i\neq 1$, $m_j\neq 1$ of n, m respectively. This can be seen easily using the representation of H with the differentials of the first kind on C. Thus, taking $k=\boldsymbol{Q}(\zeta, \eta)$, our conditions (A), (B), and also (c), are satisfied, so the conclusion of Theorem 3 holds

5) The proof is almost evident. Cf. the work quoted in footnote 4).

6) Cf. the work quoted in footnote 4).

7) The conjecture of Hasse for curves defined by equations of this type, where n, m are not supposed to be coprime, was first proved by A. Weil.

for J. Then, Theorem 2 gives the nature of class field over $\boldsymbol{Q}(\zeta, \eta)$ obtained by complex multiplication of A.

EXAMPLE 2.[8] Let D be the projective line with the universal domain $\boldsymbol{C}$ (complex number field), and D^x be the abstract variety obtained from D by omitting two points 0 and ∞. D^x is a non-complete commutative group variety with the law of multiplication $(a, b) \to a \cdot b$ (usual product of complex numbers). The subgroup E of all points with finite orders is just the group of all roots of unity. Hence the subgroup E_q of all points with order q^n $(n=0, 1, 2 \cdots)$ is isomorphic to the additive group of q-adic numbers modulo 1, for every rational prime q.

Let $\boldsymbol{Q}^{(q)}$ be the field obtaind from $\boldsymbol{Q}$ by adjunction of all roots of unity in E_q, and $\mathfrak{G}_q$ be the Galois group of $\boldsymbol{Q}^{(q)}$ over $\boldsymbol{Q}$. Then let $\overline{\boldsymbol{Q}} = \bigcup \boldsymbol{Q}^{(q)}$ and $\mathfrak{G}$ be the Galois group of $\overline{\boldsymbol{Q}}$ over $\boldsymbol{Q}$. It is clear that any σ in $\mathfrak{G}$ permutes points in E_q among themselves. Thus we get a representation μ_q of $\mathfrak{G}$ into the unit group U_q of $\boldsymbol{Q}_q$, as a transformation of q-adic numbers mod. 1, continuity being thereby evident. It is also clear that the subfield of $\overline{\boldsymbol{Q}}$ corresponding to the kernel of μ_q is just $\boldsymbol{Q}^{(q)}$. This shows that $\boldsymbol{Q}^{(q)}$ is abelian over $\boldsymbol{Q}$ for each q, hence $\overline{\boldsymbol{Q}}$ is also abelian over $\boldsymbol{Q}$. Now, it is evident that the system $\{\mu_q\}$ is a simple regular local system for $\overline{\boldsymbol{Q}}$ over $\boldsymbol{Q}$, with $\Sigma = \boldsymbol{Q}$, $\pi(p) = p$, $\nu = 1$. The representation $\tilde{\psi}$ of $J_{\boldsymbol{Q}}$ corresponding to this system is given by $\tilde{\psi}(\tilde{\mathfrak{a}}) = V(\tilde{\mathfrak{a}})^{-1}$, where $V(\tilde{\mathfrak{a}})$ denotes the volume of $\tilde{\mathfrak{a}}$. Next, in the notation of §2, the index of B' in C_0 is 1, because the infinite part $\tilde{\psi}(\tilde{\mathfrak{a}})$ of the idèle $\Psi(\tilde{\mathfrak{a}})$ is always positive. As the representation Ψ_2 is clearly *onto* in this case, Theorem 2 shows that $\overline{\boldsymbol{Q}}$ is just the maximal abelian extension of $\boldsymbol{Q}$; this is nothing but the classical theorem of Kronecker on the abelian extensions of the rational number field. We remark furthermore that Hasse's ζ-function of D^x is

$$\zeta_{D^x}(s) = \frac{\zeta(s-1)}{\zeta(s)},$$

which is also the ζ-function of regular local system $\{\mu_q\}$.

The fact that Kronecker's theorem can be interpreted by our theory of regular local systems is based on the fact that in the case of $\boldsymbol{Q}$ all representations of idèle class group can be obtaind from regularly algebraic-valued "Grössencharakter". On the contrary, this is not the case for real quadratic field and presumably for all totally

8) The idea of using non-complete group varieties to construct class fields over real number fields is due to A. Weil.

real fields of higher degree: In real quadratic fields, no *proper* "Grösisosencharakter" is algebraic-valued. This seems to suggest that $\boldsymbol{Q}$ has quite a special position among totally real fields also from the viewpoint of the theory of complex multiplication.

A Letter to André Weil

Tokyo, Feb. 14, 1956

Dear A. Weil.

I send you via air mail two manuscripts of my papers:

[1] On a certain relation between L-functions.

[2] Regular local system for an abelian extension.

[1] contains the theorem I told you at Tokyo Univ. last October, and also one formalism. I have defined namely "local system", extracting the properties of l-adic representations on abelian varieties, to which you called our attention last September. Then, L-functions with algebraic-valued "Grössencharaktere" and Hasse's ζ-functions of abelian varieties can be unified in this formalism, though there is no essential theorem. Concerning the definition, it is desirable to dispense with the condition (iv), [1], §2, which is indeed the case in regular local system ([2], §1).

[2] deserves much attention, I think. It gives the second proof of Hasse's conjecture for ζ-functions of abelian varieties with sufficiently many complex multiplications, following the idea you have explained at the Symposium last year. I proved namely a theorem, which can be formulated in short as follows: Given a system of l-adic representations of idèle class group for all l, satisfying certain concordant conditions (of the same type as those of the ring $\mathcal{A}(A)$ of an abelian variety A), then, there is a usual representation (into the complex number field) of idèle class group, concordant with these l-adic representations. But it is almost evident that abelian varieties with sufficiently many complex multiplications give rise to such systems of l-adic representations. This proves the conjecture of Hasse in this case.

As you shall see, this proof is more natural and more simple than my first proof. Moreover, we can dispense with the theory of differential forms on abelian varieties.

This last fact seems to me much interesting. In fact, please notice the following: Although almost all properties of elliptic functions have already been obtained by function-theoretic, by algebraic-geometric and by arithmetic methods, we have no key to open, nor any way to approach the mystery of ζ-functions of general elliptic curves without complex multiplications. Eichler's result means little for this general problem, because, unless we can presuppose the type of functional equations which these ζ-functions would satisfy, we cannot find the type of automorphic functions, by which the elliptic functions in question should be uniformized. Thus the essential part of the problem, it seems to me, must be of purely number-theoretic nature: The arithmetic of normal extensions obtained by the division of periods of these elliptic functions. As you shall see in [2], the continuity of l-adic representations determines almost all properties of the endomorphism $\tilde{\pi}(\mathfrak{P})$, in abelian case. As this continuity also holds trivially in general case of normal extensions, we have a right to expect that it gives some general properties of the endomorphism $\tilde{\pi}(\mathfrak{P})$, none of which is known at present except for the "Riemann hypothesis".

Now, there is one more problem. Last summer, you have made us notice to the possibility of building the theory of "complex" multiplications for real number field, by the use of non-complete commutative group varieties. My theory of regular local systems can dispose of the simplest case, namely Kronecker's theorem on rational number field, which is not the case, however, for real number fields of higher degree. This makes clear the distinction of rational number field from general real number fields. Of course, your idea might still be powerful enough. But it seems to me that non-algebraic-valued "Grössencharaktere" must be discussed from quite a different standpoint, although I cannot imagine anything about it at present.

I should be very happy if you are kind enough to read my manuscripts and to give me some advice, and also some prophecies, about these matters.

(Lemma 2, §1, [1] needs no proof, and some other proofs may also become more short and more elegant. But in any way, I think it does not matter too much.)

Next month, Mr. Y. Akizuki is to come back to Japan. I expect, it would be quite interesting to hear from him the "proof" of Riemann hypothesis, which must have been born from the bubble of the beer.

Yours truly,
Yutaka TANIYAMA

L-functions of Number Fields and Zeta Functions of Abelian Varieties

Journal of the Mathematical Society of Japan,
Vol. 9, pp. 330–366 (July, 1957)

(Received Feb. 20, 1957)

Introduction.

It was found out in several cases that Hasse's zeta functions of algebraic curves or of abelian varieties over an algebraic number field can be expressed by Hecke's L-functions with "Grössencharaktere" of that field.[1] It deserves our attention that these phenomena have always presented themselves in connection with arithmetic of abelian extensions of that number field. However, since the relation of Hasse's functions with abelian extension was not so direct in the proofs of these results, which have been done from different angles, it would be desirable to clarify the relation between Hasse's functions and abelian extensions attached to abelian varieties in question, treating all cases from a unified point of view. This is the first problem. In pursuing this problem, I have succeeded in obtaining a new interpretation of Hasse's functions in general, and in characterizing under which particular conditions the above phenomena take place.

On the other hand, "Grössencharaktere" can be interpreted as charakters of idèle class groups, so it seems natural that they have some connection with abelian varieties related to abelian extensions of the basic fields. However, as class field theory shows, it is not the idèle class group, but the factor group of it by the connected component of the identity, that can be interpreted by the Galois group of the maximally abelian extension. The above phenomena suggest conversely the possibility of an interpretation of characters of idèle class group by something connected with abelian extensions. To find out such an interpretation is a problem, first proposed by A. Weil [4], which seems no less important than the above.

I shall solve this last problem in this paper for a special type

1) Weil [9], Deuring [1], Taniyama [3]. There are also cases first treated by Eichler, where this does not hold.

of characters, called characters of type (A_0) by A. Weil [4]. The problem of interpretation of characters which are not of type (A_0) has a quite different feature, and will be entirely left open. The characters of type (A_0) correspond to certain representations of the Galois group of the maximally abelian extension of the basic field, and can be characterized by some properties of these representations. On the other hand, we can compute the zeta function of an abelian variety with the help of certain representations of the Galois group of the field obtained by division of periods of this abelian variety. When this varity has sufficiently many complex multiplications, these representations have the properties characterizing characters of type (A_0). This gives a proof of the conjecture of Hasse in case of complex multiplications, under somewhat weaker conditions than in my former paper [3]. This theory reveals more intimate relation of Hasse's functions, "Grössencharaktere" of type (A_0) and abelian extensions.

Moreover, the above relation between representations of the Galois groups and of the general Hasse's functions gives a new relation between these functions and the zeta functions of infinitely many finite extensions of the basic field. This relation holds for a little more general type of L-functions, including L-functions with characters of type (A_0), and may be considered to express, in a sense, the decomposition law of prime ideals in the infinitely many fields attached to these functions.

All this slows furthermore that Hasse's zeta functions of general abelian varieties are closely connected with the infinite normal, non-abelian extensions obtained by division of periods of these varieties. Hence these functions may have a quite different nature from those in our special cases, and we still stand far from the solutions of Hasse's conjecture in the general case, although these normal extensions have some remarkable properties as expressed in our axioms below.

The main method used in the present paper is due to A. Weil [4], where he has shown that we can associate to every character of type (A_0) a system of local representations (i.e. representations into $\mathfrak{P}$-adic completions of a number field) of the idèle class group. This idea of local representation is a quite adequate one, because, first, we can pass from idèle class group to the Galois group very naturally by its means, and second, the Galois group and $\mathfrak{P}$-adic unit group have similar topologies, while the usual character of the Galois group is necessarily of finite order. Moreover, we can connect these local representations with "l-adic representations" of the ring of endomorphisms of abelian varieties, which is the essential base of the proof of conjecture of Hasse in the present paper.

In §1, I shall give a characterization of characters of idèle class groups of type (A_0). In §2, the above mentioned relation of L-functions with characters of type (A_0) and an infinite product of zeta functions of number fields will be given. In §3, I shall first reformulate the result in §1 in a form which may be applied directly to the proof of conjecture of Hasse. This reformulation allows moreover a generalization, and an infinite product relation like that in §2 for the generalized L-function will be obtained. In §4, it will be shown that this generalization contains the case of Hasse's zeta functions of general abelian varieties. §4 implies furthermore the proof of the conjecture of Hasse in case of sufficiently many complex multiplications mentioned above.

Notations and terminologies. Basic results assumed to be known.

The following notations and terminologies will be used throughout the paper, often without references. As to the basic concepts discussed here, the reader is referred to Weil's papers [4], [6]. Terminologies and basic notations concerning algebraic geometry used in §4 will be the same as those of so-called Weil-school in algebraic geometry. As to basic results recalled in §4, see Weil [7], [8], Shimura [2] and Taniyama [3].

$\boldsymbol{Q}$ denotes the rational number field, $\boldsymbol{R}$ the real number field, $\boldsymbol{C}$ the complex number field and $\boldsymbol{Z}$ the ring of rational integers. $|\alpha|$ denotes the usual absolute value of a complex number α. σ_0 denotes the complex conjugate automorphism of $\boldsymbol{C}$, or of any subfield of $\boldsymbol{C}$: $\sigma_0\alpha=\bar{\alpha}$, $\alpha\in\boldsymbol{C}$.

If M is a square matrix, $\det M$ denotes the determinant of M. E denotes always a unit matrix. The degree of E will be clear from the context.

All groups treated in this paper are considered as topological groups with their proper topologies, maybe discrete. The words isomorphism, homomorphism, representation of groups are accordingly used in the sense of topological groups. In particular, *representation* means an algebraic homomorphism which is continuous. The word *character* is used *in the wider sense*, i.e. representation into multiplicative group $\boldsymbol{C}^*$ of $\boldsymbol{C}$.

Let k be a field. Then k^* denotes the multiplicative group of all non-zero elements of k. $\bar{k}$ denotes the algebraic closure of k, and A_k the maximal abelian extension of k in $\bar{k}$. If k' is a Galois extension of k (finite or infinite), then $G(k'/k)$ denotes the Galois group of k' over k, endowed with Krull's topology. In particular, we write $G_k=G(A_k/k)$: the Galois group of the maximal abelian extension of k over k. If k' is a finite extension of k, $N_{k'/k}$ denotes the relative norm from k' to k, and $[k':k]$ the degree of k' over k.

Algebraic number field is always considered as contained in $\boldsymbol{C}$.

Let k be an algebraic number field of finite degree. Then $H(k)$ denotes the set of all isomorphisms of k into $\boldsymbol{C}$. If k' is a finite extension of k, $H(k'/k)$ denotes the set of all isomorphisms of k' into $\boldsymbol{C}$ over k. N denotes always the absolute norm of ideals. For any α in k^*, (α) denotes the principal ideal of α. Let $\mathfrak{m}$ be an integral ideal of k. Then $G(\mathfrak{m})$ denote the group of all ideals of k, prime to $\mathfrak{m}$. Let K be also an algebraic number field, then the *Galois closure* of k and K means the smallest absolutely normal field in $\boldsymbol{C}$, containing k and K. Let k' be any normal extension of k, and ψ^* be a representation of the group $G(k'/k)$ into some group. The kernel of ψ^* being a closed subgroup of $G(k'/k)$, it corresponds to a subfield of k' containing k, by Galois theory. k' being as above, let $\mathfrak{P}$ be a prime divisor in k' of a prime ideal $\mathfrak{p}$ of k. Then the *decomposition group* of $\mathfrak{P}$ over k (consisting of all σ in $G(k'/k)$ such that $\sigma\mathfrak{P}=\mathfrak{P}$) is denoted by $G(\mathfrak{P})$. We shall denote by $\sigma_{\mathfrak{P}}$ any one of the *Frobenius automorphisms* of $\mathfrak{P}$ over k, i.e. $\sigma_{\mathfrak{P}}$ is an element in $G(\mathfrak{P})$ inducing on the residue field of k' mod. $\mathfrak{P}$ the automorphism $\xi\to\xi^{\mathrm{N}\mathfrak{p}}$. If $G(k'/k)$ is abelian, we can write $\sigma_{\mathfrak{p}}$ instead of $\sigma_{\mathfrak{P}}$. *Inertia group* of $\mathfrak{p}$ over k is the subgroup of $G(\mathfrak{P})$ consisting of all σ in $G(\mathfrak{P})$ which induce the identity automorphism on the residue field. Then $\mathfrak{p}$ is said to be *unramified* in k' if the group $G(\mathfrak{P})$ operates faithfully on the residue field of k' mod. $\mathfrak{P}$, i.e. the inertia group of $\mathfrak{P}$ is the identity. ψ^* being as above, $\mathfrak{p}$ is unramified in the field corresponding to the kernel of ψ^* if and only if $\psi^*(\sigma_{\mathfrak{P}})$ does not depend on the choice of $\sigma_{\mathfrak{P}}$ for any one of $\mathfrak{P}$. Let S be a set of ideals in k. Then the word *density* of prime ideals in S is used in Kronecker's sense, i.e. it means the limit

$$\Delta(S)=\lim_{s\to 1+0}\left(-\sum \mathrm{N}\mathfrak{p}^{-s}/\log(s-1)\right),$$

if this limit exists, the sum $\sum \mathrm{N}\mathfrak{p}^{-s}$ being taken over all prime ideals $\mathfrak{p}$ in S. The density of the set of all prime ideals of the first degree is 1. Also, the density of prime ideals in each ideal class modulo an integral ideal $\mathfrak{m}$ ("Strahlklasse" mod. $\mathfrak{m}$) is definite, and equal for all classes. Now, Tschebotareff's density theorem asserts that, for all finite normal extension k'' of k, and for any element σ in $G(k''/k)$, the density of the prime ideals $\mathfrak{p}$ of k of the first degree such that σ is a Frobenius automorphism of a prime divisor of $\mathfrak{p}$ in k'', is definite and positive. The word *almost all*, used for a set of ideals in k, means "all but a finite number of". Then *the set of all* $\sigma_{\mathfrak{P}}$ *for all prime divisors* $\mathfrak{P}$ *of almost all prime ideals* $\mathfrak{p}$ *of the first degree in* k *is everywhere dense in* $G(k'/k)$, as is immediately seen from Tschebotareff's density theorem.

$\mathfrak{p}, \mathfrak{l}, \mathfrak{h}_i$ *denote always prime ideals* in some algebraic number fields, and *corresponding latin letters* p, l, h_i *denote rational primes divisible respectively by* $\mathfrak{p}, \mathfrak{l}, \mathfrak{h}_i$, unless the other indications are explicitely given.

Let k be as above, and v be a valuation of k. Equivalent valuations will be considered as the same. k_v denotes the completion of k with respect to v, and v is considered to be extended to k_v. If v is discrete, we use $\mathfrak{p}$ to denote either the corresponding prime ideal in k, or valuation ideal in k_v, or equivalence class of v, and write $k_{\mathfrak{p}}$ instead of k_v. In particular, $\boldsymbol{Q}_p$ denotes a p-adic number field. The normalized exponential valuation corresponding to $\mathfrak{p}$ is denoted by $\nu_{\mathfrak{p}}$. Then, an element α in $k_{\mathfrak{p}}$ such that $\nu_{\mathfrak{p}}(\alpha)=0$ is called a *unit* in $k_{\mathfrak{p}}$, or $\mathfrak{p}$-*unit*. All $\mathfrak{p}$-units form a multiplicative group, the $\mathfrak{p}$-unit group, in $k_{\mathfrak{p}}^*$, which is denoted by $U_{\mathfrak{p}}$. Any element α such that $\nu_{\mathfrak{p}}(\alpha)\geqq 0$ is called $\mathfrak{p}$-integral. We define similarly units and integral elements in $\overline{\boldsymbol{Q}}_p$. *All congruences are used in the sense of valuation theory.*

Idèle group of k is denoted by I_k. There is a canonical isomorphism of the multiplicative group k^* into I_k, which is denoted by ι. The *principal idèle group* ιk^* is denoted by P_k. The canonical isomorphism of $k_v{}^*$ into I_k is denoted by ι_v, and also by $\iota_{\mathfrak{p}}$ if v corresponds to $\mathfrak{p}$. C_k denotes the *idèle class group* I_k/P_k of k. D_k denotes the connected component of 1 in C_k. Then the factor group $C'_k=C_k/D_k$ is compact and totally disconnected. *Class field theory assures now the existence of canonical isomorphism of* C'_k *onto the Galois Group* G_k *of* A_k *over* k. By this isomorphism, the image of $\iota_{\mathfrak{p}}(U_{\mathfrak{p}})$ into C'_k (by the natural homomorphism) corresponds to the inertia group of $\mathfrak{p}$, and the image of $\iota_{\mathfrak{p}}(\pi_{\mathfrak{p}}U_{\mathfrak{p}})$ into C'_k corresponds to the set of all the Frobenius automorphisms $\sigma_{\mathfrak{p}}$ of $\mathfrak{p}$, where $\pi_{\mathfrak{p}}$ denotes a $\mathfrak{p}$-prime element in k (i.e. $\nu_{\mathfrak{p}}(\pi_{\mathfrak{p}})=1$).

If an idèle a in I_k is written as $a=(a_v)$, a_v denotes the v-components of a. Let k' be a finite extension of k and $a'=(a'_{v'})$ be in $I_{k'}$. Then the *norm* $N_{k'/k}(a')$ of a' is defined by $N_{k'/k}(a')=(a_v)\in I_k$ such that $a_v=\prod N_{(v')}(a'_{v'})$, where the product is taken over all extensions v' of v to k', and $N_{(v')}$ denotes the norm of $k'_{v'}$ into k_v. (a) will denote the ideal of an idèle $a=(a_v)$ defined by $(a)=\prod\limits_{\mathfrak{p}} \mathfrak{p}^{\nu_{\mathfrak{p}}(a_{\mathfrak{p}})}$. Then for any a in I_k, the positive real number

$$||\,a\,||=N((a))^{-1}\prod_{\lambda}|\,a_{v_\lambda}|^{\eta_\lambda}$$

is called the *volume* of a, where v_λ runs over all Archimedean v, and k_{v_λ} is identified with $\boldsymbol{R}$ or $\boldsymbol{C}$ as the case may be, and $\eta_\lambda=[k_{v_\lambda}:\boldsymbol{R}]$. Then we have $\|\iota\alpha\|=1$ for any α in k^*, so we can speak of volumes of idèle classes. C_k^0 denotes the subgroup of C_k of all elements with volume 1, then C_k is isomorphic to the direct product $\boldsymbol{R}\times C_k^0$ of C_k^0 and the additive group of $\boldsymbol{R}$.

Representation ψ of C_k is identified with representation of I_k induced by ψ, and, if ψ takes the value 1 on D_k, it is also identified with that of $C_k'=C_k/D_k$ induced by ψ. In this latter case, ψ determines also a representation of the Galois group G_k under the identification of C_k' with G_k by class field theory. This representation of G_k is denoted by ψ^*. Now, a representation ψ of C_k is called *unramified at* $\mathfrak{p}$ if $\psi(\iota_\mathfrak{p}(U_\mathfrak{p}))=1$. When $\psi(D_k)=1$, *ψ is unramified at $\mathfrak{p}$ if and only if $\psi^*(\sigma_\mathfrak{p})$ does not depend on the choice of $\sigma_\mathfrak{p}$*, i.e. $\mathfrak{p}$ is unramified in the field corresponding to the kernel of ψ^*. ψ is called unramified at real Archimedean v if $\psi(\iota_v(-1))=$*the identity element.*

As to the following, special references are made to Weil [4]. For any integral ideal $\mathfrak{m}$ in k, $I(\mathfrak{m})$ denotes the subgroup of I_k consisting of all $a=(a_v)$ such that $a_\mathfrak{p}=1$ for all prime factors $\mathfrak{p}$ of $\mathfrak{m}$, and $a_v=1$ for all Archmedean v. Then $I^0(\mathfrak{m})$ denotes the subgroup of $I(\mathfrak{m})$ consisting of all a in $I(\mathfrak{m})$ such that the ideal $(a)=1$. Then the factor group $I(\mathfrak{m})/I^0(\mathfrak{m})$ is canonically isomorphic to the ideal group $G(\mathfrak{m})$. Now let a representation ψ of C_k be unramified at all $\mathfrak{p}$ in $G(\mathfrak{m})$. Then ψ takes the value 1 on $I^0(\mathfrak{m})$, hence ψ induces a representation of $G(\mathfrak{m})$, which is denoted by $\tilde{\psi}$. Notice that the subgroup $I(\mathfrak{m})P_k$ is everywhere dense in I_k (as is seen from approximation theorem for valuations), hence ψ is determined uniquely by $\tilde{\psi}$, Conversely, *a representation $\tilde{\psi}$ of $G(\mathfrak{m})$ into a complete group Γ can be obtained from a representation of C_k into Γ in this manner if and only if the following holds*: Given any neighbourhood V of the identity element of Γ, there is a natural number n and a positive number ε such that $\tilde{\psi}((\alpha))\in V$ for all $\alpha\in k^*$ satisfying $\alpha\equiv1$ mod. $\mathfrak{m}^n$ and, $|\sigma\alpha-1|<\varepsilon$ for all isomorphisms σ in $H(k)$.

All this holds in particular for a character ψ of C_k. Moreover, a character χ of C_k must be of the form $\chi(a)=\chi_1(a)\|a\|^\rho$, where χ_1 is a character with absolute value 1 and ρ is a uniquely determined real number, which is called the *real part* of χ. Any χ is unramified at almost all $\mathfrak{p}$. If we denote by $\mathfrak{h}_1,\cdots,\mathfrak{h}_t$ the exceptional prime ideals where χ is ramified, then, for each $\mathfrak{h}_i$, there is the smallest natural number c_i such that $\chi(\iota_{\mathfrak{h}_i}(\alpha))=1$ for all α in $k_{\mathfrak{h}_i}$ satisfying $\alpha\equiv1$ mod. $\mathfrak{h}_i^{ci}$. Then an integral ideal $\mathfrak{f}=\mathfrak{h}_1^{c1}\cdots\mathfrak{h}_t^{ct}$ is called the conductor of χ. Let $\tilde{\chi}$ be, as above, the corresponding character of $G(\mathfrak{f})$. Notice that $|\tilde{\chi}(\mathfrak{a})|=N\mathfrak{a}^{-\rho}$ for any $\mathfrak{a}$ in $G(\mathfrak{f})$. Now, there is a character $\boldsymbol{X}$ of k^* such that $\boldsymbol{X}(\alpha)=\tilde{\chi}((\alpha))$ for any α in k^* satisfying $\alpha\equiv 1$ mod. $\mathfrak{f}$. We shall then say, following A. Weil, that a character χ of C_k is *of type* (A_0) if the corresponding character $\boldsymbol{X}$ of k^* has the following form:

$$(*)\qquad \boldsymbol{X}(\alpha)=\pm\prod_{\sigma\in H(k)}\sigma\alpha^{n(\sigma)}$$

where $n(\sigma)$ are integers, and $\pm$ may depend on α. Notice that the real part ρ of such χ is a half integer, i.e. 2ρ is an integer, and also that $n(\sigma)$ must satisfy a certain condition. Conversely, if there is a character $\tilde{\chi}$ of ideal group $G(\mathfrak{m})$ such that $\tilde{\chi}((\alpha))=\boldsymbol{X}(\alpha)$ for any $\alpha\equiv 1$ mod. $\mathfrak{m}^n$ with a suitably fixed n, where $\boldsymbol{X}(\alpha)$ is of the form $(*)$ with integers $n(\sigma)$ independent of α, then $\tilde{\chi}$ *can be obtained from a character* χ *of* C_k *of type* (A_0) *in the above exposed manner.*

A character χ of C_k is said to be *of finite order* if some power of χ, say χ^n, is the unit character, i.e. $\chi^n(a)=1$ for all $a\in I_k$.

If χ is a character of C_k of type (A_0), the values $\tilde{\chi}(\mathfrak{a})$ of all $\mathfrak{a}$ in $G(\mathfrak{f})$ lie in a certain algebraic number field K of finite degree. Notice that K need not contain k. For any valuation w in K, $\tilde{\chi}$ may also be considered as a representation of $G(\mathfrak{m})$ into the completion K_w^*, where $\mathfrak{m}$ is a multiple of $\mathfrak{f}$ to be determined later. Then the above criterion shows that the representation determines a representation χ_w, such that $\tilde{\chi}=\tilde{\chi}_w$ on $G(\mathfrak{m})$, if we take $\mathfrak{m}=\mathfrak{f}$ when w is Archimedean, and $\mathfrak{m}=\mathfrak{f}l$ when w is associated with a prime ideal $\mathfrak{l}$ of K. In the latter case χ_w is written as $\chi_{\mathfrak{l}}$. From the definition, $\chi_{\mathfrak{l}}$ is unramified at each $\mathfrak{p}$ in $G(\mathfrak{f}l)$. Since $K_{\mathfrak{l}}^*$ is totally disconnected, $\chi_{\mathfrak{l}}$ takes the value 1 on the connected component D_k, so that $\chi_{\mathfrak{l}}$ is also a representation of C_k', and it determines the representation $\chi_{\mathfrak{l}}^*$ of the Galois group G_k. Moreover, since C_k' is compact, the image $\chi_{\mathfrak{l}}(C_k')=\chi_{\mathfrak{l}}^*(G_k)$ lies in the unit group $U_{\mathfrak{l}}$ of $K_{\mathfrak{l}}^*$. If w is Archimedean, χ_w is written as χ^τ or as $\chi^{\sigma_0\tau}$ with corresponding isomorphisms τ, $\sigma_0\tau$ in $H(K)$. Notice that χ^τ may also be defined by $\tilde{\chi}^\tau(\mathfrak{a})=\tau\tilde{\chi}(\mathfrak{a})$ for $\mathfrak{a}\in G(\mathfrak{f})$.

§1. Characterization of characters of type (A_0).

1. Let k be an algebraic number field of finite degree.

Let χ be a character of C_k of type (A_0), with conductor $\mathfrak{f}$, real part $-\rho$, and associated character $\boldsymbol{X}$ of k^* defined by

$$(1)\qquad \boldsymbol{X}(\alpha)=\pm\prod_{\sigma\in H(k)}\sigma\alpha^{n(\sigma)}.$$

We shall denote by K an algebraic number field of finite degree, containing all values $\tilde{\chi}(\mathfrak{a})$ of the associated character $\tilde{\chi}$ of the ideal group $G(\mathfrak{f})$. Then K' will denote the Galois closure of k and of K. Since a suitable power of ideals in k can be represented as principal ideals, the ideal $(\tilde{\chi}(\mathfrak{a}))$ in K' must have the form $(\tilde{\chi}(\mathfrak{a}))=\prod_{\sigma}\sigma\mathfrak{a}^{n(\sigma)}$ for all ideals $\mathfrak{a}$ in $G(\mathfrak{f})$, where all $\sigma\mathfrak{a}$ are considered as ideals in K'.

For any τ in $H(K')$, χ^{τ} is also of type (A_0), with the same conductor $\mathfrak{f}$ as χ. Let $-\rho'$ be the real part of χ^{τ}. Then, since $\tilde{\chi}((r))$ is rational for any rational number r satisfying $r\equiv 1$ mod. $\mathfrak{f}$, we see that $N((r))^{\rho}=|\tilde{\chi}((r))|=|\tau\tilde{\chi}((r))|=N((r))^{\rho'}$, where $N((r))$ is the norm of (r) considered as an ideal in k. This shows $\rho=\rho'$, i.e. χ and χ^{τ} have the same real part. In particular, we have $|\tau\tilde{\chi}(\mathfrak{p})|=N\mathfrak{p}^{\rho}$ for any $\mathfrak{p}$ in $G(\mathfrak{f})$ and any τ in $H(K')$. Observe finally that, if we put $n_0=\operatorname{Min}_{\sigma} n(\sigma)$, $\tilde{\chi}(\mathfrak{p})\,N\mathfrak{p}^{-n_0}$ is an algebraic integer in K for any $\mathfrak{p}$ in $G(\mathfrak{f})$.

Let $\mathfrak{l}$ be a prime ideal in K and $\chi_{\mathfrak{l}}$ the representation of C_k into the unit group $U_{\mathfrak{l}}$ in $K_{\mathfrak{l}}^*$ associated with χ. Denote by $k(\chi,\mathfrak{l})$ the subfield of A_k corresponding to the kernel of the representation $\chi_{\mathfrak{l}}^*$ of G_k induced by $\chi_{\mathfrak{l}}$. This $k(\chi,\mathfrak{l})$ is nothing but the field attached to χ and $\mathfrak{l}$ by A. Weil [4]. If $\mathfrak{p}$ is in $G(\mathfrak{f}l)$, $\chi_{\mathfrak{l}}$ is unramified at $\mathfrak{p}$, hence $\mathfrak{p}$ is unramified in $k(\chi,\mathfrak{l})$. Moreover, from the definition, we see $\chi_{\mathfrak{l}}^*(\sigma_{\mathfrak{p}})=\chi_{\mathfrak{l}}(\iota_{\mathfrak{p}}\pi_{\mathfrak{p}})=\tilde{\chi}(\mathfrak{p})$ for any $\mathfrak{p}$ in $G(\mathfrak{f}l)$, where $\pi_{\mathfrak{p}}$ denotes a $\mathfrak{p}$-prime element in $k_{\mathfrak{p}}^*$.

2. Let again k and K be algebraic number fields of finite degree, and K' be the Galois closure of k and K. Let S be a set of prime ideals of K with *positive* density δ. We shall consider a system $\{\psi_{\mathfrak{l}}^*\}$ of representations $\psi_{\mathfrak{l}}^*$ of the Galois group G_k into $U_{\mathfrak{l}}\subset K_{\mathfrak{l}}^*$, where $\mathfrak{l}$ runs through all prime ideals in S. We denote by $k(\psi,\mathfrak{l})$ the subfield of A_k corresponding to the kernel of $\psi_{\mathfrak{l}}^*$. Now, we assume that the following four conditions (CA_I)—(CA_{IV}) are satisfied:

(CA_I) There is an integral ideal $\mathfrak{m}$ of k with the property that $\mathfrak{p}$ is unramified in $k(\psi,\mathfrak{l})$ for any $\mathfrak{p}$ in $G(\mathfrak{m})$ and for any $\mathfrak{l}$ in S such that $\mathfrak{p}$ lies in $G(\mathfrak{m}l)$.

This means that the value $\psi_{\mathfrak{l}}^*(\sigma_{\mathfrak{p}})$ is independent of the choice of $\sigma_{\mathfrak{p}}$ for each $\mathfrak{p}$ in $G(\mathfrak{m}l)$.

(CA_{II}) $\psi_{\mathfrak{l}}^*(\sigma_{\mathfrak{p}})$ belongs to K and is independent also of $\mathfrak{l}$ in S such that $\mathfrak{p}$ belongs to $G(\mathfrak{m}l)$.

We shall denote this common value of $\psi_{\mathfrak{l}}^*(\sigma_{\mathfrak{p}})$ by $\tilde{\psi}(\mathfrak{p})$, and also put $\tilde{\psi}(\mathfrak{a})=\prod_{\mathfrak{p}}\tilde{\psi}(\mathfrak{p})^{c_{\mathfrak{p}}}$ for any ideal $\mathfrak{a}=\prod_{\mathfrak{p}}\mathfrak{p}^{c_{\mathfrak{p}}}$ in $G(\mathfrak{m})$.

(CA_{III}) We have

$$|\tau\tilde{\psi}(\mathfrak{p})|=N\mathfrak{p}^{\rho}$$

for all $\mathfrak{p}$ in $G(\mathfrak{m})$ and for all τ in $H(K)$, where ρ is a fixed half integer independent of $\mathfrak{p}$ and of τ.

(CA_{IV}) There is a natural number n_0 such that $\tilde{\psi}(\mathfrak{p})\,N\mathfrak{p}^{n_0}$ are algebraic integers in K for all $\mathfrak{p}$ in $G(\mathfrak{m})$.

Finally we impose one more condition, which is a temporary one:

(A) The principal ideal $(\tilde{\psi}(\mathfrak{p}))$ in K' can be expressed in the form

$$(2) \qquad (\tilde{\psi}(\mathfrak{p}))=\prod_{\sigma\in H(k)}\sigma\mathfrak{p}^{n(\sigma,\mathfrak{p})},$$

for all $\mathfrak{p}$ in $G(\mathfrak{m})$, where $n(\sigma,\mathfrak{p})$ are rational integers, which *may depend on* $\mathfrak{p}$ as well as σ. Notice that $n(\sigma,\mathfrak{p})$ need not be uniquely determined by $\mathfrak{p}$ and σ, unless $\mathfrak{p}$ is of the first degree.

As we have seen in **1**, the system $\{\chi_{\mathfrak{l}}^{*}\}$ obtained from a character χ of C_k of type (A_0) satisfies these conditions, with $\mathfrak{m}=\mathfrak{f}$, $\tilde{\psi}(\mathfrak{p})=\tilde{\chi}(\mathfrak{p})$ and S=*the set of all prime ideals in* K. Our aim is now to prove the converse, i.e. to show that our conditions (CA_I)—(CA_{IV}) characterize the character of type (A_0), in the language of the Galois group G_k.

If we take the natural number n_0 in the condition (CA_{IV}) and put $\varpi=\tilde{\psi}(\mathfrak{p})N\mathfrak{p}^{n_0}$ for any $\mathfrak{p}$ in $G(\mathfrak{m})$, then ϖ, and also $\bar{\varpi}$, are algebraic integers in k. From the condition (CA_{III}), we have $\varpi\cdot\bar{\varpi}=|\varpi|^2=N\mathfrak{p}^{2\rho+2n_0}$, hence any prime factor of ϖ must divide $N\mathfrak{p}$, so that any prime factor of $\tilde{\psi}(\mathfrak{p})$ (with positive or negative exponent) must divide $N\mathfrak{p}$. We thus see that, if the conditions (CA_I)—(CA_{IV}) are satisfied, and if k is suitably large, e.g. if $k=K'$, the condition (A) is automatically satisfied. Moreover, as we shall see later, (A) is logically dependent on (CA_I)—(CA_{IV}) for any field k. We have added this condition (A) for the convenience of the proof.

3. $\psi_{\mathfrak{l}}^{*}$ determines a representation $\psi_{\mathfrak{l}}$ of $C_k'=C_k/D_k$ into $U_{\mathfrak{l}}$. Then conditions (CA_I), (CA_{II}) are equivalent to the following: $\psi_{\mathfrak{l}}$ is unramified at each $\mathfrak{p}$ in $G(\mathfrak{m}l)$, and $\psi_{\mathfrak{l}}(\iota_{\mathfrak{p}}(\pi_{\mathfrak{p}}))=\tilde{\psi}(\mathfrak{p})$ for any $\mathfrak{p}$-prime element $\pi_{\mathfrak{p}}$ in $k_{\mathfrak{p}}^{*}$, for each $\mathfrak{p}$ in $G(\mathfrak{m}l)$.

More generally, let $\mathfrak{p}$ be prime to l. From the continuity of the representation $\psi_{\mathfrak{l}}\circ\iota_{\mathfrak{p}}$ of $k_{\mathfrak{p}}^{*}$, we see that there is a natural number c such that $\alpha\equiv 1 \bmod. \mathfrak{p}^{c}$ $(\alpha\in k^{*})$ implies $\psi_{\mathfrak{l}}\circ\iota_{\mathfrak{p}}(\alpha)\equiv 1 \bmod. \mathfrak{l}$. For such α, α^{p^n} converges to 1 in $k_{\mathfrak{p}}^{*}$ as $n\to\infty$, so that $\psi_{\mathfrak{l}}\circ\iota_{\mathfrak{p}}(\alpha)^{p^n}$ must converge to 1 in $K_{\mathfrak{l}}^{*}$ as $n\to\infty$, which is however not the case unless $\psi_{\mathfrak{l}}\circ\iota_{\mathfrak{p}}(\alpha)=1$, because $\mathfrak{l}$ does not divide p. Thus, for any $\mathfrak{p}$ in $G((l))$, there is the smallest non-negative integer $c(\mathfrak{p},\mathfrak{l})$ such that $\alpha\equiv 1 \bmod. \mathfrak{p}^{c(\mathfrak{p},\mathfrak{l})}$ implies $\psi_{\mathfrak{l}}\circ\iota_{\mathfrak{p}}(\alpha)=1$. We have clearly $c(\mathfrak{p},\mathfrak{l})=0$ if $\mathfrak{p}$ is in $G(\mathfrak{m}l)$, where the congruence $\alpha\equiv 1 \bmod. \mathfrak{p}^{0}$ indicates $\alpha\in U_{\mathfrak{p}}$. Then we put $\mathfrak{f}(\mathfrak{l})=\prod_{\mathfrak{p}\in G((l))}\mathfrak{p}^{c(\mathfrak{p},\mathfrak{l})}$. This $\mathfrak{f}(\mathfrak{l})$ is an integral ideal in k. Since the number of classes of $U_{\mathfrak{p}}$ modulo $\mathfrak{p}^{c(\mathfrak{p},\mathfrak{l})}$ is finite, the images of $\alpha\in U_{\mathfrak{p}}$ by $\psi_{\mathfrak{l}}\circ\iota_{\mathfrak{p}}$ are roots of unity in $K_{\mathfrak{l}}$. Hence we have

$$(3) \qquad \psi_{\mathfrak{l}}(a)=\varepsilon\tilde{\psi}((a)) \text{ for } a=(a_{\mathfrak{p}}) \text{ in } I((l)) \text{ such that } (a)\in G(\mathfrak{m}),$$

where ε is a root of unity in $K_{\mathfrak{l}}$ depending only on the classes of $a_{\mathfrak{p}}$ mod. $\mathfrak{p}^{c(\mathfrak{p},\mathfrak{l})}$ for prime factors $\mathfrak{p}$ of $\mathfrak{m}$ prime to l. We shall denote by $W(\mathfrak{l})$ the number of roots of unity ε which appear in (3) for some $a\in I((l))$ such that $(a)\in G(\mathfrak{m})$. We shall now show that $W(\mathfrak{l})$ are bounded for infinitely many $\mathfrak{l}$ in S.

Let $\mathfrak{h}_1,\cdots,\mathfrak{h}_m$ be all the prime factors in $\mathfrak{m}$ in k, and put $h=h_1\cdots h_m$, $c(\mathfrak{h}_i,\mathfrak{l})=c_i$. Denote then by $W'(\mathfrak{l})$ the number of ε which appear in (3) for some $a=(a_{\mathfrak{p}})\in I((l))$ such that $a_{\mathfrak{h}_i}\equiv 1$ mod. $\mathfrak{h}_i (i=1,\cdots,m)$. Clearly, $W(\mathfrak{l})/W'(\mathfrak{l})$ are bounded for all $\mathfrak{l}$ in S, i.e. $\leqq\prod_{i=1}^{m} N\mathfrak{h}_i$. If we take t large enough, we have $a_{\mathfrak{h}_i}^{h^t}\equiv 1$ mod. $\mathfrak{h}_i^{c_i}$ for $a\in I((l))$ such that $a_{\mathfrak{h}_i}\equiv 1$ mod. $\mathfrak{h}_i$ $(i=1,\cdots,m)$. This shows that $\varepsilon^{h^t}=1$ for any ε corresponding to these a, i.e. $W'(\mathfrak{l})$ divides h^t; here t may depend on $\mathfrak{l}$. Now take a natural number t_0 so large that we have $[K(\zeta):K]>2/\delta$ for a primitive h^{t_0}-th root of unity ζ (δ is the density of S). Class field theory shows that the set of prime ideals of the first degree in K, which split completely in the abelian extension $K(\zeta)$, has the definite density $[K(\zeta):K]^{-1}$. Hence there are infinitely many $\mathfrak{l}$ in S (at least with density $\delta/2$), which split in $K(\zeta)$ into prime ideals of higher degrees, that is to say, there are infinitely many $\mathfrak{l}$ in S such that $K_{\mathfrak{l}}$ do not contain ζ. For such $\mathfrak{l}$, $W'(\mathfrak{l})$ must be smaller than h^{t_0}. Thus we have seen that *the number $W(\mathfrak{l})$ are bounded for infinitely many prime ideals $\mathfrak{l}$ of the first degree in S.* We shall denote by S' a set of infinitely many prime ideals $\mathfrak{l}$ of the first degree in K such that $W(\mathfrak{l})$ are smaller than a given bound, and moreovei such that l are prime to $\mathfrak{m}$ and unramified in K'. We shall then denote by $\mathcal{E}$ the group of roots of unity generated by all roots of unity in K' and by all ε in (3) for all $\mathfrak{l}$ in S'; this $\mathcal{E}$ is clearly a finite group.

4. We want to prove now that $\tilde{\psi}(\mathfrak{a})$ has the form $\prod_{\sigma}\sigma\mathfrak{a}^{n(\sigma)}$ with $n(\sigma)$ independent of $\mathfrak{a}$. For this purpose, observe the integers $n(\sigma,\mathfrak{p})$ in (A). From (CA_{III}), we see $(\tau\tilde{\psi}(\mathfrak{p})\cdot\sigma_0\tau\tilde{\psi}(\mathfrak{p}))=N\mathfrak{p}^{2\rho}=\prod_{\sigma\in H(k)}\sigma\mathfrak{p}^{2\rho}$ for any $\mathfrak{p}$ in $G(\mathfrak{m})$ and for any τ in $H(K')$. If $\mathfrak{p}\in G(\mathfrak{m})$ is of the first degree, and unramified over $\boldsymbol{Q}$, then $\sigma\mathfrak{p}$ and $\sigma'\mathfrak{p}$ have no common prime factor in K' if $\sigma\neq\sigma'$, so that the expression for $\tilde{\psi}(\mathfrak{p})$ in (A) is unique, i.e. $n(\sigma,\mathfrak{p})$ are uniquely determined by $\sigma,\mathfrak{p}$. Hence we have

$$(4)\qquad n(\sigma,\mathfrak{p})+n(\tau^{-1}\sigma_0\tau\sigma,\mathfrak{p})=2\rho$$

for all σ in $H(k)$, τ in $H(K')$. Moreover, (CA_{IV}) shows that $n(\sigma,\mathfrak{p})\geqq -n_0$ if $n(\sigma,\mathfrak{p})$ are uniquely determined. Notice that if a set of integers $\{n(\sigma,\mathfrak{p})\mid\sigma\in H(k)\}$ satisfies the condition (4), then we have $|\tau\alpha'|=|N_{k/Q}\alpha|^{\rho}$, where $\alpha'=\prod\sigma\alpha^{n(\sigma,\mathfrak{p})}$, $\alpha\in k^*$.

For general $\mathfrak{p}$, this last relation maybe does not hold. But in any case, let $\mathfrak{p}_1,\cdots,\mathfrak{p}_g$ be all the prime factors of p in K', and K_i the decomposition field of $\mathfrak{p}_i$ over $\boldsymbol{Q}$. Then, a suitable power of $\mathfrak{p}_1$, say $\mathfrak{p}_1^f$, can be represented as principal ideal (π_1) with π_1 in $K_1:\mathfrak{p}_1^f=(\pi_1)$. Clearly, for any τ in $H(K')$, $\tau\mathfrak{p}_1=\mathfrak{p}_1$ if and only if $\tau\pi_1=\pi_1$. It is also clear that, if $\tau\mathfrak{p}_1=\mathfrak{p}_i$, then $\pi_i=\tau\pi_1\in K_i$ does not depend on the choice of such τ, and $\mathfrak{p}_i^f=(\pi_i)$. Now, let $\mathfrak{p}=(\mathfrak{p}_1\cdots\mathfrak{p}_j)^e$ be the prime decomposition of $\mathfrak{p}$ in K'. If we put $\pi=(\pi_1\cdots\pi_j)^e$ accordingly, π belongs to k, $\mathfrak{p}^f=(\pi)$ and $|\tau\pi'|=\mathrm{N}\mathfrak{p}^{f\rho}$ for any τ in $H(K')$, where $\pi'=\prod_{\sigma}\sigma\pi^{n(\sigma,\mathfrak{p})}$ and $n(\sigma,\mathfrak{p})$ are integers in any expression of $\tilde{\psi}(\mathfrak{p})$ in (A). Thus $\tilde{\psi}(\mathfrak{p}^f)\pi'^{-1}=\eta$ is a unit in K', and we have $|\tau\eta|=1$ for any τ in $H(K')$, so that η is a root of unity in K', i.e. $\tilde{\psi}(\mathfrak{p}^f)=\eta\pi'=\eta\prod_{\sigma}\sigma\pi^{n(\sigma,\mathfrak{p})}$ with a root of unity η in K'.

Let now $\{n_\nu(\sigma)\,|\,\sigma\in H(k)\}$ be systems of rational integers $n_\nu(\sigma)\geqq -n_0$, satisfying

$$(5)\qquad n_\nu(\sigma)+n_\nu(\tau^{-1}\sigma_0\tau\sigma)=2\rho$$

for any $\sigma\in H(k)$ and $\tau\in H(K')$. Since $n_\nu(\sigma)$ must be $\leqq 2\rho+n_0$, the number of these systems is finite. Let then T_ν be the set of all *principal* prime ideals $\mathfrak{p}$ of the first degree in $G(\mathfrak{m})$, unramified over $\boldsymbol{Q}$, such that $n(\sigma,\mathfrak{p})=n_\nu(\sigma)$ for all σ, and put $T=\bigcup_\nu T_\nu$. From what we have remarked above, these T_ν are mutually disjoint, and T contains almost all principal $\mathfrak{p}$ of the first degree in k. We shall denote by $\langle T_\nu\rangle$ the subgroup of $G(\mathfrak{m})$ generated by $\mathfrak{p}$ in T_ν. Then the finiteness of the number of T_ν brings forth the following result:

For any natural number n, $S((n))$ denotes the subgroup of $G((n))$ consisting of all $\mathfrak{a}$ in $G((n))$ representable as $\mathfrak{a}=(\alpha)$ with α in k^* satisfying $\alpha\equiv 1$ mod. n, then the factor group $\mathfrak{S}(n)=G((n))/S((n))$ is the "Strahlklassengruppe" modulo (n) in k. We denote by $G_0((n))$ the set of all principal ideals in $G((n))$, and put $\mathfrak{S}_0(n)=G_0((n))/S((n))$. Then we denote by $\mathfrak{T}_\nu(n)$ the image of the subgroup $\langle T_\nu\rangle\cap G((n))$ into $\mathfrak{S}_0(n)$ by the natural homomorphism. We shall denote moreover by $\mathfrak{S}_0'(n)$ the multiplicative group of all the prime residue classes mod. (n) in k. There is a natural homomorphism of $\mathfrak{S}_0'(n)$ onto $\mathfrak{S}_0(n)$, whose kernel consists of all classes containing a unit in k. We shall then denote by $\mathfrak{T}_\nu'(n)$ the inverse image of $\mathfrak{T}_\nu(n)$ in $\mathfrak{S}_0'(n)$. Assume now for a moment that, for each ν, there are infinitely many prime numbers $l_i^{(\nu)}$ $(i=1,2,\cdots)$ such that $\mathfrak{T}_\nu(l_i^{(\nu)})\neq\mathfrak{S}_0(l_i^{(\nu)})$, i.e. $\mathfrak{T}_\nu'(l_i^{(\nu)})\neq\mathfrak{S}_0'(l_i^{(\nu)})$. Let $m_t^{(\nu)}$ be a product of a finite number t of different $l_i^{(\nu)}$, say, $m_t^{(\nu)}=l_1^{(\nu)}\cdots l_t^{(\nu)}$. Then, since $\mathfrak{S}_0'(m_t^{(\nu)})$ is isomorphic to a direct product $\mathfrak{S}_0'(l_1^{(\nu)})\times\cdots\times\mathfrak{S}_0'(l_t^{(\nu)})$, and since the index $[\mathfrak{S}_0'(l_i^{(\nu)}):\mathfrak{T}_\nu'(l_i^{(\nu)})]\geqq 2$, we have $[\mathfrak{S}_0(m_t^{(\nu)}):\mathfrak{T}_\nu(m_t^{(\nu)})]=[\mathfrak{S}_0'(m_t^{(\nu)}):\mathfrak{T}_\nu'(m_t^{(\nu)})]\geqq 2^t$. Hence, if we denote by $T_t^{(\nu)}$ the set of prime ideals in the inverse image of $\mathfrak{T}_\nu(m_t^{(\nu)})$ in $G(m_t^{(\nu)})$), the density of $T_t^{(\nu)}$ is definite and is at most 2^{-t}.

Notice that the union $\bigcup_{\nu} T_t^{(\nu)}$ contains almost all principal prime ideals of the first degree in k, for any $t=1,2,\cdots$, hence it has a non-zero definite density. But the number of sets T_ν being finite, the density of $\bigcup_{\nu} T_t^{(\nu)}$ must become arbitrarily small if we take t suitably large, which is a contradiction. Thus we have proved that there is at least one ν, for which $\mathfrak{T}_\nu(l)=\mathfrak{S}_0(l)$ hold for all but a finite number of prime number l.

We shall take a fixed one T_ν with this property, and denote this T_ν, $\langle T_\nu\rangle$, $n_\nu(\sigma)$ by T_0, $\langle T_0\rangle$, $n(\sigma)$ respectively. With these $n(\sigma)$ we put

$$(6) \qquad \boldsymbol{X}(\alpha)=\prod_{\sigma\in H(k)} \sigma\alpha^{n(\sigma)}$$

for $\alpha\in k^*$. Since $n(\sigma)$ satisfy (5), we have $|\tau\boldsymbol{X}(\alpha)|=|N_{k/Q}\alpha|^\rho$ for any $\tau\in H(K')$. Clearly this $\boldsymbol{X}$ is a representation of k^* with discrete topology into $(K')^*$.

5. We need a new topology of k^*. Let v be a valuation of $\boldsymbol{Q}$ and $v_1,\cdots,v_g$ be all the extensions of v to k. Then the weakest topology of k stronger than each topology of k determined by $v_i\,(i=1,\cdots,g)$ will be called the v-*topology* of k. This topology is metrisable, and makes k a topological field. Then the completion k_v of k with respect to the v-topology is an algebra over $\boldsymbol{Q}_v$, and is isomorphic to the direct sum $k_{v_1}+\cdots+k_{v_g}$ as topological algebras. Hence the canonical isomorphism of $k_{v_1}^*\times\cdots\times k_{v_g}^*$ into I_k determines an imbedding ι_v of k^* into I_k. Then ι_v is bicontinuous with respect to v-topology of k and the topology of $\iota_v(k^*)$ induced by that of I_k. Let w be an extension of v to K'. Then $\alpha\to\sigma\alpha(\alpha\in k^*)$ is continuous with respect to v-topology of k and the topology of K' determined by w, for any σ in $H(k)$, hence the mapping $\boldsymbol{X}$ defined in (6) is a representation of k^* into K'^* with respect to these topologies. When v is determined by a prime number l, we speak of l-topology, and write k_l, ι_l instead of k_v, ι_v. Notice that an element α in k^* is near to 1 with respect to l-topology if and only if $\alpha\equiv 1 \mod. l^i$ with a high power l^i of l. We put finally $(\alpha)_l=(\iota_l\alpha)$ for α in k^*, i.e. $(\alpha)_l=\mathfrak{q}_1^{\nu_1(\alpha)}\cdots\mathfrak{q}_g^{\nu_g(\alpha)}$, where $\mathfrak{q}_1,\cdots,\mathfrak{q}_g$ denotes all the prime divisors of l in k and $\nu_i=\nu_{\mathfrak{q}_i}$.

Now take a prime ideal $\mathfrak{l}$ in K in the set S, such that l is prime to $\mathfrak{m}$. Put then

$$\Psi_{\mathfrak{l}}(\alpha)=\psi_{\mathfrak{l}}(\iota_l(\alpha))\cdot\tilde{\psi}((\alpha)_l)^{-1},\ \alpha\in k^*.$$

This $\Psi_{\mathfrak{l}}$ is a representation of k^* with l-topology into $K_{\mathfrak{l}}^*$, as we have $\tilde{\psi}((\alpha)_l)=1$ for $\alpha\equiv 1 \mod. l$. Moreover, from (3) in 3, and from $\psi_{\mathfrak{l}}(\iota(\alpha))=1$, we see

$$(7) \qquad \Psi_{\mathfrak{l}}(\alpha)=\pm\varepsilon\tilde{\psi}((\alpha))^{-1},$$

for any α prime to $\mathfrak{m}$, where ε is a root of unity depending only on the class of α mod. $\mathfrak{f}(\mathfrak{l})$, and $\pm$ depends on the signatures of α at real primes. This shows in particular that $|\tau\Psi_{\mathfrak{l}}(\alpha)|=|N_{k/Q}\alpha|^{-\rho}$ for such α, and for any τ in $H(K')$. Notice that, if $\mathfrak{l}$ is in S', ε belongs to the finite group $\mathcal{E}$ (defined at the end of **3**). Observe also that, if the ideal (α) belongs to $\langle T_0\rangle$, the principal ideal $(\Psi_{\mathfrak{l}}(\alpha))=(\tilde{\psi}((\alpha)))^{-1}$ must be $=\prod_{\sigma}\sigma(\alpha)^{-n(\sigma)}$. Hence, for these (α), $\Psi_{\mathfrak{l}}(\alpha)\boldsymbol{X}(\alpha)$ is a unit in K', all conjugates of which have the same absolute value 1, i.e., it is a root of unity ε' in K':

$$(8) \qquad \Psi_{\mathfrak{l}}(\alpha)=\varepsilon'\boldsymbol{X}(\alpha)^{-1},\ \alpha\in k^*,\ \text{such that}\ (\alpha)\in\langle T_0\rangle.$$

We shall now prove that (8) holds for *any* α prime to $\mathfrak{m}$. For this purpose, let $\mathfrak{p}$ be any prime ideal in $G(\mathfrak{m})$. Take a number π in k^* such that $\mathfrak{p}^f=(\pi)$ with a natural number f, and $\tilde{\psi}(\mathfrak{p}^f)=\eta\pi'=\eta\prod_{\sigma}\sigma\pi^{n(\sigma,\mathfrak{p})}$ with a root of unity η in K' (cf. 4). Put $\pi''=\boldsymbol{X}(\pi)\pi'^{-1}$, and assume for a moment that π'' is not a root of unity. Then, none of a finite number of elements $\pi''-\varepsilon$ for all $\varepsilon\in\mathcal{E}$ should not be 0. The set S' (at the end of **3**) being infinite, we can find an $\mathfrak{l}$ in S' such thet l is prime to π and to all these $\pi''-\varepsilon$, and the image $\mathfrak{T}_0(l)$ of $\langle T_0\rangle\cap G((l))$ into $\mathfrak{S}_0(l)$ is equal to $\mathfrak{S}_0(l)$ (cf. the end of **4**). From the last assumption, there is a principal ideal (α) in $\langle T_0\rangle$ such that $\alpha\pi\equiv 1$ mod. l. π being prime to $\mathfrak{m}$, we see from (7) and (8)

$$\Psi_{\mathfrak{l}}(\alpha\pi)\boldsymbol{X}(\alpha\pi)=\varepsilon'\Psi_{\mathfrak{l}}(\pi)\boldsymbol{X}(\pi)=\pm\varepsilon'\varepsilon\tilde{\psi}(\mathfrak{p}^f)^{-1}\boldsymbol{X}(\pi)=\pm\varepsilon'\varepsilon\eta^{-1}\pi'',$$

where $\varepsilon',\varepsilon,\eta$ belong to the group $\mathcal{E}$, hence $\varepsilon_0=\pm\varepsilon'\varepsilon\eta$ also lies in $\mathcal{E}$. Let $\mathfrak{l}'$ be a prime factor of $\mathfrak{l}$ in K'. Since $\alpha\pi\equiv 1$ mon. l, the number $(\alpha\pi)^{l^t}$ converges to 1 with respect to l-topology as $t\to\infty$, so that the value $(\varepsilon_0\pi'')^{l^t}=\Psi_{\mathfrak{l}}((\alpha\pi)^{l^t})\cdot\boldsymbol{X}((\alpha\pi)^{l^t})$ must converge to 1 in $K'_{\mathfrak{l}'}$, because $\Psi_{\mathfrak{l}}$, $\boldsymbol{X}$ are representation of k^* with l-topology into $K'_{\mathfrak{l}'}$. But as we have $\beta^{N\mathfrak{l}'}\equiv\beta$ mod. $\mathfrak{l}'$ for any $\mathfrak{l}'$-integral β in $K'_{\mathfrak{l}'}$, this convergence of $(\varepsilon_0\pi'')^{l^t}$ must imply that $\varepsilon_0\pi''\equiv 1$ mod. $\mathfrak{l}'$, i.e. $\mathfrak{l}'$ must divide $\pi''-\varepsilon_0^{-1}$. This contradicts however the assumption that l is prime to all $\pi''-\varepsilon$ for $\varepsilon\in\mathcal{E}$. We have thus proved that $\pi''=\boldsymbol{X}(\pi)\pi'^{-1}$ is a root of unity, hence $(\tilde{\psi}(\mathfrak{p}^f))=\prod_{\sigma}\sigma\mathfrak{p}^{f\cdot n(\sigma)}$. In order words, *integers* $n(\sigma,\mathfrak{p})$ *in the condition* (*A*) *may be taken as* $n(\sigma)$, *independent of* $\mathfrak{p}$. Thus we have arrived at the result aimed at the beginning of 4.

This implies in particular that (8) holds for any α in k^*, prime to $\mathfrak{m}$, i.e. $\Psi_{\mathfrak{l}}(\alpha)\boldsymbol{X}(\alpha)=\varepsilon'_{\mathfrak{l}}(\alpha)$, and $\varepsilon'_{\mathfrak{l}}(\alpha)$ is a root of unity in K'. Denote by $\mathcal{E}_0$ the (finite) group of roots of unity in K'. Then, $\Psi_{\mathfrak{l}}(\alpha)\boldsymbol{X}(\alpha)$ induces a representation into $\mathcal{E}_0$ of the subgroup of k^* with induced l-topology, consisting of all α prime to $\mathfrak{m}$. Hence there is a power l^c of l such that $\alpha\equiv 1$ mod. l^c implies $\varepsilon'_{\mathfrak{l}}(\alpha)=1$ for an α prime to $\mathfrak{m}$.

On the other hand, we see from (7) that $\Psi_{\mathfrak{l}}(\alpha)=\tilde{\psi}((\alpha))^{-1}$ for totally positive (i.e. positive at each real prime) α satisfying $\alpha\equiv 1$ mod. $\mathfrak{f}(\mathfrak{l})$. Thus $\tilde{\psi}((\alpha))=\boldsymbol{X}(\alpha)$ for any totally positive $\alpha\equiv 1$ mod. $\mathfrak{f}(\mathfrak{l})l^c$. Since $\tilde{\psi}$, $\boldsymbol{X}$ does not depend on $\mathfrak{l}$, this is true for any $\mathfrak{l}$ in S. Hence, if we denote by $\mathfrak{f}$ the greatest common divisor of all $\mathfrak{f}(\mathfrak{l})l^c$ for $\mathfrak{l}\in S$, then we have $\tilde{\psi}((\alpha))=\boldsymbol{X}(\alpha)$ for any totally positive α prime to $\mathfrak{m}$ satisfying $\alpha\equiv 1$ mod. $\mathfrak{f}$, i.e. $\tilde{\psi}((\alpha))=\pm\boldsymbol{X}(\alpha)$ for any α in k^* satisfying $\alpha\equiv 1$ mod. $\mathfrak{m}\mathfrak{f}$. As was recalled in "Notations and terminologies$\cdots$," this shows that there is a character χ of C_k of type (A_0), such that the associated character of k^* is exactly equal to our $\boldsymbol{X}$, and the associated ideal character $\tilde{\chi}$ of $G(\mathfrak{m})$ is exactly equal to $\tilde{\psi}$. Then, since $\psi_{\mathfrak{l}}$ and $\chi_{\mathfrak{l}}$ coincide on the dense set $I(\mathfrak{m})P_k$ of I_k, we have $\psi_{\mathfrak{l}}=\chi_{\mathfrak{l}}$ for each $\mathfrak{l}\in S$. Thus we have obtained the desired result, under the assumption of the condition (A).

6. Now, we shall drop this condition (A), and assume that $\{\psi_{\mathfrak{l}}^*\}$ satisfies only (CA_I)—(CA_{IV}). Let k' be an absolutely normal field of finite degree containing k and K (e.g. $k'=K'$). If we put $\psi'_{\mathfrak{l}}(a')=\psi_{\mathfrak{l}}(N_{k'/k}a')$ for an idèle a' in $I_{k'}$, then $\psi'_{\mathfrak{l}}$ is a representation of $C_{k'}$ into $U_{\mathfrak{l}}$, since we have $N_{k'/k}P_{k'}\subset P_k$ from the definition of $N_{k'/k}$. Let $\mathfrak{P}$ be a prime ideal of k' such that $N_{k'/k}\mathfrak{P}=\mathfrak{p}^f$ is prime to $\mathfrak{m}$, and π be a $\mathfrak{P}$-prime element in $k'_{\mathfrak{P}}$. Then, from the definition, $\psi'_{\mathfrak{l}}$ is unramified at $\mathfrak{P}$ and $\psi'_{\mathfrak{l}}(\iota_{\mathfrak{P}}(\pi))=\psi_{\mathfrak{l}}(\iota_{\mathfrak{p}}(N_{(\mathfrak{P})}\pi))=\psi(\mathfrak{p})^f$, where $N_{(\mathfrak{P})}$ denotes the norm of $k'_{\mathfrak{P}}$ into $k_{\mathfrak{p}}$. This shows that the system $\{\psi_{\mathfrak{l}}'^*\}$ of representations $\psi_{\mathfrak{l}}'^*$ of $G_{k'}$ determined by $\psi'_{\mathfrak{l}}$ satisfies the conditions (CA_I)—(CA_{IV}), with $\tilde{\psi}'(\mathfrak{P})=\tilde{\psi}(N_{k'/k}\mathfrak{P})$ and the same ρ. Moreover, it satisfies (A) automatically. Hence, from the above results (in **5**), we have $(\tilde{\psi}'(\mathfrak{P}))=\prod_{\tau\in H(k')}\tau\mathfrak{P}^{n'(\tau)}$ with integers $n'(\tau)$ independent of $\mathfrak{P}$. In particular, for any automorphism φ of k', we have $(\tilde{\psi}'(\varphi\mathfrak{P}))=\prod_{\tau}\tau\varphi\mathfrak{P}^{n'(\tau)}$. When φ is in $G(k'/k)$, we have moreover $\tilde{\psi}'(\varphi\mathfrak{P})=\tilde{\psi}(N_{k'/k}\varphi\mathfrak{P})=\tilde{\psi}(\mathfrak{p})^f=\tilde{\psi}'(\mathfrak{P})$, so that $\prod_{\tau}\tau\varphi\mathfrak{P}^{n'(\tau)}=\prod_{\tau}\tau\mathfrak{P}^{n'(\tau)}$. If we take as $\mathfrak{P}$ a prime ideal of the first degree prime to $\mathfrak{m}$, and unramified over $\boldsymbol{Q}$, then we see $n'(\tau)=n'(\tau\varphi)$ for any τ in $H(k')$ and φ in $G(k'/k)$. This means that $n'(\tau)$ is determined uniquely by the isomorphism σ of k into k' induced by τ, so we may write $n'(\tau)=n(\sigma)$. Then, for *any* $\mathfrak{p}$ in $G(\mathfrak{m})$, we have

$$\tilde{\psi}(\mathfrak{p})^f=\tilde{\psi}'(\mathfrak{P})=\prod_{\sigma\in H(k)}\prod_{\varphi\in G(k'/k)}\sigma\varphi\mathfrak{P}^{n(\sigma)}=\prod_{\sigma\in H(k)}\sigma\mathfrak{p}^{fn(\sigma)}.$$

Thus we have shown that $\tilde{\psi}(\mathfrak{p})=\prod_{\sigma}\sigma\mathfrak{p}^{n(\sigma)}$ with integers $n(\sigma)$, and the condition (A) is always satisfied for the system $\{\psi_{\mathfrak{l}}^*\}$ satisfying (CA_I)—(CA_{IV}). Summing up, we have obtained the following result:

THEOREM 1. *Let k be an algebraic number field of finite degree,* *and $\{\chi_{\mathfrak{l}}^*\}$ be the system of representations of the Galois groups $G_k = G(A_k/k)$ determined by a character χ of C_k of type* (A_0). *Then, the conditions (CA_I)—(CA_{IV}) in* **2** *are satisfied with $\psi_{\mathfrak{l}}^* = \chi_{\mathfrak{l}}^*$, $\mathfrak{m} = \mathfrak{f}$, $\tilde{\psi}(\mathfrak{p}) = \tilde{\chi}(\mathfrak{p})$ and $S =$ the set of all prime ideals. Conversely, let K be another algebraic number field of finite degree, and S be a set of prime ideals in K with positive density. If a system $\{\psi_{\mathfrak{l}}^*\}$ of representations $\psi_{\mathfrak{l}}^*$ of G_k into $U_{\mathfrak{l}} \subset K_{\mathfrak{l}}^*$, $\mathfrak{l}$ running through S, satisfies the conditions (CA_I)—(CA_{IV}), then there is one (and only one) character χ of C_k of type* (A_0) *with the property that $\psi_{\mathfrak{l}}^*$ is exactly equal to the representation $\chi_{\mathfrak{l}}^*$ associated with χ for each $\mathfrak{l}$ in S.*

§2. L-functions with characters of type (A_0).

7. Notations will be the same as in **1**, §1. First we shall observe the smallest possible field K_0 among K's, i.e. the field K_0 generated over $\boldsymbol{Q}$ by all values $\tilde{\chi}(\mathfrak{p})$ for $\mathfrak{p}$ in $G(\mathfrak{f})$. Notice that $\tilde{\chi}(\mathfrak{p}) \cdot \sigma_0 \tilde{\chi}(\mathfrak{p}) = \tau \tilde{\chi}(\mathfrak{p}) \cdot \sigma_0 \tau \tilde{\chi}(\mathfrak{p}) = \mathrm{N}\mathfrak{p}^{2\rho}$ is a rational number, hence, for any τ in $H(K_0)$, $\tau(\tilde{\chi}(\mathfrak{p}) \sigma_0 \tilde{\chi}(\mathfrak{p})) = \mathrm{N}\mathfrak{p}^{2\rho}$. From this we see that $\tau \sigma_0 \tilde{\chi}(\mathfrak{p}) = \sigma_0 \tau \tilde{\chi}(\mathfrak{p})$ holds for any τ in $H(K_0)$ and any $\mathfrak{p}$ in $G(\mathfrak{f})$. Thus, K_0 is a totally imaginary quadratic extension of a totally real field, if K_0 is not a real field. When K_0 is real, we have $\tilde{\chi}(\mathfrak{p}) = \pm \mathrm{N}\mathfrak{p}^{\rho}$, and then, comparing ideal decomposition of both sides for $\mathfrak{p}$ of the first degree, we see that ρ is an integer, hence all $\tilde{\chi}(\mathfrak{p})$ are rational and $K_0 = \boldsymbol{Q}$.

Hereafter, until the end of this §2, we shall impose on χ the following condition:

(I) χ is of infinite order, and all values $\tilde{\chi}(\mathfrak{p})$ for $\mathfrak{p}$ in $G(\mathfrak{f})$ are algebraic integers in K_0. Moreover, where $K_0 = \boldsymbol{Q}$, all $\tilde{\chi}(\mathfrak{p})$ are positive.

Clearly, the first condition in *(I)* holds if and only if all $n(\sigma) \geqq 0$ and $\rho > 0$. The second condition implies that $\prod_{\tau \in H(K)} \{\tau \tilde{\chi}(\mathfrak{p})^n - 1\} > 0$ in any case ($n > 0$), since either K_0 is totally imaginary or $K_0 = \boldsymbol{Q}$ and $\tilde{\chi}(\mathfrak{p}) = \mathrm{N}\mathfrak{p}^{\rho} > 1$. Notice that, in the former case, K is also totally imaginary.

Remember that the representation $\chi_{\mathfrak{l}}$ associated with χ is unramified at each prime ideal $\mathfrak{p}$ in $G(\mathfrak{f}l)$, and $\tilde{\chi}(\mathfrak{p})$ are $\mathfrak{l}$-units for such $\mathfrak{p}$. Notice also that, when $\mathfrak{p}$ divides l, $\tilde{\chi}(\mathfrak{p})$ is prime to $\mathfrak{l}$ if and only if $\sigma\mathfrak{p}$ are prime to $\mathfrak{l}$ (as ideals in K') for all σ in $H(k)$ satisfying $n(\sigma) > 0$. Let now $\mathfrak{q}_1, \cdots, \mathfrak{q}_r$ be all prime factors of l in k such that $\tilde{\chi}(\mathfrak{q}_i)$ is divisible by $\mathfrak{l}$, and $\mathfrak{q}_1', \cdots, \mathfrak{q}_s'$ be all the remainning prime factors of l in k, i.e. such that $\tilde{\chi}(\mathfrak{q}_i')$ is a $\mathfrak{l}$-unit. Then, for any σ satisfying $n(\sigma) > 0$, $\sigma(\mathfrak{q}_1 \cdots \mathfrak{q}_r)$ is divisible by $\mathfrak{l}$, while $\sigma(\mathfrak{q}_1' \cdots \mathfrak{q}_s')$ is prime to $\mathfrak{l}$. Hence, for an element α in k^*, $\alpha \equiv 1 \bmod. (\mathfrak{q}_1 \cdots \mathfrak{q}_r)^i$

implies $\prod_{\sigma} \sigma\alpha^{n(\sigma)} \equiv 1 \bmod. \mathfrak{l}^i$. This shows that there is a representation $\chi'_{\mathfrak{l}}$ of C_k into $U_{\mathfrak{l}}$, determined by the character χ of $G(\mathfrak{f}\mathfrak{q}_1\cdots\mathfrak{q}_r)$. $\chi_{\mathfrak{l}}$ and $\chi'_{\mathfrak{l}}$ being the same on the dense subgroup $P_k I(\mathfrak{f}l)$ of I_k, we see $\chi'_{\mathfrak{l}}=\chi_{\mathfrak{l}}$. Since $\chi'_{\mathfrak{l}}$ is unramified at $\mathfrak{q}'_i$, $\chi_{\mathfrak{l}}$ *is unramified at any* $\mathfrak{p}$ *in* $G(\mathfrak{f})$, *either prime to* l *or not, such that* $\tilde{\chi}(\mathfrak{p})$ *is prime to* $\mathfrak{l}$.

Let now U_0 be the direct product $\prod_{\mathfrak{l}} U_{\mathfrak{l}}$ of $\mathfrak{l}$-unit groups in $K^*_{\mathfrak{l}}$ for all $\mathfrak{l}$ in K, and ω be a character of U_0. If we denote by $\omega_{\mathfrak{l}}$ the character induced by ω on $U_{\mathfrak{l}}$, considered as a subgroup of U_0, then $\omega_{\mathfrak{l}}=1$ for almost all $\mathfrak{l}$. More precisely, there is an integral ideal $\mathfrak{b}=\mathfrak{l}_1^{m_1}\cdots\mathfrak{l}_t^{m_t}$ of K such that $\omega_{\mathfrak{l}}=1$ for $\mathfrak{l}$ prime to $\mathfrak{b}$, and that $\omega_{\mathfrak{l}_i}(\alpha)=1$ for any $\alpha\equiv 1 \bmod. \mathfrak{l}_i^{m_i}$. In this case, ω will be called *definable modulo* $\mathfrak{b}$. If ω is definable modulo $\mathfrak{b}$, ω is also definable modulo any multiple of $\mathfrak{b}$. Notice that the number of characters of U_0 definable modulo $\mathfrak{b}$ is exactly equal to the Euler function $\varphi(\mathfrak{b})=\mathrm{N}\mathfrak{b}\prod_i(1-(\mathrm{N}\mathfrak{l}_i)^{-1})$, i.e. the number of prime residue classes mod. $\mathfrak{b}$, and the values of such characters are $\varphi(\mathfrak{b})$-th roots of unity.

Assigning to each idèle a in I_k an element $\chi_0(a)=(\chi_{\mathfrak{l}}(a))$ of U_0 with $\mathfrak{l}$-components $\chi_{\mathfrak{l}}(a)$, we obtain a representation χ_0 of C_k, or of C'_k, into U_0. Put $\chi_\omega=\omega\circ\chi_0$. Then χ_ω is a character of C_k of finite order, so that it determines a cyclic extension $k(\chi,\omega)$ of k, by class field theory. The compositum of these $k(\chi,\omega)$ for all characters ω of U_0 is equal to the subfield of A_k corresponding to the kernel of the representation χ_0^* of G_k, and this field is nothing but the abelian extension $k(\chi)$ of k attached by χ by A. Weil [4].

It is trivial to notice that, when ω is definable modulo $\mathfrak{b}$, χ_ω is unramified at each $\mathfrak{p}$ in $G(\mathfrak{f})$ such that $\tilde{\chi}(\mathfrak{p})$ is prime to $\mathfrak{b}$. Thus, if we denote by $[\mathfrak{b}]$ the product of all $\mathfrak{p}$ in $G(\mathfrak{f})$ such that $\tilde{\chi}(\mathfrak{p})$ is not prime to $\mathfrak{b}$, then χ_ω determines a character $\tilde{\chi}_\omega$ of the ideal group $G(\mathfrak{f}[\mathfrak{b}])$ in k. However, this $\tilde{\chi}_\omega$ is in general not primitive, i.e. it may be extended to a larger group than $G(\mathfrak{f}[\mathfrak{b}])$ in some cases.

8. Let $L_{\mathfrak{b}}(s;\omega)=\prod_{\mathfrak{p}\in G(\mathfrak{f}[\mathfrak{b}])}(1-\tilde{\chi}_\omega(\mathfrak{p})\mathrm{N}\mathfrak{p}^{-s})^{-1}$ be Hecke's L-function in k with this character $\tilde{\chi}_\omega$ of $G(\mathfrak{f}[\mathfrak{b}])$, and put

$$L_{\mathfrak{b}}(s)=\prod_{\omega} L(s;\omega), \tag{9}$$

where ω runs over all the $\varphi(\mathfrak{b})$ characters of U_0 definable modulo $\mathfrak{b}$. Put furthermore

$$L_\chi(s)=\prod L_{\mathfrak{b}}(s), \tag{10}$$

$\mathfrak{b}$ running over all integral ideals of K. We must now examine the absolute convergence of this infinite product in some right half s-plane.

First we observe $L_{\mathfrak{b}}(s)$. For each $\mathfrak{p}$ in $G(\mathfrak{f}[\mathfrak{b}])$, we shall denote by $f(\mathfrak{p},\mathfrak{b})$ the smallest natural number such that $\tilde{\chi}_\omega(\mathfrak{p}^{f(\mathfrak{p},\mathfrak{b})})=1$ for all

ω definable modulo $\mathfrak{b}$, i.e. such that $\tilde{\chi}(\mathfrak{p})^{f(\mathfrak{p},\mathfrak{b})}\equiv 1$ mod. $\mathfrak{b}$. Such number exists certainly since $\mathfrak{b}$ is prime to $\tilde{\chi}(\mathfrak{p})$. Then, a well-known relation between characters of a finite abelian group shows:

$$L_{\mathfrak{b}}(s)=\prod_{\mathfrak{p}\in G(\mathfrak{f}[\mathfrak{b}])}\prod_{\omega}(1-\tilde{\chi}_{\omega}(\mathfrak{p})N\mathfrak{p}^{-s})^{-1}=\prod_{\mathfrak{p}}(1-N\mathfrak{p}^{-f(\mathfrak{p},\mathfrak{b})s})^{-\varphi(\mathfrak{b})/f(\mathfrak{p},\mathfrak{b})}$$

We shall now put, for any $\mathfrak{p}$ in $G(\mathfrak{f})$,

$$L^{(\mathfrak{p})}(s)=\prod_{\mathfrak{b}}(1-N\mathfrak{p}^{-f(\mathfrak{p},\mathfrak{b})s})^{-\varphi(\mathfrak{b})/f(\mathfrak{p},\mathfrak{b})},$$

where $\mathfrak{b}$ runs over all integral ideals in K such that $\mathfrak{p}$ belongs to $G(\mathfrak{f}[\mathfrak{b}])$, i.e. $\tilde{\chi}(\mathfrak{p})$ is prime to $\mathfrak{b}$. Then, we shall evaluate the positive term series

$$\text{(11)}\qquad \sum\nolimits_0=\sum_{\mathfrak{b}}\frac{\varphi(\mathfrak{b})}{f(\mathfrak{p},\mathfrak{b})\,(N\mathfrak{p})^{f(\mathfrak{p},\mathfrak{b})\sigma}}<\sum_{n=1}^{\infty}(\sum\nolimits'\varphi(\mathfrak{b}))N\mathfrak{p}^{-n\sigma}$$

where σ denotes the real part of the complex number s, and the sum $\sum'\varphi(\mathfrak{b})$ in parenthesis in the right hand side is taken over all $\mathfrak{b}$ such that $\mathfrak{p}$ belongs to $G(\mathfrak{f}[\mathfrak{b}])$ and $f(\mathfrak{p},\mathfrak{b})$ divides n. These conditions are equivalent to $\tilde{\chi}(\mathfrak{p})^n\equiv 1$ mod. $\mathfrak{b}$, so that $\mathfrak{b}$ runs over all integral divisors of $\tilde{\chi}(\mathfrak{p})^n-1$. Notice that $\tilde{\chi}(\mathfrak{p})^n-1$ is a non-zero integers in K, from the condition (I). Hence, if we denote by $\Phi(\mathfrak{p},n)$ this sum $\sum'\varphi(\mathfrak{b})$, we have $\Phi(\mathfrak{p},n)=|N_{K/Q}(\tilde{\chi}(\mathfrak{p})^n-1)|$. From the remark below the condition (I), we see therefore

$$\Phi(\mathfrak{p},n)=N_{K/Q}(\tilde{\chi}(\mathfrak{p})^n-1)=\prod_{\tau\in H(K)}\{\tau\tilde{\chi}(\mathfrak{p})^n-1\}.$$

In particular, $\Phi(\mathfrak{p},n)$ have the same order of magnitude as $N\mathfrak{p}^{nd\rho}$ for $n\to\infty$, where $d=[K:\boldsymbol{Q}]$. This implies that the series in the right hand side in (11) converges absolutely for $\sigma>d\rho$. Hence the infinite product for $L^{(\mathfrak{p})}(s)$ converges absolutely for $\sigma>d\rho$. We see moreover that *the infinite product*

$$\prod_{\mathfrak{p}}\prod_{\mathfrak{b}}(1-N\mathfrak{p}^{-f(\mathfrak{p},\mathfrak{b})s})^{-\varphi(\mathfrak{b})/f(\mathfrak{p},\mathfrak{b})}$$

converges absolutely for $\sigma>d\rho+1$. Consequently we can change the order of product, and we have

$$L_{\chi}(s)=\prod_{\mathfrak{p}\in G(\mathfrak{f})}L^{(\mathfrak{p})}(s).$$

9. Putting $u=N\mathfrak{p}^{-s}$, we have

$$\frac{d}{du}\log L^{(\mathfrak{p})}(s)=\sum_{n=1}^{\infty}\Phi(\mathfrak{p},n)u^{n-1}$$

$$=\sum_{n=1}^{\infty}\Big(\prod_{\tau\in H(K)}(\tau\tilde{\chi}(\mathfrak{p})^n-1)\Big)u^{n-1}$$

$$=\sum_{t=0}^{d}(-1)^{d-t}\sum_{i_1\cdots i_t}\sum_{n}(\tau_{i_1}\tilde{\chi}(\mathfrak{p})\cdots\tau_{i_t}\tilde{\chi}(\mathfrak{p}))^n u^{n-1},$$

or

$$L^{(\mathfrak{p})}(s)=\prod_{t=0}^{d}\prod_{i_1\cdots i_t}(1-\tau_{i_1}\tilde{\chi}(\mathfrak{p})\cdots\tau_{i_t}\tilde{\chi}(\mathfrak{p})N\mathfrak{p}^{-s})^{(-1)^{d-t+1}},$$

where $\{\tau_{i_1}, \cdots, \tau_{i_t}\}$ runs over all combinations of all isomorphisms $\tau_1, \cdots, \tau_d$ in $H(K)$.

Put now $\tilde{\chi}_{i_1\cdots i_t}(\mathfrak{a}) = \tau_{i_1}\tilde{\chi}(\mathfrak{a})\cdots\tau_{i_t}\tilde{\chi}(\mathfrak{a})$ for ideals $\mathfrak{a}$ in $G(\mathfrak{f})$, then $\tilde{\chi}_{i_1\cdots i_t}$ is a character of $G(\mathfrak{f})$, associated with the character $\chi_{i_1\cdots i_t} = \chi^{\tau_{i_1}}\cdots\chi^{\tau_{i_t}}$ of C_k of type (A_0). We have thus proved the following theorem:

THEOREM 2. *Let χ be a character of C_k of type* (A_0), *satisfying the additional condition* (I). *Let $L_{\mathfrak{b}}(s)$ be the function defined in* (9). *Then the infinite product of $L_{\mathfrak{b}}(s)$, taken over all integral ideals in K, can be expressed by L-functions with conjugates-product characters $\chi_{i_1\cdots i_t}$ of χ in the following manner*:

$$\prod_{\mathfrak{b}} L_{\mathfrak{b}}(s) = \prod_{t=0}^{d} \prod_{i_1\cdots i_t} L(s, \chi_{i_1\cdots i_t})^{(-1)^{d-t}},$$

where $d = [K:\boldsymbol{Q}]$, $\{i_1, \cdots, i_t\}$ runs over all combinations of $1, \cdots, d$ and $L(s, \chi_{i_1\cdots i_t}) = \sum_{\mathfrak{p}\in G(\mathfrak{f})} (1 - \tilde{\chi}_{i_1\cdots i_t}(\mathfrak{p}) N\mathfrak{p}^{-s})^{-1}$.

Notice that this function $L(s, \chi_{i_1\cdots i_t})$ may also be imprimitive, i.e. the conductor of $\chi_{i_1\cdots i_t}$ may be a proper divisor of $\mathfrak{f}$. For example, when $t=0$, $L(s, 1)$ is the zeta-function of k, from which those factors due to prime divisors of $\mathfrak{f}$ are omitted.

Remark also that, if we denote by $k(\mathfrak{b})$ the compositum of cyclic extensions $k(\chi, \omega)$ for all ω definable modulo $\mathfrak{b}$, then $L_{\mathfrak{b}}(s)$ is the zeta function of $k(\mathfrak{b})$, raised to the power $\varphi(\mathfrak{b})/[k(\mathfrak{b}):k]$ and from which those factors due to prime divisors of $\mathfrak{f}[\mathfrak{b}]$ are omitted. Hence Theorem 2 may be considered, in a sense, to express the decomposition-law of prime ideals of k in $k(\mathfrak{b})$ for infinitely many $k(\mathfrak{b})$, in the lauguage of associated zeta-functions.

§3. Reformulation and Generalization.

10. We use the same notations as before, and consider a character χ of C_k of type (A_0), not necessarily satisfying the condition (I) in 7.

l being a rational prime, the completion K_l of K may be considered as an algebra of degree $d = [K:\boldsymbol{Q}]$ over $\boldsymbol{Q}_l$. Then the representation $\tilde{\chi}$ of $G(\mathfrak{f})$ into K^* determines uniquely a representation χ_l of C_k into the multiplicative group of regular elements in K_l. Clearly, this χ_l is also a representation of $C_k' = C_k/D_k$. Let now $\xi \to R_l(\xi)$ be a regular representation of K_l with respect to a fixed basis of K_l over $\boldsymbol{Q}_l$, which is certainly a continuous mapping from K_l with l-topology into the full linear group of degree d over $\boldsymbol{Q}_l$. If we put $M_l(a) = R_l(\chi_l(a))$, M_l is a representation of C_k into that group. Since $\chi_l(a)$ are l-units, M_l is a *l-adic unimodular* representation, i.e. a representation with l-adic integral coefficients and l-adic unit determinants. M_l being also a representation of C'_k, it induces an l-adic unimodular representation M_l^* of the Galois group G_k. Then the field $k(M, l)$ corresponding to the kernel of M_l^* is the compositum of fields $k(\chi, \mathfrak{l})$ for all prime divisors $\mathfrak{l}$ of l in K.

11. We now propose to characterize the character of C_k of type (A_0) by properties of representation M_l^* thus obtained. Let $\{M_l^*\}$ be a system of l-adic unimodular representations M_l^* of the Galois group $G(\bar{k}/k)$ of $\bar{k}$ over k, with the same degree d for all rational primes l. We shall denote by $k(M, l)$ the subfield of $\bar{k}$ corresponding to the kernel of M_l^*. We shall also denote by $\mathfrak{P}$ any one of prime divisors in $\bar{k}$ of a prime ideal $\mathfrak{p}$ of k, and by $\bar{G}(\mathfrak{a})$ the set of *all* prime divisors $\mathfrak{P}$ of $\mathfrak{p}$ for all $\mathfrak{p}$ in $G(\mathfrak{a})$, $\mathfrak{a}$ being an integral ideal of k. Now, assume that the following conditions (CA'_I)—(CA'_V) are satisfied:

(CA'_I) There is an integral ideal $\mathfrak{m}$ in k with the property that $\mathfrak{P}$ is unramified in $k(M, l)$ for any $\mathfrak{P}$ in $\bar{G}(\mathfrak{m}l)$ and for any l.

This means that the matrix $M_l^*(\sigma_{\mathfrak{P}})$ is independent of the choice of Frobenius automorphism $\sigma_{\mathfrak{P}}$, for each $\mathfrak{P}$ in $\bar{G}(\mathfrak{m}l)$.

We denote by $\boldsymbol{Q}'$ the field consisting of all matrices of the form $rE, r \in \boldsymbol{Q}$, E being the unit matrix.

(CA'_{II}) The matrices $M_l^*(\sigma_{\mathfrak{P}})$ for all $\mathfrak{P}$ in $\bar{G}(\mathfrak{m}l)$ generate over $\boldsymbol{Q}'$ a semi-simple commutative algebra $\mathcal{A}_l$ of finite degree over $\boldsymbol{Q}$. Moreover, the correspondence $M_l^*(\sigma_{\mathfrak{P}}) \leftrightarrows M_q^*(\sigma_{\mathfrak{P}})$ for all $\mathfrak{P}$ in $\bar{G}(\mathfrak{m}lq)$ determines an isomorphism of $\mathcal{A}_l$ onto $\mathcal{A}_q$, for all pair (l, q) of rational primes.

This implies in particular that the characteristic equation of $M_l(\sigma_{\mathfrak{P}})$ for $\mathfrak{P}$ in $\bar{G}(\mathfrak{m}l)$ must have rational coefficients.

(CA'_{III}) The characteristic equation of $M_l(\sigma_{\mathfrak{P}})$ has rational coefficients and is independent of l for all l such that $\mathfrak{P}$ belongs to $\bar{G}(\mathfrak{m}l)$.

We shall denote by $\varpi_1(\mathfrak{p}), \cdots, \varpi_d(\mathfrak{p})$ all the characteristic roots of $M_l^*(\sigma_{\mathfrak{P}})$ (with proper multiplicities) for such l. (These characteristic roots are certainly determined only by $\mathfrak{p}$, not depending on the choice of $\mathfrak{P}$.)

(CA'_{IV}) We have

$$|\varpi_i(\mathfrak{p})| = \mathrm{N}\mathfrak{p}^{\rho} \qquad (i=1, \cdots, d)$$

for all $\mathfrak{p}$ in $G(\mathfrak{m})$, where ρ is a half integer independent of $\mathfrak{p}$ and of i.

(CA'_V) There is a natural number n_0 such that $\varpi_i(\mathfrak{p})\mathrm{N}\mathfrak{p}^{n_0}$ are algebraic integers for all $\mathfrak{p}$ in $G(\mathfrak{m})$ and for all i.

It is trivial to notice here that the system $\{M_l^*\}$ obtained from a character χ of type (A_0) satisfies these conditions, where algebras

$\mathcal{A}_l$ in (CA'_{II}) are isomorphic to the field K_0 generated over $\boldsymbol{Q}$ by all $\tilde{\chi}(\mathfrak{p})$.

In general, we see from (CA'_{II}) that $M_l{}^*(\sigma_{\mathfrak{P}})$ commutes with $M_l{}^*(\sigma_{\mathfrak{P}'})$ for all $\mathfrak{P}, \mathfrak{P}'$ in $\bar{G}(\mathfrak{m}l)$. Since such $\sigma_{\mathfrak{P}}$'s are everywhere dense in $G(\bar{k}/k)$, the image $M_l{}^*(G(\bar{k}/k))$ must be an abelian group, i.e. the field $k(M, l)$ is an abelian extension of k. The system $\{M_l{}^*\}$ can therefore be considered as a system of representations of *abelian group* $G_k = G(A_k/k)$, and we may write $\sigma_{\mathfrak{p}}$ instead of $\sigma_{\mathfrak{P}}$.

Since commutative semi-simple algebra is a direct sum of fields, condition (CA'_{II}) implies also that all matrices $M_l{}^*(\sigma_{\mathfrak{p}})$ can be decomposed (in $\boldsymbol{Q}_l$) simultaneously into direct sum:

$$\begin{pmatrix} M_{l,1}^*(\sigma_{\mathfrak{p}}) & & \\ & \ddots & \\ & & M_{l,t}^*(\sigma_{\mathfrak{p}}) \end{pmatrix}$$

for all $\mathfrak{p} \in G(\mathfrak{m}l)$, in such a way that all $M_{l,i}^*(\sigma_{\mathfrak{p}})$ with a fixed i generate over $\boldsymbol{Q}'$ a field of finite degree over $\boldsymbol{Q}$. Then, $\sigma_{\mathfrak{p}}$ being dense in G_k, all matrices $M_l{}^*(\sigma)$, $\sigma \in G_k$, can be decomposed correspondingly into the direct sum of $M_{l,1}^*(\sigma), \cdots, M_{l,t}^*(\sigma)$, and, for any one of fixed i, the system $\{M_{l,i}^*(\sigma)\}$ satisfies all conditions (CA'_I)—(CA'_V).

It is hence sufficient to consider the case where the algebras $\mathcal{A}_l$ in (CA'_{II}) are isomorphic to an algebraic number field K of finite degree. We can fix here an isomorphism μ_l of $\mathcal{A}_l$ onto K in such a way that $\mu_l(M_l(\sigma_{\mathfrak{p}})) = \mu_q(M_q(\sigma_{\mathfrak{p}}))$ for all $\mathfrak{p}$ in $G(\mathfrak{m}lq)$ and for all pairs (l, q). We shall write this common value $\mu_l(M_l(\sigma_{\mathfrak{p}}))$ by $\tilde{\psi}(\mathfrak{p})$. Clearly, all $M_l{}^*(\sigma_{\mathfrak{p}})$, hence also all $M_l{}^*(\sigma)$, can be transformed (in $\bar{\boldsymbol{Q}}_l$) simultaneously into diagonal forms:

$$M_l{}^*(\sigma) = \begin{pmatrix} \psi_{l,1}^{**}(\sigma) & & \\ & \ddots & \\ & & \psi_{l,d}^{**}(\sigma) \end{pmatrix}.$$

Then, $\tilde{\psi}(\mathfrak{p}) \to \psi_{l,i}^{**}(\sigma_{\mathfrak{p}}) = \varpi_i(\mathfrak{p})$ for $\mathfrak{p}$ in $G(\mathfrak{m}l)$, gives an isomorphism $\mu_{l,i}$ of K into $\bar{\boldsymbol{Q}}_l$, hence it determines also a prime divisor $\mathfrak{l}_i$ of l in k. If we put $\psi_{\mathfrak{l}_i}^* = \mu_{l,i}^{-1} \circ \psi_{l,i}^{**}$, $\psi_{\mathfrak{l}_i}^*$ is a representation of G_k into $K_{\mathfrak{l}_i}^*$. In this way, we obtain a system of representations $\psi_{\mathfrak{l}}^*$ of G_k into $K_{\mathfrak{l}}^*$, which satisfies evidently all conditions (CA_I)—(CA_{IV}) in §1. Hence $\{\psi_{\mathfrak{l}}^*\}$ corresponds to a character χ of C_k of type (A_0). It is also evident that our $\{M_l{}^*\}$ can be obtained from this χ in the manner described in **10.** Thus theorem 1 can be reformulated in the following form:

THEOREM 1′. *Let $\{M_l^*\}$ be a system of l-adic unimodular representations M_l^* of the Galois group $G(\bar{k}/k)$ with the same degree d for all rational primes l. Then $\{M_l^*\}$ satisfies the conditions (CA'_I)—(CA'_V)*

if and only if it is a direct sum of systems $\{M_{l,j}^*\}$ *of representations of the abelian group* G_k, *each obtained from a character* χ_i *of* C_k *of type* (A_0) *in the manner described in* **10.**

12. Let now our character χ of type (A_0) satisfy the condition (I) in **7**, and M_l^* be as is **10**. Consider then the matrix $M_l^*(\sigma_{\mathfrak{q}})$ for prime ideal $\mathfrak{q}$ *dividing* l. Let $\mathfrak{l}_i$ be, as in **11**, the prime divisor of l in K determined by the imbedding of K into $\overline{\boldsymbol{Q}}_l$ given by $\tilde{\chi}(\mathfrak{p})\rightarrow\varpi_i(\mathfrak{p})$ $=\psi_{l,i}^{**}(\sigma_{\mathfrak{p}})$. Then, $\varpi_i(\mathfrak{q})$ is a unit in $\overline{\boldsymbol{Q}}_l$ if and only if $\tilde{\chi}(\mathfrak{q})$ is prime to $\mathfrak{l}_i$. When that is so, $\chi_{\mathfrak{l}_i}$ is unramified at $\mathfrak{q}$, i.e. $\mathfrak{q}$ is unramified in $k(\chi, \mathfrak{l}_i)$, and we have $\chi_{\mathfrak{l}_i}(\sigma_{\mathfrak{q}})=\tilde{\chi}(\mathfrak{q})$, or $\psi_{l,i}^{**}(\sigma_{\mathfrak{q}})=\varpi_i(\mathfrak{q})$ (as was seen in **7**). Changing the numbering for each $\mathfrak{q}$ if necessary, we assume that $\varpi_1(\mathfrak{q}),\cdots,\varpi_r(\mathfrak{q})$ are units and $\varpi_{r+1}(\mathfrak{q}),\cdots,\varpi_d(\mathfrak{q})$ lie in the valuation ideal in $\overline{\boldsymbol{Q}}_l$. Notice that, although we may have $\mathfrak{l}_i=\mathfrak{l}_j$ for some $i\neq j$, the sets $\{\mathfrak{l}_1,\cdots,\mathfrak{l}_r\}$ and $\{\mathfrak{l}_{r+1},\cdots,\mathfrak{l}_d\}$ are mutually disjoint. Hence, for each $\mathfrak{q}$ dividing l, the representation M_l^* is decomposed in $\boldsymbol{Q}_l$ into the direct sum of representations M_l' and M_l'' of respective degrees r, $d-r$, in such a way that the characteristic roots of $M_l'(\sigma_{\mathfrak{q}})$ are exactly $\varpi_1(\mathfrak{q}),\cdots,\varpi_r(\mathfrak{q})$. It is then clear that $\mathfrak{q}$ is unramified in the subfield of A_k corresponding to the kernel of M_l'.

13. Hereafter, until the end of §3, we shall consider a general system $\{M_l^*\}$ of l-adic representations of $G(\bar{k}/k)$ with the same degree d for all l, and use the same notations as in the beginning of **11**. Assume that $\{M_l^*\}$ satisfies the conditions (CA'_I), (CA'_{III}), (CA'_{IV}) with $\rho>0$, (CA'_V) with $n_0=0$, but not necessarily (CA'_{II}). Then all $\varpi_i(\mathfrak{p})$ are algebraic integers different from roots of unity. Assume furthermore the following conditions (B_I), (B_{II}) are satisfied:

$$(B_I) \qquad \prod_{i=1}^{d}\varpi_i(\mathfrak{p})>0 \text{ for all } \mathfrak{p} \text{ in } G(\mathfrak{m}).$$

We may consider that $\varpi_i(\mathfrak{p})$ are contained in $\overline{\boldsymbol{Q}}_l$. Then, for any one of prime divisors $\mathfrak{q}$ in k of l, we change if necessary the numbering of $\varpi_i(\mathfrak{q})$ so that $\varpi_1(\mathfrak{q}),\cdots,\varpi_r(\mathfrak{q})$ are units, while $\varpi_{r+1}(\mathfrak{q}),\cdots,$ $\varpi_d(\mathfrak{q})$ lie in the valuation ideal of $\boldsymbol{Q}_l$, where $r=r(\mathfrak{q})\leqq d$ and $r(\mathfrak{q})$ may depend on $\mathfrak{q}$. Denote by $\mathfrak{Q}$ any prime divisor of $\mathfrak{q}$ in $\bar{k}$. Then our next condition reads:

(B_{II}) Restricted to the decomposition group $G(\mathfrak{Q})$ of $\mathfrak{Q}$ over k, M_l^* can be transformed in $\boldsymbol{Q}_l$ into the form

$$M_l^*(\sigma)=\begin{pmatrix} M_l'(\sigma) & 0 \\ * & M_l''(\sigma)\end{pmatrix}$$

simultaneously for all σ in $G(\mathfrak{Q})$, where M_l', M_l'' have respective degrees r, $d-r$. Moreover, if we denote by $k(M_l')$ the subfield of $k(M, l)$ corresponding to the kernel of $M_l'(\sigma)$, $\mathfrak{q}$ is unramified in

$k(M_l')$, and the characteristic roots of $M_l'(\sigma_{\mathfrak{Q}})$ are exactly $\varpi_1(\mathfrak{q}),\cdots,\varpi_r(\mathfrak{q})$.

Clearly the condition (B_{II}) is or is not satisfied irrespective of the choice of prime divisors $\mathfrak{Q}$ of $\mathfrak{q}$. As was seen in **12**, the system $\{M_l^*\}$ obtained from a χ satisfying (I) sasisfies also these conditions (B_I), (B_{II}). Our aim is now to generalize the result in §2 to our system $\{M_l^*\}$.

We shall first consider the group $U_l(d)$ of all l-adic unimodular matrices of degree d, and the direct product $U(d)=\prod_l U_l(d)$ for all l. $b=l_1^{c_1}\cdots l_t^{c_t}$ being any natural number, we shall denote by $U^{(b)}(d)$ the subgroup of $U(d)$ consisting of all $M=(M_l)$ such that $M_{l_i}\equiv E$ mod. $l_i^{c_i}$ $(i=1,\cdots,t)$. Then the set $\{U^{(b)}(d)\}$ $(b=1,2,\cdots)$ forms a fundamental system of neighbourhoods of the identity in $U(d)$, so that any representation of $U(d)$ into full linear group over $\boldsymbol{C}$ is essentially a representation of the factor group $\mathfrak{U}_{(b)}=U(d)/U^{(b)}(d)$. We shall denote by λ_b the natural homomorphism of $U(d)$ onto $\mathfrak{U}_{(b)}$. Notice that $\mathfrak{U}_{(b)}$ may be considered as the group of all matrices of degree d over the residue ring $\boldsymbol{Z}/b\boldsymbol{Z}$ of rational integers modulo b, with determinants having inverses in the ring $\boldsymbol{Z}/b\boldsymbol{Z}$.

Denote by $\mathfrak{L}_b$ a vector space of dimension d over $\boldsymbol{Z}/b\boldsymbol{Z}$, then $\mathfrak{U}_{(b)}$ can be considered as a transformation group of $\mathfrak{L}_b$. We shall call a vector in $\mathfrak{L}_b$ *proper* when the ideal (in $\boldsymbol{Z}/b\boldsymbol{Z}$) generated by all components of it is equal to whole $\boldsymbol{Z}/b\boldsymbol{Z}$. Let now $\mathfrak{x}_0$ be any fixed one of proper vectors in $\mathfrak{L}_b$. Clearly, $\mathfrak{U}_{(b)}$ transforms $\mathfrak{x}_0$ into proper vectors, and all proper vectors are obtained from $\mathfrak{x}_0$ in this way. Therefore, if we denote by $\mathfrak{B}_{(b)}$ the subgroup of all elements in $\mathfrak{U}_{(b)}$, transforming $\mathfrak{x}_0$ into itself, the cosets of $\mathfrak{U}_{(b)}$ modulo $\mathfrak{B}_{(b)}$ correspond one-to-one to all proper vectors in $\mathfrak{L}_b$. Let now $\tilde{D}_b$ be the representation of $\mathfrak{U}_{(b)}$, "induced" (in the sense of the theory of group characters) by the unit representation (i.e. all values are 1) of $\mathfrak{B}_{(b)}$, and $\tilde{\delta}_b$ the character of $\tilde{D}_b$, i.e. the characters induced by the unit character of $\mathfrak{B}_{(b)}$. Clearly, $\tilde{\delta}_b$ is a non-negative valued rational character independent of the choice of proper $\mathfrak{x}_0$. Moreover, it is evident that $\tilde{D}_b$ is the representation of $\mathfrak{U}_{(b)}$ as a permutation group of all proper vectors in $\mathfrak{L}_b$. We shall put here $D_b=\tilde{D}_b\circ\lambda_b$, $\delta_b=\tilde{\delta}_b\circ\lambda_b$, which are respectively representations and character of $U(d)$.

For the later use, we must compute the sum $\sum_{j=0}^{\infty}\delta_{l^j}(M)$ for $M=(M_q)$ in $U(d)$, such that no characteristic root of M_q is 1. Then, for a suitably large i, $\lambda_{l^i}(M)$ fixes no proper vectors in $\mathfrak{L}_{l^i}$, so that $\delta_{l^i}(M)=0$. Clearly $\delta_{l^j}(M)=0$ for all $j\geqq i$. Now the set of all vectors in $\mathfrak{L}_{l^i}$, whose all components are divisible by $l^j (j\leqq i)$ can be identified with $\mathfrak{L}_{l^{i-j}}$, hence $\mathfrak{L}_{l^i}$ may be considered, under this identification, as a direct union of all proper vectors in $\mathfrak{L}_{l^j}$ for $j=0, 1, \cdots, i$. The sum $\sum_{j=1}^{\infty} \delta_{l^j}(M)=\sum_{j=1}^{i} \delta_{l^j}(M)$ is then equal to the number of all vectors in $\mathfrak{L}_{l^i}$, left invariant by the transformation $\lambda_{l^i}(M)$, hence *it is equal to the highest power of l dividing* $\det(M_l-E)$.

14. We shall now come back to our l-adic representation M_l^* of $G(\bar{k}/k)$. Putting $M^*(\sigma)=(M_l^*(\sigma))$ for $\sigma \in G(\bar{k}/k)$, we obtain a representation M^* of $G(\bar{k}/k)$ into $U(d)$. Then, $\lambda_b \circ M^*$ is a representation of $G(\bar{k}/k)$ into $\mathfrak{U}_{(b)}$, and the field $k(M, \lambda_b)$ corresponding to the kernel of $\lambda_0 \circ M^*$ is a finite normal extension of k.

For prime ideal $\mathfrak{p}$ of k in $G(\mathfrak{m})$, *prime to* b, we put

$$\psi_b(\mathfrak{p}^n)=\delta_b(M^*(\sigma_{\mathfrak{P}}^n)),$$

the right hand side being clearly independent of the choice of $\sigma_{\mathfrak{P}}$, and also of prime divisors $\mathfrak{P}$ of $\mathfrak{p}$ in $\bar{k}$. This ψ_b coincides on $G(\mathfrak{m}b)$ with the ideal character of k associated with the character $\tilde{\delta}_b$ of the Galois group $G(k(M, \lambda_b)/k)$ in the sense of Artin's theory of L-functions.

When $\mathfrak{p}$ divides b, we define $\psi_b(\mathfrak{p}^n)$ in a little different manner. Put $b=p^c \cdot b_0=p^c l_1^{c_1} \cdots l_t^{c_t}$, where $l_i \neq p$. Let U' be the direct product $U_p(r) \times \prod_{l \neq p} U_l(d)$, where $r=r(\mathfrak{p})$ is as in (B_{II}). Then define $U'^{(b)}$ as the subgroup consisting of all $M=(M_p', M_l)$ of U' satisfying $M_{l_i} \equiv E$ mod. $l_i^{c_i}$ and $M_p' \equiv E$ mod. p^c. The factor group $\mathfrak{U}_{(b)}'=U'/U'^{(b)}$ can be considered as a transformation group of the space $\mathfrak{L}_b'=\mathfrak{L}_{b_0}+\mathfrak{L}_{p^c}'$, where $\mathfrak{L}_{b_0}$ is as before and $\mathfrak{L}_{p^c}'$ is a vector space of dimension r over $\boldsymbol{Z}/p^c\boldsymbol{Z}$. Now a representation $\tilde{D}_b'$ and a character $\tilde{\delta}_b'$ of $\mathfrak{U}_{(b)}'$ and a character δ_b' of U' are defined in the same way as $\tilde{D}_b$, $\tilde{\delta}_b$, δ_b. Then, for σ in $G(\mathfrak{P})$, we put $M'(\sigma)=(M_p'(\sigma), M_l^*(\sigma)) \in U'$, where $M_p'(\sigma)$ is a matrix defined in (B_{II}). Here, we put

$$\psi_b(\mathfrak{p}^n)=\delta_b'(M'(\sigma_{\mathfrak{P}}^n))$$

for $\mathfrak{p}$ in $G(\mathfrak{m})$ *dividing* b, the right hand side being independent of the choice of $\sigma_{\mathfrak{P}}$ and of $\mathfrak{P}$ by the condition (B_{II}).

Notice that, if b, b' are coprime natural numbers, then the space $\mathfrak{L}_{bb'}$ is isomorphic to the direct sum $\mathfrak{L}_b+\mathfrak{L}_{b'}$, and the representation $D_{bb'}$ is equivalent to the tensor product $D_b \otimes D_{b'}$, hence we have $\delta_{bb'}=\delta_b\delta_{b'}$. The same is also true for $\delta_{bb'}'$ for any $\mathfrak{p}$ dividing b or b'. This shows that $\psi_{bb'}(\mathfrak{p}^n)=\psi_b(\mathfrak{p}^n)\psi_{b'}(\mathfrak{p}^n)$ for any $\mathfrak{p}$ in $G(\mathfrak{m})$. From this we see $\sum_{b=1}^{\infty} \psi_b(\mathfrak{p}^n)=\prod_l (\sum_{i=0}^{\infty} \psi_{l^i}(\mathfrak{p}^n))$. On the other hand, from our first assumption, no characteristic root $\varpi_i(\mathfrak{p})$ of $M_l^*(\sigma_{\mathfrak{P}})$, and of $M_l'(\sigma_{\mathfrak{P}})$, is root of unity. Hence, from the remark at the end of **13**, $\sum_{i=1}^{\infty} \psi_{l^i}(\mathfrak{p}^n)$

is the highest power of l dividing $\det(M_l^*(\sigma_{\mathfrak{P}}^n)-E)=\prod_{i=1}^{d}(\varpi_i(\mathfrak{p})^n-1)$ or $\det(M_l'(\sigma_{\mathfrak{P}}^n)-E)=\prod_{i=1}^{r}(\varpi_i(\mathfrak{p})^n-1)$ according as $\mathfrak{p}$ is prime to l or not. But we have $\prod_{i=r+1}^{d}(\varpi_i(\mathfrak{p})^n-1)\equiv(-1)^{d-r} \bmod.\ \mathfrak{p}$ from (B_{II}), so that the highest power of p dividing $\prod_{i=1}^{r}(\varpi_i(\mathfrak{p})^n-1)$ is equal to that dividing $\prod^{d}(\varpi_i(\mathfrak{p})^n-1)$; we have therefore

$$\sum_{\mathfrak{b}=1}^{\infty}\psi_{\mathfrak{b}}(\mathfrak{p}^n)=\left|\prod_{i=1}^{d}(\varpi_i(\mathfrak{p})^n-1)\right|=\prod_{i=1}^{d}(\varpi_i(\mathfrak{p})^n-1)$$

for all $\mathfrak{p}$ in $G(\mathfrak{m})$; here the second equality follows from (B_I), as $|\varpi_i(\mathfrak{p})|=N\mathfrak{p}^{\rho}>1$.

15. With our "character" $\psi_{\mathfrak{b}}$, we define a "L-function" $L_{\mathfrak{b}}(s)$ as follows:

$$\log L_{\mathfrak{b}}(s)=\sum_{\mathfrak{p}\in G(m)}\sum_{n=1}^{\infty}\psi_{\mathfrak{b}}(\mathfrak{p}^n)n^{-1}\mathrm{N}\mathfrak{p}^{-ns}. \tag{12}$$

Observe that this $L_{\mathfrak{b}}(s)$ is different from Artin's L-series attached to the group character $\tilde{\delta}_{\mathfrak{b}}$ only by components due to prime factors of $\mathfrak{m}\mathfrak{b}$. Then, we form again an infinite product

$$L_M(s)=\prod_{\mathfrak{b}=1}^{\infty}L_{\mathfrak{b}}(s), \tag{13}$$

and auxiliary series for $\mathfrak{p}$ in $G(\mathfrak{m})$,

$$\log L^{(\mathfrak{p})}(s)=\sum_{n=1}^{\infty}\sum_{\mathfrak{b}=1}^{\infty}\psi_{\mathfrak{b}}(\mathfrak{p}^n)n^{-1}\mathrm{N}\mathfrak{p}^{-ns}=\sum_{n=1}^{\infty}\Big(\prod_{i=1}^{d}(\varpi_i(\mathfrak{p})^n-1)\Big)n^{-1}\mathrm{N}\mathfrak{p}^{-ns}.$$

In the same way as in §2, we see that the series

$$\sum_{\mathfrak{p}\in G(m)}\sum_{n=1}^{\infty}\sum_{\mathfrak{b}=1}^{\infty}\psi_{\mathfrak{b}}(\mathfrak{p}^n)n^{-1}\mathrm{N}\mathfrak{p}^{-ns}$$

converges absolutely in some right half s-plane, and we have

$$L_M(s)=\prod_{\mathfrak{p}\in G(m)}L^{(\mathfrak{p})}(s).$$

Again by the same calculation as in §2, we obtain a similar result:

THEOREM 3. *Let $\{M_l^*\}$ be a system of representations of $G(\bar{k}/k)$ with the same degree d for all l, satisfying the conditions (CA'_I), (CA'_{III}), (CA'_{IV}) with $\rho>0$, (CA'_V) with $n_0=0$, (B_I) and (B_{II}). Let $L_{\mathfrak{b}}(s)$ be the functions defined in* (12) *with $\psi_{\mathfrak{b}}$ obtained from M_l^* in the above exposed manner. Then we have the following relation:*

$$\prod_{\mathfrak{b}=1}^{\infty}L_{\mathfrak{b}}(s)=\prod_{t=0}^{d}\prod_{i_1\cdots i_t}L(s,\varpi_{i_1\cdots i_t})^{(-1)^{d-t}}$$

between these $L_{\mathfrak{b}}(s)$ and "L-functions" $L(s,\varpi_{i_1\cdots i_t})$ defined by

$$L(s,\varpi_{i_1\cdots i_t})=\prod_{\mathfrak{p}\in G(m)}(1-\varpi_{i_1}(\mathfrak{p})\cdots\varpi_{i_t}(\mathfrak{p})\mathrm{N}\mathfrak{p}^{-s})^{-1},$$

where $\varpi_1(\mathfrak{p}),\cdots,\varpi_d(\mathfrak{p})$ are all the characteristic roots of $M_l^*(\sigma_{\mathfrak{P}})$.

Observe that, when M_l^* is obtained from a character χ of type

(A_0) satisfying (I), we have

$$\prod_{\mathfrak{b}|a} L_{\mathfrak{b}}(s)=\prod_{b|a} L_b(s)$$

for any natural number a, where $L_{\mathfrak{b}}(s)$ are functions defined in (9) §2. (This is an immediate consequence of the above considerations.) Hence Theorem 3 is indeed a generalization of Theorem 2.

16. As the reader may have noticed, Theorem 3 still holds if we weaken the last condition in (B_{II}) (on characteristic roots of $M'(\sigma_{\mathfrak{Q}})$) to the following one: For each n, the highest power of l dividing $\det(M_l'(\sigma_{\mathfrak{Q}}^n)-E)$ and that dividing $\prod_{i=1}^{d}(\varpi_i(\mathfrak{q})^n-1)$ are equal, where the degree of M_l' may be arbitrary. But this generalization is somewhat an apparent one, as may be seen from the following lemma:

LEMMA. Let $P(X)=X^r+\sum_{i=1}^{r}a_iX^{r-i}$, $Q(X)=X^s+\sum_{j=1}^{s}b_jX^{s-j}$ be two polynominals in $\boldsymbol{Q}_p[X]$ with p-adic integral coefficients a_i, b_j, and let $\{\omega_1,\cdots,\omega_r\}$, $\{\eta_1,\cdots,\eta_s\}$ be respectively all the roots of $P(X)=0$, $Q(X)=0$ in $\bar{\boldsymbol{Q}}_p$ (with proper multiplicities). Let $P(X)=P_0(X)P_1(X)$, $Q(X)=Q_0(X)Q_1(X)$ be decompositions of $P(X)$, $Q(X)$ in $\boldsymbol{Q}_p[X]$ into products of polynominals $P_i(X)$, $Q_i(X)$, with highest coefficients 1 $(i=0,1)$, such that all roots of $P_0(X)$, $Q_0(X)$ are units, while all roots of $P_1(X)$, $Q_1(X)$ lie in the valuation ideal, in $\bar{\boldsymbol{Q}}_p$. Denote by $\|\xi\|$ the normalized valuation of ξ in $\bar{\boldsymbol{Q}}_p$. Let finally a finite number of elements $\alpha_1,\cdots,\alpha_m$ of $\bar{\boldsymbol{Q}}_p$ be given. Now, assume that the relation $\|\prod_i F(\omega_i)\|=\|\prod_j F(\eta_j)\|$ holds for any polynominal $F(T)=\sum_{i=0}^{t}c_iT^{t-i}$ with rational integral coefficients c_i such that $\|c_0\|=1$, $\|c_t\|=1$, satisfying $F(\alpha_i)\neq 0$ for $i=1,\cdots,m$. Then we have $P_0(X)=Q_0(X)$. On the other hand, if the relation $\|\prod_i F(\omega_i)\|\leqq\|\prod_j F(\eta_j)\|$ holds for any $F(T)$ with rational integral coefficients, satisfying $F(\alpha_i)\neq 0\,(i=1,\cdots,m)$, then $Q(X)$ divides $P(X)$.

This lemma is a generalization of lemma 12, n°68 in Weil's book [8]. We shall give a proof of the first part. The second part is proved similarly.

PROOF. $F(T)$ being as above, put $A(F)=\prod_i F(\omega_i)$, $B(F)=\prod_j F(\eta_j)$. Then, $\|A(F)\|$ and $\|B(F)\|$ are continuous functions of coefficients $c_0,\cdots,c_t$ of $F(T)$ with respect to the p-adic topology of $\boldsymbol{Q}$. Let $G(T)=\sum_{i=0}^{t}d_iT^{t-i}$ be a polynomial in $\boldsymbol{Z}[T]$ such that $\|d_0\|=\|d_t\|=1$, and $G(\alpha_i)=0$ for some α_i. Then, take $F(T)$ as above, and put $G_n(T)=G(T)+p^nF(T)$, which is certainly in $\boldsymbol{Z}[T]$ and $\|d_0+p^nc_0\|=\|d_t+$

$p^n c_t\|=1$ if $n\geqq 1$. Moreover, $G_n(\alpha_j)\neq 0\ (j=1,\cdots,m)$ if we take n large enough, as $F(\alpha_j)\neq 0$ by assumption. Thus we have $\|A(G_n)\|=\|B(G_n)\|$ for large n, and, since $G_n(T)$ converges to $G(T)$ as $n\to\infty$ (in p-topology), we have $\|A(G)\|=\|B(G)\|$. Hence we can drop the condition that $F(\alpha_j)\neq 0$. Now, it is clear that $\|A(F)\|=\|B(F)\|$ holds for all $F(T)$ with rational coefficients $c_0,\cdots,c_t$, whose denominators are prime to p, such that $\|c_0\|=1$, $\|c_t\|=1$. Since such rational numbers are everywhere dense in the valuation ring $\boldsymbol{O}_p$ of $\boldsymbol{Q}_p$, and since rational numbers c such that $\|c\|=1$ are everywhere dense in the unit group of $\boldsymbol{Q}_p$, we have $\|A(F)\|=\|B(F)\|$ for all $F(T)$ in $\boldsymbol{O}_p[T]$ such that $\|c_0\|=\|c_t\|=1$. Then, let β be any unit in $\overline{\boldsymbol{Q}}_p$ with degree t over $\boldsymbol{Q}_p$, and $F(T)=T^t+\sum_{i=1}^{t}c_iT^{t-i}$ be the irreducible polynominal of β over $\boldsymbol{Q}_p$. Then $F(T)\in\boldsymbol{O}_p[T]$ and $\|c_t\|=1$. If we put $\varphi(\beta)=\|\prod_i(\beta-\omega_i)\|$, $\psi(\beta)=\|\prod_j(\beta-\eta_j)\|$, we have $\|A(F)\|=\varphi(\beta)^t$, $\|B(F)\|=\psi(\beta)^t$, so that from the relation $\|A(F)\|=\|B(F)\|$ follows $\varphi(\beta)=\psi(\beta)$. Let now α be any one of ω_i, or η_j, which is a unit in $\overline{\boldsymbol{Q}}_p$. Let d, e be the multiplicities of α in $\{\omega_1,\cdots,\omega_r\}$, $\{\eta_1,\cdots,\eta_s\}$ respectively. Put then $\lambda=\prod(\alpha-\omega_i)$, $\mu=\prod(\alpha-\eta_j)$, where products are taken for $\omega_i\neq\alpha$, and $\eta_j\neq\alpha$, with their proper multiplicities. Choose β so that $\beta\equiv\alpha$ mod. p^n with sufficiently large n, then β is a unit in $\overline{\boldsymbol{Q}}_p$. If n is large enough, we have $\|\alpha-\omega_i\|=\|\beta-\omega_i\|$ and $\|\alpha-\eta_j\|=\|\beta-\eta_j\|$ for $\omega_i\neq\alpha$, $\eta_j\neq\alpha$. Hence we see $\varphi(\beta)=\|\lambda\|\ \|\beta-\alpha\|^d$ and $\psi(\beta)=\|\mu\|\ \|\beta-\alpha\|^e$, so that $\|\beta-\alpha\|^{d-e}=\|\mu/\lambda\|$. Since λ, μ are independent of n and of β, and since we can take β, so that $\|\beta-\alpha\|$ becomes arbitrarily small, we must have $d=e$. As this holds for any root α of $P_0(X)=0$, or $Q_0(X)=0$, the first part of the lemma is proved.—

To verify the last assertion in the condition (B_{II}), it is sufficient from this lemma to show the relation $\|\prod_{i=1}^{d}F(\varpi_i(\mathfrak{q}))\|=\|\det F(M_l'(\sigma_{\mathfrak{Q}}))\|$ for all polynominals $F(T)=\sum_{i=0}^{t}c_iT^{t-i}$ in $\boldsymbol{Z}[T]$ such that $\|c_0\|=\|c_t\|=1$, satisfying $F(\alpha_i)\neq 0$ for a finite number of elements $\alpha_1,\cdots,\alpha_m$ in $\overline{\boldsymbol{Q}}_l$. Because, as $M_l'(\sigma_{\mathfrak{Q}})$ is l-adic unimodular, all the characteristic roots of it are units in $\overline{\boldsymbol{Q}}_l$, hence if this relation holds, they coincide with those part of $\{\varpi_1(\mathfrak{q}),\cdots,\varpi_d(\mathfrak{q})\}$ which are units in $\overline{\boldsymbol{Q}}_l$, taking multiplicities into account.

§4. Application to abelian varieties. The conjecture of Hasse.

17. Let A be an abelian variety of dimension n defined over a field κ with characteristic p (may be 0 or a prime number). If $\mathfrak{g}$ is any set of points on A, we shall understand by *the field generated* *by* $\mathfrak{g}$ *over* κ, the smallest field in $\bar{\kappa}$ containing κ, over which all points in $\mathfrak{g}$ are rational. We shall denotes by $\mathrm{g}(m;\mathrm{A})$ the group of all points on A, whose orders divide m. If m is prime to p, $\mathrm{g}(m;\mathrm{A})$ have exactly m^{2n} points, while $\mathrm{g}(p^i;\mathrm{A})$ have p^{ir} points, where r is an integer independent of i, and we have $0\leqq r\leqq n$. Hence the field generated by $\mathrm{g}(m;\mathrm{A})$ over κ is a finite algebraic extension of κ. We shall denote furthermore by $\mathfrak{g}(l;\mathrm{A})$ the group of all points on A, whose orders are some powers of a rational prime l, i.e. $\mathfrak{g}(l;\mathrm{A})=\bigcup_{i=1}^{\infty}\mathrm{g}(l^i;\mathrm{A})$. Then $\mathfrak{g}(l;\mathrm{A})$ is isomorphic to a direct sum of $2n$ or r additive groups $(\boldsymbol{Q}/\boldsymbol{Z})_l$ of l-adic numbers modulo 1, according as $l\neq p$ or $l=p$. When we fix an isomorphism of $\mathfrak{g}(l;\mathrm{A})$ onto the direct sum of $(\boldsymbol{Q}/\boldsymbol{Z})_l$, we shall speak of "$l$-adic coordinates" of the group $\mathfrak{g}(l;\mathrm{A})$. Since any automorphism σ in $G(\bar{\kappa}/\kappa)$ permutes points of $\mathrm{g}(m;\mathrm{A})$ among themselves, σ induces an automorphism of the group $\mathfrak{g}(l;\mathrm{A})$. Hence the group $G(\bar{\kappa}/\kappa)$ can be represented, as a transformation group of $\mathfrak{g}(l;\mathrm{A})$, with l-adic coordinates of it. We shall denote by $M_l^*(\sigma)$ this representation matrix, which is clearly l-adic unimodular, and of degree $2n$ or r according as $l\neq p$ or $l=p$. It should be noticed that this representation M_l^* is certainly continuous, that is to say for any natural number i, there is a finite algebraic extension κ' of κ such that $M_l^*(\sigma)\equiv E \bmod. l^i$ for all σ in $G(\bar{\kappa}/\kappa')$. (We have only to take as κ' the field generated by $\mathrm{g}(l^i;\mathrm{A})$ over κ.) Denote then by $\kappa(\mathrm{A},l)$ the subfield of $\bar{\kappa}$ corresponding to the kernel of M_l^*. This $\kappa(\mathrm{A},l)$ is clearly equal to the field generated by $\mathfrak{g}(l;\mathrm{A})$ over κ. We shall denote finally by $\kappa(\mathrm{A})$ the compositum of all fields $\kappa(\mathrm{A},l)$.

Let us recall here some properties of endomorphisms of abelian varieties. (cf. Weil [8]).

$\mathcal{A}(\mathrm{A})$ denotes as usual the ring of endomorphisms of A, and $\mathcal{A}_0(\mathrm{A})$ denotes the tensor product $\mathcal{A}(\mathrm{A})\otimes\boldsymbol{Q}$. $\mathcal{A}_0(\mathrm{A})$ is a semi-simple algebra of degree at most $4n^2$ over $\boldsymbol{Q}$, and $\mathcal{A}(\mathrm{A})$ is an order of $\mathcal{A}_0(\mathrm{A})$. Moreover, if $\mathcal{A}_0(\mathrm{A})$ contains a commutative semi-simple algebra C of degree $2n$ over $\boldsymbol{Q}$, the commutor of C in $\mathcal{A}_0(\mathrm{A})$ coincides with C itself (cf. Weil [5] p. 12). Since any endomorphism μ of A induces an endomorphism of the group $\mathfrak{g}(l;\mathrm{A})$, $\mathcal{A}(\mathrm{A})$ can be represented with l-adic coordinates of $\mathfrak{g}(l;\mathrm{A})$, and the representation matrix $M_l(\mu)$ is an l-adic integral matrix of degree $2n$ or r according as $l\neq p$ or $l=p$. This representation M_l is faithfull for any $l\neq p$. Moreover, for $l\neq p$, the characteristic equation of $M_l(\mu)$ has rational integral coefficients, and is independent of l. When μ is defined over κ, we put $\nu(\mu)=[\kappa(x):\kappa(\mu x)]$ if this degree is finite, and put $\nu(\mu)=0$ in other case, where x denotes a generic point of A over κ. Then we have $\nu(\mu)=\deg M_l(\mu)$ for $l \neq p$. Moreover, $\nu(\mu)\neq 0$ if and only if the kernel of μ is a finite group. We call μ *separable* if the field $\kappa(x)$ is separable over $\kappa(\mu x)$. If $\nu(\mu)\neq 0$ and μ is separable, $\nu(\mu)$ is equal to the number of points b in the kernel of μ, hence the highest power of l dividing $\nu(\mu)$ is exactly equal to the number of points b in $\mathfrak{g}(l; \mathrm{A})$ such that $\mu b=0$. But this number of points is equal to the highest power of l dividing $\det M_l(\mu)$, as the definition of M_l shows. Hence if μ is separable and $\nu(\mu)\neq 0$, the highest power of l dividing $\nu(\mu)$ and that dividing $\det M_l(\mu)$ are the same also for $l=p$. Finally, we can extend this representation M_l to the algebra $\mathcal{A}_0(\mathrm{A})$ in the obvious manner.

When κ is the finite field of p^f elements, the mapping $\xi\rightarrow\xi^{p^f}$ is an automorphism of the universal domain, which leaves all elements in κ invariant. Hence this automorphism determines an endomorphism π_{A} of A, defined over κ. From the definition, $\kappa(x)$ is purely inseparable of degree p^{fn} over $\kappa(\pi_{\mathrm{A}}x)$. We see furthermore that, for any μ in $\mathcal{A}(\mathrm{A})$, μ is separable if and only if μ is prime to π_{A}, i.e. the left ideal $\mathcal{A}(\mathrm{A})\mu+\mathcal{A}(\mathrm{A})\pi_{\mathrm{A}}$ in $\mathcal{A}(\mathrm{A})$, generated by μ and π_{A}, is equal to $\mathcal{A}(\mathrm{A})$. Notice that, if any endomorphism μ is defined over κ, we have $\mu\pi_{\mathrm{A}}=\pi_{\mathrm{A}}\mu$. Let now $\varpi_1(\mathrm{A}),\cdots,\varpi_{2n}(\mathrm{A})$ be the characteristic roots of the matrix $M_l(\pi_{\mathrm{A}})$ for $l\neq p$, then the zeta-function $Z_{\mathrm{A}}(u)$ of A over κ is of the form:

$$Z_{\mathrm{A}}(u)=\prod_{t=0}^{2n}\prod_{i_1\cdots i_t}(1-\varpi_{i_1}(\mathrm{A})\cdots\varpi_{i_t}(\mathrm{A})u)^{(-1)^{t+1}}$$

(cf. Taniyama [3]).

18. When the field κ is an algebraic number field k of finite degree, the system of representations M_l^* (for all l) of the group $G(\bar{k}/k)$ defined as above satisfies the conditions stated in Theorem 3, as we shall show in the following.

First we shall consider Frobenius automorphisms $\sigma_{\mathfrak{P}}$ over k of prime divisors $\mathfrak{P}$ in $\bar{k}$. For this purpose, we use the reduction of A modulo $\mathfrak{p}$, $\mathfrak{p}$ denoting the prime ideal in k divisible by $\mathfrak{P}$ (cf. Shimura [2], Taniyama [3]). Then, for almost all $\mathfrak{p}$, the variety $\mathrm{A}(\mathfrak{p})$ obtained from A by the reduction modulo $\mathfrak{p}$ is also an abelian variety, defined over the residue field $k(\mathfrak{p})$ of $p^f=N\mathfrak{p}$ elements. In this case, $\mathfrak{p}$ is said to be *non-exceptional* for A. We can extend the process of reduction mod. $\mathfrak{p}$ to that of reduction mod. $\mathfrak{P}$. If $\mathfrak{p}$ is non-exceptional, this latter process of reduction mod. $\mathfrak{P}$ induces a homomorphism of the group $\mathfrak{g}(l; \mathrm{A})$ *onto* $\mathfrak{g}(l; \mathrm{A}(\mathfrak{p}))$, which is an isomorphism for $l\neq p$. It induces also an isomorphism of the ring $\mathcal{A}(\mathrm{A})$ into $\mathcal{A}(\mathrm{A}(\mathfrak{p}))$. Notice that, if we take another prime divisor $\mathfrak{P}'$ of $\mathfrak{p}$, these isomorphisms of $\mathfrak{g}(l; \mathrm{A})$ or of $\mathcal{A}(\mathrm{A})$ will be thereby altered in general, unless all points or endomorphisms in question are defined over k. We shall fix here the l-adic coordinates of $\mathfrak{g}(l; A)$ and of $\mathfrak{g}(l; A(\mathfrak{p}))$, by which our representations M_l^*, M_l are defined. Now, since any $\sigma_{\mathfrak{P}}$ induces on the residue field $\bar{k}(\mathfrak{P})$ the automorphism $\xi \to \xi^{pf}$ (over $k(\mathfrak{p})$), we see $M_l^*(\sigma_{\mathfrak{P}}) = M_{l,\mathfrak{P}}^{-1} M_l(\pi_{A(\mathfrak{p})}) M_{l,\mathfrak{P}}$ for any $l \neq p$, where $M_{l,\mathfrak{P}}$ denotes the transformation matrix between the original l-adic coordinates of $\mathfrak{g}(l, A(\mathfrak{p}))$ and that of $\mathfrak{g}(l; A(\mathfrak{p}))$ induced by the l-adic coordinates of $\mathfrak{g}(l; A)$ by the reduction mod. $\mathfrak{P}$. Here notice that $\pi_{A(\mathfrak{p})}$ is determined uniquely by $\mathfrak{p}$, and $M_{l,\mathfrak{P}}$ is determined uniquely by $\mathfrak{P}$, so that $M_l^*(\sigma_{\mathfrak{P}})$ does not depend on the choice of $\sigma_{\mathfrak{P}}$. This means that $\mathfrak{p}$ is unramified in the field $k(A; l)$ for any $l \neq p$, if $\mathfrak{p}$ is non-exceptional for A. Hereafter, we shall write $\pi_{\mathfrak{p}}, \varpi_i(\mathfrak{p})$ instead of $\pi_{A(\mathfrak{p})}, \varpi_i(A(\mathfrak{p}))$.

Denote by $\mathfrak{m}$ the product of all prime ideals $\mathfrak{p}$ in k, which are *not* non-exceptional for A. As such $\mathfrak{p}$ are finite in number, $\mathfrak{m}$ is certainly an integral ideal of k. Then, we have seen that, for any $\mathfrak{p}$ in $G(\mathfrak{m}l)$, $\mathfrak{p}$ is unramified in $k(A; l)$, i.e. the condition (CA'_I) is satisfied in our case. As was recalled above, the characteristic equation of $M_l(\pi_{\mathfrak{p}})$, hence also that of $M_l^*(\sigma_{\mathfrak{P}})$, have rational integral coefficients independent of $l \neq p$, which proves the condition (CA'_{III}). Now the so called Riemann hypothesis for the congruence zeta function of a curve (Weil [7]) shows that we have $|\varpi_i(\mathfrak{p})| = p^{f/2} = N\mathfrak{p}^{1/2}$ for $i = 1, \cdots, 2n$ (cf. Taniyama [3], §3). This implies the condition (CA'_{IV}) with $\rho = 1/2 > 0$. Recall also that $\varpi_i(\mathfrak{p})$ are algebraic integers, from which the condition (CA'_V) follows with $n_0 = 0$. Recall moreover that $\prod_i \varpi_i(\mathfrak{p}) = \det M_l(\pi_{\mathfrak{p}}) = \nu(\pi_{\mathfrak{p}}) > 0$, which is nothing but the condition (B_I). We have therefore only to verify the condition (B_{II}).

We must therefore consider $M_p^*(\sigma)$ for σ in the decomposition group $G(\mathfrak{P})$ of $\mathfrak{P}$ over k. Recall that the group $g(p^i; A)$ has exactly p^{2in} elements, and the group $g(p^i; A(\mathfrak{p}))$ has p^{ir} elements, where $r = r(\mathfrak{p})$ is independent of i. Moreover, by the reduction modulo $\mathfrak{P}$, $g(p^i; A)$ is mapped onto $g(p^i; A(\mathfrak{p}))$ for all i. Hence, if $\mathfrak{g}_{\mathfrak{P}}(p; A)$ denotes the kernel of the homomorphism of $\mathfrak{g}(p; A)$ onto $\mathfrak{g}(p; A(\mathfrak{p}))$ determined by the reduction mod. $\mathfrak{P}$, $\mathfrak{g}_{\mathfrak{P}}(p; A)$ is isomorphic to the direct sum of $2n - r$ groups $(\boldsymbol{Q}/\boldsymbol{Z})_p$, and is a direct component of $\mathfrak{g}(p; A)$. In other words, there is a subgroup $\mathfrak{g}'_{\mathfrak{P}}(p; A)$ of $\mathfrak{g}(p; A)$, mapped isomorphically onto $\mathfrak{g}(p; A(\mathfrak{p}))$ by the reduction mod. $\mathfrak{P}$, such that $\mathfrak{g}(p; A) = \mathfrak{g}'_{\mathfrak{P}}(p; A) + \mathfrak{g}_{\mathfrak{P}}(p; A)$ (direct sum). Clearly, this kernel $\mathfrak{g}_{\mathfrak{P}}(p; A)$ is left invariant as a whole by any σ in $G(\mathfrak{P})$. Thus, if we take p-adic coordinates in $\mathfrak{g}(p; A)$ according to the direct decomposition $\mathfrak{g}(p; A) = \mathfrak{g}'_{\mathfrak{P}}(p; A) + \mathfrak{g}_{\mathfrak{P}}(p; A)$, $M_p^*(\sigma)$ must have the form:

$$M_p^*(\sigma) = \begin{pmatrix} M'_p(\sigma) & 0 \\ * & M''_p(\sigma) \end{pmatrix}$$

for any σ in $G(\mathfrak{P})$. Here, $M''_p(\sigma)$ is of degree $2n - r$, and is a representation of $G(\mathfrak{P})$ with p-adic coordinates in $\mathfrak{g}_{\mathfrak{P}}(p;\mathrm{A})$, while $M_p'(\sigma)$ is equal to the representation of the Galois group of the residue field $\bar{k}(\mathfrak{P})$ over $k(\mathfrak{p})$ (induced by $G(\mathfrak{P})$) with p-adic coordinates in $\mathfrak{g}(p;\mathrm{A}(\mathfrak{p}))$ determined by those of $\mathfrak{g}_{\mathfrak{P}}'(p;\mathrm{A})$ by the above isomorphism. Then, just as above, we see that $M_p'(\sigma_{\mathfrak{P}})$ does not depend on the choice of $\sigma_{\mathfrak{P}}$, so that $\mathfrak{p}$ is unramified in the subfield $k(M_p')$ of $k(\mathrm{A};p)$ corresponding to the kernel of M_p'.

Here, we shall use the lemma in **16**. $\|\alpha\|$ denotes as there the normalized valuation in the field $\boldsymbol{Q}_p$. Then, as was recalled in **17**, for any separable endomorphism μ of A such that $\nu(\mu)\neq 0$, we have $\|\det \mathrm{M}_p(\mu)\|=\|\nu(\mu)\|$. Let now $F(T)=\sum_{i=0}^{t} c_i T^{t-i}$ be any polynominal with rational integral coefficients such that $\|c_0\|=\|c_t\|=1$. It is clear that $F(\pi_{\mathfrak{p}})=\sum c_i\pi_{\mathfrak{p}}^{t-i}$ belongs to $\mathcal{A}(\mathrm{A}(\mathfrak{p}))$. If $\nu(F(\pi_{\mathfrak{p}}))=0$, then $\det(M_l(F(\pi_{\mathfrak{p}})))=0$ for any $l\neq p$, hence we have $F(\varpi_i(\mathfrak{p}))=0$ with some $\varpi_i(\mathfrak{p})$. Hence, if $F(\varpi_i(\mathfrak{p}))\neq 0$ for $i=1,\cdots,2n$, then $\nu(F(\pi_{\mathfrak{p}}))\neq 0$. Moreover, since $\|c_t\|=1$, $F(\pi_{\mathfrak{p}})$ is prime to $\pi_{\mathfrak{p}}$, so that $F(\pi_{\mathfrak{p}})$ is separable. We have therefore, for any $F(T)$ such that $F(\varpi_i(\mathfrak{p}))\neq 0$, $\|\det F(\mathrm{M}_p(\pi_{\mathfrak{p}}))\|=\|\det \mathrm{M}_p(F(\pi_{\mathfrak{p}}))\|=\|\nu(F(\pi_{\mathfrak{p}}))\|=\|\det M_l(F(\pi_{\mathfrak{p}}))\|=\|\prod_{i=1}^{2n} F(\varpi_i(\mathfrak{p}))\|$, where $l\neq p$. From the remark following the lemma, we thus see that the characteristic roots of $M_p'(\sigma_{\mathfrak{P}})=M_p(\pi_{\mathfrak{p}})$ are exactly those characteristic roots of $\mathrm{M}_l(\pi_{\mathfrak{p}})$, which are units in $\bar{\boldsymbol{Q}}_p$, taking multiplicities into account. This completes the verification of the condition (B_{II}).

Hasse's zeta function $\zeta_{\mathrm{A}}(s)$ of A over k is defined by

$$\zeta_{\mathrm{A}}(s)=\prod_{\mathfrak{p}\in G(\mathfrak{m})} Z_{\mathfrak{p}}(s),$$

where $Z_{\mathfrak{p}}(s)$ denotes the zeta function $Z_{\mathrm{A}(\mathfrak{p})}(u)$ of $\mathrm{A}(\mathfrak{p})$ over $k(\mathfrak{p})$ with $u=\mathrm{N}\mathfrak{p}^{-s}$. Then we have

$$\zeta_{\mathrm{A}}(s)=\prod_{\mathfrak{p}\in G(\mathfrak{m})}\prod_{t=0}^{2n}\prod_{i_1\cdots i_t}(1-\varpi_{i_1}(\mathfrak{p})\cdots\varpi_{i_t}(\mathfrak{p})\mathrm{N}\mathfrak{p}^{-s})^{(-1)^{t+1}}.$$

This shows that $\zeta_{\mathrm{A}}(s)$ is nothing but the function $L_M(s)$ defined by (13) in **15**, for our system $\{M_l^*\}$ of representations with l-adic coordinates in $\mathfrak{g}(l;\mathrm{A})$. Thus we have seen that $\zeta_{\mathrm{A}}(s)$ can be expressed as an infinite product of "L-functions" determined as in §3.

19. We shall give an application of Theorem 1′ to the proof of the conjecture of Hasse for $\zeta_{\mathrm{A}}(s)$ in case of complex multiplication.

Let A be as in **18**, and we make following assumptions:

(CM_I) The algebra $\mathcal{A}_0(\mathrm{A})$ contains a commutative semi-simple subalgebra C of degree $2n$ over $\boldsymbol{Q}$.

(CM_{II}) Any endomorphism in $C\cap\mathcal{A}(\mathrm{A})$ is defined over k.

Let a prime ideal $\mathfrak{p}$ in k be non-exceptional for A. By the reduction modulo $\mathfrak{p}$, C is mapped isomorphically into the algebra $\mathcal{A}_0(\mathrm{A}(\mathfrak{p}))$, and any endomorphism in this image of C is defined over $k(\mathfrak{p})$. Hence

$\pi_{\mathfrak{p}}$ commutes with all elements in the image of C, which is a commutative semi-simple subalgebra of degree $2n$ of $\mathcal{A}_0(\mathrm{A}(\mathfrak{p}))$, so that $\pi_{\mathfrak{p}}$ must be contained in this image. That is to say, there is an element π in C, mapped to $\pi_{\mathfrak{p}}$ by the reduction modulo $\mathfrak{p}$. Now, $\mathfrak{P}$ being a prime divisor of $\mathfrak{p}$ in $\bar{k}$, we have $M_l^*(\sigma_{\mathfrak{P}})=M_{l,\mathfrak{P}}^{-1}M_l(\pi_{\mathfrak{p}})M_{l,\mathfrak{P}}$ as was seen in **18**, and it is clear that $M_l(\pi)=M_{l,\mathfrak{P}}^{-1}M_l(\pi_{\mathfrak{p}})M_{l,\mathfrak{P}}$. We have therefore $M_l^*(\sigma_{\mathfrak{P}})=M_l(\pi)$. Remember that M_l is a faithfull representation of $\mathcal{A}_0(\mathrm{A})$, so that all matrices $M_l(\sigma_{\mathfrak{P}})$ for all $\mathfrak{P}$ in $G(\mathfrak{m}l)$ generate over $\boldsymbol{Q}'$ an algebra $\mathcal{A}_l$, which is isomorphic to the subalgebra of C generated over $\boldsymbol{Q}$ by all π (for all $\mathfrak{p}$ in $G(\mathfrak{m}l)$), and an isomorphism is given by $M_l^*(\sigma_{\mathfrak{P}})\leftrightarrows\pi$. Hence the condition (CA'_{II}) in **11**, is satisfied. Since all other conditions (CA') have been verified in **18**, Theorem 1′ shows now that our system $\{M_l^*\}$ is obtained from a finite number of characters $\chi_1,\cdots,\chi_s$ of C_k of type (A_0). This implies in particular that, for any combination $(i_1,\cdots,i_t)$ of $1,\cdots,2n$, there is a character $\chi_{i_1\cdots i_t}$ of C_k of type (A_0) such that $\chi_{i_1\cdots i_t}(\mathfrak{p})=\varpi_{i_1}(\mathfrak{p})\cdots\varpi_{i_t}(\mathfrak{p})$ for any $\mathfrak{p}$ in $G(\mathfrak{m})$. Hence we have proved the following theorem.

THEOREM 4. *Let* A *be an abelian variety defined over an algebraic number field k of finite degree, satisfying the conditions (CM_I) and (CM_{II}). Then the zeta function $\zeta_{\mathrm{A}}(s)$ of* A *over k can be expressed in the following form*

$$\zeta_{\mathrm{A}}(s)=\prod_{t=0}^{2n}\prod_{i_1\cdots i_t}L(s\,;\chi_{i_1\cdots i_t})^{(-1)^t}$$

with L-functions $L(s,\chi_{i_1\cdots i_t})$ attached to characters $\chi_{i_1\cdots i_t}$ of C_k of type (A_0), *with defining module* $\mathfrak{m}$

$$(i.\,e.\; L(s\,;\chi_{i_1\cdots i_t})=\prod_{\mathfrak{p}\in G(\mathfrak{m})}(1-\chi_{i_1\cdots i_t}(\mathfrak{p})N\mathfrak{p}^{-s})^{-1}).$$

In particular, the conjecture of Hasse holds for our $\zeta_{\mathrm{A}}(s)$.

Remark that the corresponding result holds for a complete non-singular curve C defined over k, if a jacobian variety J of C and a canonical mapping of C into J are defined over k, and the conditions (CM_I), (CM_{II}) are satisfied for A=J. This is immediately seen from the known relation of zeta function of C and the characteristic roots of $M_l(\pi_{\mathfrak{p}})$. (cf. Weil [8] n°69; as to the explicite formula for the zeta function of C, see Taniyama [3], §4, Theorem 1′).

The author once obtained the same result by another method, in case where C is a field of degree $2n$, and k contains all algebraic conjugates of C. Under these assumptions, all endomorphisms in $C\cap\mathcal{A}(\mathrm{A})$ is defined over k. (cf. Taniyama [3]. §4.)[2] But even in

2) A passage in the proof of proposition 3 in [3] (p. 38, l. 22) would indicate that $N\mathfrak{P}'=(\iota\pi_{\mathfrak{P}'})\cdot\overline{(\iota\pi_{\mathfrak{P}'})}$ determines the ideal decomposition of $(\iota\pi_{\mathfrak{P}'})$, which is in fact not the case, since both $\mathfrak{p}_i$ and $\bar{\mathfrak{p}}_i$ may divide $\iota\pi_{\mathfrak{P}'}$. It is easy however to amend this point and obtain the desired result. See author's forthcoming paper in collaboration with G. Shimura.

this case, our present condition (CM_{II}) is in general much weaker than this former condition in [3]. Moreover, although the case of general C could also be treated by the method in [3] if k is sufficiently large, our present method would be preferred as more direct and giving more insight.

20. We shall give an example, first treated by A. Weil by a quite different method.

Let C be a plane algebraic curve defined by an equation

$$\alpha x^n + \beta y^m = 1,$$

where n, m are natural numbers and α, β are non-zero algebraic integers. Denote by d the greatest common divisor of m and n, and put $m = m_1 d$, $n = n_1 d$. Then the genus g of C is given by $2g = (n-1)(m-1)-(d-1)$. If we put $\omega_{ij} = x^i y^{j-n+1} dx$, the set $\{\omega_{ij}\}$ for all $i \geqq 0$, $j \geqq 0$ such that $(i+1)n_1 + (j+1)m_1 \leqq n_1 m_1 d - 1$ form a base of the space D(C) of all differentials of the first kind of C. Let ζ, η be respectively n-th and m-th roots of unity. Then the correspondence $\mu_0 : (x, y) \to (\zeta x, \eta y)$ of C onto itself induces an endomorphism μ of a jacobian variety J of a complete non-singular model of C. If we denote by $S(\mu_0)$ the representation matrix of μ_0 with respect to the base $\{\omega_{ij}\}$ of D(C), $S(\mu_0)$ is a diagonal matrix with diagonal elements $\zeta^{i+1}\eta^{j+1}$. Hence, if n and m are coprime, we see immediately that μ (for all ζ and η) generate over $\boldsymbol{Q}$ a commutative semi-simple algebra of degree $2g$. We can assume that J is defined over the field $\boldsymbol{Q}(\alpha, \beta)$. Then μ is defined over the field $\boldsymbol{Q}(\alpha, \beta, \zeta, \eta)$. Hence our Theorem is applicable to this case taking $k \supset \boldsymbol{Q}(\alpha, \beta, \zeta, \eta)$ so that a canonical mapping is defined over k.[3)]

Bibliography

[1] M. Deuring, Die Zetafunktionen einer algebraischen Kurven von Geschlecht Eins, Nachr. Akad. Wiss. Göttingen, 1953, 85–94.

[2] G. Shimura, On complex multiplications, Proc. International Symposium on algebraic number theory, Tokyo-Nikko, 1955, 23–30.

[3] Y. Taniyama, Jacobian varieties and number fields, ibid. 31–45.

[4] A. Weil, On a certain type of characters of the idèle-class group of an algebraic number-field, ibid. 1–7.

[5] A. Weil, On the theory of complex multiplication, ibid. 9–22.

[6] A. Weil, Sur la théorie du corps de classes, Journ. Math. Soc. of Japan, **3** (1951), 1–35.

[7] A. Weil, Sur les courbes algébriques et les variétés qui s'en déduisent, Paris, 1948.

[8] A. Weil, Variétés abéliennes et courbes algébriques, Paris, 1948.

[9] A. Weil, Jacobi sums as "Grössencharaktere", Trans. Amer. Math. Soc., **73** (1952), 487–495.

---

3) It may be not difficult to obtain the same result by this method in case where n, m are not coprime, considering also this $S(\mu_0)$. But I have not examined this case in detail.

Distribution of Positive 0-cycles in Absolute Classes of an Algebraic Variety with Finite Constant Field

Scientific Papers of the College of General Education, University of Tokyo, Vol. 8, pp. 123-137 (1958)

(Received October 16, 1958)

Introduction

On the behaviour of the zeta function of an algebraic variety over a finite field, there is a conjecture of Lang, which states as follows: $Z(u), Z'(u)$ denoting respectively the zeta functions of a variety and of its Albanese variety, both defined over a finite field with q elements, the dimensions of the variety and of the Albanese variety being r, r' respectively, $Z(q^{-r}u)$ and $Z'(q^{-r'}u)$ have the same behaviour in the disk $|u|<q$, i.e. the quotient $Z(q^{-r}u)/Z'(q^{-r'}u)$ has neither pole nor zero in the disk $|u|<q$. The principal aim of this note is to prove a theorem which is in close relation to this conjecture. We shall prove namely that the distribution of positive 0-cycles of a given degree m at each (absolute) cycle class is approximately the same for all classes, provided m is greater than $2r'+2$; more precisely: all the positive 0-cycles of degree m in any one class make an irreducible system of dimension $mr-r'$, and the zeta functions of these systems have the same behaviour in the disk $|u| < q^{-(mr-r'-1)}$, for any $m \geqq 2r'+2$.

If we assume that a part of Weil's conjecture [9] holds, the conjecture of Lang follows immediately from our theorem. Our theorem has some additional applications, which will be exposed at the end of this note.

The principle of counting rational points used here is the same as was used by Lang in his paper [3], so that it might be possible to omit the detail concerning this part. For the convenience of the reader, however, some repetitions of Lang's paper will not be avoided. Some results concerning the covering of algebraic varieties contained in Lang [2], [3], and those concerning the specialization of cycles and the reduction of constant-field contained in Shimura [6], will be often used without references.

Notations

In this note, we use the following notations. k denotes the finite field with q elements, and k_ν denotes the extension of k of degree ν over k, ν being a natural number. The Galois group of k_ν over k will be denoted by $\mathfrak{g}_\nu$. If R denotes a subset of an algebraic variety defined over the algebraic closure $\bar{k}$ of k, then R_{k_ν} denotes the set of all points in R, which are rational over k_ν, and the number of points in R_{k_ν} is denoted by $N_\nu(R)$. R_{k_1}, $N_1(R)$ will also be denoted simply by R_k, $N(R)$.

The words "variety" or "curve" are used only for absolutely irreducible variety or curve. 0-cycles will be often called simply as "cycles". If R denotes a variety, dim R denotes the dimension of R. $R^{(m)}$ denotes an m-fold symmetric product of R, which will be assumed to be defined over the same field of definition as R, and to be imbedded in some projective space if R is imbedded in a projective space.

If α: $R \to A$ is a rational map of R to an abelian variety A, then α determines in a natural way a mapping of $R^{(m)}$ to A, which will be denoted by α_m. Clearly α_m is defined over any common field of definition for R, A and α.

The number of rational points over k_ν of the projective space of dimension ρ will be denoted by $\kappa_\rho^{(\nu)}$. We have clearly

$$\kappa_\rho^{(\nu)} = (q^{\nu(\rho+1)}-1)/(q^\nu-1)$$

Throughout this note, V denotes a normal variety defined over k, imbedded in a projective space $\mathcal{P}$. r, n denote the respective dimensions of V, $\mathcal{P}$. Z denotes the set of all the multiple points of V. Clearly Z is a k-closed set of V, any component of which has dimension at most equal to $r-2$.

1. Let f: $U \to V$ be an unramified abelian covering of V, defined over k, with Galois group G. If a point Q of V is not contained in Z, the inverse image $f^{-1}(Q)$ of Q consists of exactly n distinct points, which are simple on U, where $n=[U:V]$, (the degree of the covering). When Q is rational over k_ν, the cycle $T_\nu(Q)=\sum_{\sigma \in \mathfrak{g}_\nu} Q^\sigma$ is rational over k, and the Artin symbol $(T_\nu(Q), U/V)$ is defined and uniquely determined by Q. χ being any (simple) character of G, the L-function of V over k with character χ is defined by

$$(1) \qquad \frac{d}{du} \log L_V(u, \chi) = \sum_{\nu=1}^{\infty} c_\nu(\chi) u^{\nu-1}$$

with

$$(2) \qquad c_\nu(\chi) = \sum_Q \chi((T_\nu(Q), U/V))$$

where the sum $\sum_Q$ in (2) is taken over all Q in $(V-Z)_{k_\nu}$[1].

Any point a of $V^{(m)}$ corresponds canonically to a 0-cycle of degree m of $\mathscr{P}$ supported by V and conversely. The cycle corresponding to a will be denoted by $\psi(a)$, which is rational over k_ν if and only if a is rational over k_ν. Denote by X_m the set of all points a of $V^{(m)}$ such that $\psi(a)$ is not a cycle of V, then X_m is a k-closed set of $V^{(m)}$. It is easy to see that, if z denotes a generic point of a component of X_m over k, then $\psi(z)$ consists of exactly m distinct points, at least one of which is a multiple point of V. This shows that z is a multiple point of $V^{(m)}$, so that any component of X_m is a multiple subvariety of $V^{(m)}$. a being any point in $(V^{(m)}-X_m)_k$, the Artin symbol $(\psi(a), U/V)$ is uniquely determined. We then put

$$\chi'(a)=\chi((\psi(a), U/V)).$$

With this "character" χ', we define the L-function of $V^{(m)}$ over k by

$$\frac{d}{du}\log L_{V^{(m)}}(u, \chi')=\sum_{\mu=1}^{\infty}d_\mu(\chi', m)u^{\mu-1} \tag{3}$$

with

$$d_\mu(\chi', m)=\sum_a \chi'(a), \tag{4}$$

where the sum $\sum_a$ in (4) is taken over all a in $(V^{(m)}-X_m)_{k\mu}$. $L_V(u, \chi)$ may be expressed as

$$L_V(u, \chi)=\sum_{m=0}^{\infty}d_1(\chi', m)u^m$$

(putting $d_1(\chi', 0)=1$), because we have clearly $L_V(u, \chi)=\sum_{\mathfrak{a}}\chi((\mathfrak{a}, U/V))$ $u^{\deg \mathfrak{a}}$, the sum $\sum_{\mathfrak{a}}$ being taken over all positive cycles $\mathfrak{a}$ of V, rational over k.

On the other hand, we see from (1),

$$L_V(u, \chi)=\exp(\log L_V(u, \chi))=1+(\sum c_\nu(\chi)u^\nu/\nu)+\frac{1}{2!}(\sum c_\nu(\chi)u^\nu/\nu)^2+\cdots$$

Comparing the coefficients of u^m, we obtain

(5)

$$d_1(\chi', m)=\frac{c_m(\chi)}{m}+\frac{1}{2!}\sum_{\substack{(\rho_1, \rho_2)\\ \rho_1+\rho_2=m}}\frac{c_{\rho_1}(\chi)}{\rho_1}\cdot\frac{c_{\rho_2}(\chi)}{\rho_2}+\frac{1}{3!}\sum_{\substack{(\rho_1, \rho_2, \rho_3)\\ \rho_1+\rho_2+\rho_3=m}}\frac{c_{\rho_1}(\chi)}{\rho_1}\cdot\frac{c_{\rho_2}(\chi)}{\rho_2}\cdot\frac{c_{\rho_3}(\chi)}{\rho_3}$$
$$+\cdots+\frac{1}{m!}c_1(\chi)^m.$$

If we take k_μ as the basic field (instead of k), then we have in the same way,

1) This definition is somewhat different from the one given by Lang [3]. Our theorem is of birational character, so that we might as well use Lang's definition.

(5′)

$$d_\mu(\chi', m) = \frac{c_{\mu m}(\chi)}{m} + \frac{1}{2!} \sum_{\substack{(\rho_1, \rho_2) \\ \rho_1+\rho_2=m}} \frac{c_{\mu\rho_1}(\chi)}{\rho_1} \cdot \frac{c_{\mu\rho_2}(\chi)}{\rho_2} + \cdots + \frac{1}{m!} c_\mu(\chi)^m.$$

2. $\mathcal{P}^*$ denoting the dual projective space of $\mathcal{P}$, the $(r-1)$-fold product $\mathcal{P}^* \times \cdots \times \mathcal{P}^*$ of $\mathcal{P}^*$ is denoted by Γ. Let $v=(v_1, \cdots, v_{r-1})$ be a point of Γ, with i-th "coordinate" v_i in $\mathcal{P}^*$, and L_{v_i} the hyperplane of $\mathcal{P}$ represented by v_i. We then put $L_v = \bigcap_i L_{v_i}$. As is well known, there is a k-closed set Y of Γ, different from Γ, with the property that, if a point v of Γ in not contained in Y, then the intersection $C_v = L_v \cap V$ is a non-singular curve, without any common point with Z, and the intersection product $L_v \cdot V$ is defined and equal to C_v. Let u be a generic point of Γ over k, and v be any point in $\Gamma - Y$. Then C_v is the uniquely determined specialization of C_u over the k-specialization $u \to v$, and the arithmetic genera of C_u and of C_v are the same. Since C_u and C_v are non-singular, they have the same (geometric) genus. Hence C_v has the same genus for all v in $\Gamma - Y$, which will be denoted by g.

Since the covering $f: U \to V$ is unramified at any point in $V-Z$, and since C_v is entirely contained in $V-Z$ for any v in $\Gamma - Y$, the irreducible components $W_1, \cdots, W_\rho$ of $f^{-1}(C_v)$ are non-singular and mutually disjoint, and all points of W_i are simple on U. Moreover, W_i are unramified coverings of C_v, and we have $\sum [W_i : C_v] = [U : V]$ (cf. Lang-Serre [4], Lemme 1). From the theorem of Bertini, $f^{-1}(C_u)$ has only one component with multiplicity 1, i.e. $f^{-1}(C_u)$ is a curve, for a generic point u of Γ (cf. Lang [3], p. 395). Since $f^{-1}(C_v)$ is the uniquely determined specialization of $f^{-1}(C_u)$ over the k-specialization $u \to v$, u being generic and v in $\Gamma - Y$, the principle of non-degeneracy of Zariski shows that $f^{-1}(C_v)$ is connected. Hence $f^{-1}(C_v)$ must have only one component, say W_v, which is a covering of C_v of degree $[U: V]$. Thus we have $f^{-1}(C_v) = W_v$ for any v in $\Gamma - Y$. It is evident that the Galois group of W_v over C_v is canonically isomorphic to G, the Galois group of U over V.

3. When a character χ of G is equal to the unit character χ_1, we know the following (cf. Lang [3]):

$$c_\nu(\chi_1) = N_\nu(V) - N_\nu(Z),$$

$$N_\nu(Z) < C_1 q^{\nu(r-2)},$$

$$|N_\nu(V) - q^{\nu r}| < C_2 q^{\nu(r-\frac{1}{2})},$$

$$\left| \frac{1}{\kappa_n^{(\nu) r-1}} \sum_{v \in Y} N_\nu(V \cap L_v) \right| < C_3 q^{\nu(r-\frac{1}{2})},$$

where C_1, C_2, C_3 are constants, independent of ν, which are naturally independent of the covering $f: U \to V$.

In the case where χ is not equal to the unit character χ_1, we put, for all v in Γ,

$$e_\nu(\chi, v)=\sum_Q \chi((T_\nu(Q), U/V)),$$

where the sum $\sum_Q$ is taken over all points Q in $(L_v \cap (V-Z))_{k_\nu}$ (hence in $(C_v)_{k_\nu}$ if v is in $\Gamma - Y$). Since the number of L_v corresponding to a rational v over k_ν and containing a given point Q in V_{k_ν}, is exactly equal to $\kappa_n^{(\nu)r-1}$, we have

$$c_\nu(\chi)=(\kappa_n^{(\nu)})^{-(r-1)}\left(\sum_v{}_1 e_\nu(\chi, v)+\sum_v{}_2 e_\nu(\chi, v)\right)$$

where the sums $\sum_v{}_1, \sum_v{}_2$ are taken respectively over all v in $(\Gamma - Y)_{k_\nu}$ and over all v in $(Y)_{k_\nu}$. Hence we have

$$|c_\nu(\chi)| < (\kappa_n^{(\nu)})^{-(r-1)} \left|\sum_v{}_1 e_\nu(\chi, v)\right| + C_3 q^{\nu(r-\frac{1}{2})},$$

because $|e_\nu(\chi, v)| \leqq N_\nu(L_v \cap V)$ holds for any v. On the other hand, χ may be considered as a character of the Galois group of the covering $W_v \to C_v$ for any v in Γ-Y, and the logarithmic differential of the L-function of this covering with this character χ may be expressed as

$$\frac{d}{du}\log L_{C_v}(u, \chi)=\sum_{\nu=1}^{\infty} e_\nu(\chi, v)u^{\nu-1}$$

Since the covering $W_v \to C_v$ is abelian and unramified, $L_{C_v}(u, \chi)$ is a polynomial of degree $2g-2$ in u, whose zeros are all situated in the circle $|u|=q^{-\frac{1}{2}}$ (cf. Weil [8]).

Thus we have

$$|e_\nu(\chi, v)| \leqq (2g-2)q^{\frac{1}{2}\nu},$$

so that

$$|c_\nu(\chi)| < (2g-2)q^{\frac{1}{2}\nu}(\kappa_{n+1}^{(\nu)}/\kappa_n^{(\nu)})^{r-1} + C_3 q^{\nu(r-\frac{1}{2})}.$$

Since it is evident that the inequality

$$|\kappa_{n+1}^{(\nu)}/\kappa_n^{(\nu)} - q^\nu| < C_4$$

holds with a constant C_4 independent of ν, we obtain the following evaluation for $c_\nu(\chi)$, for any character χ different from χ_1, that

$$(6) \qquad |c_\nu(\chi)| < C_5 q^{\nu(r-\frac{1}{2})},$$

where the constant C_5 is independent of ν, χ and of the covering $f: U \to V$.

From (5′) and (6), we see, for $\chi \neq \chi_1$,

$$|d_\mu(\chi', m)| < q^{\mu m(r-\frac{1}{2})}\left(\frac{C_5}{m}+\frac{1}{2!}\sum_{\substack{(\rho_1, \rho_2) \\ \rho_1+\rho_2=m}} \frac{C_5^2}{\rho_1\rho_2}+\cdots+\frac{C_5^m}{m!}\right).$$

Hence we have

$$(7) \qquad |d_\mu(\chi', m)| < C_{6, m} q^{\mu m(r-\frac{1}{2})},$$

where $C_{6,m}$ is a constant independent of μ, χ and the covering, but may depend on m. For the unit character χ_1, we have evidently

(7') $$d_\mu(\chi'_1, m)=N_\mu(V^{(m)}-X_m).$$

4. Let $\alpha: V\rightarrow A$ be a rational map of V to an abelian variety A. We say that α is composed of an isogeny of A, if there is an isogeny γ of abelian variety A' to A, and a rational map β of V to A' such that $\alpha=\gamma\circ\beta$. If α cannot be composed of any non-trivial isogeny of A, then the mapping $\alpha: V\rightarrow A$ is said to be maximal. Suppose now that $\alpha: V\rightarrow A$ is maximal, and that α, A are defined also over k. Then, the class field theory assures that, $\lambda: B\rightarrow A$ being any separable isogeny of an abelian variety B to A, defined over k[2], the inverse image $\alpha^*(B)$ of B/A by α is a variety, and that the canonical projection $\alpha^*(B)\rightarrow V$ is an unramified abelian covering of V, defined over k, with Galois group canonically isomorphic to the kernel of λ (cf. Lang [2], Serre [5]). We now take k_μ as the basic field, and consider the separable isogeny $\pi_\mu: A\rightarrow A$ defined by $\pi_\mu(x)=x^{q^\mu}-x$[3], which is clearly defined over k_μ, since the kernel of π_μ is exactly equal to A_{k_μ}. We denote by U_μ the inverse image of this isogeny $\pi_\mu: A\rightarrow A$ by α, and by G_μ the Galois group of the covering $U_\mu\rightarrow V$. Then the canonical isomorphism of G_μ to A_{k_μ} is given by the reciprocity map:

$$(\mathfrak{a}, U_\mu/V)\rightarrow\alpha(\mathfrak{a})=\sum_{Q\in\mathfrak{a}}\alpha(Q),$$

where $(\mathfrak{a}, U_\mu/V)$ denotes the Artin symbol of a rational cycle $\mathfrak{a}$ over k_μ (of V). By this isomorphism, any character χ of A_{k_μ} determines a character of G_μ, which is the unit character of G_μ if and only if χ is the unit character. We denote this character of G_μ (induced by χ) also by the same letter χ. The mapping α_m of the symmetric product $V^{(m)}$ to A is defined at each point in $V^{(m)}-X_m$, and may be expressed as

$$\alpha_m(a)=\alpha(\psi(a))$$

for any point a in $V^{(m)}-X_m$.

Denote now by Λ_m the graph of α_m in $V^{(m)}\times A$, and put $W'(m, y)=\mathrm{pr}_{V^{(m)}}(\Lambda_m\cap V^{(m)}\times y)$, $W(m, y)=W'(m, y)\cap(V^{(m)}-X_m)$, for any point y in A. Clearly $W'(m, y)$ is a k-closed set of $V^{(m)}$, and $W(m, y)$ consists exactly of all points a in $V^{(m)}-X_m$ such that $\alpha_m(a)=y$. Hence $W(m, y)$ and $W(m, y')$ are mutually disjoint for any different points y, y'. Observe furthermore that any irreducible component of $W(m, y)$ has a dimension at least equal to $mr-t$.

Now, from the definition of $d_\mu(\chi', m)$, we see

$$d_\mu(\chi', m)=\sum_b\chi(b)N_\mu(W(m, b)),$$

2) We say that $\lambda: B\rightarrow A$ is defined over k if λ, B, A are defined over k and all points in the kernel of λ are rational over k.

3) x^{q^μ} denotes the image of x by the automorphism $\xi\rightarrow\xi^{q^\mu}$ of the universal domain.

for any character χ of $A_{k\mu}$, where the sum $\sum_b$ is taken over all b in $A_{k\mu}$. Thus we have

$$N_\mu(A)N_\mu(W(m,b))=\sum_\chi \bar{\chi}(b)d_\mu(\chi',m)$$

where $\bar{\chi}$ denotes the complex conjugate character of χ, and the sum $\sum_\chi$ is taken over all characters of A_{k_μ}. From this and from (7), (7′) we see

$$(8)\quad |N_\mu(A)N_\mu(W(m,b))-N_\mu(V^{(m)}-X_m)|<C_{6,m}N_\mu(A)q^{\mu m(r-\frac{1}{2})}.$$

Putting $t=\dim A$, we have $|N_\mu(A)|<C_7q^{\mu t}$ with a constant C_7 depending only on t, so that the above inequality (8) may be written as

$$(9)\quad |N_\mu(A)N_\mu(W(m,b))-N_\mu(V^{(m)}-X_m)|<C_{6,m}C_7q^{\mu(m(r-\frac{1}{2})+t)}$$

(b in $A_{k\mu}$). On the other hand, b, b' being any two points in $A_{k\mu}$, we see form (8)

$$(10)\qquad |N_\mu(W(m,b))-N_\mu(W(m,b'))|<2C_{6,m}q^{\mu(mr-\frac{1}{2}m)}.$$

Here notice that $N_\mu(X_m)<C_{8,m}q^{\mu(mr-2)}$ holds with a contant $C_{8,m}$ independent of μ.

5. Suppose now that $m\geqq 2t+1$ $(=2\dim A+1)$, then

$$m\left(r-\frac{1}{2}\right)+t\leqq mr-\frac{1}{2}.$$

Hence the inequality (9) shows that, b being in $A_{k\mu}$,

$$N_\mu(A)N_\mu(W(m,b))=N_\mu(V^{(m)})+O\left(q^{\mu(mr-\frac{1}{2})}\right)$$

for $\mu\rightarrow\infty$. On the other hand, we have

$$N_\mu(A)=q^{\mu t}+O\left(q^{\mu(t-\frac{1}{2})}\right) \text{ and } N_\mu(V^{(m)})=q^{\mu mr}+O\left(q^{\mu(mr-\frac{1}{2})}\right)$$

for $\mu\rightarrow\infty$ (cf. Lang [3]), so that we must have

$$N_\mu(W(m,b))=q^{\mu(mr-t)}+O\left(q^{\mu(mr-t-\frac{1}{2})}\right)$$

for $\mu\rightarrow\infty$. Clearly, this implies that the set $W(m,b)$ has only one component, i.e. $W(m,b)$ is a variety, and the dimension of $W(m,b)$ is exactly equal to $mr-t$. Notice here that, if we take ν sufficiently large, there is a positive cycle $\mathfrak{a}'$ of degree m of V, consisting of exactly m distinct points, rational over $k_{\mu\nu}$, such that $\alpha(\mathfrak{a}')=b$. Since the point a' of $V^{(m)}$ corresponding to this $\mathfrak{a}'$ is simple on $V^{(m)}$, and is contained in $W(m,b)$, so the variety $W(m,b)$ is simple on $V^{(m)}$. On the other hand, any component of $W'(m,b)$ other than $W(m,b)$ must be contained in X_m, so that it must be a multiple subvariety of $V^{(m)}$, as all components of X_m are multiple on $V^{(m)}$. Hence the above obtained result shows that the intersection product $\Lambda_m\cdot(V^{(m)}\times b)$ is defined, and that $\alpha_m^{-1}(b)=\mathrm{pr}_{V^{(m)}}\Lambda_m\cdot(V^{(m)}\times b)$ is equal to a ($(mr-t)$-dimensional) cycle $\rho_b\overline{W}(m,b)$ with a natural number ρ_b, for all b in $A_{k\mu}$, where $\overline{W}(m,b)$ means the closure of $W(m,b)$ in $V^{(m)}$. Since μ

may be taken arbitrarily, this result holds for all b algebraic over k. Let now y be any point of A, not necessarily algebraic over k. If $\overline{W}(m, y)$ has at least two distinct components, then we can find a point b of A, algebraic over k, such that these distinct components have distinct specializations over the k-specialization $y \to b$, and also that $\overline{W}(m, b)$ is the uniquely determined specialization of $\overline{W}(m, y)$ over $y \to b$. This contradicts however the above result that $\overline{W}(m, b)$ is a variety. In the same manner we see that $\dim \overline{W}(m, b) = mr - t$. Thus the intersection product $\Lambda_m \cdot (V^{(m)} \times y)$ is defined and $\alpha_m^{-1}(y) = \rho_y \overline{W}(m, y)$ for any y in A. Let now y' be a generic point of A over k and y be again any point of A. From these results we see that the cycle $\alpha_m^{-1}(y)$ is the uniquely determined specialization of the cycle $\alpha_m^{-1}(y')$ over the k-specialization $y' \to y$, and $\deg \alpha_m^{-1}(y') = \deg \alpha_m^{-1}(y)$, where deg denotes the degree of a cycle of $V^{(m)}$ considered as a cycle of the ambient projective space.

Remark. Let $\beta: V \to B$ be a rational mapping of V to an abelian variety B, both defined over k. Then β can be expressed with an isogeny $\gamma: A \to B$ in the form $\beta = \gamma \circ \alpha$ such that $\alpha: V \to A$ is a maximal mapping of V to an abelian variety A. We may suppose that A, α, γ are all defined over a finite algebraic extention k' of k. It is then evident that $\beta_m = \gamma \circ \alpha_m$, and that

$$\beta_m^{-1}(y) = \alpha_m^{-1}(\gamma^{-1}(y))$$

for any y in B. Hence, if we denote by h the number of distinct points in the kernel of γ, $\beta_m^{-1}(y)$ has exactly h distinct components: $\overline{W}(m, y_1), \cdots, \overline{W}(m, y_h)$, where $y_1, \cdots, y_h$ denote all the h distinct points of A such that $\gamma(y_i) = y$. It is also evident that, if we put $\beta_m^{-1}(y) = \sum_i \rho'_{y_i} \overline{W}(m, y_i)$, then $\deg \rho'_{y_i} \overline{W}(m, y_i)$ are the same for all y_i.

6. Suppose now that $m \geqq 2t+2$, then $m\left(r - \frac{1}{2}\right) + t \leqq mr - 1$, hence we see from (9)

$$N_\mu(V^{(m)} - X_m) = N_\mu(A) N_\mu(W(m, b)) + O(q^{\mu(mr-1)}). \tag{11}$$

This implies that the behaviour of the zeta function over k of $V^{(m)}$ in the disk $|u| < q^{-(mr-1)}$ is determined by those of the zeta function over k of A in the disk $|u| < q^{-(t-1)}$ and of the zeta function over k of $W(m, b)$ in the disk $|u| < q^{-(mr-t-1)}$, for any b in A_k. On the other hand, from (10) we see that the quotient of the zeta function over k of $W(m, b)$ by that of $W(m, b')$ has neither pole nor zero in the disk $|u| < q^{-(mr - \frac{1}{2})}$ for any b, b' in A_k.

Summing up, we have proved the following theorem:

THEOREM. *Let V be a projective normal variety of dimension r, defined over the finite field k with q elements, and let $\alpha: V \to A$ be a rational map of V to an abelian variety A. Suppose that α is* *maximal, and A, α are defined over k. Denote by α_m the mapping of the m-fold symmetric product $V^{(m)}$ to A, determined in a natural way from α, and denote by Λ_m the graph of α_m. Then the inverse image* $\alpha_m^{-1}(y)=\mathrm{pr}_{V^{(m)}}(\Lambda_m\cdot(V^{(m)}\times y))$ *is defined, and has only one component W(m, y), whose dimension is equal to* $mr-\dim A$, *and* $\deg\alpha_m^{-1}(y)$ *are the same, for all y in A, provided* $m\geqq 2\dim A+1$.

Moreover, if we denote, for any point b of A, rational over k, the number of positive 0-cycles $\mathfrak{a}$ of degree m of V, rational over k_μ, satisfying $\alpha(\mathfrak{a})=b$, by M(m, b), then we have the following inequality:

$$|M_\mu(m,b)-M_\mu(m,b')|<C_{0,m}q^{\mu(mr-\frac{1}{2}m)}$$

for any two points b, b' in A, rational over k_μ, where $C_{0,m}$ denotes a constant, which does not depend on μ (but may depend on m).

7. We now assume that the following conjecture holds, which is a part of Weil's conjecture on zeta functions of varieties (cf. Weil [9]):

(I) Let W be an algebraic variety defined over the finite field with q elements, and s the dimension of W. Then, i) the function $Z_W(u)$ of W over k is regular in the disk $|u|<q^{-(s-1)}$ except for only one pole of the first order at $u=q^{-s}$, and ii) all zeros of $Z_W(u)$ in this disk are situated in the circle $|u|=q^{-(s-\frac{1}{2})}$.

If this conjecture holds, then it is clear that the number of zeros of $Z_W(u)$ in the disk $|u|<q^{-(s-1)}$ is finite. If we denote by $\{\beta_1^{-1},\cdots,\beta_l^{-1}\}$ the set of all these zeros, each written as often as its order, then we have $|\beta_i|=q^{(s-\frac{1}{2})}$, and

$$\frac{d}{du}\log[(1-q^{s}u)Z_W(u)/\prod_i(1-\beta_iu)]$$

is everywhere regular in our disk.

Apply now this conjecture to our variety V. Let $\{\omega_1^{-1},\cdots,\omega_\lambda^{-1}\}$ be the set of all zeros in the disk $|u|<q^{-(r-1)}$, of $Z_V(u)$, hence also of $L_V(u,\chi_1)$, each written as often as its order. Put

$$R(u)=(1-q^ru)L_V(u,\chi_1)/\prod(1-\omega_iu);$$

then $\dfrac{d}{du}\log R(u)$ is everywhere regular in the disk $|u|<q^{-(r-1)}$. Now, it is easy to see that between the L-function $L_V^{(\mu)}(u,\chi_1)$ over k_μ and the L-function $L_V(u,\chi_1)$ over k there is a relation: $L_V^{(\mu)}(u^\mu,\chi_1)=\prod_\zeta L_V(\zeta u,\chi_1)$, where the product $\prod_\zeta$ is taken over all the μ-th roots of unity. Hence, we have

$$L_V^{(\mu)}(u,\chi_1)=R^{(\mu)}(u)\prod(1-\omega_i^\mu u)/(1-q^{\mu r}u),\ \text{with}\ R^{(\mu)}(u^\mu)=\prod R(\zeta u).$$

If we put now

$$\log R(u)=\sum c'_m u^m,$$

then the above relation shows that

$$\log R^{(\mu)}(u)=\sum \mu c'_{\mu m}u^m.$$

Since $\log R(u)$ is regular in $|u|<q^{-(r-1)}$, we have

$$|c'_m|<C_9 q^{m(r-1+\varepsilon')}$$

with an arbitrarily fixed positive number ε', and with C_9 independent of m. Put now $R^{(\mu)}(u)=1+\sum_{m=1}^{\infty}d'_{\mu,m}u^m$. Then, comparing the coefficients of u^m in the both sides of

$$1+\sum_{m=1}^{\infty}d'_{\mu,m}u^m(=R^{(\mu)}(u))=\exp\left(\sum_{j=1}^{\infty}\mu c'_{\mu j}u^j\right)$$

we have

$$|d'_{\mu,m}|<C_{10,m}q^{\mu(r-1+\varepsilon)},$$

where ε is an arbitrarily fixed positive number, and $C_{10,m}$ is independent of μ. Thus we see

$$\begin{aligned}L_V^{(\mu)}(u,\chi_1)&=(1-q^{\mu r}u)^{-1}R^{(\mu)}(u)\prod(1-\omega_i^{\mu}u)\\&=\left(1+\sum_{\rho=1}^{\infty}q^{\mu r\rho}u^{\rho}\right)\left(1+\sum_{\rho=1}^{\infty}d'_{\mu,\rho}u^{\rho}\right)\prod_i(1-\omega_i^{\mu}u).\end{aligned}$$

On the other hand, we have evidently

$$L_V^{(\mu)}(u,\chi_1)=1+\sum_{m=1}^{\infty}d_{\mu}(\chi'_1,m)u^m$$

so that we obtain

$$\text{(12)}\qquad d_{\mu}(\chi'_1,m)=q^{\mu mr}-\sum\omega_i^{\mu}q^{\mu(m-1)r}+O(q^{\mu(mr-1+\varepsilon)})$$

for $\mu\to\infty$, since $|\omega_i^{\mu}|=q^{\mu(r-\frac{1}{2})}$ from our assumption (I). From this and from the definition (3), $L_{V^{(m)}}(u,\chi'_1)$ may be expressed as

$$L_{V^{(m)}}(u,\chi'_1)=(1-q^{mr}u)^{-1}R_m(u)\prod(1-\omega_i q^{(m-1)r}),$$

where $R_m(u)$ is regular and has no zero in the disk $|u|<q^{-m(r-1+\varepsilon)}$. As $d_{\mu}(\chi'_1,m)=N_{\mu}(V^{(m)}-X_m)$, we see from (9), (12)

$$q^{\mu mr}-\sum\omega_i^{\mu}q^{\mu(m-1)r}=N_{\mu}(A)N_{\mu}(W(m,b))+O(q^{\mu(mr-1+\varepsilon)}),$$

for $\mu\to\infty$. We know that

$$N_{\mu}(A)=\prod_{i=1}^{2t}(1-\pi_i^{\mu})=q^{\mu t}-\sum_{i=1}^{2t}\pi_i^{\mu}q^{\mu(t-1)}+O(q^{\mu(t-1)}),$$

where π_i are certain algebraic integers with absolute value $q^{\frac{1}{2}}$, so that the set $\{\pi_1,\cdots,\pi_{2t}\}$ must be a subset of $\{\omega_1 q^{-(r-1)},\cdots,\omega_\lambda q^{-(r-1)}\}$. If we put $\{\omega_1 q^{-(r-1)},\cdots,\omega_\lambda q^{-(r-1)}\}=\{\pi_1,\cdots,\pi_{2t}\}\cup\{\omega'_1 q^{-(r-1)},\cdots,\omega'_{\lambda-2t}q^{-(r-1)}\}$. then we have

$$N_{\mu}(W(m,b))=q^{\mu(mr-t)}+\sum\omega_i'^{\mu}q^{\mu((m-1)r-t)}+O(q^{\mu(mr-t-1+\varepsilon)}).$$

From these we obtain the following consequence: (In the following, $(A(V),\varphi)$ denotes the Albanese variety of V, defined over k. Notice that φ: $V\to A(V)$ is maximal. $Z_V(u)$, $Z_{V^{(m)}}(u)$ denote the zeta functions over k of V, $V^{(m)}$ respectively.)

1) Under the assumption (I), the quotient $Z_{V^{(m)}}(q^{-mr}u)/Z_V(q^{-r}u)$ has neither pole nor zero in the disk $|u|<q$.

2) Under the assumption (I), if $Z_V(u)$ has no zero in the circle $|u|<q^{-(r-1)}$, then the Albanese variety of V must be trivial.

3) Under the assumption (I), assume moreover that the number λ of zeros of $Z_V(u)$ in $|u|<q^{-(r-1)}$ is equal to $2\dim A(V)$ (this is also a part of Weil's conjecture), then the conjecture of Lang holds for V, and moreover, the unique component variety of $\varphi_m^{-1}(b)$ has trivial Albanese, for all b in A, algebraic over k, hence also for all b in A.

Now, it has been proved recently by S. Koizumi[4] that $\varphi_m^{-1}(b)$ is a variety with trivial Albanese, if m is sufficiently large. Using this result, we have

4) Under the assumption (I), the conjecture of Lang holds for all varieties if and only if it holds for varieties with trivial Albanese, and if this conjecture holds, the unique component of $\varphi_m^{-1}(b)$ has trival Albanese for all $m\geqq 2\dim A(V)+2$.

8. We shall conclude this note by giving an application of our theorem in $n°6$.

In the following, let V_0 be a normal variety without singularities of codimension 1, and assume that V_0 is imbedded in a projective space $\mathcal{P}_0$, where the field of definition K for V_0 may be arbitrary. Without loss of generality, we may assume that K is finitely generated over the prime field. Put $r=\dim V_0$, $n=\dim \mathcal{P}_0$. Let α_0: $V_0\to A_0$ be a rational map of V_0 to an abelian variety A_0, both defined over K, and assume that α_0 is maximal.

$\mathfrak{p}$ being any discrete valuation of K, we dnote by $\widetilde{X}$ the geometric object obtained from an object X (defined over K) by the reduction mod. $\mathfrak{p}$.

First, notice that, by a similar reasoning as was used in $n°5$, making now use of the reduction mod. $\mathfrak{p}$ (instead of K-specialization $y\to b$) with a suitable discrete valuation $\mathfrak{p}$ of K with finite residue field, we can show that the first half of our theorem holds, provided we can find a $\mathfrak{p}$ such that $\tilde{\alpha}_0$: $\widetilde{V}_0\to\widetilde{A}_0$ is maximal.

Suppose for a moment that, for a generic point y_0 of A_0 over K, the cycle $\alpha_{0m}^{-1}(y_0)$ of $V_0^{(m)}$ has l distinct components, where m is supposed to be greater than $2\dim A_0+1$. Let b_0 be any point of A_0. Then we can find a discrete valuation $\mathfrak{p}$ of K with a finite residue field with the properties that $\widetilde{V}_0$ is normal, $\widetilde{A}_0$ is an abelian variety and $\tilde{\alpha}_0$ is a rational map of $\widetilde{V}_0$ to $\widetilde{A}_0$, that $\widetilde{V_0^{(m)}}=\widetilde{V}_0^{(m)}$, $\widetilde{\alpha_{0m}}=(\tilde{\alpha}_0)_m$ hold, and that $\alpha_{0m}^{-1}(y_0)$, $\alpha_{0m}^{-1}(b_0)$ go to $\tilde{\alpha}_{0m}^{-1}(\tilde{y}_0)$, $\tilde{\alpha}_{0m}^{-1}(\tilde{b}_0)$ by the reduction mod

4) unpublished.

$\mathfrak{p}$, whereby distinct components of $\alpha_{0m}^{-1}(y_0), \alpha_{0m}^{-1}(b_0)$ go to distinct components of $\tilde{\alpha}_{0m}^{-1}(\tilde{y}_0)$, $\tilde{\alpha}_{0m}^{-1}(\tilde{b}_0)$ respectively, and the degrees of these cycles are preserved (cf. Shimura-Taniyama [7]). Then $\tilde{\alpha}_0$ can be expressed with an isogeny $\nu: B \rightarrow \tilde{A}_0$ of an abelian variety B to $\tilde{A}_0$, in the form $\tilde{\alpha}_0 = \nu\beta$, where $\beta: \tilde{V}_0 \rightarrow B$ is maximal. From the remark at the end of $n^\circ 5$, the kernel of ν must contain exactly l distinct points. Then the number of distinct components of $\tilde{\alpha}_0^{-1}(\tilde{b}_0)$ must be also l, so that $\alpha_0^{-1}(b_0)$ must have exactly l distinct components. Moreover, if we put $\alpha_0^{-1}(b_0) = \sum_{i=1}^{l} \rho_{ib_0} W_i(b_0)$, with irreducible $W_i(b_0)$ and positive ρ_{ib_0}, then, from the same remark, and from our assumption on $\mathfrak{p}$, we see that $\deg(\rho_{ib_0} W_i(y_0))$ is equal to $\deg(\rho_{jy_0} W_j(y_0))$ for any i, j. Since b_0 was arbitrary, we have seen that $\alpha_0^{-1}(b_0)$ has exactly l distinct components, and $\deg(\rho_{ib_0} W_i(b_0))$ are the same for all b_0. Notice that $W^i(b_0)$, $W_i(b_0')$ cannot have any common point outside X_{0m}, where X_{0m} is defined for $V_0^{(m)}$ in the same way as X_m was defined for $V^{(m)}$, provided of course $(i, b_0) \neq (i', b_0')$. Now these results allow us to make Chow variety B_0' of all $\rho_{ib_0} W_i(b_0)$ for all $i = 1, \cdots, l$ and for all b_0 in A_0. Let x be a generic point of $V_0^{(m)}$ over K, and ξ be the Chow point of $\rho_{ib_0} W_i(b_0)$ containing x, which is clearly uniquely determined by x. Then, the correspondence $x \rightarrow \xi$ determines a rational map β_0' of $V^{(m)}$ to B_0'. Now, denoting by $\xi_i(b_0)$ the Chow point of $\rho_{ib_0} W_i(b_0)$, we denote by B_0 the set of points $(\xi_i(b_0), b_0)$ in $B_0' \times A_0$ for all b_0 in A_0 and for all i. Then the mapping $\beta_0' \times \alpha_0$ of V_0 to $B_0' \times A_0$ maps V_0 into B_0. On the other hand, the correspondence $\gamma_0; (\xi_i(b_0), b_0) \rightarrow b_0$ is a rational map of B_0 to A_0, which is evidently an unramified covering with the degree l. This shows that B_0 is an abelian variety and $\gamma_0: B_0 \rightarrow A_0$ is an isogeny (cf. Lang-Serre [4], Théorème 2), and that α_0 is composed of γ_0. This contradicts however to the maximality of α_0, unless $l = 1$. Thus the first half of our theorem holds for *any* field of definition, provided V_0 satisfies an additional assumption that it has no singularities of codimension 1. We have shown at the same time that, if the reduction mod. $\mathfrak{p}$ work well for $V_0, V_0^{(m)}, \alpha_0, \alpha_{0m}, A_0$,[5] then $\tilde{\alpha}_0: \tilde{V}_0 \rightarrow \tilde{A}_0$ connot be composed of non-trivial separable isogeny.

Now, assume that (A_0, α_0) is the Albanese variety of V_0, and put $t = \dim A_0$. Let $\zeta_1, \cdots, \zeta_t$ be a basis of the space of linear differentials of the 1st kind on A_0. Then the differentials of the 1st kind $\delta\alpha_0(\zeta_1), \cdots, \delta\alpha_0(\zeta_t)$ on V_0 are also linearly independent over K (cf.

5) We say that the reduction mod. $\mathfrak{p}$ works well for V_0, $V^{(m)}$, α_0, α_{0m}, A_0, if the following conditions are satisfied: $\tilde{V}_0$ is a normal variety without singularities of codimension 1, $\tilde{A}_0$ is an abelian variety, $\tilde{\alpha}_0$ is a rational map of $\tilde{V}_0$ to $\tilde{A}_0$, and $\widetilde{V_0^{(m)}} = \tilde{V}_0^{(m)}$, $\widetilde{\alpha_{0m}} = (\tilde{\alpha}_0)_m$ hold.

Igusa [1]). Put now $\alpha_0^{-1}(b_0)=\rho_{b_0}W(b_0)$ with irreducible $W(b_0)$, and put $\rho'_{b_0}=\rho_{b_0}/\rho_{y_0}$ with generic y_0. Clearly ρ'_{b_0} is a positive integer. Making use of the Chow variety of $\rho'_{b_0}W(b_0)$ (instead of $\rho_{b_0}W(b_0)$) and reasoning as above, we can easily see that ρ_{y_0} is a power of the characteristic of the universal domain, which contradicts to the result of Igusa, unless $\rho_{y_0}=1$. We have thus seen that $\alpha^{-1}(y_0)=W(y_0)$ for any generic y_0. Now, let $\mathfrak{p}$ be a discrete valuation of K with a finite residue field, such that the reduction mod. $\mathfrak{p}$ works well for V_0, $V_0^{(m)}$, α_0, α_{0m}, A_0, and that $\tilde{\zeta}_1,\cdots,\tilde{\zeta}_t$ forms a basis of the space of linear differentials of the 1st kind of $\tilde{A}_0$ such that $\delta\tilde{\alpha}_0(\tilde{\zeta}_i)=\widetilde{\delta\alpha_0(\zeta_i)}\neq 0$ for all i. Then it is evident that $\tilde{\alpha}_0\colon \tilde{V}_0\to\tilde{A}_0$ cannot be composed of non-trivial inseparable isogeny, so that $\tilde{\alpha}_0$ must be maximal. It is also evident, that in this case, the dimension of the Albanese variety of $\tilde{V}_0$ is not smaller than $t=\dim A_0$. On the other hand, it is well known that the Albanese variety and the Picard variety of a given normal variety have the same dimension. Hence, if we can prove that the dimension of Picard variety of V_0 is not smaller than that of $\tilde{V}_0$, then we can conclude, that, in the above case, $(\tilde{A}_0, \tilde{\alpha}_0)$ is the Albanese variety of $\tilde{V}_0$.

Let now $\mathcal{G}(V_0)$, $\mathcal{G}_a(V_0)$ and $\mathcal{G}_l(V_0)$ be respectively the group of all divisors, the group of all divisors algebraically equivalent to zero, and the group of all divisors linearly equivalent to zero, and let $\mathcal{G}(\tilde{V}_0)$, $\mathcal{G}_a(\tilde{V}_0)$, $\mathcal{G}_l(\tilde{V}_0)$ be corresponding groups for $\tilde{V}_0$. Then the Picard variety of V_0 is isomorphic to the factor group $\mathcal{G}_a(V_0)/\mathcal{G}_l(V_0)$, and the theorem of Severi-Néron states that $\mathcal{G}(V_0)/\mathcal{G}_a(V_0)$ is finitely generated. On the other hand, for the $\mathfrak{p}$ with the above properties, we have clearly $\widetilde{\mathcal{G}(V_0)}=\mathcal{G}(\tilde{V}_0)$, $\widetilde{\mathcal{G}_l(V_0)}=\mathcal{G}_l(\tilde{V}_0)$, so that the factor group $\mathcal{G}(V_0)/\mathcal{G}_l(V_0)$ is mapped homomorphically onto $\mathcal{G}(\tilde{V}_0)/\mathcal{G}_l(\tilde{V}_0)$ and $\mathcal{G}(\tilde{V}_0)/\mathcal{G}_a(\tilde{V}_0)$ must be finitely generated. Now, if the dimension of the Picard variety of $\tilde{V}_0$ is greater than that of V_0, the factor group

$$\mathcal{G}_a(\tilde{V}_0)/\widetilde{\mathcal{G}_a(V_0)}=[\mathcal{G}_a(\tilde{V}_0)/\widetilde{\mathcal{G}_l(V_0)}]/[\widetilde{\mathcal{G}_a(V_0)}/\widetilde{\mathcal{G}_l(V_0)}]$$

cannot be finitely generated, which is a contradiction. We have thus proved that, if the discrete valuation $\mathfrak{p}$ of K satisfies the above stated conditions, then $(\tilde{A}_0, \tilde{\alpha}_0)$ is the Albanese variety of $\tilde{V}_0$. This statement is still valid for valuations $\mathfrak{p}$ with infinite residue field, as will be easily seen by successive reductions of residue fields till the step cannot be continued, i.e. till the final residue field becomes a finite field.

Summing up, we have the follwing:

SUPPLEMENT TO THE THEOREM. *The first half of the theorem*

holds without any restriction on the field of definition, provided the variety satisfies the additional condition that it has no singularity of codimension 1. Moreover, using the same notation as in the theorem, we have

$$\alpha_m^{-1}(y) = \overline{W}(m, y)$$

for any generic point y of A, provided (A, α) is the Albanese *variety of V.*

COROLLARY. *Under the same assumption as above on V_0, if (A_0, α_0) is the Albanese variety of V_0, then $(\tilde{A}_0, \tilde{\alpha}_0)$ is the Albanese variety of $\tilde{V}_0$ for almost all valuations of the field of definition for V_0.*

Bibliography

[1] J. Igusa, A fundamental inequality in the theory of Picard varieties, *Proc.* **41**, *N.S.A.*, 317-320 (1955).

[2] S. Lang, Unramified class field theory over function fields in several variables, *Ann. Math.*, **64**, 285-325 (1956).

[3] ———, Sur les séries L d'une variété algébrique, *Bull. Soc. Math. France*, **84**, 385-407 (1956).

[4] S. Lang et J.-P. Serre, Sur les revêtements non ramifiés des variétés algébriques *Amer. J. Math.*, **79**, 319-330 (1957).

[5] J.-P. Serre, Groupes algébriques et thèorie du corps de classes, *Lecture note at College de France* 1957.

[6] G. Shimura, Reduction of algebraic varieties with respect to a discrete valuation of the basic field, *Amer. J. Math.*, **77**, 134-176 (1955).

[7] G. Shimura and Y. Taniyama, Complex multiplication of abelian varieties and its applications to number theory (To appear).

[8] A. Weil, *Sur les courbes algébriques et les variétés qui s'en déduisent*, (Paris, 1948).

[9] ———, Numbers of solutions of equations in finite fields *Bull. Amer. Math. Soc.*, **55**, 497-508 (1949).

Problems

These problems were distributed to the participants of Tokyo-Nikko conference on Number Theory, 1955, and published in Japanese in Sugaku, vol. 7(1956), p. 269 and p. 271.

Problem 10. Let k be a totally real number field, and $F(\tau)$ be a Hilbert moduler form to the field k. Then, choosing $F(\tau)$ in a suitable manner, we can obtain a system of Hecke's L-serieses with "Grössencharaktere" λ, which corresponds one-to-one to this $F(\tau)$ by the process of Mellin-transformation. This can be proved by a generalization of the theory of operator T of Hecke to Hilbert modular functions (cf. Herrmann).

The problem is to generalize this theory in the case where k is a general (not necessarily totally real) number field. Namely, to find an automorphic form of several variables from which L-serieses with "Grössencharatkere" λ of k may be obtained, and then to generalize Hecke's theory of operator T to this automorphic form.

One of the aim of this problem is to characterize L-serieses with "Grössen-oder Klassen-charaktere" of k ; especially to characterize the Dedekind zeta function of k in this method, which is not yet done even if k is totally real.

Problem 11. Let C be an elliptic curve of characteristic zero, whose ring $\mathscr{A}(C)$ of endomorphisms is isomorphic to the principal order of an imaginary quadratic number field Σ. Assume C is defined over the field $k=\Sigma(j)$, where j is the absolute invariant of C. Let $\mathfrak{a}$ be an ideal in Σ, $b_{\mathfrak{a}}$ a proper $\mathfrak{a}$-Teilpoint on C and $d_{\mathfrak{a}}=[k(b_{\mathfrak{a}}):k(b_{\mathfrak{a}})_0]$, where $k(b_{\mathfrak{a}})_0$ is the field generated over k by the value of Weber τ-function on C for the argument $b_{\mathfrak{a}}$. Denote by $L_{\mathfrak{a}}(s)$ the L-function of $k(b_{\mathfrak{a}})$ with the principal character mod.$\mathfrak{a}$. Then the zeta function $\zeta_C(s)$ of C over k can be expressed as $\zeta_C(s) = \prod_{\mathfrak{a}}(L_{\mathfrak{a}}(s))^{d_{\mathfrak{a}}}$, where $\mathfrak{a}$ runs through all the ideals in Σ. Therefore, we have

$$(*)\qquad \prod_{\mathfrak{a}}(L_{\mathfrak{a}}(s))^{d_{\mathfrak{a}}}=\zeta_k(s)\zeta_k(s-1)\Big[L\Big(s-\frac{1}{2},\chi\Big)\cdot L\Big(s-\frac{1}{2},\bar{\chi}\Big)\Big]^{-1}$$

where $\zeta_k(s)$ is the zeta function of k, and $L(s,\chi)$ is the L-function with "Grössencharakter" χ of k, which is essentially the character in Σ.

The problem is, first, to prove $(*)$ directly, without using the theory of algebraic curves on functions, next, to generalize this relation $(*)$ to an arbitrary number field k. Namely, does there exists a suitably defined infinite sequence of abelian extensions of k, such that the infinite product of L-functions of these extensions with suitable characters can be expressed with L-functions with "Grössencharaktere" in a simple form ?

Problem 12. Let C be an elliptic curve defined over an algebraic number field

k, and $L_c(s)$ denote the L-function of C over k. Namely,

$$\zeta_c(s) = \frac{\zeta_k(s)\zeta_k(s-1)}{L_c(s)}$$

is the zeta function of C over k. If a conjecture of Hasse is true for $\zeta_c(s)$, then the Fourier series obtained from $L_c(s)$ by the inverse Mellin-transformation must be an automorphic form of dimension -2 of some special type (cf. Hecke). If so, it is very plausible that this form is an elliptic differential of the field of that automorphic functions. The problem is to ask if it is possible to prove Hasse's conjecture for C, by going back this considerations, and by finding a suitable automorphic form from which $L_c(s)$ may be obtained.

Problem 13. Concerning the above problem, our new problem is to characterize the field of elliptic modular functions of "Stufe" N, and, especially, to decompose the jacobian variety J of this function field into simple factors, in the sense of isogeneity.

It is well known that, in case $N = q$ is a prime number, satisfying $q \equiv 3 (\text{mod } 4)$, J contains elliptic curves with complex multiplication. Is this true for general N?

Problem 28 (Tsuneo Tamagawa). Let K be an algebraic function field of one variable of genus g over a field of constant k of characteristic $p \neq 0$. Let J be the Jacobian variety of a non-singular model C of K. Let A be the Hasse-Witt matrix of K and let r be the rank of $A \cdot A^p \cdot \cdots \cdot A^{p^{g-1}}$. Show that the number of p-division points of J is equal to p^r.

Remark. For a finite field k, this assertion is true by the class field theory of K.

Problem 29. Let the notation be as in Prob. 28. The matrix A is decomposed in the form

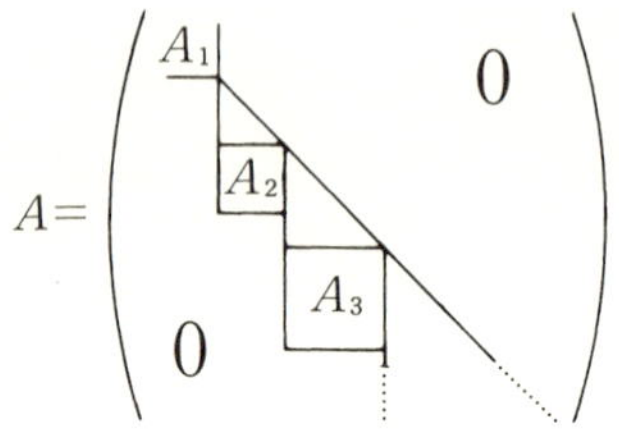

where the degrees of $A_1, A_2, A_3, \cdots$ are invariants of K (cf. Hasse-Witt). What is the meaning of these invariants for the Jacobian variety of C?

Remark. When K is defined by $y^2 = 1 - x^5$, one has the following result.

a) $A = \begin{pmatrix} 1 & 0 \\ 0 & 1 \end{pmatrix}$ if $p \equiv 1 \ (\text{mod } 5)$,

b) $A = \begin{pmatrix} 0 & 0 \\ 1 & 0 \end{pmatrix}$ if $p \equiv 2 \text{ or } 3 \ (\text{mod } 5)$,

c) $A = \begin{pmatrix} 0 & 0 \\ 0 & 0 \end{pmatrix}$ if $p \equiv 4 \ (\text{mod } 5)$,

and in the case b) the Jacobian variety is reducible. It is very plausible that a similar result holds for K defined by $y^2 = 1 - x^q$ (q a prime).

第二部

ABEL 函数体ノ n-分割ニツイテ*

WEIL ノ有限性定理ノ代数的証明

A. Weil ハ，ソノ Thèse [1] ニ於テ，有限次代数体 k 上定義サレタ curve ノ，jacobian variety ノ，k ニ関スル rational points ノナス加群ガ，有限個ノ base ヲ持ツコトヲ証明シタ．彼ノ証明ハ，本質的ニ，Θ-函数ノ理論オヨビ，代数函数ノ解析的理論ニ基イテイル．

所デ，Hasse [1] ガ指摘シタ様ニ，Weil ノ証明ノ途中デ得ラレル，Abel 函数ニツイテノ 2, 3 ノ性質ハ，Abel 多様体ノ数論ニ於テ興味アル結果ヲ有シ，又ソレ等ハ，純代数的ニ導クコトガデキル．然シ Hasse ハ此等ノコトヲ，curve ノ genus 1 ノ場合，即チ楕円函数体ノ場合ニ，示シテイルニ過ギナイ．

然シ abelian variety ノ代数幾何的理論ノ最近ノ発展ニヨリ，コレラノ結果ヲ任意ノ genus ノ場合ニ導キ，サラニ Weil ノ有限性定理ノ代数的証明ヲ与エルコトハ，極メテ容易ナコトトナツタ．此レヲ実際ニ行フコトガ，コノ小論ノ目的デアル．

§1 ニ於テ，後ニ必要トサレル結果ヲ引用シテオク．

§2 デハ，定義体ハ，ソノ有限次拡大ニヨリ満サレル様ナアル条件ヲ除ケバ，全ク任意トシ，Abel-函数体ノ，n-分割（“n-Teilung”）ノ理論ヲ扱フ．コレハ Hasse [1] ノ方法ヲ，代数幾何カラノ補助手段ヲ用イテ拡張スレバ，得ラレル．ソノ主要目標ハ，n-分割ニ際シテノ，rational point ノ分解法則ヲ，対応スル常数体ノ拡大ノ考察ヲモ含メテ，研究スルコトデアル．此処デ扱ハレルノハ，拡大ガ Kummer 拡大トナル，特別ナ場合ダケデアルガ，ソノ結果ハ興味アルモノデアル．

§3 デハ，常数体ヲ有限次代数体トシテ，Weil ノ有限性定理ノ，代数的証明ヲ与ヘル．此処デハ Hasse ノ方法ヲソノママ拡張スルコトハ出来ナイ．然シ Weil ニヨリ展開サレタ “Arithmétique sur les variétés algébriques” ガ，ソノ目的ニ対シ極メテ有効ナコトガワカル，ソシテ Weil ノ初メノ考ヘ方 [1] ニ従ツテ，証明ガ行ハレル．

§1. 準　備

我々ハ，Weil [2] ニ於ル abelian variety ノ理論ヲ用ヒル．但シ Weil ハ，curve Γ ノ jacobian variety ヲ Γ ノ定義体ノ超越拡大ノ上デ，abstract variety トシテ構成シテイル．基礎体ノ超越拡大ヲセネバナラヌコトハ，“arithmétique” ニ対シテ致命的ナ障害トナルコトガ多イ．シカシ最近，Chow [1] ハ jacobian variety ヲ基礎体ヲ拡大セズニ，射影空間内ニ構成スルコトニ成功シタ．コノ論文ハ未発表ナノデ，我々ハ松阪氏 [1] ニヨリ，同ジ条件ノ下デ構成サレタ jacobian variety ヲ用イルコトニスル．

以后，Γ ヲ，genus g ノ，non-singular curve デ，体 k 上定義サレタモノトシ，J ヲ k 上定義サレタ，Γ ノ projective jacobian variety，φ ヲ $\Gamma \longrightarrow$ J ノ canonical mapping トスル．

1°. δ ヲ J $\longrightarrow$ J ノ恒等写像，$n\delta$ ヲ，$a \longrightarrow na$ ナル J ノ endomorphism トスル．X ヲ J ノ divisor，X_a ヲ X ヲ a ダケ移動

* これは 1952～3 年東大理学部数学科の講義 “幾何学続論” の単位のために弥永教授に提出したリポートである。（編者注）

シテ (J ノ加法群ノ意味デ) 得ラレル divisor トスル. ソノトキ x ヲ J ノ k 上ノ generic point, $y=nx$ トスレバ, n ガ k ノ標数 p ト素ノトキ, $k(x)/k(y)$ ハガロア拡大デ, ソノガロア群ハ, $na=0$ ナル $a\in$ J ノナス群 $\mathfrak{g}$ ト同型デアル. ソシテ $\mathfrak{g}$ ハ位数 n^{2g} ノアーベル群ヲナシ, ソレハ n 位ノ巡回群ノ $2g$ 個ノ直積デアル. [Weil [2], Théorème 33 ノ Cor. 1, Théorème 12, 及ビソノ Cor.]

以後 $\mathfrak{g}$ ノ元ヲ $a_1, a_2, \cdots, a_N$; $N=n^{2g}$ ト書クコトニスル.

2°. A ヲ k 上定義サレタ abelian variety, f ヲ k 上定義サレタ $\Gamma \longrightarrow$ A ノ函数トスル. ソノトキ f ハ J$\longrightarrow$A ノ homomorphism λ ニ, $f=\lambda\varphi+a$, a ハ constant, ナル形デ一意ニ拡張サレル. X ヲ A ノ divisor, $\mathfrak{x}_a=\overset{-1}{f}(\mathrm{X}_a)$ トスル. 但シ a ハ右辺ガ定義サレル様ニ選ブ. $\mathfrak{m}=\sum m_i \mathrm{P}_i$ ヲ Γ ノ divisor, $\mathrm{S}[\varphi(\mathfrak{m})]=\sum m_i \varphi(\mathrm{P}_i)$ (右辺ノ $\sum$ ハ J ノ点ノ加法ヲ表ス) ト書クコトニスレバ, $\mathrm{S}[\varphi(\mathfrak{x}_a)]=\rho a+b$ ニヨリ A$\longrightarrow$J ノ homomorphism ρ ガ定義サレ, λ, X ノミニヨリ定ル. ソレヲ λ'_{X} ト書ク. λ'_{X} ハ X ニ linear ニ依ル [Weil [2], Théorème 21, Théorème 23].

ソシテ

$$\lambda'_{\mathrm{X}}=0 \Longleftrightarrow \mathrm{X}_{\lambda z} \sim \mathrm{X} \qquad \forall z \in \mathrm{J}$$

[Weil [2], Théorème 30]. ソレ故特ニ A=J, $\lambda=\delta$ トシ記号 $\mathrm{X}\equiv 0$ ヲ $\mathrm{X}_t-\mathrm{X}\sim 0$ $\forall t\in$ J ヲ意味スルトスレバ

$$\mathrm{X}\equiv 0 \Longleftrightarrow \delta'_{\mathrm{X}}=0.$$

3°. $\mathrm{M}_1, \cdots, \mathrm{M}_{g-1}$ ヲ Γ ノ k 上独立ナ generic points トシ, $\sum\varphi(\mathrm{M}_i)$ ノ k 上ノ locus ヲ Θ トスル. コノ Θ ハ以下ノ推論ニ於テ基本的ナ役割ヲ演ズル.

m_i ヲ整数, x_i ヲ J ノ点トスルトキ.

$$\sum m_i \Theta_{x_i} \sim 0 \Longleftrightarrow \begin{cases} \sum m_i=0 \text{ 及ビ} \\ \sum m_i x_i=0. \end{cases}$$

更ニ X ヲ J ノ divisor トシテ

$$\mathrm{X}\equiv 0 \Longleftrightarrow \begin{cases} \text{J ノ点 t ガアリ,} \\ \mathrm{X} \sim \Theta_t-\Theta. \end{cases}$$

此ノ t ハ $t=\mathrm{S}[\varphi(\Gamma)\cdot\mathrm{X}_a]$ $a\in$J ニヨリ一意ニ定リ, X ガ k 上 rational ナ divisor ナラバ, t モ k 上 rational デアル.

従ツテ $\mathrm{X}\equiv 0$ ナル divisor X ノナス群ヲ, $\mathrm{X}\sim 0$ ナル divisor ノナス群デ割ツタ商群ハ, J ノ点ノ加法群ニ, 上ノ対応ニヨリ同型ニ写像サレル. 此レハ jacobian variety ノ Picard variety ガモトノ jacobian variety ト同型ナコトヲ表スモノニ外ナラナイ. [Weil [2], Théorème 30 ノ Cor. 3, Théorème 32 ノ Cor. 2]

4°. X ヲ J ノ divisor, $\nu=n\delta$ トスレバ, $\overset{-1}{\nu}(\mathrm{X})$ ハ常ニ定義サレ,

$$\overset{-1}{\nu}(\mathrm{X})\equiv n^2\mathrm{X}.$$

特ニ $n>0$, X ヲ k 上 rational, $n\mathrm{X}\sim 0$ トスレバ, J ノ函数 $\varphi_{\mathrm{X}}, \psi_{\mathrm{X}}$ デ k 上定義サレ,

$$(\varphi_{\mathrm{X}})=nX, \quad \varphi_{\mathrm{X}}(nx)=\psi_{\mathrm{X}}(x)^n$$

ナルモノガ存在スル. [Weil [2], Proposition 31, 32].

5°. 我々ハ更ニ, algebraic variety 上ノ arithmetic ニツイテノ一定理ヲ引用セネバナラヌ.

V ヲ有限次代数体 k 上定義サレタ, complete, normal ナ abstract variety トシ $(x_0, x_1, \cdots, x_n)$, $(y_0, y_1, \cdots, y_m)$ ヲ各々 k 上定義サレタ, V ノ上ノ函数トシテ, F, G ヲ, コノ函数系ニヨリ定マル V カラ各々 n 次元, m 次元ノ射影空間内ヘノ写像トスル. ソシテ更ニ $(x_i)=\mathrm{X}_i-\mathrm{Z}$, $(y_i)=\mathrm{Y}_i-\mathrm{Z}$, $(\mathrm{X}_i \succ 0,\ \mathrm{Y}_i \succ 0,\ \mathrm{Z}\succ 0)$ デ, $\mathrm{X}_0, \mathrm{X}_1, \cdots, \mathrm{X}_m$ ハ共通点ヲ持タヌトスル. ソノトキ F ハ V 上スベテノ点デ定義サレ, G ハスベテノ Y_i ニハ共通デナイ V ノ点デ定義サレル.

H ヲ射影空間ニ於ル点ノ高サ (height) ヲ表ストスレバ, ソノトキ, 常数 γ ガアリ

$$H[F(\mathrm{P})] \geqq \gamma H[G(\mathrm{P})]$$

ガ，V ノ absolute algebraic ナ点 P デ，スベテノ Y_i ニハ含マレヌモノスベテニ対シ成立ツ．[Weil [3] Théorème 8].

§ 2. n-分割体

此ノ § デハ，定義体 k ハ任意トスル．ソシテ n ヲ，k ノ標数 p ト素トスル．記号ハ § 1 ニ於ルト同様トスル．但シ A ハ abelian variety ヲ表ハサズ，abel 函数体ヲ表ハストスル．

6°． モシ必要ナラ，k ヲ有限次代数拡大デオキカエルコトニヨリ，次ノ仮定ガ満サレテイルトスル．

1) $na=0$ ナル J ノ点 $a_1, \cdots, a_N$, $N=n^{2g}$ ハスベテ k 上 rational.

2) k ハ 1 ノ n 乗根ヲスベテ含ム．

A ヲ k 上定義サレタ J ノ函数ノナス体トスル．（即チ Γ ノ Abel 函数体．）J ノ endomorphism $\nu=n\delta$ ニ対応シテ，A ノ meromorphism $\psi \longrightarrow \psi^{(n)}$ ガ

$$\psi^{(n)}(x)=\psi(nx)$$

ニヨリ定ル．此ノ meromorphism ニヨル A ノ像ヲ $A^{(n)}$ デ表ハシ，A ノ n-分割体 (n-Teilungskörper) トイウ．ソノトキ 1° ニヨリ，$A/A^{(n)}$ ハ $N=n^{2g}$ 次ノ abel 拡大デ，ソノ ガロア群 ハ $\mathfrak{g}$ ト同型，即チ $(n, n, \cdots, n)$ 型デアル．ソレ故 仮定 2) ヨリ，$A/A^{(n)}$ ハ Kummer 拡大デアル．ソノ適当ナ生成元ヲ以下ニ求メヨウ．

7°． **3°** ヨリ $n\Theta_{a_i}-n\Theta \sim 0, i=1, \cdots, N$. 仮定ヨリ a_i ハ k 上 rational ダカラ，左辺ノ因子モサウデアリ，従ツテ **4°** ヨリ函数 $\varphi_i, \psi_i \in A$ ガアリ

$$(\varphi_i)=n(\Theta_{a_i}-\Theta), \quad \psi_i{}^n=\varphi_i{}^{(n)}.$$

A ノ函数ニ，次ノ同等関係ヲ導入スル．

$\omega \underset{n}{=} \pi$ in A トハ，$\omega=\pi\rho^n$, $\rho \in A$ ナルコトトスル．此レガ同等関係ヲ定メルコトハ明カ．此レニヨル class ヲ，A ノ n-class トイフコトニスル．同様ニ $A^{(n)}$ ニ於テモ n-class ガ定義サレル．ソノトキ

$$\underline{a_i+a_j=a_k \Longrightarrow \varphi_i\varphi_j \underset{n}{=} \varphi_k \quad \text{in } A}$$

ナゼナラバ，3° ニヨリ $\Theta_{a_i}+\Theta_{a_j}-\Theta_{a_k}-\Theta=0$, 故ニ $\omega \in A$ ガアリ

$$(\omega)=\Theta_{a_i}+\Theta_{a_j}-\Theta_{a_k}-\Theta,$$

ソシテ

$$\begin{aligned}(\varphi_k\omega^n) &= n(\Theta_{a_k}-\Theta)+n(\Theta_{a_i}+\Theta_{a_j}-\Theta_{a_k}-\Theta)\\ &= (\varphi_i\varphi_j)\end{aligned}$$

ソレ故，$\varphi_i\varphi_j=c\varphi_k\omega^n$, $c \in k$ ナル $c \neq 0$ ガアル．此レヨリ特ニ

$$\varphi_i(nx)\varphi_j(nx)=c\varphi_k(nx)\omega(nx)^n$$

即チ

$$(\psi_i(x)\psi_j(x))^n=c(\psi_k(x)\omega(nx))^n.$$

故ニ，x ヲ，此ノ式ノ両辺ガ $\neq 0, \infty$, ナル様ニ取レバ，c ハ k ニ於ル n 乗数ナルコトガワカル．故ニ $\varphi_i\varphi_j \underset{n}{=} \varphi_k$ in A.

次ニ $\varphi_i \underset{n}{=} 1$ in A トスレバ，$\varphi_i=\pi^n, \pi \in A$. ソレ故 $(\pi)=\Theta_{a_i}-\Theta$, 即チ，$\Theta_{a_i}-\Theta \sim 0$. 故ニ $a_i=0$.

従ツテ，φ_i ハ A ニ於ル アル n-class ト 1 対 1 ニ対応シ，$\varphi_i \longleftrightarrow a_i$ ニヨリ，φ_i ニヨリ代表サレル A ノ n-class ハ $\mathfrak{g}$ ト同型ナ群ヲナス．

而モ，$\psi_i{}^n=\varphi_i{}^{(n)}$, $\psi_i \in A$ ダカラ，A ハ $A^{(n)}$ ノ Kummer 拡大トシテ

$$A=A^{(n)}(\sqrt[n]{\varphi_i{}^{(n)}})=A^{(n)}(\psi_i)$$

ニヨリ与ヘラレル．

8°． $A/A^{(n)}$ ハ不分岐アーベル拡大デアルガ，ソノ逆モイヘル．

ソシテ又，上ニ求メタ $\psi_1, \cdots, \psi_N$ ハ $A^{(n)}$ 上一次独立デ，ソレ故 $A/A^{(n)}$ ノ base ヲナス．

ソレヲ云フタメニ，先ヅ次ノコトニ注意スル．3° ヨリ任意ノ s ニ対シ（s ハ k 上 rational).

$$\Theta_{a_i}-\Theta-(\Theta_{a_{i+s}}-\Theta_s) \sim 0$$

ダカラ此ノ左辺ヲ divisor トスル $\rho \in A$ ヲ

取レバ，$\varphi_i\rho^n \underset{n}{=} \varphi_i$ デ，$(\varphi_i\rho^n)=n(\Theta_{a_{i+s}}-\Theta_s)$. J ノ点 x ヲ任意ニ与エルトキ，s ヲ適当ニ取リ，$\Theta_{a_{i+s}}-\Theta_s$ ガ x ヲ含マヌ様ニ出来ルカラ，φ_i ノ代表スル A ニ於ケル n-class カラ $\overline{\varphi}_i$ ヲトリ，$\overline{\varphi}_i$ ガ x ニ於テ $\neq 0, \infty$ デアル様ニ出来ル．ソレニ対シ，$\bar{\phi}_i$ ヲ作レバ，明カニ

$$A = A^{(n)}(\bar{\phi}_i).$$

ソシテ $\bar{\phi}_i$ ハ x ヲ含ム J ノ任意ノ prime divisor ニ対シ整デアル．ソレ故 $\bar{\phi}_1,\cdots,\bar{\phi}_N$ ニ対スル判別式ヲ作ツテ見ル．$\mathfrak{S}$ ヲ $A/A^{(n)}$ ノ trace ヲ表ハストシテ

$$\mathfrak{S}(\bar{\phi}_i\,\bar{\phi}_j)=\begin{cases} n^{2g}\,\bar{\phi}_i\,\bar{\phi}_j & a_i+a_j=0 \\ 0 & a_i+a_j\neq 0. \end{cases}$$

ナゼナラ，$a_i+a_j=0$ ナラバ $\bar{\phi}_i\bar{\phi}_j\in A^{(n)}$ ダカラ，ソノ trace ハ $n^{2g}\bar{\phi}_i\,\bar{\phi}_j$ デアル．サモナケレバ，上ニ証明シタコトヨリ $\bar{\phi}_i\,\bar{\phi}_j\in A^{(n)}$，而モソノ $A^{(n)}$ 上ノ conjugate ハ 1 ノ n 乗根ヲ掛ケルコトニヨリ得ラレルカラ $\mathfrak{S}(\bar{\phi}_i\,\bar{\phi}_j)=0$ トナル．此レヨリ，

$$\det|\mathfrak{S}(\bar{\phi}_i\bar{\phi}_j)| = (n^{2g})^{n^{2g}}\,(\prod_i \bar{\phi}_i)^2.$$

$\overline{\varphi}_i$ ノ作リ方ニヨリ，此レハ x ニ於テ $\neq 0, \infty$，ソレ故 $\bar{\phi}_1,\cdots,\bar{\phi}_N$ ハ $A^{(n)}$ 上一次独立デ，故ニ x ヲ含ム素因子ニ対シ整ナ $A/A^{(n)}$ ノ base ヲナス．

所デ，X ヲ J ノ divisor，$nX\sim 0$ トスレバ，2° ヨリ，$X\equiv 0$ デ，ソレ故 3° ニ於ル同型対応ヨリ $X\sim\Theta_s-\Theta$；$ns=0$；故ニ s ハ a_i ノ一ツデアル．

ソレ故 $\varphi\in A$ ヲ $(\varphi)=nX$ ナル函数トスレバ

$$(\varphi)=nX\sim n(\Theta_{a_i}-\Theta)$$

ソレ故 $\varphi \underset{n}{=} \varphi_i$，故ニ $\varphi_1,\cdots,\varphi_N$ ハ A ニ於テ，ソノ divisor ガアル divisor ノ n 倍トナル様ナスベテノ函数ノナス n-class ヲ代表スル．

所デ，$K/A^{(n)}$ ヲ $A^{(n)}$ 上ノ n 次ノ不分岐巡回拡大トスレバ，此レハ Kummer 体トシテ $K=A^{(n)}(\sqrt[n]{\varphi})$ ニヨリ生成サレ，不分岐ナコトヨリ

$$(\varphi)=nX$$

ナル形デ，故ニ，今述ベタコトヨリ，K ハ $A/A^{(n)}$ ノ部分体デアル．

ソレ故 k ヲ特ニ代数的閉体トスレバ p ト素ナスベテノ n ニ対シ 6° ノ仮定 1) 2) ガ満サレルカラ，A 上ノ不分岐アーベル拡大（次数ガ p ト素ナ）ハ A ノ "n-Teilung" ニヨリ得ラレルモノ以外ニナイコトガワカル．

9°. 我々ノ目的ハ，J ノ rational point ノ研究デアツタ．Dヲ，k上 rational ナ，J ノ点ノナス群，ソノ n 倍ナル形ノ点ノナス群ヲ D^n，トスル．ソシテ商群 D/D^n ヲ考ヘ，D^n ヲ法トシテ同ジ coset ニ属スルコトヲ $\underset{n}{\sim}$ デ表ス．即チ

$s\underset{n}{\sim}s'$ トハ $s'=nr+s$, $r\in D$, ナルトキトスル．

A ニ於テ，φ_i ノ代表スル n-class ヲ Φ_i トカク．任意ノ $s\in D$ ニ対シ，8° デ述ベタ様ニ $\overline{\varphi}_i\in\Phi_i$，ガアリ，$\overline{\varphi}_i$ ハ s デ定義サレテ $\neq 0,\infty$．ソシテ Φ_i ノ函数ノ s ニ於ル，$0,\infty$ ト異ル値ハ，体 k ニ於テ，同一ノ n-class ニ属スルカラ，ソノ class ヲ $\Phi_i(s)$ ト書クコトニスル．ソノトキ，上ノ様ナ代表 $\overline{\varphi}_i$ ガ存在スルコトニヨリ，$\Phi_i(s)$ ハ実際ニ，k ニ於ル n-class ヲ表ス．

今 $D\ni s\underset{n}{\sim}0$ トスレバ，$s=nt$, $t\in D$, 故ニ，$\overline{\varphi}_i{}^{(n)}=\bar{\phi}_i{}^n$ ヨリ

$$\overline{\varphi}_i(s)=\overline{\varphi}_i(nt)=\bar{\phi}_i(t)^n$$

デ，t, $\bar{\phi}_i$ ハ k 上 rational ダカラ，$\bar{\phi}_i(t)\in k$ デ，$\bar{\phi}_i$ ノトリ方ヨリ $\bar{\phi}_i(t)\neq 0,\infty$. ソレ故 $\bar{\phi}_i(s)\underset{n}{=}1$ in k. 即チ $\Phi_i(s)=1$ (in k).

逆ニ $\Phi_i(s)=1$ (in k) トスレバ，$\overline{\varphi}_i(s)=c_i{}^n$, $c_i\in k$ デアル．ソコデ，$s=nt$ トスレバ，t ハ一般ニ k上 algebraic ナ点デアルガ，

$$\overline{\varphi}_i(s)=\bar{\phi}_i(t)^n=c_i{}^n,$$

デ，一方 1 ノ n 乗根ハスベテ k ニ含マレテイルカラ，$\bar{\phi}_i(t)\in k$ トナル．所ガ $\bar{\phi}_i$ ハ $A^{(n)}$ 上 A ヲ生成シ，$A^{(n)}$ ノ函数 $\rho^{(n)}$ ニ対シテハ $\rho^{(n)}(t)=\rho(nt)=\rho(s)\in k$ ダカラ，A ノ

スベテノ函数ガ t デ取ル値ハ k ニ属スル．故ニ t ハ k 上 rational，ソレ故 $s \underset{n}{\equiv} 0$．従ツテ

$$s \underset{n}{\equiv} 0 \Longleftrightarrow \Phi_i(s)=1(\text{in } k),\ i=1,\cdots,N.$$

ソレ故，$\Phi_i(s)$，$i=1,\cdots,N$ ノ値ト，D/D^n ノ元トガ，此ノ対応ニヨリ，一対一ニ対応スル．

所デ，$\mathfrak{g}$ ハ $(n,\cdots,n)$ 型ノ アーベル群デアツタ．ソノ生成元ヲ $a_1,\cdots,a_{2g}$ トスレバ，任意ノ $a_i\in\mathfrak{g}$ ハ $a_i=i_1a_1+i_2a_2+\cdots+i_{2g}a_{2g}$ ト表セル．ソノトキ，7° ヨリ $\bar\varphi_i \underset{n}{\equiv} \bar\varphi_1^{i_1}\cdots\bar\varphi_{2g}^{i_{2g}}$，ソシテ a_j $(j=1,\cdots,2g)$ ハ丁度 n 位ダカラ，$\bar\varphi_j^m \underset{n}{\equiv} 1 \Longleftrightarrow m$ ガ n ノ倍数，而モ $a_1,\cdots,a_{2g}$ ガ独立ナコトヨリ $\bar\varphi_1^{i_1}\cdots\bar\varphi_{2g}^{i_{2g}}=1 \Longleftrightarrow i_1,\cdots,i_{2g}$ ガスベテ n ノ倍数．結局，$\bar\varphi_i(s)$ ノ，k ニ於ル n-class ハ，$\bar\varphi_1(s),\cdots,\bar\varphi_{2g}(s)$ ノ k ニ於ル n-class ニヨリ一意ニ定ル．ソレ故，s ノ属スル D/D^n ノ coset ヲ $\bar s$ トスレバ，

$$(\Phi_1(s),\cdots,\Phi_{2g}(s)) \longleftrightarrow \bar s$$

ニヨリ D/D^n ヲ $2g$ 個ノ値 $(\Phi_1(s),\cdots,\Phi_{2g}(s))$ デ，一対一ニ表シ得ル．

10°. 上ノ一対一対応ハ，J ノ有理点ノ "n-分値" ノ分解法則ヲ表ス．即チ，$\nu=n\delta$ トシテ，$s\in D$ ニ対シ，$\nu^{-1}(s)$ ヲ考エル．此レハ，ソノ中ノ一ツヲ t トスルトキ，$\{t+a_i;\ i=1,\cdots,N\}$ ニヨリ表サレルガ，$\nu^{-1}(s)$ ガ，k 上 prime rational ナ，0 次元ノ cycle ノ和トシテ，ドノ様ニ分解サレルカガ，対応スル $(\Phi_1(s),\cdots,\Phi_{2g}(s))$ ノ，k ニ於ル n-class トシテノ値ニヨリ定メラレルノデアル．ソレ故，此ノ分解ハ，同一ノ coset $\bar s$ ニ属スル元スベテニ対シ，同ジ型デアル．

我々ハ更ニ，此ノ n-分値ニヨリ，常数体 k ガ，如何ニ拡大サレルカヲ考ヘヨウ．仮定ニヨリ a_i ハ k 上 rational ダカラ $k(t+a_i)=k(t)$．故ニ簡単ノタメニ此レヲ k_s ト書クコトニスル．明カニ $s\underset{n}{\equiv}s'$ ナラバ $k_s=k_{s'}$．故ニ k_s ハ，D ニ於ケル，s ノ D^n ヲ法トスル coset $\bar s$ ノミニシカヨラヌカラ，此レヲ $k_{\bar s}$ ト書クコトニスル．

既ニ 9° デ示シタ様ニ，$k_s \underset{n}{\equiv} k(\bar\phi_i(t))$．所ガ

$$\bar\phi_i(t+a_i)^n=\bar\phi_i(nt)=\bar\varphi_i(s)\in k$$

ダカラ，k_s ハ k ノ Kummer 拡大トシテ

$$k_s=k\,(\sqrt[n]{\bar\varphi_i(s)})$$

ニヨリ与ヘラレル．ソシテ k_s/k ノ次数ハ，対応スル $(\Theta_1(s),\cdots,\Theta_{2g}(s))$ カラ，実際ニ計算スルコトガデキル．特ニ

$$k_s=k \Longleftrightarrow s\underset{n}{\sim}0.$$

11°. 9° ニ於ル対応 $\bar s\longleftrightarrow(\Phi_1(s),\cdots,\Phi_{2g}(s))$ ハ D/D^n ノ加法及ビ vector ノ加法

$$(\Phi_1(s),\cdots,\Phi_{2g}(s))\cdot(\Phi_1(t),\cdots,\Phi_{2g}(t)) = (\Phi_1(s)\Phi_1(t),\cdots,\Phi_{2g}(s)\Phi_{2g}(t))$$

ノ意味デ，同型対応ニナル，ソシテ更ニ A/k ノ automorphism τ_s (s ニヨル translation) ガ，A ニ於ル n-class Φ_i ノ変換トシテ具体的ニ表現サレル．此レガ，Weil ノ Thèse [1] ニ implicit ニ含マレテイタ最モ重要ナ結果デアル．

J ニ於ル，s ダケノ translation ヲ T_s デ表シ，

$$\varphi^{\tau_s}(x)=\varphi(T_sx)=\varphi(x+s)$$

ニヨリ，A/k ノ automorphism τ_s ヲ定義スル．

$D\ni s$ ニ対シ $\bar\varphi_i(s)\neq0,\infty$；$\bar\varphi_i(0)\neq0,\infty$ ナル $\bar\varphi_i\in\Phi_i$ ヲ選ブ．

$(\bar\varphi_i)=n\mathrm{X}$ トスレバ，$n\mathrm{X}\sim0$ ヨリ，$\mathrm{X}\equiv0$．ソシテ $(\bar\varphi_i^{\tau_s})=n\mathrm{X}_{-s}$ デアル．$\mathrm{X}\equiv0$ ヨリ $\mathrm{X}_{-s}-\mathrm{X}\sim0$，故ニ $(\omega)=\mathrm{X}_{-s}-\mathrm{X}$ ナル $\omega\in A$ ガアル．ソノトキ

$$\bar\varphi_i^{\tau_s}=c\,\bar\varphi_i\,\omega^n,\quad c\in k,\quad c\neq0.$$

此レニヨリ特ニ

$$\bar\varphi_i^{\tau_s}(0)=c\,\bar\varphi_i(0)\,\omega^n(0)$$

即チ $\bar\varphi_i(s)=c\,\bar\varphi_i(0)\omega^n(0)$.

$0\underset{n}{\sim}0$ ヨリ 9° カラ．$\bar\varphi_i(0)\underset{n}{\equiv}1$ in k，ソレ故，$c\underset{n}{\equiv}\bar\varphi_i(s)$ in k，即チ

$$\bar\varphi_i^{\tau_s}\underset{n}{\equiv}\bar\varphi_i(s)\bar\varphi_i \quad \text{in } A.$$

或ハ，φ_i ノ属スル n-class Φ_i ニ対シ

$$\Phi_i^{\tau_s} = \Phi_i(s)\cdot\Phi_i.$$

更ニ，$r\in D$ ニ対シ

$$\Phi_i(s+r) = \Phi_i(s)\cdot\Phi_i(r).$$

即チ，始メニ述ベタ２ツノ結果ガ得ラレタ．

Weil [1] ニ於テ，此ノ結果ハ，基本第三種積分ノ，変数ト parameter トノ交換法則ニヨリ導カレテイル．ソレ故 Hasse [1] ガ指摘シタ様ニ，此レヲ，ソノ代数化ト見做スコトモ出来ルカモ知レナイ．然シ，代数的ニ，基本第三種積分ヲ定義スルコトモ出来テイナイ現在，ソレニツキ云々スルニハ，時期尚早デアル様ニ思ハレル．

尚，D/D^n ガ，丁度 $a_1,\cdots,a_N$ デ代表サレル場合，此レハ $A/A^{(n)}$ ノガロア群ニ対シ，Artin ノ reciprocity ト同ジ型ノ対応ヲ与ヘルガ，サウデナイ限リ，一般ニ τ_s ハ $A^{(n)}$ ノ元ヲ固定シナイ．故ニ此レカラ，ガロア群ノ表現ヲ得ヨウトスレバ，Hasse [1] ガ述ベタ様ナ方法ニヨラネバナラナイデアラウ．

12°. 2·3 ノ実例ヲ考ヘテミル．

(**a**) 最モ簡単ナ場合ハ，$g=1$，$n=2$，k ガ有限次代数体ノトキ，此ノトキ上ノ ψ_i ハ Jacobi ノ楕円函数ニ外ナラナイ．

即チ，$sn(u)$ ノ週期ヲ $4\mathrm{K}$, $2i\mathrm{K}'$ トシ，週期 $4\mathrm{K}$, $4i\mathrm{K}'$ ヲ持ツ，k 上定義サレタ楕円函数ノナス体ヲ考エル．週期平行四辺形内ニ於ル sn, cn, dn ノ零点ハ，各〻 $(0, 2\mathrm{K}, 2i\mathrm{K}', 2\mathrm{K}+2i\mathrm{K}')$, $(\mathrm{K}, 3\mathrm{K}, \mathrm{K}+2i\mathrm{K}', 3\mathrm{K}+2i\mathrm{K}')$, $(\mathrm{K}+i\mathrm{K}', 3\mathrm{K}+i\mathrm{K}', \mathrm{K}+3i\mathrm{K}', 3\mathrm{K}+3i\mathrm{K}')$ デ，極ハスベテ共通デ $(i\mathrm{K}', 2\mathrm{K}+i\mathrm{K}', 3i\mathrm{K}', 2\mathrm{K}+3i\mathrm{K}')$ デアル．ソレ故半週期 $0, 2\mathrm{K}, 2\mathrm{K}+i2\mathrm{K}', 2i\mathrm{K}'$ ニ対応シテ，上ノ φ_i ヲ作リ，$\psi_i^2(x)=\varphi_i(2x)$ トスレバ，ソレハ sn, cn, dn, 1 ニ外ナラナイ（常数因子ヲ除キ）．

ソレ故，上ノ考察ニ於テ，n ト素ナ素数 q ニ対シ，有理点ノ q 分値ニ対スル ψ_i ノ値ヲ考ヘルエトハ，興味深イコトデアルト思ハレル．ソシテソノ数論的性質ヲ考ヘルトキ，Jacobian variety ノ常数体ノ reduction ガ有効デアルカモ知レナイ．（常数体ノ reduction ニツイテノ簡単ナ事実ハ，玉河先生ニ，report トシテ提出サレル筈デアル．）*

一般ニ，此ノ ψ_i ハ，一次独立ナ n 位ノ Θ-函数ノ商ニ対応スルモノデアル．

(**b**) k ヲ有限体トスル．k ガ $q=p^f$ 個ノ元ヨリ成ルトスレバ，k ガ１ノ n 乗根ヲ含ムトイフ仮定ヨリ，$n\mid p^f-1$，ソレ故，k ニ於ル n-class ハ位数 n ノ巡回群ヲナス．

今 $n=l^h$，l ハ p ト素ナ素数トスル．

ソレニ対スル D ノ元ハ，明ラカニ，スベテ，位数有限デアル．ソシテ位数ガ l ト素ナ元ハ，D^n ニ含マレルカラ，D/D^n ハ，位数ガ l ノ巾デアル元ニヨリ代表サレル．所ガ，ソノ様ナ元ハ，Weil [2], [4] ニヨリ，l 進数 modulo 1 ノナス群ノ，$2g$ 個ノ直積デアル．ソレ故，上ニ見出サレタ対応

$$\bar{s} \longleftrightarrow (\Phi_1(s),\cdots,\Phi_{2g}(s))$$

ハ，此ノ場合，既ニ良ク知ラレテイル結果ヲ与ヘルニスギナイ．

又，$s \underset{n}{\simeq} 0$ デナケレバ，$\Phi_i(s)$ ハ k ニ於テ n 乗数デナイカラ，$[k_s:k]$ ハ l ノアル巾トシテ与ヘラレル．($s^{l^i}\in D^n$ ナラ $[k_s:k]=l^i$)．

(**c**) k ヲ $\mathfrak{p}$ 進体，n ヲ $\mathfrak{p}$ ト素ナ素数トスル．ソノトキ，k ニ於ル n-class ハ (n, n) 型ノアーベル群ヲナシ，$\varepsilon^i\pi^j, 0\leqq i, j<n$ ニヨリ代表サレル．但シ ε ハアル $\mathfrak{p}$ 単数，π ハ $\mathfrak{p}$ 素元．

此レヨリ，D/D^n ガ有限群デ，$(n,\cdots\cdots, n)$ ($4g$ 個) 型ノアーベル群ノ部分群ニ同型ナコトガワカル．

楕円函数体ノ場合ハ，丁度４個ノ巡回群ノ直積ニナルコトガ知ラレテイル．一般ノ場合ニモ，$4g$ 個ノ巡回群ト（ソノ部分群トデハナクテ）同型ニナルト思ハレル．

---

* このreportは発見されていない．（編者注）

§3. Weil ノ有限性定理

以后，常数体 k ヲ有限次代数体トスル．

n ヲ十分大トシ，$na=0$ ナル点，$a_1,\cdots,a_N\in$ J, $(N=n^{2g})$ ニ対シ divisors $\Theta_{a_i}+\Theta_{e-a_i}$, $i=1,\cdots,N$ ガ共通点ヲ持タヌ様ニスル，但シ e ハ J ノ任意ノ固定シタ，k 上 algebraic ナ点．以下，此ノ様ナ n ヲ一ツ固定シ，ソレニ対シ 6° ノ仮定 1)，2) ガ満サレル様ニ，k ノ有限次代数拡大ヲシ，簡単ノタメニソレヲ再ビ k ト書クコトニスル．ソシテ k 上 rational, k 上 algebraic ナコトヲ単ニ rational, algebraic トイフコトニスル．

13°. J ハ non-singular ダカラ，J ノ上ニ Weil ノ意味ノ distribution (Weil [1]) ヲ定義シテ，分解定理ガ成立スルヨウニ出来ル．

X ヲ J ノ，algebraic divisor トシ，X ニ対応スル distribution ノ，algebraic point s ニ於ル値ヲ $\omega(X,s)$ デ表ハスコトニスル．k ニ於ル イデヤル類数ハ有限ダカラ，各イデヤル類ノ代表ヲ $\mathfrak{a}_1,\cdots,\mathfrak{a}_h$ トスル．ソシテ X, s ガ rational ナラ $\omega(\mathrm{X},s)$ ハ k ノ整イデヤルダカラ，ソノ属スル類ノ代表 $\mathfrak{a}_i$ ヲ取リ，$\omega^1(\mathrm{X},s)=\omega(\mathrm{X},s)\ \mathfrak{a}_i^{-1}$ トスレバ，此レハ k ノ数ヲ値トシ，ω ト同等ナ distribution ヲ定義スル．ソレ故，初メカラ，$\omega(X,s)$ ガ k ノ数ヲ値トスルトシテ差支ヘナイ．ソノトキ，distribution ノ分解定理ヨリ，7° ニ定義サレタ函数 φ_i：

$$(\varphi_i)=n\,(\Theta_{a_i}-\Theta)$$

及ビ，J ノ rational point $s\in D$ ニ対シ

$$\varphi_i(s)=\frac{\lambda_i(s)}{\mu_i(s)}\left(\frac{\omega(\Theta_{a_i},s)}{\omega(\Theta,s)}\right)^n$$

デ，$\lambda_i(s)$，$\mu_i(s)$ ハ s ニヨラヌアル一定ノ有理整数 c_i ヲ割ル k ノ整数デアル．

φ_i ノ代リニ，同ジ n-class ニ属スル $\bar{\varphi}_i\in\Phi_i$ ヲ取レバ，$\bar{\varphi}_i=\varphi_i\,\rho^n$, $\rho\in A$ ダカラ，

$$\bar{\varphi}_i(s)=\frac{\lambda_i(s)}{\mu_i(s)}\left(\frac{\omega(\Theta_{a_i},s)}{\omega(\Theta,s)}\cdot\rho(s)\right)^n,$$

ソレ故，一ツノ i ニ対シ，$\lambda_i(s)$，$\mu_i(s)$ ノ作ル単項イデヤルニハ，有限個ノ可能性シカナイ．又 k ノ単数ハ，有限個ノ base ヲ持ツカラ，ソノ n-class ハ有限，ソレ故，$\varphi_i(s)$ ノ代表スル，k ニ於ケル n-class $\Phi_i(s)$ ニハ，有限個ノ可能性シカナイ．故ニ 9° ニ於ル同型対応ヨリ，D/D^n ガ有限群デアルコトガ分ツタ．

14°. 我々ハ，Weil [1] ニ従ツテ，"descent infini" ノ方法ヲ適用スル．ソノタメニ，J カラ，アル射影空間内ヘノ写像ヲ考ヘル．

5° ニ引用サレタ定理ヲ適用スルタメニ，7° ニ定義サレタ函数 φ_i,ψ_i，ノ divisor ヲ吟味セネバナラヌ．$\nu=n\delta$ ト書クコトニシテ，

$$(\varphi_i)=n(\Theta_{a_i}-\Theta),\ \psi_i^n=\varphi_i^{(n)} \text{ ヨリ}$$
$$(\psi_i)=\bar{\nu}^1(\Theta_{a_i}-\Theta).$$

所ガ，4° ヨリ $\bar{\nu}^1(\Theta)-n^2\Theta\equiv 0$ ダカラ，J ノ rational point e ガアリ，

$$\bar{\nu}^1(\Theta)-n^2\Theta\sim\Theta_e-\Theta.$$

ソレ故 $\pi\in A$ ヲ $(\pi)=\bar{\nu}^1(\Theta)-(n^2-1)\Theta-\Theta_e$ ナル様ニ取レバ，

$$\xi_i=\pi\,\psi_i$$

トスルトキ

$$(\xi_i)=\bar{\nu}^1(\Theta_{a_i})-(n^2-1)\Theta-\Theta_e.$$

ソノトキ J ヨリ $n^{2g}-1$ 次ノ射影空間 $\mathrm{P}^{n^{2g}-1}$ 内ヘノ写像 Ξ,Ψ ヲ

$$\Xi:\ s\to\Xi(s)=(\xi_1(s),\cdots,\xi_N(s)),$$
$$\Psi:\ s\to\Psi(s)=(\psi_1(s),\cdots,\psi_N(s)),$$

$N=n^{2g}$，デ定義スル．$\Xi(s)$ ト $\Psi(s)$ トハ，ソレガ定義サレル限リ，射影空間内ノ点トシテ，同ジ点ヲ表ス．

更ニ，r ヲ J ノアル rational point トシテ，$\mathrm{J}\longrightarrow P^{n^{2g}-1}$ ノ写像 F_r 及ビ $\mathrm{J}\longrightarrow P^{n^{2g(n^2-2)}-1}$ ノ写像 G' ヲ

$$F_r:\ s\to F_r(s)=\Xi(ns-r)=(\xi_1(ns-r),\ \cdots\cdots\cdots,\xi_N(ns-r)),$$

$$G':\ s\to G'(s)=(\cdots\cdot,\psi_{i_1}(s)\cdots\psi_{i_{n^2-2}}(s),\ \cdots\cdots\cdots\cdots)$$

ニヨリ定義スル．但シ $i_1,\cdots,i_{n^2-2}$ ハ，1，$\cdots,n^{2g}$ ノ値ヲ独立ニ動クモノトスル．ソシテ，F_r, G' ノ座標ヲ表ス函数ヲ $f_i^{(r)}, g_i'$ デ表スコトニスレバ，T_r ヲ r ニヨル translation トシテ，

$$(f_i^{(r)})=\nu^1T_r((\xi_i)))=\nu^1T_r\nu^1(\Theta_{ai})-\nu^1((n^2-1)\Theta_r+\Theta_{e+r}),$$

$$(g_i')=\sum_{k=1}^{n^2-2}\nu^1(\Theta_{a_{i_k}})-\nu^1((n^2-2)\Theta).$$

今 t_1, t_2 ヲ J ノ rational points デ $t_1+t_2=n^2r+e$ ナルモノトスレバ，3° ヨリ

$$(n^2-2)\Theta+\Theta_{t_1}+\Theta_{t_2}\sim(n^2-1)\Theta_r+\Theta_{e+r},$$

ソレ故，$\nu^1\{(n^2-2)\Theta+\Theta_{t_1}+\Theta_{t_2}-(n^2-1)\Theta_r+\Theta_{e+r}\}\sim 0$. 故ニ此ノ式ノ左辺ヲ divisor トシテ持ツ函数 $\rho\in A$ ヲトリ，

J $\longrightarrow P^{n^{2g(n-2)}-1}$ ノ写像 G ヲ

$$G:\ s\to G(s)=(\cdots,\rho(s)\psi_{i_1}(s)\cdots\psi_{i_{n^2-2}}(s),\cdots)$$

デ定義スレバ，射影空間ノ点トシテハ $G(s)$ ト $G'(s)$ トハ同ジデ（両者ガ共ニ定義サレルトキ），G ノ座標ノ函数 g_i ノ divisor ハ

$$(g_i)=\nu^1(\sum_{k=1}^{n^2-2}\Theta_{a_i}+\Theta_{t_1}+\Theta_{t_2})-\nu^1((n^2-1)\Theta_r+\Theta_{e+r}).$$

ソシテコノ § ノ始メニナサレタ仮定ヨリ，$\Theta_{a_i}(i=1,\cdots,N)$ ハ共通点ヲ持タヌカラ，$\nu^1(\Theta_{a_i})$，$T_r\nu^1(\Theta_{a_i})$，$\nu^1T_r\nu^1(\Theta_{a_i})$ モ各々，共通点ヲ持タヌ．

又 $\nu^1(\sum_{k=1}^{n^2-2}\Theta_{a_i}+\sum_{k=1}^{2}\Theta_{tk})$ ナル $n^{2g(n^2-2)}$ 個ノ divisors ノ共通点ハ $\nu^1(\Theta_{t_1}+\Theta_{t_2})$ ニ含マレル点ダケデアル．ソレ故，5° ニ引用サレタ定理ヲ写像 F_r, G ニ対シ適用デキル．

即チ H ヲ，射影空間ノ点ノ高サヲ表ストシ，$\nu^1(\Theta_{t_1}+\Theta_{t_2})$ ニ含マレヌ J ノ algebraic point s ニ対シ

$$H[F_r(s)]\geqq cH[G(s)]$$

但シ $c\neq 0$ ハ，s ニヨラヌ，アル正ノ数．

所ガ，点ノ高サハ，ソノ射影座標ノ取リ方ニヨラヌカラ，　$H[G(s)]=H[G'(s)]$，$H[\psi(s)]=H[\varXi(s)]$．ソシテ又，G' ノ定義ニヨリ　$H[G'(s)]=H[\psi(s)]^{n^2-2}$（証明ハ Weil [3] ヲ参照）．ソレ故，結局

$$\underline{H[F_r(s)]\geqq cH[\varXi(s)]^{n^2-2}.}$$

15°．13° ニヨリ D/D^n ハ有限群デアル．ソノ代表元ヲ $r_1,\cdots,r_l$ トスレバ，任意ノ $s\in D$ ニ対シ

$$s=ns_1+r_{i_1},\ s_1=ns_2+r_{i_2},\ \cdots\cdots\cdots$$

ソレ故

$$s=n^ks_k+n^{k-1}r_{i_k}+\cdots\cdots+nr_{i_2}+r_{i_1}.$$

モシ $s_k=0$ トナル k ガアレバ s ハ $r_1, r_2,\cdots,r_l$ カラ生成サレル群ニ属スル．

スベテノ s_k, $k=1,2,\cdots\cdots$ ガ 0 デナイトスル．

14° ニ定義サレタ F_r トシテ，F_{-r_i}：$F_{-r_i}(s)=\varXi(ns_1+r_{i_1})=\varXi(s)$ ヲ取ル．又 G ヲ定義スルトキノ t トシテ，$t_1=a_i$，$t_2=n^2r+e-a_i$ トシ，ソレデ定義サレル G ヲ G_i ト書クコトニスル．ソノトキ s_1 ガ $\nu^1(\Theta_{a_i}+\Theta_{n^2r+e-a_i})$ ニ含マレヌ様ニ i ヲ取レバ（ソレハ此ノ § ノ始メニナサレタ仮定カラ可能デアル），14° ヨリ s_1 ニハヨラヌ常数 c_i ガアリ

$$H[F_{-r_{i_1}}(s_1)]\geqq c_iH[\varXi(s_1)]^{n^2-2}.$$

$a_1,\cdots,a_N$ ハ有限個ダカラ，$c=\operatorname{Min}_i c_i$ トスル．c ハ尚モ r_{i_1} ニ依ルガ，$r_1,\cdots,r_l$ ハ有限個ダカラ，ソノ中カラ再ビ最小ノモノヲ取リ c トスレバ，$c>0$ デアル．

而モ $s=ns_1+r_{i_1}$ トスレバ，$F_{-r_i}(s)=\varXi(s)$ ヨリ

$$\left(\frac{1}{c}H[\varXi(s)]\right)^{\frac{1}{n^2-2}}\geqq H[\varXi(s_1)]$$

全ク同様ニシテ

$$\left(\frac{1}{c}H[\varXi(s_1)]\right)^{\frac{1}{n^2-2}}\geqq H[\varXi(s_2)],$$

此レヲ繰返セバ

$$\left(\frac{1}{c}H[\varXi(s)]\right)^{\left(\frac{1}{n^2-2}\right)^k}\geqq H[\varXi(s_k)].$$

所ガ s_k ハ rational point ダカラ $\Xi(s_k)$ ノ座標ハ k ノ数デアル．P^{n^2-1} ニ於テ k ノ数ヲ座標トシ，高サ有限ナル点ハ有限個シカナイカラ，k ヲ十分大トスレバ $\Xi(s_k)$ ハ有限個ノ点ノ一ツト一致セネバナラヌ．

所ガ写像 Ξ ハ 1対1 デアル．ナゼナラ，$s \neq s'$，$\Xi(s)=\Xi(s')$ トスレバ

$$\psi_i(s) = \alpha\,\psi_i(s') \qquad \alpha \in k,$$

ψ_i ノ中ニハ常数 1 ガアルカラ，$\alpha = 1$．故ニ

$$\psi_i(s) = \psi_i(s') \quad i = 1, \cdots, N.$$

ψ_i ハ A ヲ構成スルカラ，A ノスベテノ函数ニ対シ s, s' ニ於ル値ガ一致スルコトニナリ，$s = s'$．

ソレ故 k ヲ十分大トスレバ，$s_k = 0$ 又ハ s_k ハアル有限個ノ点ノ一ツト一致スル．故ニ D ハ有限個ノ元カラ生成サレル．

此レガ，始メニ述ベタ Weil ノ有限性定理デアル．

［注意］ 以上ノ推論ハ，J ヲ jacobian variety トセズニ，一般ノ abelian variety トスルトキモ，些少ノ変更ヲ加ヘレバ成立ツ．ソノ際，問題トナルノハ，一般ノ abelian variety デハ，ソノ Picard variety ガ，始メノ variety ト同型デハナクナリ，有限個ノ点ヨリ成ル部分群ニヨル商群ト同型ニナルコトデアル．然シ此ノコトハ，“有限性” ニ関スル問題ニ対シテハ，本質的ナ障害トナルモノデハナイ．

文　献

H. Hasse, [1]: Der n-Teilungskörper eines abstrakten elliptischen Funktionenkörpers als Klassenkörper, nebst Anwendung auf den Mordell-Weilschen Endlichkeitssatz. (Math. Zeitschr. Bd. 48 (1942), pp. 48-66).

松阪輝久 [1]: Jacobi 多様体の構成について．(数学 3巻, (1951), pp. 199~206).

A. Weil, [1]: L'arithmétique sur les courbes algébriques. (Acta. Math. 52 (1929), pp. 281-315).

[2]: Variétés abéliennes et courbes algébriques. (Actualités sci. et ind. 1064, (1948)).

[3]: Arithmetic on algebraic Varieties. (Ann. of Math. (2) 53 (1951), pp. 412-444).

[4]: Sur les fonctions algébriques à corps de constantes fini. (C.R. Acad. Paris, 210 (1940), pp. 542-595).

W.L. Chow, [1]: The Jacobian varieties of an algebraic curves; (forthcoming in the Amer. Journ.).*

* Amer. Journ. Math. 76 (1954), 453-476 (編者注)

日本数学会講演アブストラクト

十分多くは無い虚数乗法を持つ Abel 多様体の ζ 函数について

$\mathscr{A}_0(A)$ の部分環でその表現が CM 型の体の表現の直和に分解する様なものがある様な Abel 多様体 A の，Hasse の ζ 函数が Hecke の L で表せること．

(1957 年 5 月 16 日講演)

Weil 群の表現について

Weil 群 $G_{K,k}$ の表現で，$C_K{}'$ 上に (A_0) 型の指標を induce する様なものは，対応する局所表現により特性付けられ得る．これを応用して，k 上定義された Abel 多様体の，K 上の ζ 函数が Hecke の L で表せれば，k 上の ζ 函数についても同様なことが成立つことがいえる．

(1957 年 5 月 16 日講演)

代数多様体の 0 次 cycle の同等類について

V^r を有限体 $k=GF(q)$ の上の正規代数多様体，$V(n)$ を V の n 個の対称積，(A^m, f) を V の Albanese 多様体，f_n を $V(n)$ から A への標準写像とする．A の点 a に対し，$W_{(a)}=f_n{}^{-1}(a)$ とおく．(A, f) により V の 0 次 cycle の類別が定まり，類体論によりこれは V の不分岐拡大に対応する (Lang)．この拡大に対する L- 函数を計算して次の評価式を証明する ($N_\nu(X)$ は X の体 $GF(q^\nu)$ 上の有理点の個数を表わすとし，また $m=\dim A$, $r=\dim V$ とする)．$n\geqq 2m+1$ のとき，$\nu\to\infty$ に対し

$N_\nu(V(n)=N_\nu(A)N_\nu(W_{(0)})+O(q^{r(r-1/2)n})$.

このことから，$n\geqq 2m+1$ に対し，$W_{(a)}$ が A のすべての点 a に対し既約で，同一次元 $rn-m$ をもつことがわかる．定数体の reduction を使ってこの結論は任意の定義体に対して確かめられる．また上の評価式から，V の ζ-函数と A のそれとの間の関係がわかる．

(1958 年 5 月 29 日講演)

リーマン面の週期と Hasse-Witt の行列とについて

最近井草氏は次の定理を得たといわれる：E_λ を $y^2=x(1-x)(\lambda-x)$ で定義される平面楕円曲線，定義体の標数を p として，E_λ の Hasse-Witt の不変量を A_λ とすれば (A_λ は λ の多項式 (Deuring))，A_λ は次の微分方程式を満たす：

$$\lambda(1-\lambda)\frac{d^2A_\lambda}{d\lambda^2}+(1-2\lambda)\frac{dA_\lambda}{d\lambda}-\frac{1}{4}A_\lambda=0.$$

これは標数 0 における E_λ の第一種積分の週期の満たす微分方程式と同じである．ここでは，この結果が一般の genus の代数曲線の系に対しても成り立つことを証明する．すなわち，一定の genus g をもつ平面代数曲線のある系列を考え，その分岐点をパラメーターとみて，代数曲線の Hasse-Witt の行列を分岐点の函数と考える．一方，標数 0 の場合に同様のパラメーターを入れた系列を考え，この系列の曲線の満たすパラメーターについての偏微分方程式を mod p で考えたものを，対応する標数 p の系列の Hasse-Witt の行列が満たすことを証明する．

(1958 年 5 月 29 日講演)

Hilbert のモヂュラー函数について

k を n 次の総実な代数体，R を k の整数全体の作る環とする．k の valuation vector を係数とし，その行列式が k のイデール

となるような，2次の正方行列全体の作る環を V とする．そのとき，V の各元に対し，乗法子環が R を含むようなある一つの Abel 多様体（n次元の）を対応させることができる．Vの部分群で，kの元を係数とする行列全体から成るものを $k(2)$ とするとき，この対応により同型な Abel 多様体に対応する行列全体は，V のある最大 compact 部分群 C および $k(2)$ による double coset を作る．このことから，V の上で定義され，上記の double coset の上で常数値を取り，ある種の解析性の条件を満す函数として Hilbert のモヂュラー函数を定義することができる．この方法によれば Hecke の作用素，Dirichlet 級数との対応等の理論は極めて自然に構成され，Abel 多様体との関連も明白になる．同じ見地から，Stufe のある場合を論ずることも出来る．

（1958年10月28日講演）

数学 第7巻，4号 (1956 年5月)

Jacobi 多様体と数体[1)]

§1. 緒 言

虚2次体上の類体はすべて，虚数乗法を持つ楕円函数のモズルと，周期の等分値とにより生成され得るというのが，古典的な虚数乗法論の主要な結果である．所で，この理論の，より高次の体への一般化を試みたものとしては，これまで，Hecke [2]，[3] の理論があるにすぎない．ここでは，此等の理論を特別の場合として含む，一般的な虚数乗法論を作ることを目標とする．

一方，代数体上定義された代数曲線の ζ-函数に対し，Hasse は，それが全複素平面で有理型で，普通の型の函数等式を満すであろうと予想した．この予想は，$ax^m+by^n+c=0$ という形の方程式で定義される曲線に対しては Weil [8] により，虚数乗法を持つ楕円曲線に対しては Deuring [1] により，それぞれ肯定的に解決された．ここでは，上の理論の応用として虚数乗法を十分多く持つ曲線及び Abel 多様体に対し，この予想を証明する．

§2. 準 備[2)]

A を g 次元 Abel 多様体，$\mathfrak{A}(A)$ を A の endomorphism の環，$\mathfrak{A}_0(A)$ を，$\mathfrak{A}(A)$ の係数環の，有理数体への拡大により得られる algebra とする．$\mathfrak{A}(A)$ の元は，A の1次第1種微分の空間の1次変換を引き起すから，此の空間の基底 $\omega_1,\cdots,\omega_g$ を用いて，$\mathfrak{A}(A)$ を逆表現することが出来る．$S(\mu)$，$\mu\in\mathfrak{A}(A)$ を，此の逆表現の行列とする．今以上すべてのものが，有限次代数体 k の上で定義されているとする．$\mathfrak{P}$ を k の素イデアルとし，此等のものを mod. $\mathfrak{P}$ で reduce して得られるものを，対応する記号の上に～をつけて表わすことにする．

補助定理 1. ほとんどすべての（つまり有限個を除く）$\mathfrak{P}$ に対し，$\tilde{\mathrm{A}}$ は Abel 多様体で，$\mu\to\tilde{\mu}$ は，$\mathfrak{A}(A)$ から $\mathfrak{A}(\tilde{\mathrm{A}})$ の中への同型写像であり，又 $\tilde{\omega}_1,\cdots,\tilde{\omega}_g$ は $\tilde{\mathrm{A}}$ の1次第1種微分の空間の基底で，それによる $\mathfrak{A}(\tilde{\mathrm{A}})$ の逆表現を $\tilde{\mathrm{S}}(\tilde{\mu})$ と書けば，$\widetilde{\mathrm{S}(\mu)}=\tilde{S}(\tilde{\mu})$.

この補助定理の中で除かれた素因子 $\mathfrak{P}$ を，A の例外素因子という．以後 mod. $\mathfrak{P}$ で考えることは，例外でない $\mathfrak{P}$ に限ることにし，一々断らない．'ほとんどすべての' という形容詞も省略することがある．

仮定 1. 以後，$\mathfrak{A}(A)$ が，$2g$ 次の代数体 K の order と同型な環 R を含むと仮定する．

K を含む最小の正規体を $\bar{K}$ と書く．R のすべての元に対し，$\mathrm{S}(\mu)$ は，$k\cdot\bar{K}$ において，同時に対角形に変換される．そのとき，

$$\mathrm{S}(\mu)=\begin{pmatrix}\mu_1^{\sigma_1} & & \\ & \ddots & \\ & & \mu^{\sigma_g}\end{pmatrix},\quad \mu_1\in K$$

と書ける．σ_i は K から複素数体の中への同型写像で，特に σ_1 は恒等写像とする．以後 μ と μ_1 を同一視する．

補助定理 2. σ_0 を，K の複素共軛同型とすれば，$\sigma_1,\cdots,\sigma_g,\sigma_0\sigma_1,\cdots,\sigma_0\sigma_g$ は，K から複素数体の中への，$2g$ 個の同型写像すべてである．それ故とくに，A が単純ならば，K は，総実なる g 次の体 K_0 の，総虚な2次の拡大である．

系. k が $\bar{K}$ を含むとき，σ_i の k への接続の一つを再び σ_i で表すことにする．そのとき k の数 β に対して，$\mathrm{N}_{k/K}(\beta^{\sigma_1^{-1}}\cdots\beta^{\sigma_g^{-1}})$ のすべての共軛は，その絶対値が $(\mathrm{N}\beta)^{1/2}$ である．ここで N は絶対ノルムを表わす．

§3. Hasse の ζ-函数

まず，R が K の principal order であるとする．R のイデアル $\mathfrak{a}$ に対し，$\alpha a=0$, $\forall\alpha\in\mathfrak{a}$ を満す A の点 a を，A の $\mathfrak{a}$ 分点という，a が，$\mathfrak{a}$ より本当に大きいどんなイデアル $\mathfrak{b}$ に対しても $\mathfrak{b}$ 分点でないとき，本来の $\mathfrak{a}$ 分点という，$\mathfrak{a}$ 分点は $\mathrm{N}(\mathfrak{a})$ 個，本来の $\mathfrak{a}$ 分点は $\varphi(\mathfrak{a})$ 個ある，ただし $\varphi(\mathfrak{a})$ は $\mathfrak{a}$ の Euler 函数．それ故，a を本来の $\mathfrak{a}$ 分点とするとき，σ を，$k(a)/k$ の同型写像とすれば，$a^{\sigma}=\mu_{\sigma}a$ なる $\mu_{\sigma}\in R$ がある．μ_{σ} を含む mod. $\mathfrak{a}$ での剰余類を $\langle\sigma\rangle$ と書けば，$\langle\sigma\rangle$ は σ により一意に定まり，$\mathfrak{a}$ と素な剰余類である．k はすべての μ の定義体だから，$a^{\sigma\tau}=(\mu_{\sigma}a)^{\tau}=\mu_{\sigma}\mu_{\tau}a$．それ故，$\sigma\to\langle\sigma\rangle$ は，$k(a)/k$ の Galois 群から，mod. $\mathfrak{a}$ の素剰余類群の中への同型写像を与える．特に，$k(a)/k$ は Abel 拡大である．

さて，x を A の k 上の generic point（以下 G. P. と略す）とする．$\mathfrak{P}$ を k の素因子とする．そのとき，$\tilde{x}\to\tilde{x}^{\mathrm{N}\mathfrak{P}}$ は $\tilde{A}$ の endomorphism であるが，仮定 1 より，それは，R の中に逆像 $\pi_{\mathfrak{P}}$ を有する．（$\tilde{x}^{\mathrm{N}\mathfrak{P}}$ は，$\tilde{x}$ の座標の $\mathrm{N}\mathfrak{P}$ 乗を座標に持つ点を表わす．）又，例外素因子と素な k のイデアル $\mathfrak{B}=\mathfrak{P}_1\cdots\mathfrak{P}_r$ に対し $\pi_{\mathfrak{B}}=\pi_{\mathfrak{P}_1}\cdots\pi_{\mathfrak{P}_r}$ と定義する．

仮定 2. $k\supset\bar{K}$（§3 のみ）．

補助定理 3.[3]

$$(\pi_{\mathfrak{B}})=\mathrm{N}_{k/K}(\mathfrak{B}^{\sigma_1^{-1}}\cdots\mathfrak{B}^{\sigma_g^{-1}}),$$

今，上に考えたイデアル $\mathfrak{a}$ が，K の判別式の 2 倍と素であるとする．適当な自然数 F を取れば，$k(a)$ は，k 上の，F を法とする Strahl 類体に含まれる．ここで F が $\mathfrak{a}$ で割れるとしてよい．$\sigma(\mathfrak{P})$ を $k(a)/k$ での，$\mathfrak{P}$ の Frobenius 自己同型とすれば，$\widetilde{a^{\sigma(\mathfrak{P})}}=\tilde{a}^{\mathrm{N}\mathfrak{P}}=\tilde{\pi}_{\mathfrak{P}}\tilde{a}$ だから，$\pi_{\mathfrak{P}}$ は $\langle\sigma(\mathfrak{P})\rangle$ に属する．故に，$\sigma(\mathfrak{B})$ を $k(a)/k$ での，$\mathfrak{B}$ の Artin 記号とすれば $\pi_{\mathfrak{B}}\in\langle\sigma(\mathfrak{B})\rangle$．特に，$k$ の数 β に対し，$\beta\equiv1$ mod. F ならば，$\pi_{(\beta)}\equiv1$ mod. $\mathfrak{a}$．それ故補助定理 2 の系と補助定理 3 とより，$\mathfrak{a}$ についての仮定を使えば，此のような β に対しては，

$$\pi_{(\beta)}=\mathrm{M}_{k/K}(\beta^{\sigma_1^{-1}}\cdots\beta^{\sigma_g^{-1}}).$$

それ故，$\chi(\mathfrak{B})=\pi_{\mathfrak{B}}/|\pi_{\mathfrak{B}}|$ は k の量指標である．

さて，A の，Hasse の ζ-函数は例外でない $\mathfrak{P}$ に対する $\tilde{A}$ の合同 ζ-函数の無限積として定義される．所が後者は，$\tilde{A}$ の，$\tilde{k}$ 上 f 次の点の個数 N_f から計算される[4]，所が N_f は，A の，$\tilde{\pi}_{\mathfrak{P}}^f-1$ 分点の個数，故に $N_f=\mathrm{N}(\pi_{\mathfrak{P}}^f-1)$．以上をまとめて，

定理 1. 仮定 1，2 の下に，A の k 上の ζ-函数 $\zeta_A(s)$ は，次のように表わせる．

$$\zeta_A(s)=\Phi(s)\prod_{\mu=1}^{2g}\prod_{i_1\cdots i_\mu}L\left(s-\frac{\mu}{2},\chi^{\rho_{i_1}}\cdots\chi^{\rho_{i_\mu}}\right)^{(-1)^\mu}$$

ここで χ は上に定義された量指標，ρ_i は K の $2g$ 個の同型写像で，積は，一定の μ に対しては，ρ_i の，μ 項の組合せすべてにわたる．L は k の L-函数，$\Phi(s)$ は，簡単な形の有理函数で，例外素因子に由来する[5]．

系. C を，有限次代数体 k 上定義された代数曲線で，その Jacobi 多様体が，上の定理の条件を満すものとする．そのとき C の ζ-函数に対しても，Hasse の予想が成立つ．即ちそれは，量指標の L-函数により表せる．

§4. 不分岐拡大体の構成

仮定 3. 以後 R は K の principal order とする．$\bar{K}$ の Galois 群を G とし，K に対応するその部分群を H とする．$\sum_{i=1}^{g}H\sigma_i\sigma=\sum_{i=1}^{g}H\sigma_i$ を満す σ 全体の作る，G の部分群を，G^* とする．K^* を，G^* に対応する，$\bar{K}$ の部分体とする．そのとき次の形の分解が成立つ：$\sum_{i=1}^{g}H\sigma_i=\sum_{j=1}^{s}\tau_jG^*$．今，$K^*$ のイデアル $\mathfrak{b}$ に対し，$\Gamma(\mathfrak{b})=\mathfrak{b}^{\tau_1^{-1}}\cdots\mathfrak{b}^{\tau_s^{-1}}$ とおけば $\Gamma(\mathfrak{b})$ は K のイデアルになる．そのとき k の素イデアル $\mathfrak{P}$ で，$\mathrm{N}_{k/K^*}=\mathfrak{P}=\mathfrak{p}^f$ なるものに対し，$(\pi_{\mathfrak{P}})=\Gamma(\mathfrak{p})^f$．

仮定 4. K^* の，ほとんどすべての 1 次素イデアル $\mathfrak{p}$ に対し，$\Gamma(\mathfrak{p})$ は，K の，どんな真部分体のイデアルにもならない．

補助定理 4. 以上の仮定が成立てば，Aは単純である．又 K^* の1次の $\mathfrak{p}$ の，k における素因子 $\mathfrak{P}$ に対し，$\tilde{A}$ も単純である．したがって $\mathfrak{A}(A)$, $\mathfrak{A}(\tilde{A})$ は共に R に同型である．

補助定理 5. xをkの上の G.P. とする．R のイデアル $\mathfrak{a}$ に対し，A から，或る Abel 多様体 B への準同型 $\lambda_{\mathfrak{a}}$ で，$k(\lambda_{\mathfrak{a}}x)=\cup_{\alpha\in\mathfrak{a}}k(\alpha x)$ を満すものがある．此の $\lambda_{\mathfrak{a}}, B$ は同型を除き一意に定り，k 上で定義され得る．そして，$\lambda_{\mathfrak{a}}A\cong\lambda_{\mathfrak{b}}A$ は，K で $\mathfrak{a}\sim\mathfrak{b}$ なること，即ち，$\mathfrak{a}=\mathfrak{b}(\beta)$ なる K の数 β があることと同等である．

$\mathfrak{p}$ を，K^* の1次素イデアル，$\mathfrak{P}$ を$\tilde{k}$における その素因子とすれば，明かに $\tilde{k}(\tilde{x}^p)=\tilde{k}(\lambda_{\Gamma\widetilde{(\mathfrak{p})}}\tilde{x})$. ただし $p=N\mathfrak{p}$.

ここで $B=\lambda_{\mathfrak{a}}A$ として，$\mathfrak{A}(A)$ と $\mathfrak{A}(B)$ との同型対応を，$\lambda_{\mathfrak{a}}\mu=\mu^*\lambda_{\mathfrak{a}}$ なる $\mu\in\mathfrak{A}(A)$ と $\mu^*\in\mathfrak{A}(B)$ とを対応させることにより固定する．そのとき，$\mathfrak{A}(A)$ のイデアル $\mathfrak{b}$ の像を $\mathfrak{b}^*$ と書けば，明かに，$\lambda_{\mathfrak{b}^*}\lambda_{\mathfrak{a}}A\cong\lambda_{\mathfrak{a}\mathfrak{b}}A$.

一方，kの同型写像 σ に対し，AからA^σへの準同型 λ があると仮定する．σ は $\mathfrak{A}(A)$ から $\mathfrak{A}(A^\sigma)$ への同型 $\bar{\sigma}:\mu\to\mu^{\bar{\sigma}}$ を引き起す．故に，$\mu^{\bar{\sigma}}\to\mu^*$ は，$\mathfrak{A}(A^\sigma)$ の自己同型である．所が，

補助定理 6. 仮定4の下に，$\mathfrak{A}(A)$ の自己同型 Σ に対し，表現 $S(\mu^\Sigma)$ と $S(\mu)$ とが同じ特性根を持てば，Σ は実は恒等置換である．

系. $S(\mu^{\bar{\sigma}})$ が $S(\mu)$ と同じ特性根を持てば，$\mu^{\bar{\sigma}}=\mu^*$.

ここで，解析的理論を援用しなければならない．Endomorphism 環が R と同型で，同一の表現 $S(\mu)$ を許容するような，互いに同型でない Abel 多様体を $A_1,\cdots,A_h$ とすれば，Riemann 行列の理論より，A_i の間には，上への準同型が存在し，さらに，$A_i=\lambda_{\mathfrak{a}_i}A$ と取れる．今此等の中で，単因子がすべて1である Riemann 形式を持つものを $A_1,\cdots,A_{h'}$ とすれば，これらも，その中の一つ A により，$A_i=\lambda_{\mathfrak{a}_i}A$ と書けるが，そのとき $\mathfrak{a}_i$ は K のイデアルで，K の，指数2の実体 K_0（補助定理2を見よ）へのノルム $\mathfrak{a}_i\bar{\mathfrak{a}}_i$ が，K_0 の狭い意味での主類に属するものの作る，Kのイデアル類すべてにわたる．此のような類の作る群を $\mathfrak{C}$ と書けば，$\mathfrak{C}$ の位類は，それ故 h' である．

簡単のため k/K^* を正規とし，その Galois 群を $\mathfrak{G}$ とする．$\sigma\in\mathfrak{G}$ に対し，$\mathfrak{A}(A^\sigma)$ はRと同型で，又 A^σ もすべての単因子 1 なる Riemann 形式を持つから，$A^\sigma\cong\lambda_{\mathfrak{a}(\sigma)}A$ と書け，これにより，$\mathfrak{a}(\sigma)$ を含むイデアル類$[\sigma]$が一意に定まる．勿論 $[\sigma]\in\mathfrak{C}$. 又$\tau\in\mathfrak{G}$ に対し，$A^{\sigma\tau}\cong\lambda^\tau_{\mathfrak{a}(\sigma)}\lambda_{\mathfrak{a}(\tau)}A=\lambda_{\mathfrak{a}(\sigma)^{\bar{\tau}}}\lambda_{\mathfrak{a}(\tau)}A$. 所が，$\tau$ は，K^* の上の自己同型だから $S(\mu^{\bar{\tau}})$ は $S(\mu)$ と同じ固有値を持つ，故に補助定理6の系より $\mu^{\bar{\tau}}=\mu^*$, 従って $\mathfrak{a}(\sigma)^{\bar{\tau}}=\mathfrak{a}(\sigma)^*$, 結局 $A^{\sigma\tau}\cong\lambda_{\mathfrak{a}(\tau)\mathfrak{a}(\sigma)}A$. 故に $\sigma\to[\sigma]$ は，$\mathfrak{G}$ から $\mathfrak{C}$ の中への準同型写像である．その kernel を $\mathfrak{H}$ とし，$\mathfrak{H}$ に対応する，k の部分体を $\hat{K}$ とすれば，$\hat{K}/K^*$ の Galois 群は $\mathfrak{C}$ の部分群と同型，故に Abel 群である．

xを A の k 上の G.P. とする．$\mathfrak{p}$ を K^* の1次素イデアルとし，$\sigma(\mathfrak{p})$ を，$\mathfrak{p}$ の，$\hat{K}/K^*$ での Frobenius 自己同型とすれば，$\mathfrak{p}$ の，k での素因子 $\mathfrak{P}$ で reduction して，$\tilde{k}(\widetilde{x^{\sigma(\mathfrak{p})}})=\tilde{k}(\tilde{x}^p)=\tilde{k}(\lambda_{\Gamma\widetilde{(\mathfrak{p})}}\tilde{x})$, 即ち $\widetilde{A^{\sigma(\mathfrak{p})}}\cong\lambda_{\Gamma\widetilde{(\mathfrak{p})}}\tilde{A}$. $\mathfrak{A}(\tilde{A})\cong R\cong\mathfrak{A}(A)$（補助定理 4）より，これから，$\Gamma(\mathfrak{p})\in[\sigma(\mathfrak{p})]$. 特に $\sigma(\mathfrak{p})^f=1$ は $\Gamma(\mathfrak{p})^f\sim1$ と同等．それ故，K^* のイデアル $\mathfrak{b}$ で，$\Gamma(\mathfrak{b})$ が K で単項イデアルになるものすべての作る群をIとすれば，

定理 2. $\hat{K}$は，イデアル群Iに対する，K^*の類体である．

特に，$\mathfrak{C}$ の各類が，$\Gamma(\mathfrak{b})$ なる形のイデアルで代表されるときは，$[\hat{K}:K^*]=h'$. 此れは，‘類方程式’の既約性を意味する．

§5. 分点の体における分解法則

K^* のイデアル数の系を一つ選び[6]，イデ

アル $\mathfrak{b}$ を表すイデアル数を $\hat{\beta}$ と書く. $\Gamma(\hat{\beta})=\hat{\beta}^{\tau_1^{-1}}\cdots\hat{\beta}^{\tau_s^{-1}}$ は $\mathfrak{b}$ により1の冪根因子を除き定まり，従って，本質的には，イデアル数の系の選び方によらない．絶対値を比べれば，$\pi_{\mathfrak{P}}=\varepsilon'\Gamma(\hat{\tilde{\omega}})^f$ がわかる．ただし $\mathfrak{P}$ は k の素イデアル，$N_{k/K^*}\mathfrak{P}=\mathfrak{p}^f$, $\mathfrak{p}=(\hat{\omega})$, 又 ε' は1の冪根である．

K に含まれる1の冪根の集合を E と書き，Aの点 a に対し，$Ea=\sum_{\varepsilon\in E}(\varepsilon a)$ なる，0次元の cycle を定義する．そして $k(a)_0$ を，kを含み，その上で Ea が有理的になる最小の体とする．

$\mathfrak{a}$ を K のイデアル，aを，A の，本来の $\mathfrak{a}$ 分点とする．K^* の素イデアル $\mathfrak{p}$ が k で f 次に分解するとする．一方 f_0 を合同式 $\varepsilon''\Gamma(\hat{\tilde{\omega}})^{f_0}\equiv\varepsilon \bmod. \mathfrak{a}$ が成立つ最小の自然数とする．ここに ε'' は1の冪根，$\varepsilon\in E$. 今 F を，f と f_0 との最小公倍数とすれば，$\pi_{\mathfrak{P}}^{F/f}=\varepsilon'^{F/f}\Gamma(\hat{\tilde{\omega}})^F\equiv\varepsilon_2(\mathfrak{a})$, $\varepsilon_2\in E$, だから，Ea は $\pi_{\mathfrak{P}}^{F/f}$ で不変，故に $\widetilde{Ea}$ は $\tilde{k}$ 上高々 F/f 次，故に K^* 上高々 F 次である．$\widetilde{Ea}$ の $\widetilde{K}^*$ 上の次数は，$\mathfrak{P}$ の $k(a)_0/k$ での分解の次数に等しい．それを F_0 とすれば，それ故 $F_0\leqq F$. 一方定義より，$\tilde{\pi}_{\mathfrak{P}}^{F_0/f}\tilde{a}=\tilde{\varepsilon}\tilde{a}$ なる $\varepsilon\in E$ があるから，$\pi_{\mathfrak{P}}^{F_0/f}\equiv\varepsilon(\mathfrak{a})$. 即ち $\varepsilon'\Gamma(\hat{\omega})^{F_0}\equiv\varepsilon(\mathfrak{a})$. 又 F_0 が f の倍数であることは明かだから $F\leqq F_0$, 故に $F=F_0$. 故に，$\varepsilon'\Gamma(\hat{\beta})\equiv\varepsilon \bmod. \mathfrak{a}$, $\varepsilon\in E$ を満すイデアル $(\hat{\beta})$ の作る群を $I_{\mathfrak{a}}$ と書けば，

定理 3. $k(a)_0$ は，k と，イデアル群 $I_{\mathfrak{a}}$ に対する，K^* の類体との合成体である．

系. A が $\hat{K}$ 上で定義されていれば，$\hat{K}(a)_0$ は，$I_{\mathfrak{a}}$ に対する，K^* の類体である．

註

1) 二三の事情のため，この題名は内容にそぐわないものになった．妥当な題名は，‘Abel 多様体と数体’であろう．

2) 以下に現われる概念の定義，主要性質については Weil [6], Shimura [5], Hecke [4] その他を見られたい．

3) 補助定理2はこの補助定理を使って証明される．従ってこれを先に書いた方が良いかもしれない．

4) Weil [7].

5) R が principal order でないときは，principal order のときに帰着される．

6) 定義は Hecke [4].

文　献

[1] M. Deuring, Die Zetafunktionen einer algebraischen Kurven vom Geschlecht Eins, Nachr. Akad. Wiss. Göttingen, (1953), 85-94.

[2] E. Hecke, Höhere Modulfunktionen und ihre Anwendung auf die Zahlentheorie, Math. Ann., 71 (1912), 1-37.

[3] E. Hecke, Über die Konstruktion relativ ABEL-scher Zahlkörper durch Modulfunktionen von zwei Variablen, Math. Ann., 74 (1913), 465-510.

[4] E. Hecke, Eine neue Art von Zetafunktionen und ihre Beziehung zur Verteilung der Primzahlen, II, Math. Z., 6 (1920), 11-51.

[5] G. Shimura, Reduction of algebraic varieties with respect to a discrete valuation of the basic field, Amer. J. Math., 77 (1955) 134-176.

[6] A. Weil, Variétés abéliennes et courbes algébriques, Paris, 1948.

[7] A. Weil, Number of solutions of equations in finite fields, Bull. Amer. Math. Soc., 55 (1949), 497-508.

[8] A. Weil, Jacobi sums as ‘Grössencharaktere’, Trans. Amer. Math. Soc., 73 (1952), 487-495.

（本研究は文部省研究助成金（昭 29 年）によるものである）

虚数乗法に関する非公式討論会[1]

会は非専門家による質問から始まる．（以下，Artin を A., Deuring を D., Weil を W. と略記する.）

A．非可換な group variety の等分を考えて，非 Abel 拡大の場合を扱うことはできないか？

W．Abel 多様体で既に非 Abel 拡大が得られる．虚数乗法のない楕円曲線の，週期の n 等分から生ずる拡大体の Galois 群は，

$$\begin{pmatrix} a & b \\ c & d \end{pmatrix} \bmod n,\ (ad-bc, n)=1,$$

a, b, c, d は有理整数 mod n，という行列群の部分群に同型だ．定義体が1の n 乗根を含むときは，さらに $ad-bc\equiv 1 \pmod n$ という条件のついた部分群に含まれる．特に modulus j を超越元とし，複素数体に j を添加した体を常数体とすれば，Galois 群は，丁度この後の群全体（すなわち modular 群 mod n）になる[2]．面白いのは週期の p^N–分点全部[3] を添加した体で，その Galois 群は p–進行列による表現をもつ[4]．一方 ζ–函数は Abel 拡大を精密に調べるもう一つの方法で[5]，このことは，Frobenius 置換を通して結びついているが，今のところ全く神秘的に見える．詳しくいうと，Eを，代数体k上定義された楕円曲線，その週期の m–分点をkに添加した体を k_m とし，

$$k^{(p)} = \bigcup_n k_{p^n}$$

とする．p, q を二つの素数，l を p, q と違う素数とする[6]．$G^{(p)}$ を，$k^{(p)}/k$ の Galois 群とすれば，

$$G^{(p)}=GL(Z_p, 2),\ G^{(q)}=GL(Z_q, 2)^{7)}$$

と同一視できる[8]．さて，ほとんどすべての l は $k^{(p)}/k, k^{(q)}/k$ で不分岐である[9]．この拡大に関する，l の Frobenius 置換を，$M_l^{(p)} \in G^{(p)}$, $M_l^{(q)} \in G^{(q)}$ とする．その特性方程式は，Frobenius 置換のとり方によらない[10]．ここで E を mod l で考えれば，$M_l^{(p)}$ の特性方程式が，$M_l^{(q)}$ のそれと等しいことがわかる[11]，ここで注目すべきことは．普通無限次代数拡大の理論には，有限次拡大の性質をひとまとめに表現するだけの意味しかないのだが，今の場合には無限次ということが本質的であることだ．有限次で切って考えると，p–進近似と q–進近似との間には何の関係もないから，特性方程式の間にも，どんな関係もみられない．

A．それではその特性方程式を explicit に計算することが大切だと思う．

W．ζ–函数の分子だから計算はすぐできる[12]．

D．虚数乗法がなくても同じだ．

W．（志村に），variable modulus[13] の理論について考えていることを話せ．

志村．Kronecker の仕事を再構成しただけだ．

W．そういっただけでは，ほかの人にはわからない．

志村．E を，variable modulus の楕円曲線とし，定義体を有理数体 Q とする[14]．従って modulus j は変数だ．今，$k=Q(j)$ とし，$p \neq 2, 3$ とすれば k_p/k の Galois 群は

$$\begin{pmatrix} a & b \\ c & d \end{pmatrix} \bmod p$$

なる行列の群で[15]，k_p/k での p の分解，分岐の次数もわかる．またその分解群，惰性群も計算できた．また Kronecker の合同関係式[16] もでる．

W．Totally imaginary でない体の上の類体の構成を考えようとすれば，コンパクトでない，可換な group variety を考えるこ

とが必要になる．例えば，有理数体の上の類体は，指数函数の週期の等分によってできる．虚数乗法と，このような場合との両方を含む理論ができることが望ましい．それはまた，一般の代数体上の量指標の理論と関係がある[17]．

志村．（D. に），今日，講演の内容を変えたが[18]，始めやる予定だった話を聞きたい．

D．Weierstrass の $\wp$ 函数を考える．これに対し，

$$\wp(pu\,;\,p\omega_1,\omega_2)\equiv\wp(u,\omega_1,\omega_2)^p \mod p$$

が成立つ．この式の意味は，左辺を，$\wp(u, \omega_1, \omega_2)$ の有理函数として表わしたときの合同を表わすのだが，また一方，両辺の差を modular 函数として Fourier 展開したときの係数が全部 p で割れるということにもなる[19]．

次に K を，虚数乗法をもつ楕円函数体とし，その虚数乗法の環が，ある虚 2 次体 Σ の principal order と同型であるとする．Σ の整イデアル $\mathfrak{a}$ に対し，K の部分体 $K^{\mathfrak{a}}$，K から $K^{\mathfrak{a}}$ への同型写像 $\varphi(\mathfrak{a})$ がある[20]．これは，常数体 k の元をも動かす，すなわち k に，自己同型 $(\Omega/\mathfrak{a})$ を引き起す．du を K の第 1 種微分とすれば，

$$(du)^{\varphi(\mathfrak{a})}=\theta(\mathfrak{a})du$$

という形の式が成り立つ[21]．$\mathfrak{a}$ が，単項イデアル (α) のときは，$\varphi(\mathfrak{a})=\alpha$, $\theta(\mathfrak{a})=\alpha$ となるが[22]，一般の場合には，$(\theta(\mathfrak{a}))=\mathfrak{a}$ は，$\mathfrak{a}$ の，単項イデアルとしての表現である．次に $\mathfrak{b}$ を別のイデアルとする．

$$K\to K^{\mathfrak{ab}}=\varphi(\mathfrak{a})\varphi(\mathfrak{b})K$$

を考えて

$$(du)^{\varphi(\mathfrak{b})\varphi(\mathfrak{a})}=\theta(\mathfrak{b})^{(\Omega/\mathfrak{a})}\cdot\theta(\mathfrak{a})du$$

となる．ここで $\varphi(\mathfrak{a})$ は，$\mathfrak{a}$ により Σ の単数を除いて決るから，

$$\theta(\mathfrak{b})^{(\Omega/\mathfrak{a})}\cdot\theta(\mathfrak{a})=\theta(\mathfrak{ab})\varepsilon$$

となる．ここで ε は k の単数であるだけでなく，Σ の単数でさえもある．これは単項イデアル定理のより強い形である[23]．

ついでに，私はこの間，Fricke の，“Lehrbuch der Algebra, III” の中で，古典的方法による単項イデアル定理の証明をみつけた[24]．この本は Weber と違って誤りも少く，良い本なのだが，余り読まれていない．非常に難かしいからだろう．とにかくも‥‥[25]．

A．（D. に），単項イデアル定理についての予想というのは，一般的にいえば，代数体 k のイデアル $\mathfrak{a}$ を表わす，k の絶対類体 K の数 $\theta(\mathfrak{a})$ を適当に取れば，その coboundary が k に属するようにできるということか？

D．そうだ．

W．虚数乗法論の一般的テクニックを話そう．A を虚数乗法をもつ Abel 多様体とする．まず定義体を下げること[26]．

Kummer 多様体[27] に対し，A は一意には決らないが，簡単のために，A が，最小定義体 k で定義されているとする．σ を k の自己同型とし，A, A^σ に対応する R–module を $\mathfrak{m}, \mathfrak{m}'$ とする．A は $\mathfrak{m}$ の類と，$\sigma_1,\cdots,\sigma_n$ とによって決り，逆に A は $\mathfrak{m}$ の類を決める[28]．しかし $A\to(\mathfrak{m})$ という対応は解析的関係で，代数的には定式化できない[29]．代数的に考えるには，次のようにやる．まず

$$\lambda_{\mathfrak{a}}: A\to A^\sigma$$

なる準同型 $\lambda_{\mathfrak{a}}$ がある．さらに，A から A^σ の上への準同型の全体は R–module を作る．この module の類が A と A^σ との関係を代数的に表わすのである[30]．

だが，完全な説明をするのには，polarized Abel 多様体が必要になる．A, A^σ は，それを polarize する因子により，射影空間 P の中に imbed されるから[31]，始めから P の中に入っているとして差支えない．A から A^σ への準同型の module は，今いったように，A と A^σ とを較べるには適当だが，polarized Abel 多様体には向かない[32]．μ を，A から A^σ の上への一つの準同型とし，$\mathfrak{D}$, $\mathfrak{D}'$ を，A, A^σ を polarize する因子類とする．$\mathfrak{D}$ と $\mu^{-1}(\mathfrak{D}')$ とを比較するのだ．所で，A 上の各

因子類は，k_0 の，或る総正な数 η から作られた $\sqrt{-\eta}$ に対応する[33]．今 $\mathfrak{D}, \mu^{-1}(\mathfrak{D}')$ が，$\sqrt{-\eta}$，$\sqrt{-\eta_1'}$ に対応するとすれば

$$\sqrt{-\eta_1'}\big/\sqrt{-\eta}=\xi$$

は総正な k_0 の数で μ により決るから $\xi(\mu)$ と書ける．さて，A から A^σ への準同型の module $\mathfrak{N}$ は R-module（1次元の）だから，k_0 上2次元，故に $\mu\in\mathfrak{N}$ は，$\mathfrak{N}$ の k_0 上の基底に関する成分 μ_1, μ_2 $(\in k_0)$ により表わせる：$\mu=(\mu_1, \mu_2)$．そのとき，$\xi(\mu)$ は，この μ_1, μ_2 の2次形式になる．これは，虚数乗法と2次形式との古典的な関係の一般化である[34]．

志村，谷山．(W.に)，我々の方法で，Abel 拡大がどの程度構成できると思うか？

W．それについては既に予想を話したはずだ．Abel 多様体に対応する量指標から作られる拡大体と一致するだろうというのだ[35]．

W．Abel 多様体の modulus について話そう．与えられた次元の Abel 多様体全体では，algebraic family を作らない．Riemann 行列の古典論から，単因子 $e_1, \cdots, e_n$ が決る[36]．簡単なのはそれが全部1のときで，その全体は，modulus の一つの family を作り，丁度 Siegel の modular 函数に出てくるものになる[37]．一般に，$e_1, \cdots, e_n$ が同じもの全体は，それぞれ一つの family を作る．それには Siegel の modular 群を適当に変えたものが対応する[38]．Confort が何かやっているが，彼の仕事には大した価値はない．Siegel の modular 函数そのものでさえ，arithmetical な目的に使えるかどうかは，今の所何ともいえない．

W．(谷山に)，Hasse の ζ-函数についての，これからの計画を話せ．

谷山．別にない．ただ，虚数乗法のある楕円曲線と modular 函数との関係[39]から類推して，Hecke の作用素の理論[40]を使って考えようと思っている．

W．楕円函数は全部，modular 函数で一意化されると思うか？

谷山．Modular 函数だけでは駄目だろう．別の特別な型の automorphic function も必要だと思う．

W．もちろんそれで或るものはできるだろう．しかし一般の場合は，今までとは全く違い，全く神秘的に見える[41]．だが差当り，Hecke の作用素を使うことは有効だ．Eichler が，Hecke の理論を応用したが，虚数乗法のない楕円曲線の或るものはその中に含まれる[42]．無限に多くのそのような楕円曲線が……．

D．違う！有限個しかわかっていない．(これに続き W．と D．との間に二三のやりとりがあったが早口で聞きとれない．)

W．(谷山に)，このような問題と Ramanujan の予想[43]との間にどのような関係があると思うか？

谷山．Ramanujan の予想は Eichler が証明したと思うが[44]．

W．私のいっているのは本来の予想，つまり $\varDelta(\tau)$ の場合だ．

谷山．それについては考えたことはない．

森川．(W．に) 高次の singular relation をどう思うか？

W．それは一体何だ？

森川．Riemann 行列 (E, T) に対する singular relation とは

$$(E, T)\begin{pmatrix} A & B \\ C & D \end{pmatrix}=A(E, T)$$

なる形[45]のものだが，代数多様体の，r 次の第1種微分の週期行列

$$(E, T^{(r)})$$

に対し同じ形の関係が成り立つとき，それを higher singular relation というのだ[46]．

Chevalley．(W.に)，さっき定義体をさげる云々といったが，あの言葉の定義は何だ？

W．V を k 上定義された代数多様体，k_0 を k の部分体で，k/k_0 が分離的代数拡大であるとする．σ, τ を k/k_0 の同型とすると

き，V^σ/k^σ から V^τ/k^τ [47] への双有理対応 $f_{\sigma\tau}/k^\sigma \cup k^\tau$ があり，すべての組 (σ, τ, ρ) に対して

$$\begin{array}{ccc} V^\sigma & \xrightarrow{f_{\sigma\tau}} & V^\tau \\ & f_{\rho\sigma}\searrow \quad \swarrow f_{\tau\rho} & \\ & V^\rho & \end{array}$$

が good diagram であり，またさらに $f_{\sigma\lambda,\tau\lambda} = (f_{\sigma,\tau})^\lambda$ であるとする．そのとき，k_0 で定義された V_0, V から V_0 への双有理対応 f/k があり，従って f^σ/k^σ は V^σ から V_0 への双有理対応になるわけだが，すべての対 (σ, τ) に対して

$$\begin{array}{ccc} & V_0 & \\ f^\sigma \nearrow & & \nwarrow f^\tau \\ V^\sigma & \xrightarrow[f_{\sigma,\tau}]{} & V^\tau \end{array}$$

が good diagram になる．始めのような条件の下でこのような V_0 が存在することを，定義体が k_0 まで下げられるというのだ．これは V の，自分自身への双有理対応がただ一つ（すなわち恒等写像）しかないときが最も簡単だ．k/k_0 が正則拡大のときにも似たような定理が成り立つ[48]．

（ここで散会，その後誰かが W. に，Ramanujan の予想とは何かと聞いたので，）

W. $\varDelta(\tau)$ を Weierstrass の判別式，つまり

$$\varDelta(\tau) = q\prod_{\nu=1}^{\infty}(1-q^\nu)^{24}, \quad q = e^{2\pi i\tau}$$

とし，q の羃で整頓して，

$$\varDelta(\tau) = \sum_{n=1}^{\infty}\tau(n)q^n.$$

これに対する Dirichlet 級数[49] は，そのとき，次のような Euler 積に分解される[50]．

$$\sum_{n=1}^{\infty}\frac{\tau(n)}{n^s} = \prod_p(1-\tau(p)p^{-s}+p^{11-2s})^{-1}.$$

この各因子を，p^{-s} の2次方程式と考えたとき，これが虚根をもつというのが Ramanujan の予想だ．これは合同 ζ-函数に対する Riemann 予想に似た形をもっている[51]．

ここで有理点が問題だ．つまり有理数体 Q の上で定義された多様体 V で，mod p で考えたときの有理点の個数が，何かしら $\tau(p)$ と関係があるようなものが存在するか？対称の性質，合同性····，函数等式が····，この V は 23 次でなければならぬ．決定的なことは······[52]．

註

1）この記事は筆者のノートと記憶とだけにもとづいて書かれたものであるから，記憶違い，誤解もあることと思う．読者は前以て，このことを承知しておいていただきたい．

2）週期の n 分点は (n, n) 型 Abel 群をつくり，Galois 群はその自己同型を引き起すから，このような行列で表現できるのである．詳しいことは，例えば Weber の代数学の3巻に書いてある．(§ 63)

3）p を固定し N を自然数全体を動かしたもの全部．

4）すぐ下に説明されている．

5）W. の1日目, 3日目の講演を参照．

6）W. は素数といったが実は，l は，pq を割らない k の素イデアルのつもりなのである．以下そのつもりで読んでいただきたい．

7）$GL(Z_p, 2)$ は p–進整数環 Z_p 上の2次の行列群．

8）p^N–分点全体を p–進数 mod 1 を係数とする，2次元のベクトル空間で表わし，その変換群として表現するのである．Deuring [1] § 2 または Weil [13] n° 31 参照．

9）l が p, q を割らないとき, E を mod l で考えたものがやはり楕円曲線ならば，E の p^n, q^n–分点はすべて，mod l で考えたとき，位数が小さくなることはない．不分岐性はこのことからすぐに証明できる．

10）l の，相異る素因子に対応する Frobenius 置換は，Galois 群の内部自己同型によって互に移りうるから．

11）mod l で考えることにより, Frobenius 置換は，局所的に，E mod l の $x \to x^{\mathrm{Norm}\, l}$ なる endomorphism により与えられ，その特性方程式は，$M_l^{(p)}$, $M_l^{(q)}$ のそれと等しく

なる．Weil [13] Th. 36 参照．

12) Weil [13] p. 138，または Hasse [6] § 11 参照．

13) E の modulus j を変数として，E を，u と j とを超越元とする代数曲面として取り扱う理論．Kronecker [11] 参照．

14) 曲面と考えた E は，Q 上定義されうる．Weber の τ 函数の方程式を使えばよい．Deuring [2] 参照．

15) 記号は W. と同じ．

16) Kronecker [11], p. 439，式 (64)．

17) W. の 1 日目の講演参照．有理数体のときは，量指標：$\chi(n)=|n|$ との関係は明白である．

18) D. は単項イデアル定理について話すはずであったが ζ-函数の話に変更になった．

19) このことは D. の東京での講義で，くわしく証明された．

20) x を K の generic point: $K=k(x)$; とするとき，$K^{\mathfrak{a}}=\bigcup_{\alpha\in\mathfrak{a}} k(a\mathfrak{a})$ と定義する．Deuring [1] 参照．このとき，K と $K^{\mathfrak{a}}$ とは抽象体としては同型だが，$\mathfrak{a}$ が単項イデアルでなければ，函数体としては同型にならない．従って $\varphi(\mathfrak{a})$ は，"verallgemeinierte Meromorphism" で常数体 k の元を動かすのである．Deuring [3] 参照．

21) $(du)^{\varphi(\mathfrak{a})}$ は，微分 du の同型 $\varphi(\mathfrak{a})$ による $K^{\mathfrak{a}}$ への像の，写像 $\varphi(\mathfrak{a})$ の微分による逆像で，故に K の微分となるから，このように書ける．$\theta(\mathfrak{a})$ はそのとき，k の数となる．

22) 最初の α は K の，自身の中への自己同型 (Meromorphism)，第二の α は Σ の数．

23) 実際に証明を遂行するためには，次の三つの補題が必要である：(i) 与えられた singular modulus j と，$\Sigma(j)$ の与えられた素イデアル $\mathfrak{P}$ とに対し，modulus j をもつ楕円曲線 E で，$\Sigma(j)$ 上定義されるものが存在して，mod $\mathfrak{P}$ での reduction が E に対しうまく行く．(ii) $\Sigma(j)$ で定義され，modulus j をもつ楕円曲線 E で，E から E^{σ} への準同型 λ_{σ} が，$\Sigma(j)$ で定義されるようなものが存在する．ここで σ は $\Sigma(j)$ の任意の自己同型である．(iii) 任意の楕円曲線 E 上に，第 1 種微分 ω があり次の性質をもつ：mod $\mathfrak{P}$ での reduction が E に対しうまく行くようなすべての $\mathfrak{P}$ に対し ω を mod $\mathfrak{P}$ で考えたものはやはり第 1 種微分である．

D. は，(i), (ii) は証明できたが，(iii) は modulus j が，或る合同式を満す場合にしか証明できなかったということである．(この註は志村の話による.)

24) p. 362 にある．

25) 以下略．詳細は Fricke の本を見よ．

26) これについては後に再び話がある．

27) 以下に使われる概念，記号と，その主要性質とについては，W. の 3 日目の講演参照．

28) W. の 3 日目の講演参照．

29) A から代数的に決るのは R だけで，$\mathfrak{m}$ は決らない．たとえば A, A^{σ} は同じ R をもつが，どの $\mathfrak{m}$, $\mathfrak{m}'$ がそれに対応するかはわからないのである．

30) $\lambda_{\mathfrak{a}}$ なる記号については，志村，谷山の講演参照．もちろん σ には適当なる条件が必要である．R が principal order のときには，$\mathfrak{a}^{-1}$ の類が，$A\to A^{\sigma}$ の準同型の module の類と同じになる．一般の場合には $\mathfrak{a}^{-1}$ は作れないことに注意．

31) W. の 3 日目の講演および Weil [16] 参照．この imbedding が到る所双正則になるように，因子を十分大きくとっておくのである．

32) この module は A と A^{σ} とだけで決るから，因子類の違いは反映されない．

33) W. の 3 日目の講演参照．$\sqrt{-\eta}$ は $K=k_0(\sqrt{-\eta})$, $\mathfrak{J}\sigma_i(\sqrt{-\eta})>0\ (i=1,\cdots,n)$ なる性質をもつとしてある．ここで K は R の商体，k_0 は K の，指数 2 の総実なる部分

体である．W．はこのような $\sqrt{-\eta}$ を使って，Riemann 形式 $E(p(\lambda), p(\lambda'))=T_r(\sqrt{-\eta}\lambda\bar{\lambda}')$ を作っている．所で Riemann 形式は，A の，≡での同等による因子類と1対1に対応するのである．Weil [14] 参照．

34) A から A^σ への準同型 μ は，$\mu(p(\mathfrak{m}))\subset p(\mathfrak{m}')$ なる，C^n の複素1次変換 μ で与えられる（記号は W．3日目の講演を見よ）．いいかえれば，$\mu\mathfrak{m}\subset\mathfrak{m}'$ なる数 $\mu\in K$ で代表される．簡単のため，R が principal order のときに説明する．$\mu\mathfrak{m}=\mathfrak{m}'\mathfrak{a}$ からイデアル $\mathfrak{a}$ が決るが，そのとき $\mu=\lambda_{\mathfrak{a}}$ と書けることは容易にわかる．そして A から A^σ への準同型 μ の module は $\mathfrak{m}^{-1}\mathfrak{m}'$ となるから，$\mathfrak{a}$ と逆の類に属するわけである．今 $\mathfrak{D}$, $\mathfrak{D}'$ に対応する数を $\sqrt{-\eta}$, $\sqrt{-\eta'}$ とする．すなわち $E(p(\lambda),\ p(\lambda'))=T_r(\sqrt{-\eta}\lambda\bar{\lambda}')$, $\lambda, \lambda'\in\mathfrak{m}$, $E'(p(\nu),\ p(\nu'))=T_r(\sqrt{-\eta'}\nu\bar{\nu}')$, $\nu,\ \nu'\in\mathfrak{m}'$; $\mu^{-1}(\mathfrak{D}')$ に対応する Riemann 形式 E_1' は，

$$E_1'(p(\lambda), (\lambda'))=E'(\mu p(\lambda), \mu p(\lambda'))$$

により与えられるから，E_1' に対応する $\sqrt{-\eta_1'}$ に対し

$$T_r(\sqrt{-\eta_1'}\lambda\bar{\lambda}')=T_r(\sqrt{-\eta'}\mu\lambda\cdot\overline{\mu\lambda'})$$

となる．これから，$\sqrt{-\eta_1'}=\sqrt{-\eta'}\mu\bar{\mu}$ が出る．結局，

$$\xi(\mu)=(\sqrt{-\eta'}/\sqrt{-\eta})\mu\bar{\mu}.$$

ここで μ を，成分 μ_1, μ_2 で表わせば，$\xi(\mu)$ は2次形式になるわけである．

特に A が楕円曲線のときを考えて見る．このときには，Riemann 形式は本質的に一つしかないから，

$$E=E'=\begin{pmatrix} 0 & 1 \\ -1 & 0 \end{pmatrix}$$

として差支えない．$\mathfrak{m}=(\omega_1, \omega_2)$, $\mathfrak{m}'=(\omega'_1, \omega'_2)$ と，有理整数環上の基底で表わす．さて，$\lambda=x_1\omega_1+x_2\omega_2$, $\lambda'=y_1\omega_1+y_2\omega_2$ に対し，$E(p(\lambda),\ p(\lambda'))=x_1y_2-x_2y_1=T_r(\sqrt{-\eta}\lambda\bar{\lambda}')$ より，$\sqrt{-\eta}=(\omega_1\bar{\omega}_2-\bar{\omega}_1\omega_2)^{-1}$ なることが計算される．同様にして $\sqrt{-\eta'}=(\omega'_1\bar{\omega}'_2-\bar{\omega}'_1\omega'_2)^{-1}$．それ故 $\sqrt{-\eta'}/\sqrt{-\eta}=N(\mathfrak{m}\cdot\mathfrak{m}'^{-1})$ となる（N は絶対ノルム）．結局 $\xi(\mu)=(\mu\bar{\mu})/N(\mathfrak{m}'\mathfrak{m}^{-1})$．それ故 μ を $\mathfrak{m}'\mathfrak{m}^{-1}$ の基底で表わせば，$\xi(\mu)$ が，イデアル類 $\mathfrak{m}'\mathfrak{m}^{-1}$ に対応する2次形式であることがわかる．

一般の場合には余り簡単ではない．$\mathfrak{m}, \mathfrak{m}'$ が共に，k_0 の整数環上の相対基底 (ω_1, ω_2), (ω'_1, ω'_2) をもつ場合には，E と E' との単因子が同じとすれば類似の計算によって，

$$\xi(\mu)=\mu\bar{\mu}N_{K/k_0}(\mathfrak{m}'\mathfrak{m}^{-1})$$

が成り立つことがわかる．一般の $\mathfrak{m}, \mathfrak{m}'$ に対しては，この場合を仲介にすれば，同じ式が証明できるであろう．しかしこれからすぐに，$\xi(\mu)$ が $\mathfrak{m}'\mathfrak{m}^{-1}$ に対する2次形式であるとはいえない．なぜならどの場合にも，$\mathfrak{m}'\mathfrak{m}^{-1}$ が k_0 上に相対基底をもたないのが普通だからである．W．はこの点，何か感違いをしているのではないかと思われる．

35) W．の1日目，3日目の講演参照．またここで見られるような量指標の性質と Hasse の予想との関係については谷山の近刊の論文参照．

36) Riemann 形式 E は歪対称だから

$$\begin{pmatrix} 0 & D \\ -D & 0 \end{pmatrix}$$

なる標準形にできる．ここに D は整係数対角行列．その対角元 $e_1, \cdots, e_n$ を，E の単因子という．そのとき Riemann 行列は (D, T) なる形に変換される．T は対称で虚数部が正定符号．この T を modulus と考えるのである．これは polarized Abel 多様体に対応する概念である．W．の3日目の講演参照．

37) Siegel [12] の終りにくわしく説明されている．

38) Siegel の modular 群は，I を単位行列として

$${}^t\begin{pmatrix} A & B \\ C & D \end{pmatrix}\begin{pmatrix} 0 & I \\ -I & 0 \end{pmatrix}\begin{pmatrix} A & B \\ C & D \end{pmatrix}=\begin{pmatrix} 0 & I \\ -I & 0 \end{pmatrix}$$

（t は転置行列を表わす）なる関係を満す整係数行列 A, B, C, D による1次変換から成るが，ここで I の代りに D^{-1} とするのであ

る．Confort の仕事については，筆者は何も知らない．

39) Hecke［7］参照．

40) Hecke［9］,［10］参照．

41) W. は公開講演のときにも同じことをいっている．

42) Eichler［4］,［5］参照．ただし話題になっている，楕円曲線の例は書いてない．Stufe 11, 17, 19 などのときにそうなる．

43) 最後に説明がある．

44) Eichler は，-2 次元の，ある型の形式に対し，対応する予想を証明したのである．Eichler［4］,［5］参照．この［5］で彼は，Ramanujan の予想が，Hasse の höhere Differential と関係があるといっている．

45) A, B, C, D は n 次整係数，Λ は n 次複素係数行列．このような変換は T を，すなわち Reimann 形式を変えないから，普通の虚数乗法よりせまい概念で，丁度 polarized Abel 多様体の自身への変換になる（古典論の principal transformation である）．

46) これに対する W. の答は記憶にない．高次微分の週期行列を考えることは，いわゆる高次 Jacobi 多様体（Weil［15］参照）を考えることであるが，W. は，別の機会に，高次 Jacobi 多様体には，複素解析的でない（従って代数的でない）要素が入って来るので，考えるのを止めたと語っている．

47) 一般に V/k は，V が k 上定義されていることを示す簡略記法である．f/k も同様．

48) この定理は Chow, Lang, 松阪（輝）によるという話である．

49) 一般に，$\sum_{n=1}^{\infty} a(n)q^n$ なる modular 形式に対し $\sum_{n=1}^{\infty} \dfrac{a(n)}{n^s}$ なる Dirichlet 級数を対応させるので，両者は Mellin 変換により結びついている．Hecke［8］参照．

50) これも Ramanujan の予想の一部だが，既に解決されている．証明は，たとえば Hecke［9］にある．

51) -2 次元の，ある型の形式に対しては，この二つの予想は実際，本質的に同等である．Eichler［4］,［5］．

52) この最後の話は，散会後の騒がしさも手伝って，ほとんど聞きとれなかった．近くにいた人の話では，W. は何かいいかけて，まだ話す時期ではないと思ったらしく，ニヤニヤ笑って止めてしまったという．

文　献

［1］ M. Deuring, Die Typen der Multiplikatorenringe elliptischer Funktionenkörper, Abh. Math. Sem. Univ. Hamburg, **14** (1941), 197–272.

［2］ M. Deuring, Invarianten und Normalformen elliptischer Funktionenkörper, Math. Z. **47** (1941), 47–56.

［3］ M. Deuring, Die Struktur der elliptischen Funktionenkörper und die Klassenkörper der imaginären quadratischen Zahlkörper, Math. Ann. **124** (1952), 393–426.

［4］ M. Eichler, Quaternäre quadratische Formen und die Riemannsche Vermutung für die Kongruenzzetafunktion, Arch. Math. **5** (1954), 355–366.

［5］ M. Eichler, La théorie des corréspondences des corps des fonctions algébriques et leurs application dans l'arithmétiques. (lecture note, Nancy, 1954)

［6］ H. Hasse, Abstrakte Begründung des komplexen Multiplikation und Riemannsche Vermutung in Funktionenkörpern, Abh. Math. Sem. Univ. Hamburg, **10** (1934), 325–348.

［7］ E. Hecke, Bestimmung der Perioden gewisser Integrale durch die Theorie der Klassenkörper, Math. Z. **28** (1928), 708–727.

［8］ E. Hecke, Uber die Bestimmung Dirichletscher Reihen durch Ihre Funktionalgleichung, Math. Ann. **112** (1936), 664–700.

［9］ E. Hecke, Über die Modulfunktionen und die Dirichletschen Reihen mit Eulerschen Produktentwicklung, I, Math. Ann. **114** (1937), 1–28.

［10］ E. Hecke, 〃 , II, Math. Ann. **114** (1937), 316–351.

［11］ L. Kronecker, Zur Theorie der elliptischen Funktionen, Werke, IV., Leipzig-Berlin,

1929, 345-495.
[12] C. L. Siegel, Über die analytische Theorie der Quadratischen Formen, I, Ann. of Math. **36** (1935), 527-606.
[13] A. Weil, Variété abélienne et courbes algébriques, Paris, 1948.
[14] A. Weil, Théorémes fondamentaux de la théorie des fonctions theta (Séminaire Bourbaki, Paris, 1949).
[15] A. Weil, On Picard varieties, Amer. J. Math. **74** (1952), 865-894.
[16] A. Weil, On the projective embedding of abelian varieties, forthcoming somewhere.

問　　題

問題 11. kを総実な代数体，$F(\tau)$ をk上の Hilbert modular form とする．$F(\tau)$ を適当にえらぶと，量指標 λ の Hecke の L-級数の体系が得られて，この $F(\tau)$ と Mellin 変換により，1対1に対応する．このことは Hecke の作用素 T の理論を Hilbert modular 函数に拡張することによって証明される．(cf. Hermann)

問題はこの理論を（必ずしも総実でない）一般の代数体kに拡張することである．即ち，kの量指標 λ の L-級数が得られる如き多変数の automorphic form を見出して Hecke の作用素 T の理論をこの automorphic function に拡張するのである．

この問題の目的の一つは，kの量または類指標の L-級数を特性づけることにある．現在ではkが総実の場合にもまだできていない．

問題 12. C を代数体k上で定義された楕円曲線としk 上 C の L-函数を $L_C(s)$ とかく：

$$\zeta_C(s)=\zeta_K(s)\zeta_K(s-1)/L_C(s)$$

は k 上 C の zeta 函数である．もし Hasse の予想が $\zeta_C(s)$ に対し正しいとすれば，$L_C(s)$ より Mellin 逆変換で得られる Fourier 級数は特別な形の－2次元の automorphic form でなければならない．(cf. Hecke) もしそうであればこの形式はその automorphic function の体の楕円微分となることは非常に確からしい．

さて，C に対する Hasse の予想の証明は上のような考察を逆にたどって，$L_C(s)$ が得られるような適当な automorphic form を見出すことによって可能であろうか．

問題 13. 問題 12 に関連して，次のことが考えられる．“Stufe” N の楕円モジュラー函数体を特性づけること，特にこの函数体の Jacobi 多様体 J を isogenus の意味で単純成分に分解すること．また $N=q=$素数，且 $q\equiv 3$ (mod. 4) ならば，J が虚数乗法をもつ楕円曲線をふくむことはよく知られているが，一般の N についてはどうであろうか．

問題 29. 記号は問題 28* の通りとする．行列Aは，

$$A=\begin{pmatrix} \boxed{A_1} & & & 0 \\ & \boxed{A_2} & & \\ & & \boxed{A_3} & \\ 0 & & & \ddots \end{pmatrix}$$

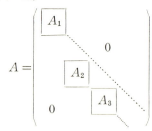

の形に変形されるが，そのとき $A_1, A_2, A_3, \cdots$ の寸法はKの不変数である．(cf. Hasse-Witt) この不変数はCの Jacobi 多様体に対してどのような意味を有するであろうか．

註. Kが $y^2=1-x^5$ で定義される場合は次のことがいえる．

a) $A=\begin{pmatrix} 1 & 0 \\ 0 & 1 \end{pmatrix}$ ($p\equiv 1$ mod.5),

b) $A=\begin{pmatrix} 0 & 0 \\ 1 & 0 \end{pmatrix}$ ($p\equiv 2$ または 3 mod. 5),

c) $A=\begin{pmatrix} 0 & 0 \\ 0 & 0 \end{pmatrix}$ ($p\equiv 4$ mod. 5),

しかも b) の場合は多様体は可約である．類似の結果が $y^2=1-x^q$ (q は素数) で定義されるKのときも成立することは，極めて確からしい．

* **問題 28.** Kを標数$p\neq 0$なる定数体k上の種数 g なる 1 変数代数函数体とする．K の non-singular モデル C の Jacobi 多様体を J, A を K の Hasse-Witt 行列とする．また $A\cdot A^p\cdots\cdot A^{p^{g-1}}$ の階数を r とする．このとき J の p-分点の数が p^r であることを証明せよ．

註. k が有限体の場合は K の類体論より主張は正しい．（玉河恒夫）

A. Weil の印象

冷い風と雨の吹きつける夜の羽田からWeil の乗った日航機が飛立ったとき，これで遂にサヨナラだと思った．実際，彼が日本にいると考えただけで，僕達は何となく落着けなくなるのだった．彼の与えた印象は，余りにも強過ぎたのである．

あの精悍な感じ，波のうねるような話し方，何でもずけずけいう遠慮なさ，それだけでもう十分である．だがこれは単に一面に過ぎない．

彼の数学上の活動が多方面にわたっていることは周知の通りであるが，その方法も多様である．埋もれた多くのアイデアの発掘，新しい概念の構成による，時の大問題への挑戦（もちろん常に成功とはいえないが），多くの実験による新しい事実の発見，乃至はそれ自身の法則性を持つ函数の育成など，一口に理論家といっても，その幅は極めて広い．

それは数学には限らない．文学，詩，絵画その他芸術全般に，異常に強い関心を持っている．‘良いものはすべて好きになる権利がある．’滞日の時間の多くは，各地の美術館その他の見学に費された．また，日本の多くの美術書を買い集め，或は片隅の個人のコレクションまで訪れるなど，その熱心さは驚くべきものがある．そして，‘もちろん数学も芸術の一部だ’と彼はいうのである．

だが Weil の才とエネルギーをもってしても，数学全般に，或は芸術全般に精通することは不可能であろう．彼の理解が時として極めて常識的，皮相的になるとしても，それは止むを得まい．

否，そもそも彼は，豊富な常識の持主なのである，この‘特異なる性格’，‘我儘で附合いにくい天才’が，人の及ばぬ高みに，深く澄んだ自由の空気を呼吸しながら摑っている氷河から離れて，忽ちにして常識円満な社会人となるのを，僕は何度か見た．‘人生は数学ではない．常に妥協が必要だ．’と彼はいう．ここで常識とは，現実を現実として認め，そのまま受け入れる妥協の精神にほかならない．もちろん彼は，妥協するとき黙ってはいない．その叫びが時として誤解を招くのであろう．

年をとると老婆心が強くなり，後進の育成に意を用いるようになるのは，どこの国でも同じとみえる．滞日中彼は常に，若い人々を励まし，指導し，引き立てようと心を配っていたように思われた．‘今度は君達が予言する番だ．’討論の際，彼はよくそういった．アイディアを明確に提出し（正確に定式化することではない），それを具体的に根拠づけるという討論形式を，僕は彼から始めて学んだ．だがアイディアは時として，目に見えるなんらの根拠もなしに，‘無から’生ずることもある．そのようなものは，遂に彼の網にはかからぬのであろうか．

時には学術会議のお嬢さん方をアイスクリームのみにさそう彼である．そのお嬢さん方の話では二人の子供の，非常に良い父親であるという．故国に帰ることを，‘感傷的’といいながらも，機会あればと秘かに望んでいるように思われるのである．

彼の顔は，見る角度により，時により，またそこに現れる心情により，極めて様々な印象を与える．だが，額に深く刻まれた皺には，戦争と亡命，時には投獄されたこともあるという生活の労苦が潜んでいるのであろうか．彼の‘我儘’のかげにある思いやりも，このようにして生れたのかも知れない．

大勢の中にいるとき，彼は意気軒昂として当るべからざるものがある．大抵の人はその毒気に当てられてしまう．しかし，二人きりで話しているとき，ふと話がとぎれて沈黙が座を包むことがある．そんな時に現れる彼の素顔には，思い上りの影など微塵もない，一種の淋しさが浮ぶのである．それは人が己を空しくして他に心を用いるときに現れる，あの謙虚な淋しさなのである．

数学の歩み（月報）第1巻

A. Weil をめぐって

A. Weil は，C. L. Siegel を除けば，恐らく世界第一の現役数学者であろう．シカゴ大学教授．彼は歯に衣を着せない．その批判は辛辣である．その公明な卒直さが広い視野，高い見識と相俟って Bourbaki の運動の一つの推進力となるのであろうが，温厚な大先生方には余り評判が宜しくない．然し，それを一概に排斥しないだけの自由な空気がなかったならば，数学は窒息してしまったであろう．

彼の視野の広さと見識の高さ，それは例えば「数学の将来」や，1950 年の congress に於ける講演に見られる．時には，余りにも大きく飛躍し，余りにも大胆に推測し，ハッタリではないかと思われる箇所もなくはない．然し，凡眼を以て，天才の思想を云々するのは危険であろう．

周知の通り，Bourbaki は，Weil を中心として，フランスに於る analysis の余りにも強固な伝統に反抗して生れた．然しその外見上の modernism の故に，彼等の持つ classic の素養の深さを見落してはなるまい．嘗て数学とは，抽象化であり，公理系であり，無矛盾の体系であると云われた時代があった．然し 30 年代の陶酔から醒めた「現代」数学は此の形式的規定を甚だ不十分なものと感じている．無矛盾らしく見える抽象的体系の中で「意味」あるものは何か？

それは，classic な諸結果を抽象し統一し見通しよく再構成し得るものでなければならぬ．その構成と展開，それが現代数学の任務であると云われる．而も，一つの抽象的基盤の上のみに閉ち籠った分野は，スコラ化する危険に曝される．（アメリカに於る一部の学派には，既にその徴候が見られる．）Bourbaki の仲間の伝統に培われた見識が此の危険を克服していることに注意せねばなるまい．

とにかくも，Weil は，此の意味で，現代数学を代表する．そこに，彼の秘密があり，又限界も見出される．

彼の思考は明晰判明であるが，その論文は表現は簡潔，援用手段は豊富，方法は強引，甚だ読みづらい．今その中から，最も興味ある二，三を取出して見よう．不定方程式の理論（Thèse），代数幾何に関する三部作，類体論と L-函数の理論……そこには，共通な著しい特徴があることに気附くであろう．彼は先づ classic な理論の中から，本質的なもの，その keystone を，鋭く見抜く．何が，如何に抽象され，一般化さるべきか？　これが第一の問題である．次には，此の計画を実行に移さねばならぬ．そこには勿論，重大な障害が山積する．大抵の数学者は，そこで挫折するか，迂路を取る．然し彼は，始めの計画を変えない．障害を一つ一つ，強引に捩ち伏せる．此の腕力の強さと息の長さ，それが彼の第二の才能である．単なる抽象以上に出た彼の業績の深遠さは，此処に由来するのである．

だが，才人は万に走る．彼は余りにも多くのことに手を着けるため，一つの問題を十分に深く追求しない恨みがある．彼の重要な諸結果が繊細さを欠くのも，之れに基くのであろう．

更に一つ問題が残る．成程，確実な基礎は

得られ，見通しより一般化は成し就げられた．然しただそれだけではないか？ 此れは現代数学そのものに対する疑問である．我々はいつまでも 19 世紀の脛を齧っているべきなのか？ 全く新しい分野，予期されぬ展開，幾つかの部門の形式的類似性を越えた深い関連，それ等は最早存在しないのであろうか？ Weil の方法では此の新天地を開拓することは不可能なのである．然し，之れは，世紀の天才が，時を得て始めて為し得る様なものなのであろう．現代数学の枠の中に於てさえも，未だ為されねばならぬことが余りにも多い現在，それ以上を求めるのは当を得たものとは云えぬかも知れない．我々は寧ろ，第二，第三の Weil を必要とするのではなかろうか．

さきに， Weil の腕力について述べた．彼より遙かに独創的な Siegel は，その点でも彼を凌駕する．独創的な深みに達するには，綺麗事が好きで腕力の弱い，我国の多くの数学者にとっては正に，頂門の一針と云うべきであろう．　（1号，1953 年7月）

類体の構成について

虚二次体上の類体の構成に関する Kronecker の青春の夢の拡張を企てる時第一に，基礎にとるべき体の選択が問題になる．虚二次体の場合には，その整数の base を周期とする楕円函数体の乗法子環が始めの体と同型である事が，本質的に重要である．従ってここでは，或る abelian variety の乗法子環と同型になる体を考えるべきであろう．

一般に g 変数の Θ 函数の周期行列となり得る，g 行 $2g$ 列の行列 $\Omega=(\omega_{ij})$ を，genus g の Riemann matrix と云う．abelian variety A は或る Ω に属する Θ 函数から構成されるが，此の Ω を A の Riemann matrix と云う．そのとき，g 次正方複素行列 Λ, $2g$ 次正方有理行列 B で

(1)　$$\Lambda\Omega=\Delta B$$

を満すものがあるとき，変換 (1) を Ω の或は A の乗法子と云う．Ω の乗法子環が，division algebra になる条件は，Ω が pure なること，即ちより小さい genus を持つ二つの Riemann matrix の直和に分解されぬことである．それ故我々は pure な Ω のみを考えることにする．そのとき，その乗法子環 D は，有理数体上高々 $2g$ 次の division algebra で而も "Rosati の anti-automorphism" を有するから，Rosati に従い，その center は total real な体であるか，又はその虚二次拡大であることを結論できる．

我々の興味は，D が可換体なる場合にあるから，まず D を，g 次の total real な体 F とする．そのとき対応する Ω として genus g のものが存在し，g 個の essential parameters に依る．楕円函数からの類推で，D の次数最大のもののみを考えようとすれば，此の場合の Ω の parameter は，更に特殊化されねばならぬ．

結局，第二の可能性，即ち F の虚二次拡大 K を乗法子環とする，genus g の pure Riemann matrix の存在のみが問題である．一般に genus g の pure Riemann matrix が，K に同型な乗法子環を持てば Lefschetz により，それは

$$(2)\quad \Omega=\begin{pmatrix} \omega_1^{(1)}\cdots\cdots\omega_{2g}^{(1)} \\ \omega_1^{(2)}\cdots\cdots\omega_{2g}^{(2)} \\ \cdots\cdots\cdots\cdots \\ \cdots\cdots\cdots\cdots \\ \omega_1^{(g)}\cdots\cdots\omega_{2g}^{(g)} \end{pmatrix}$$

なる形に変換され得る．ここに $\omega_1,\cdots,\omega_{2g}$ は K の base. 又 K から複素数体内への $2g$ 個の同型写像の中，複素共軛なものから一つづつ

選んで $\sigma_1, \cdots, \sigma_g$ とし $\sigma_i\omega_j=\omega_j^{(i)}$ とする．そこで逆にKの base $\omega_1, \cdots, \omega_{2g}$ から (2) により Ω を作るとき，それが pure Riemann matrix になるか否かを考える．今，この Ω が Riemann matrix になると仮定し，(1) を満す Λ を考える．Ω の各行が互いに共軛なことより，Λ の各行は，Λ の第1行の共軛の適当な順列であることがわかる．それ故 Λ はその第1行により全く定まる．所で今，

$\sigma_1K=\sigma_2K$ とすれば $\Lambda=\begin{pmatrix}\lambda_1\lambda_2\,0\cdots\cdots0\\ \cdots\cdots\cdots\cdots\\ \cdots\cdots\cdots\cdots\end{pmatrix}$, $\lambda_1\in\sigma_1K$, $\lambda_2\in\sigma_2K$, なる Ω は明かに Ω の乗法子を定め，それ故 Ω の乗法子環は有理数体上少くとも $4g$ 次，故に此のときは，Ω が Riemann matrix になるとしてもそれは pure ではない．特にKがガロア体なら，(2) により対応する Ω は "maximal singular" になる．そこで，K, Fの最小ガロア体を $\tilde{K}, \tilde{F}$ として

$$(3)\qquad [\tilde{K}:\tilde{F}]=2g$$

なる Kのみが問題である．所が (2) によってかかる K に対応する Ω が，実際に pure Riemann matrix になることは，容易に証明できる．そのとき Ω は乗法子として，

$$(4)\qquad \Lambda=\begin{pmatrix}\lambda^{(1)} & & 0\\ & \ddots & \\ 0 & & \lambda^{(g)}\end{pmatrix}\qquad \lambda\in K$$

により与えられるものを有する．かかるものが既に $2g$次の体を作るから，Ω の乗法子は此れ以外にない．故にその乗法子環は実際に K と同型になる．それ故 Kronecker の青春の夢をかかる体に拡張することがまず試みられるべきである．

所で，かかる K に対する Ω は，始めに考えた，F に対する Ω の parameter を特殊化して得られるから，後者に対して考えられる Hilbert の modular 函数の理論が応用できることは注目に値する．Hecke がその Dissertation に於て企てたのは，正しく此の様なことの $g=2$ の場合に外ならない．そこに既に，我々の条件 (3) が“考える4次体とその共軛とが8次の体を合成する”と云う形で表れている．Hilbert, Hecke が，如何にして，此等の制限を必要とするに至ったかは明かでないが，以上の考察と考え合せるとき，それ等が非常に本質的なものであることがわかるであろう．　　　　（2号，1953年10月）

月報 No. 2 所載の「類体の構成について」には誤りがあるから訂正する．即ち，行列 (ω_{ij}) の乗法子を定める行列 Λ は，その第一行により決るが，第一行を此の様に与えても，乗法子を定める Λ を作り得るとは限らない．此れは，選ばれた同型写像 σ_i を組合せて，複素共軛写像が生ずるかも知れないことによる．此の間の事情は，Fをガロア体として考えて見れば簡単にわかる．そのとき，Fが真の虚部分体を含めば，対応する Ω が pure であり得ぬのは明かだが，そうでない時は一概には云えない．此の見地から，或る種の円分体又はその部分体を基礎に取ることも興味あるかも知れない．此れに対しては Hecke の流儀は適用できないが，Λ の3又は4乗根の体に対する特殊な方法が参考になるであろう．

筆者の健康上の理由により，此れ以上調べることができないのは残念である．

（3号，1953年12月）

数学の歩み（月報）第2巻

モヅル函数と合同ζ-函数

今度の国際数学者会議で，M. Eichler は，極めて注目すべき講演を行った．それは，楕円モヅル形式に対応するヂリクレ級数のオイラー積展開に関する Hecke の理論を，所謂 Hasse のL函数に関する Weil の結果の方向に結び付けたもので，整数論の，２つの離れた分野の間に一つの架け橋の作る点から，又 Hecke の理論の意味の探求のために，Hasse のL函数の理論の進歩のためにも，喜ばしい貢献である．以下それを簡単に紹介して見よう．

Stufe q のモヅル函数で，モヅル変換

$$\tau \to \frac{a\tau+b}{c\tau+d}, \quad c\equiv 0\ (q), \quad \left(\frac{a}{q}\right)=1$$

により不変で，そのフーリエ展開の係数が，有限個の素数のみを分母に持つ有理数なるものすべてのなす代数函数体をKとすれば，Kの常数体kは有理数体である．素数pを法としてKを reduce すれば，有限個のpを除き，Kと同じ種数gを有する，標数pの素体 k_p 上の函数体 K_p を得る．K, K_p の常数体 k, k_p を代数的閉体に拡大して得られる函数体を $\bar{K}$, $\bar{K}_p$ で表すことにする．さて，此のK_p の ζ-函数 $\zeta_p(s)$ の，上に除外されたp以外の素数にわたる積 $\Pi'_p\zeta_p(s)$ を，函数体Kの，Hasse の L 函数と云うので，その研究が当面の目的である．

簡単のために，Kの函数が上の条件の中，$\left(\frac{a}{q}\right)=1$ を除いたモヅル変換でも不変である場合のみを考える．所でKの第一種微分は，-2次元の整の Spitzenform で，それのフーリエ係数は，Kの函数のそれと同じ性質を有する．その一次独立なg個を $S_i(\tau)$, $i=1,\cdots,g$ とする．そのとき，p を q と異る素数として，Hecke の作用素 $T(p)$ の，第一種微分に対する作用は，

$$S_i(\tau)\,|\,T(p)=pS_i(p\tau)+\frac{1}{p}\sum_{\nu=0}^{p-1}S_i\left(\frac{\tau+\nu}{p}\right)$$

と書け，従って $S_i(\tau)\,|\,T(p)$ も，-2次元の Spitzenform で，更に始めの条件を満すモヅル変換で不変だから，Kの第一種微分である．故に

$$S_i(\tau)\,|\,T(p)=\sum_{k=1}^{g}t_{ik}(p)\,S_k(\tau)$$

と書け，$T(p)$ は，Kの第一種微分の一次変換を引き起す．その行列 $(t_{ik}(p))$ は，$\bar{K}$ のある乗法子 $\tau(p)$ の，第一種微分による表現の行列である．此の乗法子 $\tau(p)$ は明かに，変換：$\varphi(\tau)\to\varphi(p\tau)$;$\varphi(\tau)\to\varphi((\tau+\nu)/p)$ の"和"である．所で有限個の素数 p を除外すれば，modulo pでの reduction により，$\bar{K}$ の乗法子環は，$\bar{K}_p$ の乗法子環の中に同型に写像される．それによる $\tau(p)$ の像を $\tau(p)_p$ と書く．又 $\bar{K}_p$ は，$\varphi_p\to(\varphi_p)^p$ により定義される乗法子 π_p を有する．Rosati の逆自己同型による π_p の像を $\pi_p{}^*$ とすれば，$\pi_p\cdot\pi_p{}^*=p$ は周知であるが，更に

$$\pi_p+\pi_p{}^*=\tau(p)_p$$

が成立つ．此れは

$$\varphi(p\tau)\equiv\varphi(\tau)^p(\mathrm{mod}\ p)$$

より，変換 $\varphi(\tau)\to\varphi(p\tau)$ を mod. p で考えればπ_pとなり，従って，$\varphi(\tau)\to\varphi((\tau+\nu)/p)$ なる変換の，$\nu=0,\cdots,p-1$ についての和が，mod. p で考えて $\pi_p{}^*$ になることからわかる．所で Weil の理論により，此の π_p のある行列表現の特性多項式から，K_p の ζ-函数$\zeta_p(s)$ を計算できる．此の表現は，もとの函数体Kで考えれば，第一種微分による表現を組み合せて得られるものと同等で，その特性方程式は，mod. p での reduction によ

り不変である．従って

$$\pi_p\pi_p{}^*=p, \qquad \pi_p+\pi_p{}^*=\tau(p)_p$$

より，

$$\zeta_p(s)=\frac{\det(E_g-p^{-s}(t_{ik}(p))+p\cdot p^{-2s}E_g)}{(1-p^{-s})\ (1-p^{1-s})}$$

(det (　) は行列式，E_g は g 次単位行列)

一方，Hecke の理論に於る，作用素 $T(p)$ の本来の意義は，第一種微分 $\sigma_i(s)$ に対応するヂリクレ級数 $d_i(s)$ の，オイラー積展開の存在（行列の積の形に書かれる）：

$$(d_1(s),\cdots,\ d_g(s))=$$
$$(1,\cdots,1)\prod_p{}'(E_g-p^{-s}(t_{ik}(p))+p^{1-2s}E_g)^{-1}$$

にある．此れと，上の，$\zeta_p(s)$ の式とより，K の Hasse の L 函数と，此のヂリクレ級数 $d_i(s)$ の体系との関連は明かであろう．此れより特に前者は，全 s-平面で有理型で，普通の型の函数等式を満すことがわかる．

極めて特殊な型の函数体に対する，此の驚くべき結果から，幾つかの示唆を読み取ることも出来よう．それを追求して行くことにより，代数的整数論の中心問題に行き当るかも知れない．とにかくも，徐々に集積され形成されて来た一つの流れが，明確な形となって現れようとするとき，それは先ず，幾つかの個所に，此の様な，小さい美しい形で結晶するものなのである．

(2号，1954 年 10 月)

モヂル函数と合同 ζ 函数（補遺）

2巻2号の紹介記事「モヂル函数と合同 ζ-函数」には，重大な書き洩しがあるので補足したい．

そのために先ず，Ramanujan の予想を明かにする．Weierstrass の楕円函数の判別式

$$\varDelta(\tau)=e^{2\pi i\tau}\prod_{m=1}^{\infty}(1-e^{2\pi im\tau})^{24}$$
$$=\sum_{n=1}^{\infty}c(n)e^{2\pi in\tau}$$

の係数 $c(n)$ が乗法的構成法則を持つこと，即ち，オイラー積の展開により，

$$\sum c(n)n^{-s}=\prod_p(1-c(p)p^{-s}+p^{11-2s})^{-1}$$

から定められること，更にこの大さは，

$$c(p)<2p^{\frac{11}{2}}$$

なることを嘗って Ramanujan は，経験的に予想した．此の第一の予想は間もなく Mordell により証明され，後に，オイラー積を持つヂリクレ級数についての Hecke の理論により，究極的な形で一般化された．

所で第二の予想，係数の絶対値の評価は，上のオイラー積の p-因子を，p^{-s} の二次式と見る時，それが実根を持たないことを意味する．此の形ではそれは合同 ζ-函数についての所謂 Riemann 予想と，形式的に極めて類似していることを，Petersson が注意している（1940 年）．然し，此の前紹介した講演に於て，Eichler は，-2 次元の或る種のモヂル形式に対しては，此の両者が本質的な関連をも有することを示したのである．前々号の記事に戻り，そこに於る，独立な g 個の形式

$$S_i(\tau)=\sum\sigma_i(\mathrm{n})e^{2\pi in\tau}$$

を適当に取れば，対応するヂリクレ級数 $d_i(s)$ がオイラー積を持つ様に出来る．

$$d_i(s)=\sum_n\sigma_i(n)n^{-s}$$
$$=\prod_p{}'(1-\sigma_i(p)p^{-s}+p^{1-2s})^{-1}$$

その時 $\sigma_i(p)$ は $\tau(p)_p$ の固有値になるが，前に説明した様に，$\pi+\pi^*=\tau(p)_p$ で，π, π^* の固有値は，Riemann 予想より絶対値 $\sqrt{p}$ を有する．それ故

$$|\sigma_i(p)|\leq 2\sqrt{p}$$

此れは，-2 次元の形式 $S_i(\tau)$ に対する Ramanujan の予想の一般化に外ならない．(前の $\varDelta(\tau)$ は -12 次元なることに注意) 此れが彼の講演の主要目的の一つであろうと思われる．

序でながら，$\varDelta(\tau)$ の 3, 4, 6 乗根に対しては，それを或る虚2次体の Hecke の量指標 (Grössencharakter) の L-級数に対応する ϑ-級数で表すことにより，Ramanujan の予想の一般化が証明される (Schöneberg).

又此の前述べた Hasse の L-函数については，虚数乗法を持つ楕円函数体のそれは，或る代数体の Hecke の量指標の L-級数二つの積の逆数になることを，Deuring が証明している．これは，彼の虚数乗法論の一つの応用である．此の様な，種々の概念の，注目すべき相互関連の，深い根拠は，まだ知られていない．然しその中心には Hecke の理論が位置している．彼により開かれた領域は最近益々その重要性が認められて来た様に思われる．紹介者の不勉強と，問題認識とのズレとのために此の前の記事が甚だ不十分であった事をこの機会にお詫びしたい．

（4号，1955 年 3 月）

投　　書

人間の生活を支え豊かにするものとして工業技術があり，その基盤として自然科学一般がある．その一部として又科学そのものの基礎として数学があり，それ故数学を進歩発展させるのは有意義である．数学が統一ある有機体である以上，純粋数学の研究も人類の幸福に役立つものである．成程立派な見解です．だがそれが空々しく響くのは何故でせう．或人は，日本に於ける科学と工業の分離について述べます．つまり科学は空に蒸散し技術は輸入に頼る，科学は尊重されず，科学者は国民の支持を感じない．戦時中は必要上，一部の科学は政策的に尊重，奨励されました．その頃学生生活を送った人々が科学に強い情熱を感じ得るのは，或程度このためかも知れません．だが今や科学は無用の長物と化したかの様です．ましてや余りにも迂遠な数学など，やりたい人はやればいいさ，と言うわけです．それだからこそ，つまりすべてが植民地化された現在こそ，科学の殿堂だけでも毅然として独立を維持すること，民族的誇りの一つの依り所をそこに築くこと，それは素晴らしいことではないか！　そう，日本国民の名誉のために，とおっしゃるのですね．フルハシ，ハシヅメが世界記録を樹立し，ユガワがノーベル賞を，コダイラがフィールド賞を貰う，それは劣等感に打ち挫かれた国民精神を感奮興起させる．非常に結構です．自国の高い文化を誇ること，それに愛着し，それを更に高める努力をすること，それは自然でもあり美しくもあります．その文化が真に国民の中に根を下しているならば．だが，普段は見向きもせず考えても見ない人々，輸入した問題と取組み外国で業績を挙げた人々を，何か賞を貰ったと言うだけで担ぎ廻り誇りとする．成程 12 才の少年にふさわしい無邪気さかも知れません．

日本人の持つ非合理性，日本社会に於る資本主義と前近代性との奇妙な混淆，その上に重なる植民地政策，此のジャングルを切り抜けるには，科学的合理的精神の涵養以外に道はない．キチンとした理科，数学教育こそはその目的に最も適うものである．数学をちゃんとやっている研究室があり，その雰囲気に育って教師が教育に当る，それは理想的ではないか．その通りです．もしそれが求められているならば．だが教育も馬に水を飲ませることは出来ません．一般に信じられている所では，曽て合理主義の担い手であった資本主義は，この国にあっては，合理主義の徹底を恐れ，時には植民地政策に順応することにより自らを守ろうとしています．而も勤労大衆も合理主義に救いを求めようとはしていない．表立って求められていない此の合理主義を叩き込むには，人は啓蒙家にならなければなりません．だが，フェルマアの問題を研究しながら啓蒙家になることは出来ません．純粋数学が啓蒙の光となり得た時代は，遙か昔のことなのです．合理主義的啓蒙それ自身に対する疑問は，ここでは一応措くとしても．

＊　＊　＊

趣味として数学をやると云う人があります．大変結構な趣味です．だが人は，単なる趣味に一生を捧げはしません．つまり，此の言葉は様々なニュアンスを以て語られているのです．数学の持つ純粋さ，曖昧を許容しない明確さ，一種の精神的な高み，その様なものに精神の依り所を求めようとする人，学問に対する漠然とした尊敬，何かに徹する生活の魅力から，学の中の学たる数学に没頭しようとする人もあるでしょう．人間精神の名誉のために透明で稀薄な空気を呼吸しつつ，高い氷河に攫るのも良いし，数学者は詩人であり，その論文は芸術作品であるとも考えられましょう．誰かが云ったように，詩を理解しないものは真の数学者ではない．——或はアカデミズムへの憧れから象牙の塔に立て籠り，或る人は暇潰しのために，他の人は蒐集癖，骨董趣味から，etc. etc. とにかく何でも結構です．暇があり，余裕のある人は，どんな生活を選ぼうと自由でありませんか．類体論の読者が，マラルメの読者の何千分の一であろうと，独得な記号の下に，著者にしか理解出来ぬ“詩”をものにしようと，煩雑な計算，記号的，形式的推論により，詩作と云うよりは大量生産の名にふさわしい論文が山積しようと．浮世の俗物のことなど，顧慮する必要がどこにあるのでしよう．その俗物共も，あなた方に対しては極めて寛容です．実際，あなた方の生活は素晴らしいものですし，合理主義的啓蒙とか，資本主義の矛盾とか喚き立てて，現在の秩序を乱し，社会を転覆しかねない不逞の輩の尻馬に乗るおそれもなく，おまけに，求めようともせず，考えても見なかった何か素晴らしい名誉を，人間精神に与えてくれるかもしれないのですからね．

＊　＊　＊

自分の生きていたしるしを世の中に遺すために，死の床にあって，私は此れだけのことをしてきた，私の生活は無意味でなかったこと，満足の中に眼を閉じられるように，数学の殿堂に自分の鑿の跡を残そうとする人もあります．山小舎の壁に落書きをし，木の幹に自分の名を彫りつけるのも同じ心理によるのでしょう．死の床にあって，守銭奴は子孫に遺し得る金高に満足の笑を浮べ，政治家は自分の当選回数を，将軍は自分の灰にした町の数を思い浮べる．或る人は自分の書き写した経典の山を，他の人は自分のだました女の数々を‥‥何と素晴しい“しるし”が世に遺ることでしょう．人間の事業がすべて空々しいと云うのではありません．それに“価値”を与えるのは，何か別な原則によるのだと云うのです．

何も死の床に限りません．私は此んなに頭が良いんだ．私は此んなに能力がある．私はこんなにいろんなことをした．それは私が前に証明しました．それより此の方が簡単な証明です．此れは私の最大発見定理です．私はどんな論文でも理解できる．私は此れだけの本を書いた．私は学位を6つ取った．私は何とか大学の教授だ．私は大物だ．‥‥‥別に数学でなくとも‥‥私はこんなに将棋が強い．私は柔道6段だ．私はこんなにお酒がのめる．私は金庫破りの名人だ．泥棒の親分だ．私は妾を6人持っている．私は神の使いだ．私は神様だ！

わかりました．わかりました．わかりました．たゞ僕によくわからないのは，そんなに頭がよく，能力があり，何でも証明できるあなたに取って，あなたの能力の証明だけは難かしいらしく，その証明にあなたの一生を要すると思っていらっしゃるらしいのは何故かと云うことです．

（2号，1954 年 10 月，署名T）

整　数　論*

S. S. S. では，昨年秋，各部門に亘って日本の数学の発展をしらべる事を企画し，整数論，函数論，位相数学，微分方程式論，確率統計論の各分野でその方面の人が分担して概括的なことをしらべました．それらは，S. S. S. 例会で数度に亘って紹介されました．その後も，その内容は各グループ諸個人の間で検討され，深められました．それらの成果を月報に連載する事は，可成り以前から予定されており，今度整数論の部が実現した事は，喜びにたえません．この試みは日本でも始めての事であり，今後共討論と研究を通じて，日本数学の発展の方向を見出して行くようなものが作られて行く事を希望します．

（編　集　部）

×　×　×

昨年来，S. S. S. 数論グループで，日本に於ける整数論の歴史につき討論されたが，必ずしも意見の一致を見たわけではない．本稿は，その際の筆者の立場にもとづき書かれたものである．尚史実については河田先生にも簡単にお伺いしたが，それ等は筆者の考えに従い取捨した．

本文中では一切の敬称を省略する．

序　論

整数論は数学の女王といわれる．それは女王の様に美しいが，又女王の様にかよわくて，独りで生きて行くことは出来ない．それが美しく成長するためには，数学の他の分野の精髄を吸収することが必要である．一方それ等の分野も，数論への応用によってその武器を鍛えられ，新たな力により豊かにされる．整数論の意義は，その美しさ，神秘的な深さの外に，数学の活動性への此の様な，鼓舞の力にも存するのである．即ち 18 世紀末に学として誕生して以来，整数論は，順次，複素数の理論，イデヤル論，ζ-函数，モヅル函数，楕円函数，代数函数，アーベル函数などの解析函数論，方程式と有限群の理論，抽象代数学，位相群論，抽象的代数函数論及び代数幾何学，多変数函数論，代数的位相幾何学その他と交渉し，特にこれ等の幾つかは，その発生の直接の由来を整数論に負うているのである．

此の様な事情の故に，整数論から出発して，関連ある他の部門の研究へと入って行く人が多いのは不思議ではない．それは所謂逃避ではなく，数学全体から見て喜ばしい結果をもたらすことが多い．例えば抽象代数学，多変数函数論の発展など，その様な例は枚挙に暇ないが，それ等は勿論，或は代数学であり或は解析学であって，最早整数論ではない．整数論を代数学の一部と見なす俗説の存在を考慮すれば，此の区別を強調する必要があろう．然らば何を整数論と云うのか？　本稿では，二三の理由により，有理数体，代数的数体，代数函数体の arithmetic 及び主としてその研究のための手段（ζ-函数，イデール等）をも含めて，（広い意味での）整数論と呼ぶことにする．

環　境

日本で整数論が生れつつあった 19 世紀末に於ける，世界の大勢を概観することも興味があるであろう．整数論は，一世紀の間に急速に成長して，数学の中心部門の一つとなっていた．ガウスに始まる代数体の理論は，イデヤル論，円分体の理論，虚数乗法論などに支えられて大きく成長し，ヒルベルトの Bericht に結晶した．又デリクレ，エルミット，ミンコフスキーなどにより，二次形式，デオファ

* これは SSS の各専門別グループが執筆した“日本における数学の発展”シリーズの第1部としてかかれたものである．

ンタス近似，不定方程式論などが進められ，数の幾何学が生れる．此等は，ヂリクレの ζ-函数，クライン，フルヴィッツの楕円モヂゥル函数論などにより解析的にも研究される．勿論此等の函数は，代数体研究の手段でもあって楕円函数論と共に，クロネッカー，ウエーバーの虚数乗法論に，一つの頂点を見出す．一方，代数函数論では，ヤコービの逆問題の解決を目指して，リーマン，ワイヤストラスのアーベル及び一般函数論が生れたが，その代数的取扱いも成功し，又イタリーでは，幾何学的な研究方法が生れつつあった．又代数学では，不変式論，方程式とそのガロア群の理論，多項式環のイデヤル論など，多項式の理論全盛で，以上の研究と結びつきながら発展していった．

20世紀に入ってから，代数体の理論は，ヘンゼルの p-進数体の理論から高木の類体論へと飛躍した．代数学では，ウェダーバーンの代数系の理論，シュタイニッツの体論を経て，抽象化の傾向が目立って来た．代数函数論は，アルチンの合同 ζ-函数の理論によって新たな生命を吹き込まれ，解析数論では，ハーディ，リットルウッド，ランダウ，ジーゲル，ヘッケその他の人々により多様な深い理論が発展されて行くのである．

とにかくも，話をもとに戻し，日本に於ける歴史に入ろう．

歴　史

1898年，高木貞治がドイツに行って，50年の遅れを感じたと云う話は有名である．我々の歴史も此の辺から始めることにしよう．

1　高木はヒルベルトの下で50年の遅れを取り戻し，ガウスの数体の上の類体の構成の論文をまとめて帰朝し，ここに，日本の整数論発展の基盤が築かれた．その後第一次大戦中にまとめた類体論を，20年に発表し，整数論の発展に決定的な影響を与えるに至った．即ち，その整理と，派生的諸問題とは，最近に至るまで，世界整数論界の，一つの中心課題をなしていたのである．高木はその後，相互法則，円体の理論などについての成果も得ている．又，高木から約 10 年程後の，竹内端三は，1の3乗根の体の上の類体の構成，その他，代数体の整数論についての業蹟を上げたが，後に函数論の講座を受持って，整数論から離れるに至った．そして日本の整数論は，主として高木の弟子達により築かれて行くのである．

30年代に至るまで，仙台を中心に，別な流派が栄えていた．即ち，藤原松三郎を中心に柴田，深沢，森本などにより，連分数論，ヂオファンタス近似論などにつき多くの結果が発表されて来た．これは主としてミンコフスキーの系統のもので，而もその方法は幾何学的なものに限られ，題材もそれほど広範囲にはわたらなかった．この流派はいつしか衰え，仙台では後に他の流派が生れるに至るのである．

2　20年代に入ってから，高木の下より，優れた学者が出始め，30年代の全盛期を来すに至るのであるが，そこには自ずと，幾つかの流れが見られる．

類体論は順次世界の注目を惹き始める．特にハッセは，25年に Bericht を書いてその重要性を強調した．我が国の研究者も多くは此の Bericht，後には又シュヴアレーの Thèse などによって，類体論を勉強したのである．

さて，弥永昌吉は，アルチンの下で，単項イデヤル定理を簡単化及び一般化し，その際 Geschlechtmodul の理論等を作ったが，後に位相数学に転じた．高木の弟子ではないが，淡中忠郎も，単項化定理を扱い，又 p-群に対する Einbettungsproblem* を解決したが，後やはり位相群論に入って行った．此の様な転向は，当時華々しく生れつつあった抽象数学の影響によるものであろう．

原註*　与えられた代数体の上に，与えられた有限群をガロア群に持つ正規拡大を作る問題即ち此頃，ドイツを中心に，抽象数学が盛んになり，第五問題を目指す位相群論，局所類体論への応用により更新された代数系の理論などを中心として，急激に発展した．ネーターの所から，此の様な気風を，我が国に持ち帰ったのは正田建次郎の功績であろう．この線に沿い，中山正，次いで秋月康夫が，所謂秋月-中山の Homomorphismus を導入して，局所類体論に一見地を開いたが，間もなく抽象代数学その他のものに向って行った．又守屋美賀雄は，エルブランなどと並んで，無限次代数体の局所的並びに，大局的類体論を構成し，又，賦値論，抽象的代数函数論などにも貢献した．守屋と共に北大に行った稲葉栄次も，代数函数論，イデヤル類群論，Einbettungs-problem などを扱った．又京都の園正造は，ネーターと独立に公理的イデヤル論を作ったが，其一派は正田，中山などと同様に，代数学の研究へと入って行った．

さて，此れ程代数的でない線を進んだ弟子達もいる．菅原正夫は，虚数乗法による類体の構成に就いて二三の結果を発表したが，その際のモヅル函数の研究を通して，ジーゲルの影響の下に，多変数函数論へと進んだ．此の様に解析学に向ったのは，我国では珍らしい例である．黒田成勝は，整数論に於てはイデヤル類群，単数群，類数の計算等我が国の一般的潮流から多少離れた面で活動した．又森島太郎は，終始フェルマーの問題と取組み，円体の類数の計算を続け，最近遂にその解決に達したと噂されている．

以上の流派とは独立して，解析数論の流れがある．末綱恕一は高木の最初の弟子であるが，ランダウの下で解析数論を学び，帰朝後多くの論文を発表した．中でも，ハッセと共同の，約数問題についてのものが有名である．然し其後基礎論に転向した．竜沢周雄はその直系で，やはり多くの論文があり，伊関兼四郎と共に，一つの流れをなしている．

所で 30 年代の中頃には特に整数論が流行し，稲葉，中山，竜沢，河田，森その他の人が出たのも此の時期である．此の頃，菅原，森，荒又など，所謂一高グループが生れ，末綱の影響の下に，主に解析数論を研究した．又解析数論のグループは，戦後山本，江田などを中心に金沢にも生れた．

30年代は又，各地に大学が新設され，新生の数学教室で活発な研究が行われた時期であるが，その中心題目は，腰の重い整数論よりは寧ろ，新興の抽象数学であった．名古屋などに整数論が根を下すのは，少し後のことになる．

3　第二次大戦の激化に伴い，研究成果は急激に減少する．終戦後数年，生活も落着き海外の分献も手に入るようになって，我が国の学界も活気を取戻すに至る．そして整数論に於て，我々は，二つの新しい学説を耳にしたのである．即ち，代数系のコホモロジー論と代数幾何学と．後者については，井草準一がヴェイユの考えに従って，合同 ζ-函数のリーマン予想の別証を与えたほか，主としてヴェイユの影響の下に，京都に於て一流派を成すに至るが，整数論に関心を持つ秋月，井草，河合などの努力にも拘らず，その成果は主に代数幾何プロパーに関するものである．又コホモロジーにより，類体論に新しい見方が開かれたため，中山，淡中など，此の面から整数論に戻ってくる人もあった．そして仙台に，一つのグループが生れ，名古屋でも，中山，黒田のもとに，若い研究者が集り，コホモロジー論その他につき，多くの結果が得られた．又河田敬義は，戦後，整数論に戻り，無限次代数拡大の理論，アーベル拡大の一般論などを作った．

然し，此等がすべてではない．岩沢健吉も戦後，整数論に入って，ζ-函数をイーデルにより説明する新見地を開いたが，それは少し別な関連から，ヴェイユにより深められ，その際提出された ζ-函数の問題は，佐武一郎の，無限次拡大の分岐の理論に基き，玉河恒夫により解かれる．又その，コホモロジーの面は，中山その他により研究された．これとは別だが，遠山啓は，ヴェイユの超アーベル函数論を研究し，戦後にその成果をまとめて発表した．又岩沢は，渡米後，可解的拡大の理論その他の結果を得ている．

その他，最近に至って，日本の整数論界は若い人達の参加により急激に賑かになるが，それ等の詳論は，既に歴史の範囲を超えるであろう．

議 論

整数論は，深い伝統に根ざす学である．伝統のない土地にそれが移し植えられるとき，屢々，その一部面のみに片寄り，遂には生命を失うに至る．それ自身充足する程の，広い視野と強い伝統に基く学派が形成されるまでは，従って，外国からの，絶えざる刺戟，影響が必要とされるのである．

仙台を中心に，数の幾何が盛んであったことは前に述べた．筆者は，その由来は知らない．然しその後の歴史が示す様に，此れは，二次形式論，不定方程式論の，解析的，幾何-代数的理論と相俟って始めて興味ある発展を遂げたのであって，方法に於ても，対象に於ても局限され，全体的結合を欠いていた我が国の学派が衰退したのは，寧ろ自然であるとも云えよう．

この点，類体論に於ては，少しく異る．高木の理論は，世界の学界をリードするものであったため，アルチン，ハッセ，シュヴァレー，エルブランなど，有能な学者がそれに参加し，我が国の研究者は，或は彼等と相助け，或はその影響の下に，研究を進めて行った．そして類体論に関する限り，殆んど充足した伝統を作るに至ったかに見えるのである．然し海外の学界との交流も一歩あやまれば，彼等の伝統の上に研究を進めるという形となり，自らの足下に強固な地盤を作るという地道な努力はなおざりにされ勝ちになる．この弱点は，はからずも，第二次大戦中に露呈されたと云えよう．

この傾向の表れとして，取附き易い，抽象-位相代数学の全盛となり，30 年代の類体論の主目標が整理，代数化であったことと相俟って，整数論から代数学に転ずる人が多くなるのである．戦後の整数論も，主としてコホモロジーと云う，極めて代数的な理論に代表され，而も再び代数学の方向に転化しつつある．勿論整数論からの転向は，世界的現象であり，数学全体に対する，整数論の有用性を例証するものではあるが，その方向が代数にのみ向くこと，或は寧ろ，代数的興味からのみ整数論を見ると云う傾向は，必ずしも世界的とは云えない．例えば，抽象的代数函数論は，アルチン以来，合同 ζ-函数論を軸として発展して来たものであるにも拘らず，我が国では，主として賦値論及び類体論の立場から取扱われていたのである．これは，整数論的伝統の差異とでも云うものであろうか？

勿論我が国には，解析数論の流れも存在する．筆者の知識の不足の故に，その発展の方向を論評できないのは遺憾であるが，唯それは，解析数論全般を網羅すると云うには程遠いものであると思われる．此の面では，そもそも我が国に根を下すに至らなかった理論も少くはない．而もそのうちの幾つかは，尚広い前途を有しているのである．

結 び

我々は今何処に立っているか？　我々は何処へ行くべきなのか？　筆者の小論は曲りなりにもこの問に答え得たとは思われないが，敢て，将来への道を尋ねて見よう．

整数論は本来，自然数の arithmetic である．代数学のみならず，解析学，幾何学の成果をそれに応用しうるゆえんも此処にあり，それが活き活きとして枯渇しない生命を有するのも，自然数と云う「もの」に根ざすからである．どの様に発展した理論の中に於ても整数論が問題である限り，此の立場は忘れらるべきではない．勿論それは，その原始的な形で固執さるべきものではないが．

この故に，その進歩が，理論の理論的展開に依存する度合いは，他の分野に比べてより小である．即ち，整数論の発展のためには，常に新しいイデーが必要である．他の数学の分科では，一つの新しい方法，新しい見方が生れるとき，そこから，数多くの成果を得ることも可能であろう．それはいわば，平坦な面上を拡がる発展である．然し整数論における主要な発展は，新しいイデーに依ることが，多く，飛躍的である．先人の生み出した新しい見地に安住するとき，それは整数論以外のものに転化するに至る．整数論に於ては，師と別な道を歩くことにより偉大になる人が多いことは，世界の歴史の示す通りで，師弟相伝の，固定した学派の生れにくいのも，上の様な理由にもとづくのであろう．事実数論に関する限り，偉い先生は，寧ろ必要ないとも云える．

然し，イデーは虚無から造られるのではない．それを支える基盤は，過去の理論の畜積，数多くの実例，問題などを含む，幅の広い伝統であり，その発生の機縁は活潑な而も落着いた雰囲気である．具体的な素材，当面の問題なしに，どの様なイデーが生れ得るであろうか．我が国の様に，この伝統の浅い場合には，前に云った通り，外国の伝統の成果によりかかり，海外から絶えず刺戟を受けることが必要なのは，歩むを得ないことではあるが，それに満足することなしに，自らのうちに強い伝統を育てる様にしなければ，整数論の大いなる発展は，結局は望めないであろう．過去の畜積を身につけた偉い先生は，この様な点から我が国ではやはり必要とされるのではなかろうか．

類体論の整理を終り，世界の整数論界は，一つの転換期を迎え，我が国でも，新たな働きが目立って来た．我々は，我が国に成長した，及び成長しなかった各理論を広く展望し，各々の個性に従い，各々のイデーを持って，相協力して前進すべきであろう．もし整数論の発展が望ましいとするならば．

（最近筆者は非常に消耗している上に，編集部から無暗にせき立てられたため，甚だ不満足な形のまま発表することにする．事実の誤認，見落しも少なくないことと思われるが，読者諸賢の叱正に俟ちたい．又別な立場の人からは，反論，補説が提出されることと思う.）

(5 号，1955 年 5 月)

空々しさということ

主観的にはどうあろうとも，客観的に見て肉食することは栄養になる．数学をやることは結局有用である．それは本当かも知れません．然し，ふしだらで投げやりな生活を続ける病人が，好きなカツレツを食べる時だけ，栄養学説を引合いに出し，又役に立った数学の方向を検討してその旅を進もうとはしない数学者が，自分の好きなことをやりながら有用性の歴史的解明を持出すとしたら，その言葉はやはり空々しく響き，客観的にもどんなものでしょう．例えその学説が如何に真であっても，成程，時々はそんな生活を送りながら恢復する病人もあり，好きでやっただけの数学が役に立つことも起る．100 万円必要な人が宝くじを買い，特等に当る人もいるのだからと考えて安心する．実際，当る人がいると言うのは，客観的事実ではありませんか！（此れは清水氏の所論に対する直接の反論ではない）

(5 号，1955 年 5 月，署名T)

数学の歩み（月報）第3巻

虚数乗法と合同ζ-函数

編集部から，筆者の最近の研究について書く様にとのことであるが，順序として歴史的な説明から始める．

1. 楕円函数の虚数乗法と類体の構成

Kronecker により提出された問題，所謂彼の“青春の夢（Jugendtraum）”は19世紀末の数学の重要課題の一つであった．“虚2次体の上のアーベル方程式は，特異週期の楕円函数の変換方程式で尽くされるであろう．”その解決のための努力は，楕円函数，楕円モズル函数の理論の詳細な研究となり，更には Hilbert による類体論の構想にまで発展したのであるが，その完全な解決のためには，高木の類体論の出現を必要としたのである．

所でその函数論的な研究は2つの面を持つ．モズル函数論として，又楕円函数論としての面を．最初から判然と分れていたのではない．然し後に至って，此の差異は重大になる．主流は，モズル函数の面に偏るに至り，後 Hasse により，此の面での理論は究極的に整理されて（[8]，[9]）問題はその終極に達したかに見えた．

然し間もなく，予想外の展開が起る．Artin [1] により提出された，合同ζ-函数に対する Riemann 予想の類似（以下単にリーマン予想と呼ぶ）の証明を目指し，抽象的代数函数論が急激に発展する．ここで Hasse は，虚数乗法論を応用して，最初の成功をおさめたのである [10]．彼の考えは次の通り：有限体上の楕円函数体 $\bar{K}$ のζ-函数は，基礎体における有理点の個数により定る．今 $\bar{K}$ がある代数体 k 上の，虚数乗法を持つ楕円函数体 K から，k の素イデヤル $\mathfrak{p}$ で reduce して（定義方程式を mod $\mathfrak{p}$ で考えて）生ずるとする．（所謂常数体の reduction）．その時，K の点が，mod $\mathfrak{p}$ で $\bar{K}$ の有理点になることは，その点を k に添加した体で $\mathfrak{p}$ が一次に分解することに外ならない．後者は，類体の構成の際の分解法則から判定出来る．かくて，k において norm $(\mathfrak{p})=q=\pi\bar{\pi}$ とするとき，問題の有理点は K の週期の $\pi-1$ 分値に丁度対応する．その個数は norm $(\pi-1)$ である．此れから，合同ζ-函数は次の形になる．

$$\zeta(s)=\frac{(1-\pi q^{-s})(1-\bar{\pi} q^{-s})}{(1-q^{-s})(1-qq^{-s})}$$

ここで $|\pi|=|\bar{\pi}|=q^{1/2}$ であることが，此の場合のリーマン予想に外ならない．

然し Hasse はすぐに，此の方法は逆行可能であることに気づいた．即ち，上記の函数体 $\bar{K}=k(x,y)$ の変換

(1) $(x,y)\to(x^q,y^q)$

が $\zeta(s)$ の分子から得られる2次方程式

$$(X-\pi_1)(X-\pi_2)=0$$

を満すことを直接証明する（此の方程式が相異なる実根を持たないことがリーマン予想である）．そのとき此の π を掛ける虚数乗法を mod $\mathfrak{p}$ で考えたものが，変換（1）になるわけだが，それがフロベニウス置換の型であることから，週期の $\pi-1$ 分値の体に於ける分解法則が得られる [11]．リーマン予想の，此の方針による完全な証明は [12] でなされた．

此れは重要な転機であって，ここに再び，虚数乗法論が，楕円函数の問題として，今度は代数的に，扱われることになった．此の考えに従い，類体の構成の理論を，純代数的に基礎づけたのが Deuring [4]，[5] である．そこでは，常数体の reduction の理論 [3] が本質的に利用される．

2. リーマン予想と Hasse の ζ 函数

Hasse の論文が出た以上，（楕円函数でなく）一般の代数函数体に対するリーマン予想の証明の論点を見出すことは，さして困難ではなかったに違いない．間もなく，Weil は，証明の計画を説明している [19]．然しそれを実行するには，重大な方法論的障碍を克服せねばならなかった．即ち，此の場合には無力である代数函数論に代って，代数幾何が利用されるのだが，後者は，久しく，イタリーの人々の優れた直観に依存していて，厳密な証明を欠いていた．だが標数が 0 でないと直観も怪しくなる．彼は先ず，その厳密な基礎付けから始めなければならなかった [20]．続いてリーマン予想を証明し [21]，更にアーベル多様体の理論，特に，その乗法子環の理論（此れは虚数乗法に対応する）を展開した [22]．その際，変換（1）と，合同 ζ-函数の零点との間の，Hasse に於けると同じ関連が，此の場合にも確かめられたのである．此のことは重大な意味を持つ．

だが，代数函数論も，自己の名誉を主張し得た．即ち，此れより以前，Davenport と Hasse [2] とは

(2)　　$y^n=1-x^m$

なる形の方程式で定義される代数函数体に対し，リーマン予想を証明した．此の体は，ある有理函数体上の，都合のよい形の巡回拡大体になり，その合同 ζ-函数を，ガロア群の指標の L-函数によって，一次因子に分解で来るのである．その各々の零点は，一般的なガウスの和（所謂ヤコビの和）で表せる．ここでリーマン予想は，ヤコビの和の絶対値の性質から，直接の結果となる．所が此のヤコビの和は，絶対値だけでなく，そのイデヤル分解，符号の決定まで行われていた（Stickelberger [18]）．それは円分体の理論である．ここでもリーマン予想は，代数的整数論に極めて接近している．

だが [18] の結果はリーマン予想だけではない．ここに Hasse の ζ-函数が登場する．ある代数体 k 上，(2) で定義される函数体の Hasse の ζ-函数とは，(2) を k の素イデヤル $\mathfrak{p}$ を法として得られる有限体上の函数体の ζ-函数体の函数を，すべての $\mathfrak{p}$ につき掛けたものである．[18] の結果より，その各成分の零点の算術的性質はすべて解っている．それを用いて Weil [23], [24] は，此の Hasse の ζ-函数が Hecke の量指標の L-函数 [15] を用いて表わせることを証明した．Hasse の ζ-函数は，ここに忽ちにして注目の的となり間もなく Deuring [6] は，虚数乗法を持つ楕円函数体に対し，類似の結果を証明した．最近の Eichler の結果 [7] も，同じ線に沿うものである（月報 2 巻 2 号の紹介参照）．

虚数乗法論，ζ-函数，代数体の理論などの此の様な相互関連は，何を意味するのであろうか？　それは誰しも抱く疑問であろう．

3. 虚数乗法の拡張．Hecke の研究

話を昔に戻そう．Hilbert は 1900 年，有名なパリ講演の中で，Kronecker の青春の夢の，任意の代数体への拡張を，“整数論と函数論とに於ける，最も重要な問題の一つ”と呼んだ．整数論的な面は勿論類体論であるが函数論として彼は，多変数のモズル函数（所謂ヒルベルトのモズル函数）を考えていたに違いない．それは又，アーベル函数の虚数乗法論でもある．

とにかくも，此のモズル函数を用いて，Hecke は，類体の構成を研究した [13]，[14]．それは極めて困難な道であった．然も彼の得た結果はいささか意外であった．ある 4 次体を乗法子環に持つ必ずしも一つではない函数体から出発し，その不変量の対称函数により生成される体を研究する．そのとき，それは，別な 4 次体の上の類体，それも絶対類体ではなくその部分体，が得られるのである．此の様な食い違いの意味は何であろう

か．すべてのアーベル体を得ることは不可能なのか？　事態は錯雑し，その上 Hecke の論文自身，甚だ難解なので，後を追うものも現れず，問題はいつか，数学の主流から離れるに至った．

然し，楕円函数の虚数乗法論の新展開，代数幾何学の整備，特にアーベル多様体とその乗法子環の理論の発展などは，Hecke の研究を，新たな角度から見直すことを，現実の問題に戻した様に思われた．筆者は以前から，その様な意図を抱いていたが，その後志村［17］による，代数多様体の，常数体の reduction の理論が出たので，必要な道具立てはすべて整った．後は，第一歩を，どの方向に踏み出すかと云うことだけが問題であったのである．

4. 虚数乗法と合同 ζ 函数

然しその方向は明白である様に思われなかった．筆者は多少戸惑っていたのである．然し Hecke の扱った函数体に対する Hasse の ζ-函数を考えている中に，その証明の際のキー・ポイントから，上記 Hecke の結果が，自然に出て来ることがわかった．今から考えれば此れは当然であって，1°，2° で説明した様に，中心的な問題は常に，合同 ζ-函数の零点の算術的性質を決定することにあったのである．一方では此れから，Hasse の ζ-函数の理論が出，他方では類体の構成の理論が可能になるのである．此の様な応用の契機は，合同 ζ-函数の零点と，変換 (1) との間の，既に説明された，Hasse-Weil による関連である．

詳細な説明は省略し，主要な結果の得られる筋道を書いて見よう．ここで，証明の方針は明白であっても，方法論的な困難の伴う場合もあるのだが，それもすべて，主として Weil による此れまでの理論の枠の中で，比較的簡単に片付けられるのである．

先ず，合同 ζ-函数の零点のイデヤル分解を求める．此れは，変換 (1) により，微分が 0 になることを利用して，乗法子環の微分による表現を mod $\mathfrak{p}$ で考えることにより決定されるのである．此の結果を利用して，最大特異アーベル多様体の週期行列の性格と，此の多様体が，単純であることの判定条件 (Lefschetz［16］) が，後で必要とされるだけは，代数的に再証明される．(［16］ではトポロジーと ϑ-函数論とを用いる．) 次に，やはりイデヤル分解を用いて，適当な条件の下に，函数体の不変量の生成する体に於ける分解法則が求まる．それにより，此の体は乗法子環の体とは一般に違う，ある (explcit に定まる) 体の上の，ある定った分解法則を持つ類体であることがわかる．Hecke の結果は，此の特別な場合で，そこでの不思議な食い違いは，一般の場合にも，どこまでも付きまとうものであることがわかる．又此の結果を，(2) で決る函数体に応用すれば，n，m が素数のとき，Stickelberger［18］による円分体のイデヤル類の間の関連が一部分出てくる．

週期の等分値の体の場合も同様で，此のときも，期待される Strahl 類体の，一定の性格を持つ部分体が得られることがわかる．このときの証明の根拠は，楕円函数の場合のそれと全く同じである．

此等の分解法則が得られれば，それに基いて，合同 ζ-函数の零点の符号を決定することは，原則的には容易である．然し実際には Hecke の函数体の場合及び，(2) の函数体の場合に，極めて厄介な道具立てを用いて決定することが出来るに過ぎない．然し此れは，本質的に，遙かに容易に決定し得るものと思われる．とにかくもそれにより，此の場合にも，その Hasse の ζ-函数が Hecke の量指標の L-函数を用いて表せることがわかるのである．此れは Hecke の函数体の場合には新しい結果で (2) の函数体 (n, m が素) のときは，Weil の結果の，より一般的な関連の中での別証を与えるものである．此れは無意味ではないと思う．それにより，色々な事柄がより密接に結び合っていることが明かになるから．

5. 二三の問題

此の研究はまだ完結していない．何よりもアルチンの相互法則を与えること，類方程式の代数的性質の研究が欠けている．此等に対しては，かなり重大な難点があり，今の所種数2の函数体に対してしか，解決の見込みはない．更に Hasse の ζ-函数の問題が残されていることは上に書いた通りである．

又，4° で書いた範囲内でも，その結果は未だ十分に一般的でも十分に精密でもない．更に実際に利用される不変量の，函数論的な性格（モズル函数？ としての）についてはまだ何もわかっていない．

その上，Hilbert の問題，すべてのアーベル体を得ること，は，まだ答えられていないのである．上記の結果は寧ろ，アーベル函数のみを用いての類体の構成の限界を明らかにするものと云えよう．もし Kronecker の青易の夢の拡張を更に追求するとすれば，全く別な函数を探さなければならないであろうその可能性については，筆者は何も知らない．

他方，特異週期の函数体の Hasse の ζ-函数が，Hecke の量指標の L-函数で表わせることが，次第にわかって来たわけだが，此のことの持つ意味は依然謎のままである．

Eichler [7] の結果は，それ以外のジリクレ級数も現れ得ることを示唆している．ここで，Hecke の晩年の仕事と関連がより明かとなり，色々な ζ-函数を，或程度包括する理論が出来るかも知れない．そのとき，ヒルベルトのモズル函数が再び登場するのではないかと思われる．手掛りはやはり，量指標の L-函数であろう．然し此の函数の本質さえも，まだ十分にわかっているとは云えない様に思う．

此の様な問題は皆，非常に魅力あるものである．それ等は，多くの人々を容れるだけの多面性を持っている．然し此等の研究を進めるためには，やはり強固な伝統の支えが必要であると思われる．我々は，それを築くために努力すべきではなかろうか．

文　献

[1] E. Artin: Quadratische Körper im Gebiet der höheren Kongruenzen (Math. Zeitschr. 19 (1924) I, 153–206, II, 207–246).

[2] H. Davenport und H. Hasse: Die Nullstellen der Kongruenzzetafunktionen im gewissen Zyklischen Fälle (Crelles Journ. 172 (1935) 151–182).

[3] M. Deuring: Reduktion algebraischer Funktionenkörper nach Primdivisoren des Konstantenkörpers (Math. Zeitschr. 47(1942) 643–654).

[4] ——: Algebraische Begründung der komplexen Multiplikation (Hamburger Abh. 16 (1949) 32–47).

[5] ——: Die Struktur der elliptischen Funktionenkörper und die Klassenkörper der imaginären quadratischen Zahlkörper (Math. Ann. 124 (1952) 393–426).

[6] ——: Die Zetafunktion einer algebraischen Kurven vom Geschlecht Eins (Nachr. Göttinger Akad. (1953), 85–94).

[7] M. Eichler: Modulfunktionen und Riemannsche Vermutung für die Kongruenzzetafunktion (Proc. Congr. Intern. Math. Union, Amsterdam, 1954).

[8] H. Hasse: Neue Begründung der komplexen Multiplikation I (Crelles Journ. 157 (1927) 115–139).

[9] ——: II (Crelles Journ. 165 (1931) 64–88).

[10] ——: Beweis des Analogons der Riemannschen Vermutung für die Artinschen und F. K. Schmidtschen Kongruenzzetafunktionen in gewissen elliptischen Fälle (Göttingen Nachr. (1933) 253–262).

[11] ——: Abstrakte Begründung der komplexen Multiplikation und Riemannsche Vermutung in Funktionenkörpern (Hamburger Abh. 10 (1934) 325–348).

[12] ——: Zur Theorie der abstrakten elliptischen Funktionenkörper (Crelles Journ. 175 (1936) I, 55–62, 69–88, III, 193–208).

[13] E. Hecke: Höhere Modulfunktionen und ihre Anwendung auf die Zahlentheorie (Math. Ann. 71 (1912) 1-37).

[14] ——— : Über die Konstruktion relativ Abelscher Zahlkörper durch Modulfunktionen von zwei Variablen (Math. Ann. 74 (1913) 465-510).

[15] ———: Eine neue Art von Zetafunktionen und ihre Beziehung zur Verteilung der Primzahlen II (Math. Zeitschr. 6 (1920) 11-51).

[16] S. Lefschetz: On certain numerical invariants of algebraic varieties with application to abelian varieties (Trans. Amer. Math. Soc. 22 (1921) 327-482).

[17] G. Shimura: Reduction of algebraic varieties with respect to a discrete valuation of the basic field (Amer. Journ. Math. 77 (1955) 134-176).

[18] L. Stickelberger: Über eine Verallgemeinerung der Kreisteilung (Math. Ann. 37 (1890) 321-367).

[19] A. Weil : Sur les fonctions algébriques à corps de constantes finis (C. R. Acad. Sci. Paris 210 (1940) 592-594).

[20] ———: Foundations of algebraic geometry (New York 1946).

[21] ———: Sur les courbes algébriques et les variétés qui s'en déduisent (Paris 1948).

[22] ———: Variétés abéliennes et courbes algébriques (Paris 1948).

[23] ———: Numbers of solutions of equations in finite fields (Bull. Amer. Math. Soc. 55 (1949) 497-508).

[24] ———: Jacobi sums as "Grössencharaktere", (Trans. Amer. Math. Soc. 73 (1952) 487-495).

(1号，1955 年 7 月)

シンポジウムについて

I. M. U. 主催の整数論シムポジウムが開かれる．だが，S. S. S. 始め各グループの有能な論客達は，どんな意見をも，誌上に発表していない．一体彼等は眠っているのだろうか？

とにかくも喜ぶべきことである．高木以来優れた整数論者を輩出した日本の真価が世界に認められたのである．此の機会に，日本の数学は一段と進歩し，伝統は益々強固になり，多くの先生方は，安んじて国を空けて，外国の学生の前に，日本民族の優秀性を実証することが出来る．何と素晴らしいことではないか！

だが，Barbaki 氏乃至 Berbeki 氏が論文を書けばすぐに飛び附いて，気のきいた remark に自己の才能をひらめかす，それが日本の伝統と云うものだろうか？ 成程，少し Barbaki 氏に注目される様な論文を書いて外国へ行き大きい仕事はそれから始める，伝統は外国にあれば十分なのだ．今度も，多くの駿秀が Berbeki 調に転向し，特に目立つ人々は，その伝統を強化しに，海を越えて行くことになるであろう．世界的見地に立てば，日本の伝統なんてケチな考えが間違っているに違いない，とU氏は考えたのである．

だが，とU氏は考え直した．では俺の様に頭の悪い人間はどうなるのか？ 判りもしないのに傍聴したいなんて生意気だと云われ，何も話すことがないだろうと，来日した偉い先生方にも会わされなかった．ではと奮発して勉強を始めようとしたら，廻りはガラ空きで，おまけに伝統とやらは海の後方にある．せめて盛んになった Berbeki 調でやろうとしても，どうも底が浅く，浮ついていて取りとめない．それに来年，今度は Barbaki 氏が来たら，Barbaki 調が流行するんじゃないか．先見の明を発揮してその方をやろうかとフラフラしていたら，他の人々はシムポジウムを機会に盛んに業蹟をあげているのに，お前は怠けていて何もしないから学位はやれないと来た．その上シムポジウムで数学が大いに宣伝されたのはよいが，整数論は人間理性の創作で，才能に恵まれた少数の人々が象牙の塔に閉ぢ籠って研究するものだと云うことになってしまって，だからお前の様な頭の悪いものは大学に残っていても仕方あるまいと放り出された．……‥来年春風の吹き始める頃，此んなことになるんじゃないかと心配になって来たので，此の問題をもう少し立ち入って考えて欲しいと思い，おづおづと投書する次第である． (1号，1955年7月，署名U)

代数幾何学と整数論

編集部からの注文は整数論の綜合報告と云うことであるが，此の様な大きい題目のものは，筆者の手に余ることは明かである.

そこで，今度のシムポジウムで議論された幾つかの重要な問題を説明するに止めることにした．但し，類体論関係の問題については次号に佐武一郎氏の報告が寄せられる筈なので省略することにし，又代数幾何学プロパーについては，筆者はよく知らないので，やはり省略することにした.

1. 形式所類似

代数函数体の構造と代数的数体の構造とが形式的に似ていることは，既に 19 世紀半ば以来認められていた．Dedekind-Weber による代数函数論，乃至近年の抽象的代数函数論は，此の様な根拠の上に，即ち，代数的整数論から類推に基いて築き上げられたものであるが，理論が出来てみると，函数体の方が代数体よりも遙かに簡単な構造をもっていることがわかった．そこで逆に，函数体の理論からの類推にもとづいて，代数体の理論を考え更には両者を統一する理論を作ろうとする気運が生れることは当然である．所で多変数の代数函数論を有効に取扱う方法が，代数幾何学の方法以外にないことは現在明かだから，此の統一的理論は，有理整数環，多項式環乃至一般の整域の上の代数幾何学と云う形で試みられるに至るのである．勿論此の様な試みは今に始るものでなく，既に Kronecker が，その，整数論の基礎づけの中で行い，楕円函数論の中に応用しているものである．彼は，$y^2=1-\rho x^2+x^4$ で定義される楕円函数の理論を，変数と考えた Modul ρ の多項式環の上で展開したのである．此の理論はそれ自身としても甚だ興味あるもので，つまり，楕円函数論及び楕円モヅル函数（即ち ρ）の理論を包括する，代数幾何学的理論が目標となるのである．現在此の様な楕円函数の週期の等分値に対し幾つかの結果が得られているが，その一般論の展開は将来の問題であろう.

始めの問題については，整域の上の代数幾何学については永田氏及び山崎氏の研究がある．然しながらより興味のあるのは，代数幾何学的方法の応用である．ここではリーマン予想に関することを中心に説明して見よう．合同 ζ- 函数に対するリーマン予想は Weil により説明されたが，それは，実質的に，問題の代数曲線のヤコビ多様体の乗法子環の理論に基くものである．そこで代数体に対しても，ヤコビ多様体に当るものを考えたくなるのは当然であるが，それは一体どんなものであろうか？　先づ，ヤコビ多様体とは，代数曲線の，0 次の因子類群に，アーベル多様体としての構造を入れたものであることに注意しよう．所が因子類群は，その曲線上の，或る種の fibre space の類の群として表現される．（月報 2 巻 1 号の記事参照）．そこで代数体に対し，fibre space を適当に定義することにより，何等かの類似が得られるのではないか？　此れが山崎氏の立場である．一方，又 0 次の因子類群は，函数体の不分岐アーベル拡大を統制することは，類体論により知られている．所で，曲線の不分岐拡大は又その次数だけの discrete な点を fibre にする fibre space と考えることが出来，そのときガロア群は，此の fibre space の自己同型群（リーマン面の Decktransformation の群）として表現される.

代数曲線の最大アーベル拡大のガロア群はその曲線のイーデル類群と同型だから，イーデル類群はいわば，何かある fibre space の自己同型の群と考えることが出来るわけである．此の様にして，イーデル類群を通して，ヤコビ多様体の類似を求めることも出来る．所がどの場合にも，重大な困難が生ずるのである．第一の方法では，代数体には常数体に当るものがないから，代数幾何学との類似の追求は，技術的な困難に行き当り，その有効な解決は容易でない．而も代数体には，無限遠素点（アルキメデス的賦値に由来するもの）があるため余計面倒である．そしてこの点は，第二の方法に対し致命的なのである．即ち代数体のイデール類群は，無限遠素点の存在のために，totally disconnected でなく，単位元の連結成分が存在し，従ってガロア群とは同型になり得ない．そのため，それを何かある fibre space の自己同型群として表現すれば，極めて奇妙な fibre space を考えなければならないわけだが，勿論今まで誰もそれに成功してはいない．

此の様に，イデール類群の，単位元の連結成分の存在が，現代の代数的整数論の困難のそして又面白さの，主要な源泉の一つなのである．類体論の問題に関して，此の様な困難を強引に克服して，函数体との類似をやってのけた Weil の論文があるが，その意義はまだ十分明かとは云えない．

イデール類群の問題については，後に全く別の立場から説明する筈である．

2. 不定方程式

代数幾何学と整数論との関係は，勿論両者の形式的類似の中にだけあるのではない．整数論の或る種の問題は，実は，代数幾何学の問題と考えられるのである．ここでは不定方程式についてだけ説明しよう．即ち，

$$f(x,y)=0.$$

なる不定方程式の解を求める問題は，此の方程式により定義された代数曲線の上の有理点を求める問題に外ならない．2次形式論はその特別な場合であるが，此れについては相当詳細な理論が出来ているので，普通，一般の不定方程式論とは別に取扱われている．先づ此れを説明しよう．

Gauss 以来，2次形式論が，代数的整数論の発展に与えて来た影響は大きい．その最も重要な一つとして，現在 Hasse の原理と呼ばれているものを指摘したい．此れは大雑把に云えば，或る問題が解けるための必要十分条件が，対応する問題が，すべての素点に対し局所的に解けることであると云う原理で，2次形式論に限らず，整数論の多くの分野で認められるのである．もとの形で云えば，2次形式 $f(x_1,\cdots x_n)$ に対し，$f(x_1,\cdots x_n)=a$ が有理整数解を持つ必要十分条件は，此れが実数解を持ち更に，$f(x_1,\cdots,x_n)\equiv a \pmod p$ がすべての素数 p に対し，有理解を持つことであると云うのである．

此の原理が，どの様な範囲の問題に対し成立つかを考えることは，極めて興味ある問題であろう．小野氏の研究は，此の様な点に関するものである．

然しながら我々は，更に，定量的なことを問題にすることが出来る．Hardy-Littlewood による Waring の問題の解はそれを示唆している．即ち，$x_1{}^k+\cdots\cdots+x_m{}^k=a$ なる不定方程式の整数解の個数が，此れの実数解及び対応する合同式の有理解の密度の積により近似されることを証明することにより，彼らは此の問題を解いたのである．此の原理は後に Siegel により更に深く追求された．即ち彼はある2次形式の種が，他の2次形式を表す仕方の数に対し，同様な結果を，近似式でなく等式の形で，証明したのである．

此の様にして，不定方程式の研究は，その局所的な研究と深く結びついていることがわかるのである．我々はここで，p-進体の上の代数幾何学へと導かれる．そして，p-進体は解析的な体であることから，ここでは，代数体の場合よりは問題は容易になる．例えば我々は，有理点の問題を考えるときに，p-進体上の微分方程式論を応用することも出来るわけで，実際，楕円曲線上の有理点に関する

Siegel の結果などは，此の様な方法に基いているのである．

ここで我々は再びイデールに行き当る．イデールは，代数体の，局所的理論を大局的理論に結びつける一つの重要な鍵だからである．そしてここでは，位相解析的方法が自由に応用され得るのである．例えば，2次形式論に関する Siegel の結果は，イデールの上の行列環の位相解析から導くことは出来ないであろうか？ 実際 Weil は Siegel の平均値定理を，此の様にして証明した．2次形式についての結果は，久賀氏が研究しているが完全な解決は中々難かしいようである．

所で，一般の不定方程式の理論は非常に難かしく，北欧学派の多年の研究にも拘らず楕円曲線に対してすら，満足すべき結果は何も得られていない．我々は実質的に，Weil 及び Siegel による有限性定理以上の何物をも持っていないのである．有限性定理とは，genus が1より大きい代数曲線上の有理因子類の群が有限個の生成元を持つこと及び整数点の個数が有限であると云う定理で，その証明は，一つの重要な原理を生み出したのである．此の原理によれば，代数多様体の素点の全体は，多様体の普通の点の全体と同時に，或る意味での代数的な構造を持っていることが明かになるので，虚数乗法論などに本質的に利用される，常数体の reduction の有効性の実質的な基礎を与えるのである．

それはさておき，不定方程式論の此の様な方法による研究は，最近に到り Néron の研究が現われるまで，一歩も前進しなかった．彼は，その一般に信用されていない研究（証明が誰にも理解出来ないのである）で，此の様な原理に基いて，代数多様体上の，色々な類別による因子類群の構造を研究し，特別な曲線に対し，その生成元の個数を評価しているのである．

不定方程式の問題は勿論有限体の上で考えることも出来る．その時我々は，極めて重要な理論に行き当るのである．それは，合同 ζ-函数の理論である．

3. 虚数乗法と ζ-函数

一般に，或る方程式の整数解の個数を係数とする無限級数により解析函数を定義し，その函数論的な研究から逆に，解の個数についての洞察を得るという原理が，屢々用いられる．その著しい例は，ζ-函数及びモヂュル函数である．例えば，p 個の元から成る有限体の上で定義された代数曲線の ζ-函数は

$$\log \zeta(p^{-s})=\sum_{\nu}(N_{\nu}\cdot p^{-\nu s}/\nu)$$

により定義される．ここで N_{ν} は，その代数曲線上の点で，p^{ν} 個の元を持つ体の上で有理的なるものの個数である．同様な原理により2次形式からモヂュル函数及び ζ-函数を作ることが出来るがそれは後廻しにしよう．

此の ζ-函数をめぐって，代数幾何学と，整数論との，第3の型の関連が生ずる．此れは代数多様体と，代数的数体との実質的な関連に基くもので，一般に虚数乗法論と呼ばれているものである．

本来虚数乗法論は，虚2次体の上のアーベル拡大体の構成を考えるもので，類体論の生れる基盤となったものであるが，類体論の完成した現在，その拡張を考えることは，整数論と他の分野の理論との関連を示す意味で興味のある以外には，Artin の相互法則の具体的な形が求められる位の意味しかないと考える人も多いかと思う．然しながら我々はここで，一見意外な事実に行き当るのである．（虚数乗法論の歴史とその基本的な考え方については，前号の記事で紹介したから，ここでは，2，3 の補足を加えるだけとしたい．）

即ち，代数体の上で定義されたアーベル多様体で，その虚数乗法の環が十分大きいものの，Hasse の ζ-函数（前号の記事参照）が，その代数体の，Hecke の量指標の L-函数により表わされることがわかったのである．所でアーベル多様体の ζ-函数は，有限体に

reduction した時の有理点の個数により定るから，つまり，位数有限な点を添加した体におけるイデヤルの分解法則により定るわけで，此れは類体の構成に関する理論に外ならない．ここで我々は又，イデールに行き当るのである．

Hecke の量指標が，イデール類群の指標として定義され得ることは，岩沢，Tate により証明されたことである．このとき，前に説明した様に，イデール類群の単位元の連結成分が存在することから，此の様な指標は一般に，アーベル拡大の言葉では，即ち類体論の言葉では説明できないので，それの算術的意義付けが，現在の重要な問題の一つなのである．所で，此れと全く同様にして，量指標を用いて，イデール類群の，𝔭-進体の中への表現を作ることが出来る．𝔭-進体の位相はバラバラだから此の表現は連結成分の上では 1 となり，従って本来の意味で類体論により説明出来る．即ち此の表現から，定義体の無限次アーベル拡大が得られるわけである．そのとき，此の量指標として，アーベル多様体の ζ-函数から得られるものを取れば，此のアーベル拡大は，アーベル体の週期の 𝔭- 巾分値を添加して得られる体と同じであろう．此れが今度のシムポジウムでの Weil の予想である．此の様にして，アーベル多様体の ζ-函数と，量指標の L- 函数との関連の深い根拠が見出され，同時に量指標の算術的意義付けが得られるであろう．又此れにより，虚数乗法論は，整数論に対し，何か不自然なものであるのではなく，19世紀に於てそうであったと同様に，代数体の整数論の深い神秘に導くものであることがわかるのである．此の様な理論は，志村氏，筆者及び Weil により研究されたが，勿論今後も続けられる筈である．

所で，アーベル多様体の理論からは，常に，基礎体が totally imaginary な場合の理論しか出て来ないことは明かである．実数体を扱おうとすれば，有理数体に対する指数函数の例で明かな様に，基本領域がコンパクトでない週期函数を考えることが必要になる．その様なものを用いて，実代数体の量指標の理論が解明されることが期待されている．

此の様な理論はすべて，非常に興味あるものではあるが，我々はここで立ち止ってはならない．例えば，虚数乗法を持たない楕円曲線の Hasse の ζ-函数の理論は，依然として深い神秘に閉されているのである．そして，その解明の鍵として，主として Hecke によるモヂュル函数の理論が注視を浴びるに至ったのである．

4. モヂュル函数

筆著はここで Hecke の業蹟の全面的な紹介をするつもりはない．ただ二三の重要な点の説明は絶対に必要である．前に述べた様な 2 次形式 $f(x_1, \cdots x_n)$ から，解析函数を作ることが出来る．即ち，a_ν を $f(x)=\nu$ の整数解の個数として，

$$\zeta(s)=\sum a_\nu/\nu^s,$$

$$\varphi(s)=a_0+\sum a_\nu e^{2\pi i\nu/N}$$

を作る（N は f の判別式）．そのとき $\varphi(s)$ はモヂュル函数になり，此のことは，$\zeta(s)$ が函数等式を満すことと同値である．此の様に，函数等式を満すヂリクレ級数，特に ζ- 函数の理論は，モヂュル函数の理論に飜訳することが出来る．特に，あるヂリクレ級数がオイラー積を持つこと（$\zeta(s)=\Pi_p\zeta_p(s)$ と，素数についての積に分解できること）は，対応するモヂュル函数が，ある作用素環の固有函数になることと同等である．此の作用素は，モヂュル形式の n 次の変換の trace として定義されるもので，普通 Hecke の作用素 T と呼ばれる．

所で，モヂュル群の，指数有限な部分群に属するモヂュル函数は代数函数体を作るから，此の作用素 T を，代数幾何学の言葉で説明することが出来る．最近 Eichler は，此の様な考えに従い，特別な部分群に対し，対応する函数体の Hasse の ζ-函数の函数等式を証明し

た（月報2巻2号の記事参照）そのとき，実際に計算することにより，虚数乗法を持たない幾つかの楕円曲線が，此の様な場合に含まれることがわかったので，此れにより Hasse の ζ-函数の理論は，もう一つの新しい武器を得たわけである．

一般に此の Hecke の理論は，ζ-函数，特に量指標の L-函数の理論を考える，非常に有効な手段なのである．所が一般の体の L-函数或は少し違った型の ζ-函数を考えようとすると，楕円モヅル函数だけでは不十分なことがわかり，その一般化の問題が起る．この一般化は，例えば Siegel のモヅル函数に対しては，Maass により，又 Hilbert のモヅル函数については Herrmann などにより試みられているが，未だ，整数論への興味ある応用が出来るまでには至っていない．此れは当面の最も重要な問題の一つである．

然し，楕円モヅル函数に話を限っても，重要な問題は多い．此処で，2つの考え方が生ずる．一つは代数的，他は解析的なそれである．代数的見地に立てば，楕円モヅル函数とは，楕円函数のモヅル又は週期の等分値の作る函数に外ならない．此の点から既に 1. で述べた様に，モヅルを変数と考える楕円函数論が生れる．此れは志村氏により研究されているものであるが，最近，井草氏も，此の様な見地から，興味ある結果を得たと伝えられる．

然しながらモヅル函数は又，モヅル群により不変な函数である．此の見地は直ちに，一般の automorphic function の理論へと導くが，それでは一般に過ぎる．然しながら，モヅル群の代りに整数論的な意味を持つ他の群を取ることにより，同様に整数論に対し有効な函数を得るであろう．此れが Eichler の考えであって，彼は，四元数の整域の単数群の行列表現の群を考えることにより，此の様な試みに成功している．一般に2次形式を主として考えれば此の様な立場に導かれるので，ここでは最早楕円函数による説明は不可能である．

前に述べた様に，モヅル函数又はそれに類似な性質を持つ函数で，一意化出来る代数函数に対しては，Hasse の ζ-函数を研究する現実的な道が開けるので，その意味でもこの Eichler の試みは注目に値する．然しながらいずれにせよ，どの様な代数函数が，此の様な函数により一意化出来るかと云う疑問が当然起るであろう．此れは，逆に云えば，モヅル函数体のヤコビ多様体を，その単純成分に分解する問題に外ならず，それは又第一種積分の週期を決定する問題である．ここで，Hecke が既に此の様な試みをしていることに注意したい．即ち彼は，mod. N で考えたモヅル群(所謂モジュラル群又は有限モヅル群)の Stufe N のモヅル函数体の第一種積分による表現を考えることにより，いくつかの興味ある結果を得ているのであるが，この問題の完全な解決は非常に難しいように思われる．

しかしながら，最近，アーベル多様体の理論が十分に発展して来たので，Eichler の理論と関連してこの問題を考えることは，ふたたび現実の問題となったとも云えよう．

5. 結　び

以上，我々は，現在の整数論に於て重要と思われる幾つかの問題を議論して来たのであるが，同時に，此等の問題は皆相互に関連し合っていることが明かになったと思う．だが更に次のことに注意したい．即ち，整数論に於ては，その対象は多面的，多義的で，又その方法も多様である．そのどれか一つに固執することは，一時的には有益なことも多いが結局は危険であると．

嘗て Hecke の量指標が，イデール類群の指標であることが明かとなったとき，これこそは量指標の本質を表わすものであり，イデール類群の上の位相解析的方法により，此の方面に，重要な発展がもたらされると考えられ，整数論の主流は今や位相的代数的方法に存すると思われたことがあった．然しながら此の様な方法によっては，量指標に関するどの様な発展も生れず，その様な展開は却って他の面から生じたのである．此の結果を再びイデールの言葉に書き直すことは勿論可能であろうし，それは更に深い洞察を可能にするかも知れない．ここで強調したいのは，イデールのみに固執することは余り生産的とは云えないということである．例えば Hecke の作用素 T の理論は一寸イデールでは扱えない．なぜならイデールから生ずるヂリクレ級数はすべてオイラー積を持つから．同様に，現在非常に重要になりつつある代数幾何学的方法も，それのみに固執することは，同様な運命を辿るかも知れない．どの様な方法にせよ，万能ではないのである．

モヂル函数についても同様である．歴史を重視する人々は，モヂル函数を楕円函数から切り離して考えることが出来ず，モヂル群は何か外面的，偶然的なものであると思い勝ちであるが，此の様な見方が一面的であることは既に指摘した通りである．

要するに整数論の対象を，安直に，此れはこういうものであると割り切って考えることは，成程気持の良いことには違いないが，殆んど常に誤りに陥るのである．そもそも整数論は，実在する対象に関する学問で従って，概念の面からも，また方法の面から，割切ってしまうことは不可能なのである．このような一面性に陥る危険は，多くの人々の協同により最も有効に避けることができる．S.S.S. 数論グループの意義は，この点に存すると思う．

附　記

本稿を書き終えてから，筆者は更に興味ある事項を発見した．上に述べたように，代数的な値を取る量指標から，$\mathfrak{p}$-進体，およびその直積の中への連続写像が得られる．これから，無限個の有限次アーベル拡大（基礎体上の）が得られる．この拡大体の ζ-函数の無限積が，最初に考えた量指標及びその共軛指標の L-函数により表示されることが明かとなったのである．このことは，前にのべたアーベル多様体の ζ-函数に対しては成立つことが予想されていたのであるが，これが一般に，虚数乗法の理論を用いずに証明できたのである．このようにして，量指標の L-函数が，新たな観点から説明されたわけだが，この説明がイーデルの言葉のみでいい表わされているとはいえ，その発見のためには虚数乗法論が必要であったことを注意すべきである．これは，以上の所論を裏書きするものであると思う．

（2号，1955 年 10 月）

A. Weil に接して

Weil との座談会の際には，慣例に反して記録を取らなかったので，以下書くべきことにつき一片のメモさえ残っていない．

この記事は全く少数の個人の記憶のみに基いて書かれたもので，局所的な正確さまでは保証できないが，大局的な誤りは無いと信ずる．

1. フジホテルで

10月10日夜，訪問者は K. S. Y. の3人

Gauss のように！

Weil： 日本人の数学を見ていて特に感ずることがある．日本人には，先輩や目上の人に従う様に要求する習慣があるのか．

S.S.S.：戦前，戦争中は特にそうだった．日本人のモラルの中心でさへあった．我々も小学校以来盛んにタタキ込まれた．

Weil： 特に数学では，若い人々に，手軽に出来ることをやり，余り大きいことには手をつけないようにすすめる習慣があるのか．自分の考えを押し出さず，偉い人の思想圏内で仕事をする方が安全だと忠告する習慣があるのか．

S.S.S.：大体そうだ．

Weil： 日本で本当に独創的な研究を始める人は少なかった．岩沢はその少い一例だが一方小平は，非常によく出来るにも拘らず，私や Lefschetz, Hodge などの仕事を完成する様な事にしか手を出さなかった．極く最近やっと，彼自身の考えに基く研究が出始めた．尤もこれは，岩沢が小平より優れた数学者だという意味ではない．私の云いたいのは小平の様に素晴らしい数学者が，自分のアイデアを見出すのにこんなにも遅れたことで，これはまさに驚くべきことだ．

然し，戦後，日本の若い人の間に，自分のアイデアを持って始めようとする者がふえて来た．特に君達は皆，高みを狙っているが，日本でこの様な傾向が出来たのは極く最近のことで，非常に良いことだ．

とにかくも，自分のアイデアを持って始める様に，Gauss はそうだった．君達も Gauss の様に始めろ．そうすれば間もなく君達は，自分が Gauss では無いことを発見するだろうが．それでも良い．とにかく Gauss の様にやれ．

モラルを変えるのは大変だが，数学のやり方を変えるだけなら，それ程難かしくはないだろう．

良いアイデアを

Weil： アイデアはフッと浮ぶものだ．何故浮んだか自分にも解らぬことが多い．然し良いアイデアを見出すことの出来る人が，数学の才能を持った人だ．だが，良いアイデアを少くとも2つ持った人でないと，良い数学者とは云い切れない．なぜなら，凡庸な数学者に，良いアイデアが偶然浮ぶこともあるからだ．例えば Morse がそうで，彼は非常に愚かな数学者で，数学も大して知らないのだが，変分法で一つの良いアイデアを持っている．然し2つ以上の良いアイデアが，凡庸な者に偶然学ぶ確率は非常に少く，neglisible だ．

S.S.S.：あなたの最初のアイデアは？

Weil： それは私の Thèse だ．私は先生（名前は記憶不正確故省略）から，Mordell の研究を聞いて，その論文を一晩で読み，一つのアイデアを得た．論文を返すときそれを先生に話したが，信用されなかった．私には自信があったので，多くの人々に話したが誰

も信用しなかった．然しそれに基いて研究を進めて，大体出来た頃，どこかで Siegel に会ったので，彼にそのアイディアを話したら，彼は非常に喜んで，大いに激励して呉れた．もう大体完成していたのだから，彼に激励されなくても結局は出来たろうが，とにかくも私はそのとき大変嬉しかった．

その後 20 年程経って，又，Siegel に会ったとき，Siegel がいうには，“私（Siegel）は，君（Weil）が，あのアイデアを話したとき，それに従ってうまく行くとは思わなかった．にも拘らず私は君を激励したのだ．私はいつもそうしている．なぜなら，若い人が何かアイデアを持って大問題と取組んでいるときには，非常な不安があるものなので，誰か名の知れた数学者がそれを激励しないと，途中で断念してしまうかも知れないから云々”と．私はこの話に大変感心して，それ以後，私自身そうすることにしている．

但し，Riemann 予想だけは例外で，それをやることは思い止らせたいと思うが，とにかくも，激励されたために，偶然にでもその大問題が解ければ，それは数学のために非常に喜ばしいことで，又出来なくても，それは本人の不幸たるに止るから問題はない．数学に取ってはもともとのことだ．

共鳴箱の理論

S.S.S. は又，Weil の共鳴箱の理論註)を攻撃したがこれに対し，

Weil：私は，共鳴箱はつまらんと云っているだけではない．それは又必要でもあるのだが，共鳴箱がないと，音叉は淋しいではないか．例へば Bernoulli は Leibniz の弟子だが（記憶多少アイマイ）Leibniz は，Bernoulli が現れてから，余計良い仕事をするようになった．勿論，Bernoulli は単なる共鳴箱ではなく，彼自身優れた数学者だが，ここで私の云いたいのは，優れた数学者も，共鳴箱がないと不幸で，余り良い仕事が出来ないことが多いということだ．自分の話すことを理解して呉れる共鳴箱が必要なのだ．音楽に作曲家と演奏家がある様に，数学にも新しい理論を作る人と，それを多くの人々に，上手に講義する人の両方があっても良いと思う．

色々な雑誌，例えば Crelle の創刊当時のものでも開けて見給へ．下らない論文が一杯載っていることが解るだろう．そんな論文は現在何の価値もないが，当時は必要だったのだ．なぜなら，彼等共鳴箱は，優れた数学者に共鳴するだけでなく，自分でも何か論文を書いて見たかったに違いない．その様な論文のために，紙面を広くあけておくことが必要なので，いつの世の中でも妥協が必要なのだ……．

S.S.S.：以前あなたにお渡ししたプリントの中の L 氏の説〈少数精鋭主義に反対する（月報 3 巻 1 号）.〉をどう思うか？

Weil：あれは昨夜読んだが，理論的と云うよりは感傷的な文章だ．

数学の才能

Weil：数学には，天才と云うものは確かにある．Gauss を見給へ．Galois, Abel を見給へ．数学者として生れついている人と云うものはあるのだ．そんな人が数学者にならないことは，勿論あるが，そうでない人が数学者になることはない．

勿論，才能のない数学者は多い．数学が非常に好きで，良く勉強しても一向にできない人もある．丁度，ある女が好きになっても，女の方で彼を愛して呉れない場合が多いのと同じことで，彼にとっては，不幸であるに違いないが，仕方のないことではないか．

然し，解決が無いわけではない．私の友人の……（カヴァリエ — とか云ったが，記憶不正確）は，数学が非常に好きだったが，大

原註）2 流の数学者は，彼等自身の出すことのできない音色に共鳴する共鳴箱の役割を果すに過ぎない（Weil “数学の将来” より）

したことは出来なかった．然し途中から数学史に転向し，その領域では色々良い仕事をしている．彼は自分の解答を見出したわけだ．

もう一人の友人（ロートマン）は，レジスタンスに向って，ナチに殺された．これも彼の解答であったのかも知れない．

S.S.S.：日本では才能があっても伸びないことも多い．第一先生がいない．皆アメリカへ行ってしまう．我々にとっては大きな問題だ．次に先生がいたとしても，我々に何を読むべきか，古典が何であるかさえも教へて呉れない．皆自分で発見しなければならないのだ．

Weil：私の若い頃も丁度それと同じだった．私がアーベル函数を始めたとき，フランスには，まともに ϑ- 函数を知っている人さへ一人も居なかったので，自分で色々な文献を見出さなければならなかった．今ではフランスには S.S.S. が出来て (Bourbaki のことか？) old S.S.S. が young S.S.S. に教へる様になったが，最初フランスで始めたときは君達と同じ事情だったのだ．

S.S.S.：例へば私は，代数函数論を，Weyl のリーマン面から勉強し始めたのだが，あれには ϑ-函数が書いてないので，Riemann を読むまで，Jacabi の逆問題の意味を誤解していた．

Weil：Riemann から始めるべきだ．私は Riemann から始めたので誤解しなかった．

それから，Enzyklopädie の Krazer-Wirtinger の書いたものを読んだ…….

2. Weil への手紙

Weil と S.S.S. との座談会は，10月22日（土曜日）と予定された．それに先立ち，日本数学界の特殊事情を多少知っておいて貰う方が都合が良いと考へた対策委員会は，一通の手紙を早急にしたため，海外流出の問題（月報2巻5号投書）と共に，手分けして飜訳し，タイプになぐり打ちして出来上ったのが21日午後10時，直ちにホテルに届けたが Weil はまだ外出中であった．

以下その要旨を掲げる．

A. Weil 様

来る土曜日の貴方との討論を実り多いものとするために，日本における数学研究の現実と，それに対する我々の考へとを，前以て貴方に知らせておくことは意味あることだと思う．……

先づ，日本では，経済的理由と，研究者としての地位が非常に少ないことのために，研究を続け得る人は非常に少い．一方研究を志す人の数は，それに比べれば相対的に非常に多いのである．従って幸いにチャンスを得た若干の人達を除いては，大部分は失業し，又は研究を断念して，別の職につくことになる．

第二に，チャンスを与へる権利を握っている教授層は，古い固定観念にとりつかれた，高度に保守的な人々の集りであるということである．この古い観念 —— それを L 氏は少数精鋭主義と呼んだのだが ——これは多くの良くない傾向と結果を生んで来た．

第三に，S.S.S. はこれらのすべての悪い傾向を克服し，日本の数学を正しい方向にそって発展させるために生れたものである．それ故，S.S.S. は失業研究者，未来を保証されない学生——以下これをボロボロと呼ぶことにする——それも若干の先進的な少数精鋭によって組織されている．ここで特にはっきり知っておいてもらいたいことは，信じられない程の生活上の困難と，研究の困難で S.S.S のボロボロは，決してヤケになったり，失望したりはしないで協力して数学研究を続けていること．更に日本の数学の正しい発展のためには，目覚めた少数精鋭と，ボロボロとの協力が，自然でもあり，必要でもあるということである．なぜならボロボロは，徹底的に，決して妥協することなく，従来の悪い傾向に反対して斗うことができるし，先進的な少数精鋭の協力が，それを成功させるからである．而も，今では，ボロボロが優れた研究成果をあげることが可能と思われるし，而も日本の数学の根強い発展のために，このことはどうしても必要なのである．貴方がこれまで話し合った S.S.S. の会員はすべて目覚めた少数精鋭であった．だから S.S.S. における ボロボロの存在とその役割とについて聞くのは，貴方にとってこれが始めてであると思う．

更に又，今日までの大学の数学教育の与へる多くの悪い影響についても知っておいてもらいたい．即ち，ボロボロにせよ，少数精鋭にせよ．そのためにスポイルされ，いためつけられてしまうのである．これらの点についての立入った討論こそ，我々の希望する所である．……

今度の会合では，研究の組織や体制の問題についても是非討論したい．S.S.S. では，グループによる共同研究の組織と体制を整へるということが，当面の重要な課題になっている．この点に関しては，ブルバキのゼミナールや討論の持ち方について貴方から聞くことが大きな利益をもたらすと期待している．例へば数論グループの谷山は “Weilと 話して，アイデアの討論が可能であることを知った” といっている．今後 “アイデアの討論” が，グループ研究の最高の目標となるであろうがその点では貴方に深く感謝しなければならない．然しブルバキと違う点は，S.S.S. のグループ研究の目的が，単に数学上の研究成果をあげるということの他に，ボロボロの自衛組織として役立つこと，ボロボロと精鋭との相互の理解を深める場となること，絶えず若い人達を吸収して，それを育てていくという教育的自的を持っていることは知っておいて欲しいことである．

S.S.S. に現にあり，又は作られつつあるグループは次の6つである．

その他 “数学の将来” や “大学における数学教育” にあらわれた貴方の数学思想については是非討論したい．最後に，日本の数学及び S.S.S. に対する忠告を聞くことができれば非常に嬉しい．

S.S.S. 運営委員会

3. 池の端の散歩

遂に 22 日は来た．Weil は東大数学教室で，午後3時半から，1時間，個人的な discussion を行ったが，その後開会までの約 50 分を，discussion に出席した S. T. の2人と共に，上野不忍の池の端の散歩に費した．折しも夕闇の迫る頃，晩秋の池の端には一種の詩情がただよっていたが， Weil はそれに眼もくれず，彼の意見を滔々とお説教した．S. T. の2人はその英語を理解するのに精一杯で殆んど口を挿む余裕もなかったという．

Weil： 今日出席する人の中にボロボロはどの位いるか．

S.S.S.： 少くとも半分以上はいると思う．

Weil： 今の教育が良くないという実例があるのか．

S.S.S.： 例へばある学生は正規のセミナリーに出て全然伸びられなかったが，独立に別の部門をやる様になってから始めて才能を伸ばすことができた．これはほんの一例だ．

Weil： それは制度の問題か．雰囲気の問題か．

S.S.S.： これは雰囲気の問題だ．然し制度の良くない例もある．といって二三挙げる

Weil： どんな制度にも良くない所はある‥‥そしてある制度のために，直接に被害を受けた人は，その制度を悪く云うもので，その気持は一般的に理解され得るものだ．然し同じ程度に被害を被っても，その原因は様々であることがある．例へばある人がナチの収容所に放りこまれて……（この例は，日本人には良く理解できないと考えたのか，間もなく他の例に変る）．例へばある人が，何の罪もないのに裁判にかかり，判事の誤判のため2年間禁固されたとする．又他の人は病気

のためそれだけの期間安静を強いられるかも知れないし，又，別の人は医者の誤診のためにそうするかも知れない．

どの場合でも，2年間動けなかったという事実は同じだが，その受取り方は場合々々で異り，然も何故異るかを我々は理解することが出来る．誰も細菌の悪口を云はないが，間違へた判事や医師のことは，非常に悪く考へるだろう．而もその場合でも，医師と判事とでは全く感じ方が違う．それは，そもそも判事は誤判しないことを期待されて，又そうできると考へられているからだ．所が，裁判制度はどうしても必要な制度だが，それがある限り，或程度の誤判は避けられない．一般に或る制度の欠陥を完全に無くそうとすると，非常に多くの努力が必要になり，得らるべきその成果に釣り合わない．然し制度が非常に悪くなって，幣害が続出する様になったら，改革が必要になる．どの場合にも，幣害と労力とを秤にかけて考へなければならない．

例へばフランスの一般教育制度は非常に悪くて，私は，すぐにも改革に手を着けるべきだと考へているのだが，色々な理由によってその様な改革が実際には決して行われないだろうと感じている．（いささか悲観的である．尤も彼は今フランスに居るのではない．）

こんなこともある．一般に入学者を選抜しなければならぬ以上，入学試験により選抜することは，割合に良い方法なのだが，やはり不公平が起る．フランスの試験官はその点非常に神経質で，Cartan など，採点に不公平のない様にするため非常に多くの時間をそれに費すが，私は常々，それは時間の浪費だと忠告しているのだ．なぜなら，試験を不公平にする要素は色々あって，或る学生は当日風邪を引いて頭が痛いかも知れず，他の学生は前日喧嘩してムシャクシャしているかも知れない，等々．Cartan がどんなに頑張っても，試験場の学生の風邪を直すことは出来ない……どの試験にも附きものの，試験問題を偶然良く知っていたか否かによる，重大な不公平の要素は抜きにして考へてもそうだ．だから，採点だけに或る程度以上の公平さを求めて，余りにも多くの時間を費すのは時間の浪費というものだ．

4. 座　談　会

Weil のお説教が終らぬ中に，3人は会場に達した．既に会場には，50人余りの人々が詰めかけていた．かくて会は（日本の習慣に従って！）定刻を20分程遅れて始まった．

海外流出の問題

S.S.S.：我々は，先日の共同コミュニケ（月報3巻2号参照）に感謝しているが，あの主旨を詳しく聞きたい．

Weil：あれは，一般社会並びに政府に対するものであって，その主旨は，数学者の給料を上げろということだ．

S.S.S.：我々は少し別な解釈をしていた．経済的問題も重要だが，日本に帰って来ている人もいるし，物理のように殆んど皆帰って来ている所もある．これはモラルの問題だと思う．

Weil：すべての人に英雄になる様に要求することは出来ない．或る種の人々には高いモラルを抱かせることはできない．又一方経済的欠乏の中でも良い仕事の出来る人もいるし，余裕がないと何もできないタイプの人もいる．だからすべての人に同じことを要求することは出来ない．それにモラルといっても自分の仕事にふさわしいだけの給料を貰っていると感じられない時，モラルが簡単に生れて来る筈はないではないか．

一時フランスでも同様のことが問題になり旅券の期間を制限することにより調整しようとしたが，正直な人には無暗に煩雑で，ずるい奴は結局抜け道を見附けるから，何の役にも立たなかった．然しフランスでの現象は一時的で，現在は正常に戻っている．

S.S.S.：然しあなたの見落している点がある．彼等は，日本の大学からも給料を貰っているから，後任を入れるわけには行かない．彼等は給料を貰いながら，講義も指導もしないのだ．

Weil： 君達はアメリカにいる日本の数学者について誇張された考へを持っている．彼等の中の或る人々は真に優れた数学者だが，その人達が日本に帰っても，どれだけ君達の役に立つか疑問だ．それに簡単な質問応答なら，航空便で片が附く．

S.S.S.：然し，日本には数学の伝統が余りないから，色々困ることもあると思う．（以下二，三の押問答の後に）

Weil： 君達の云うことを聞いていると，君達は皆 leadership-complex を持っているのではないかと思われて来た．今まで日本では，上の人に従うことが美徳とされていると聞いていたが，S.S.S. はそうではないと思っていた．所が実は S.S.S. もその例にもれず，君達は唯々現在の先生には，何か神秘的な理由のため服従出来ないが，一度び有力な指導者が現れれば，それに従うことにより万事うまく行くと考へている様に思われる．

（此の辺我々の趣旨が良く通じなかった）

伝統について

S.S.S.：ボロボロの役割をどう思うか．

Weil： 程度の高い講義やセミナリーに出席するだけでも十分意義がある．いつか Siegel が天体力学の講義をしたとき，学生が一人もいなくなったことがあったが，これでは良い講義も永続きしない．……更に自分で何か論文でも書ける様になれば一層良い．

大体 19 世紀始めまで，数学者は孤立して仕事をしていた．例へば Gauss は，自分の研究について，殆んど人に話さなかった．然し Jacobi は，一人では居られない性質だったが，彼の廻りには，数学者は一人も居なかったので，彼は，自分と一緒に研究すべき数学者を，自分で育て上げねばならなかった．これがドイツの大学におけるセミナーの発端で，それ以後，数学は大勢でやるものになった．

S.S.S. は君達の独特な組織だが，第一次大戦後，école normale に起きた動きが実を結べば，今の S.S.S. の様になったかも知れない．

S.S.S.：その会について聞きたい．

Weil： 会なぞ出来なかったのだ．école normale の生徒は皆アナーキスティックな傾向（政治的意味でなく）が強くて，まとまることが出来ない．

S.S.S.：日本には伝統が少いから，まとまってやることが必要なのだ．

Weil： 君達は伝統，伝統というが，余り重視するのは良くない．Gauss が数学を始めたとき，ドイツには伝統どころか数学さへも無かったのだ．その後，Riemann，Dedekind が来た時，Göttingen には Gauss がいたが，彼は他人と話さなかったので，やはり何もないのと同じだった．Riemann は若くして死に，Dedekind は他に転じたので再び Göttingen には何も無くなってしまった．所謂 Göttingen の伝統が出来たのは，1880 年頃，Klein の世代からで，君達とそんなに違はない．

然し数学的雰囲気も大切だ．私は若い頃しばらく Noether の所に居たが，余りまとまったものが得られず，多少失望して帰って来た．然し後になって見ると，そのとき，色々な数学の術語が話されるのを聞いていただけでも，非常に有効であったということがわかった．

招待者の問題

S.S.S.：今度のシムポジウムでの，国内の人々の招待の仕方は良くないと思う．若い人がもっと多く出席出来た方が有効であったと思はないか．

Weil： 然し会議には大体定った人数があ

り，出席者がそれ以上でも，以下でも会議の性格は別のものになる．

S.S.S.：だが，専門外の教授達や，アメリカから帰って来た連中が招待されている．

Weil：この様な会議を開くには金がかかるものだ．だから色々な妥協が必要になる．

例えば寄附した人々には，何かパーティーでも開いてやらなければならないだろう．(Weil はこの様なパーティーは大嫌いなのだが，会期中，欠かさず出席していた.）又，会議を組織するためには，多くの教授達の力が必要だが，それらの教授達に，組織だけ頼んで，出席はお断りと云うことは出来ない．S.S.S. が，自分達だけの力で会議を開くのなら，どの様にでも出来るだろうが．

（これでこの問題は打切り）

S.S.S.：数学の社会的機能について．

Weil：数学内部から見れば，ある論文を書いた人の社会的意識が，論文の内容に反映しないことは，その人の宗教が反映しないのと同様だ．数学の外から見れば，先づ，社会が何故数学者に給料を払っているかということが問題になる．（この問題は何故かこれ以上発展せず）

共同研究

S.S.S.：20世紀数学の特殊性は色々あるがその一つは，数学が細かく専門に分れたためそのすべてを知ることが不可能になり，共同研究が不可避的になったことであると思う．

Weil：そんなことはない．今でも，Chevalley や Dieudonné は数学全般に通じている．

S.S.S.：然し，それは，彼等が全体にわたって指導性を持っているか否かとは別だろう．

Weil：勿論それは別問題だ．だが，20世紀数学も，19世紀又はそれ以前の数学と本質的には変ってはいないことは極めて確かだ．アポロニウスの頃の本を読んで見れば，その当時の学生も現在と同じく，数学は相互に関連のない多くの部門に分れていて，そのすべてに通ずることは不可能だと思ったに違いないことがわかる．

S.S.S.：だが吾々は共同研究が必要だと思っている．そこで Bourbaki の例を聞きたい．

Weil：第一に注意することは，Bourbaki の目的は Bourbaki（教科書）を書くことで，だから本を書かなければ止めて貰うことになっている．共同研究が目的なのではない．

次はアイデアは個人の中からしか生れない． Bourbaki の集りで，共同で何かのアイデアが生れることがあり，その時は皆興奮するが，後になって見ると非常に小さいアイデアでしかない．それさへも共同で生れることは非常に稀なことなのだ．

だがアイデアは，孤立していると無くなってしまう恐れがある．そんなとき，皆に話していれば，その雰囲気の中で自然に育っていく．

S.S.S.：それも協力の成果ではないか．

Weil：そうではない．アイデアが人に話せる位にまで成熟するためには，場合により違うが，普通，数ケ月から数年の間，個人の胸の中で育くまれて行かなければならないものだ．これは全くその人だけの問題だ．

S.S.S.：我々はその後の発展をも共同研究の成果と呼びたい．

Weil：非常に明かなことは，良いアイデアとか，新しい事実の発見とかは，個人からしか生れないということだ．もし共同でその様なことが出来たら，至急電報で知らせて欲しい．そんな驚くべきニュースに対しては，電報代など安いものだ．

S.S.S：さっき，Bourbaki は本を書くための団体だと云ったが，Bourbaki に入っていれば色々目に見えない利益があるのではないか．

Weil：勿論そうだ．例へば最初は，数学全体に通じていたのは，私と Chevalley だけだったが，今では全員，大低のものに通じるようになった．その他有益なことが色々ある．

君達は共同研究に非常に熱意を持っている様に見えるから注意しておくが，共同で何かやるときには，一種のテクニックが必要だ．Bourbaki は，共同で本を書くテクニックを発見したので，それ以後は，それに従ってうまくやっている．ただ一つ，共同でアイデアを生むテクニックだけは存在しない．

最後に共同研究に対する忠告を３つ．

先づ，overorganize し過ぎないこと．

次に，常に，あらゆる種類の失望に対し備えていなければならないこと．失望は共同研究の一部分であると考へるべきである．

第三に，アイデアは集団から生まれるのではなく，個人から生まれる，という事だ．

この様なことに注意すれば，共同の成果は素晴らしいものになるであろう．

5. 二次会以後

Weil は，座談会中，盛んにお菓子をつまみ，サンドウィッチも平げたが，まだ何か食べたいというので，皆で通りへ出た．本郷の喫茶店は皆狭過ぎたので，お茶の水までゾロゾロ歩いて行く．彼は歩くのは大好きなのである．

あるレストランに落着き，ビールを飲み出すと又，話に活気が出て来る．

S.S.S.：あなたは物理学に興味を持っているか．

Weil：全然持っていない．私には物理学は理解できない．

S.S.S：フランスの応用数学が不振なのは？

Weil：それはフランスの教育制度の欠陥のためだ．昔は Fourier 始め優れた応用数学者がいたが，現在の制度では応用数学が発展することは望めない．もし制度が変れば又発展するだろう．

S.S.S.：数学には，芸術と似た所が多くあると思わないか？　私は絵に興味を持っているのだが．

Weil：似た所がある所ではない．数学も一つの芸術なのだ．

S.S.S.：絵画におけるアブストラクトと，数学のアブストラクトと関係があると思う．

Weil：名前が同じという以外に共通なものは何もない．私は両方について話すことは好きだが，一緒には出来ない．

S.S.S.：あなたは盛んにアイディアと云うがアイディアとは一体何だ？

Weil：アイディアを定義することは出来ない．丁度フォックス・テリアに，鼠は何かと聞く様なもので，彼は鼠を定義することは出来ないが，匂をかげばわかる．アイディアも，論文の中から嗅ぎ出すことが出来るものなのだ．人によって上手，下手はあるが．

S.S.S.：他人の仕事を follow した論文の中にもアイディアはあるのか？　あなたは以前小平は Hodge の仕事を follow していたと云ったそうだが．

Weil：これはフォックス・テリアの問題で，場合により違う．例えば Galois の研究の萌芽は，Lagrange やその他の人の中に見られ，彼はそれを follow したともいえるのだが，どんな貧弱なフォックス・テリアでも，彼の論文の中に，非常に優れたアイディアを嗅ぎ分けることができる．

一方小平の，調和積分についての最初の大きな論文は，Hodge の研究を follow したものだが，そこにあるのは巧妙なテクニックだけで，自分はその中に，どんなアイディアも嗅ぐことは出来ない．

S.S.S.：アイディアはどんな人に浮ぶのか？

Weil：アイディアに値する人にだ．それに値する人とは，アイディアなしに永い間研究を続けることに耐える秘訣を心得た人だ．

アイディアは，インスピレーションの様に

浮んで来るものではない．Borel が良い例で，彼は若い頃，天才だと思われていたので，アイディアがインスピレーションの様に浮んで来るものと思って，余り努力せずにそれを待っていたので，結局大したことは出来なかった．

だから才能だけでは駄目で，持ちこたえ，努力を続けて行くことの出来る性格の力も必要なのだ．又環境も重要だろう．

S.S.S.：E. Cartan の研究を現代流に書き変えることにはどのような意義があると思うか？

Weil ： どう書き変えるのか？

S.S.S.：例えば Chevalley 流に．

Weil ： 現在明かなことは，Lie 群の初等的な部門についての良い教科書が一冊もないということだ．（Weil は相当程度の高いものまで“初等的”と呼ぶ習慣があることに注意）

Weil ： Cartan の研究は非常に重要なものを色々含んでいて，神秘的に見えるものもある．（symmetric space は重要かという S.S.S. の質問に対し，かなり長い時間考えて）symmetric space これは幾何だけでなく，数論始め，様々な部門に現われる．私は以前，Siegal に，彼の研究が，実は symmetric space についての研究なのだということを教えてやらなければならなかったことがあった．

S.S.S.：Cartan の研究を現代流に書き直せば，問題の意味が明かになるし，色々有効なのではないかと思う．

Weil ： 自分は問題の意味などと云うことは皆知っているから，書き変えなくても差支えないし，又これから勉強しようとする人にとって云えば，確実なことは，現在，微分幾何と，Lie 群の初等的部分についての良い教科書が一冊もないと云うことだ（と，前の意味を繰り返す．）

その他，話は尽きなかったが，レストランの閉る時刻となったので，一同みこしを上げる．

*　　*　　*

A. S. T. の3人がホテルまで Weil を送って行くことになった．以下は，その途中及びホテルのバアの一隅で，12時頃まで続けられた話の一部である．

S.S.S ： あなたはフランスに帰りたいとは思わないか？　誰でも年をとると，母国語の話されている所に住みたくなると思うが．

Weil ： それはセンチメンタルだ．私はセンチメンタルな気持の中で生活しているのではない．

だが，私の妻と子供とは，毎年夏をフランスで過す．こんな生活は不便で不愉快だ．

然し，フランスにポジションが見つかっても，帰るかどうかは，その時の色々な条件によって左右される．一概には云えない．

人生は数学ではない．色々矛盾したものが同時に存在している．だから妥協が必要なのだ．

だが……色々な現実的な条件を無視すれば，私の夢は，フランスに住んで，時々外国を旅行することだ．（と，多少，シンミリする．）

（話はとんで，政治的なことが問題になる）

S.S.S.：あなたは共産主義をどう思うか．

Weil ：共産主義は圧制的だから嫌いだ．私はロシアを旅行したので良く知っている．

又，数学者は，共産主義者になると，色々党の仕事が多くなって，勉強出来なくなる．数学は all time job だ．十分暇がなければ出来ない．私の廻りで，何人も実例を見て知っている．

S.S.S.：然しそれと数学をやるのとどっちが良いか判らないと思う．

Weil ： 勿論それは別問題だ．ただ明かなのは，人は同時に数学者であり共産主義者であることは出来ないということだ．人はどちらかを選ばなければならない．

S.S.S.：数学の教師と研究者との関係はどうか？ 教えるのに熱心になると研究出来なくなることもあると思う．

Weil： 或る人々は，講義を，必要以上に elaborate するために多くの時間を使っている．然しそれが学生のためになるか否かは別問題だ．だが，教師と研究者との関係は連続的で，人は，その中間の，自分の好む位置に立ち止っていることが出来る．共産主義者の時の様な断絶はない．

S.S.S.：あなたの政治的信条は？

Weil： アナーキズムだ．然しこのアナーキズムというのは，論理的には矛盾なのだがこれも前に云った人生の矛盾の一つで，実際には存在するのだ．

S.S.S.：あなたは，大学における講義は重要だと思うか？

Weil：（肩をすくめて）それは，為される講義により，又，聞く学生にもよる．私について云えば，学生の頃私は講義を余り聞かなかったし，それは又，役に立つ講義でもなかった．

S.S.S.：近頃日本では，制度が変ったため学生が講義を非常に重視する様になった．だからあなたの云ったことは，彼等には良く通じなかったと思う．

（話は又変る）

Weil： 日本人が留学するとき，奥さんを国において行く人が多いのは，不思議なことだ．これは日本と欧米諸国との大きな違いの一つだ．彼等は，奥さんを連れて行くだけの金がないとか，それでも自分が外国に行くことは，自分の研究のために絶対必要なのだとか云うが．例えばフランスの若い人が，同じ様な事情の前に立たされたとすれば，彼が外国行きを断念するだろうことは，殆んど確実だ．

（最後に）

S.S.S.：あなたは火曜日に，何時の飛行機で羽田を立つのか．

Weil： 9時半だったと思う．送りに来るか？

S.S.S.：良くわからないが多分行くと思う．

Weil：politeness のためならば送りに来る必要はない．（彼は形式的なことは大低嫌いである．）

どうせ飛行場は混雑しているからと，最後のサヨナラをして外に出た3人のクタクタに疲れた頭には冷い夜の風が心地よかった．

（3号，1956 年 2 月，署名 SSS 会員）

Weyl と 整 数 論

Weyl は整数論にはそれ程深く立ち入らなかった．尤も 1916 年の，ヂオファンタス近似の領域での仕事（数の，1を法とする平等分布について）は，若き日の彼の才能の閃きを示すもので，現在でも高く評価されている．晩年彼は，Riemann 行列論，二次形式論などで再び整数論と接触する様になるが，これ等は寧ろ形式的な一般化に重点の置かれたもので，往年の輝きは最早彼から期待すべくもなかった．

然しながら，学生の頃 Hilbert の Zahlberichte に読み耽ったという彼は，生涯，整数論に対する関心は失わなかったと思われる．彼の著書 “Algebraic Theory of Numbers” (Princeton, 1940) は，この関心を例示するものということも出来るであろう．

この本は 1938–39 年 Princeton 大学での講義にもとずいて書かれた教科書であるから，数論についての独創的な見解をそこに求めてはいけない．然し教科書として見れば，何と独創的に書かれていることであろう．ここには歴史家として，歴史評論家としての Weyl

が，到る所に顔を出している．

代数的整数論の基本定理を論ずるに当り，彼は，Kronecker と Dedekind とを対比する．Hilbert が，整数論の基礎から Kronecker の名を抹殺して以来，この頃まで，整数論は Dedekind の基礎の上に，即ちイデヤル論の上に，建設されるのが常であった．そしてドイツ抽象代数学派の隆盛は，この傾向を強く支持する様に思われた．然しこの隆盛の中に既に異分子が生れつつあったのである．

30年代に盛んになったものの一つに，代数体と代数函数体とを同時に整数論の対象にしようとする傾向がある．そしてこの様な立場こそ，Kronecker の理論の出発点であった．この再生した傾向は間もなく，イデヤル論は自分に取って余り適当な道具ではないと考え始めるに到った．一方30年代末の Chevalley のイデールの出現は，類体論からイデヤルを追放するに致るのである．

時の流れに敏感な Weyl が，この様な動きを見逃がす筈がない．彼は進んで Kronecker を取り上げ，ついで Kummer のイデーを詳説し，両者の後継者である Hensel に説き及んでいる．この Hensel の p-進数こそ，まさしく，イデヤル論に対し自己の優位を主張し始めていた賦値論の母胎なのである．

そこには趣味の問題も加わる様に思われる．Kronecker は彼の整数論の中で，その“有限性の哲学”を貫いている．彼はイデヤルの様な無限集合を，本来の数学の対象と認めることは出来なかったに相違ない．一方，Dedekind にとっては，無限集合は最初から実体として扱われていた．彼は寧ろ方法の純粋性に関心があったのであろう．この様な感覚は現在では共に過去のものとなっている．実際，有限性の立場も，純粋性への偏愛も，共に固執すれば忽ち生産性を失うに到ることは明かである．然しながら，抽象代数学が未だ方法の純粋性を謳歌していた時代に Kronecker を取り上げたのは，Weyl の趣味の反映でもあるのではなかろうか．

然しながら Weyl は，この様な対比を提出したまま，それを更に追求しようとはしなかった．この様な見解の徹底的な追求は，10年後 (1950 年)．Cambridge の国際数学者会議での Weil の講演により始めて実現されるのである．この講演で Weil は，明かに Weyl の本に示唆されながらも，彼自身の豊富な知識と直観とによって遙かに広汎な，数学的に徹底した見解を提出している．その故にこの講演は，その後の整数論の発展に対し実際的に貢献することが出来たのである．これは，Weyl と Weil の数学者としての性格を象徴している出来事ではなかろうか．

（4号，1956 年 3 月）

数論グループ

一口に数論と云うが，その意味は明確ではない．簡単に云えば，人々が数論と見做すもの，それが数論である．

ギリシャに於て初期の開花を見た数論もその後の発展は散慢だった．解析学が急激な発展を続けていた時代に，数論は少数の好事家の手で，断片的，趣味的に扱われたに過ぎない．然しながらその成果の蓄積は，18 世紀末には，Gauss による飛躍的発展を準備する程にまで至っていたのである．19 世紀数学の基盤は複素数であり，数論は，函数論と共に，複素数の数学の立役者となった．同時に，解析学や，他の部門との有機的関連が強まり，数学内での数論の位置が確定した．然し，イデヤル，p 進数などの抽象概念の発生は，理論の平面的進歩の結果である計算の煩雑化，全体的な見透しの悪さと共に，第一次戦後の抽象数学的展開の原因となったのである．

この新展開は余りにも花々しかったために，数論は一時抽象数学の中に自らの姿を没したかに見えた．然し，第二次戦後の反省期に入って，数論的思想の優位が復活するに至り，前途尚洋々たるものがある．

この様に数論が，各時代の様々な新数学を武器として取り入れながら，尚一つの部門としての独立性を保っているのは，その全体が数論的思考，数論的問題意識に貫かれているためである．それが失われれば，雑多な数学的方法の展覧会場に堕するであろう．この特徴は，数論グループの運営に影響を与えないはずがない．

S. S. S. 数論グループは，1953 年夏発足した．当時は，自分達の勉強している分野につき，その大要，問題や見透しなどをお互いに語り合うことが主眼であった．会が終った後も，おそばを食べながら色々なことを話題にのせ，人数の少いせいもあってかなりなごやかな雰囲気であった．然し間もなく，立ち入った討論をするための共通の地盤のないことが痛感されるに至り，セミナリーを行って共通の知識を身につけようとする方針に変った．これは約一年続いたが結果は失敗であった．その中 55 年春となり，迫り来るシムポジウムに備えて，泥縄式にせよ，数論の知識を詰込み，整理するための研究集会が大がかりに開かれ，S. S. S. 以外の人々の参加も少くない状態になった．シムポジウム中の活躍を通して，数論グループのメンバーも増え，各自の実力も増大し，将来の発展が大いに期待されるに至ったのである．だが此の様な状態は，一方では運営の困難さをも飛躍的に増大させた．その打開のための新構想も立てられたが，事実上は，以前のセミナリー方式を惰性的に続けることになり，間もなく，二，三の偶然的な事情とからみ合って，数論グループは“冬眠”に入るに至る．

だが既に春だ．我々は冬眠から醒めなければならない．再び，新構想が必要である．その前に，失敗の原因を考えよう．数論では，その内容と方法の多様性の故に，すべての人に共通であるべき知識は非常に少いのである．例えば類体論はおろか，二次体の数論さえ知らなくても，数論は出来る．だから，全員共同のセミナリーを開けば，ある人々に取っては判り切ったことの繰返しとなり，他の人々に取ってはさして関心のない理論を義理で聞かされることになる．うまく行く筈がない．セミナリーは，類似した分野を研究する小人数のサブ・グループに分けて行われるべきである！

ではグループ全体の結合は？　我々は，個々は，個々の理論の技術的な細目についての知識を共有しなくても，その理論の構造と構成について，又多くの問題の意義，それを支える問題意識について，又理論の発展の歴史と将来の見透しについて，更には特定の数学者が理論を建てるときの態度，問題を解くときの方針などに現れる個性的な特徴について語り合い，各自の見識を広め，高めることが出来る．要するに推論の連鎖の中からアイデアを取り出し，それを語り，又語られるアイデアを理解し，討議することが出来る．これこそは数論グループ全員が集って行うに最も適した課題であり，これから毎月1回開かれる“数論グループ例会”の中心題目となるものである．又例会では，参考文献の紹介や勉強法のガイダンス，又サブ・グループのセミナリーの発足，経過の報告などは当然行われるべきである．以上が新構想の大要である．

又我々は，始めて整数論を学ぶ人々のために，Baby-Seminar を予定し，5 月下旬に開講する．これは，出来るだけ早く，数論の主要部門の二，三の大勢を把握させることを目的とし，差当り7月始めまでに，代数体の数論が紹介される．

*　　　*　　　*

最後に，数論に見られる最近の傾向につき一言したい．20 世紀後半は，新解析学が確立さるべき時期であろうと思われる．そして数論が，この面から豊かにされるであろうことは殆んど疑いがない．だが将来を云わずとも，何等かの解析的方法の応用が，数論の大勢となりつつある．既に位相群の上の調和解析さえ，まだ十分に利用されているとは云えない．対称空間乃至それに類似の空間の理論は，今後益々数論にとり重要になるであろう．加法数論，不定方程式論なども，又違った解析学の発展によって生気を取り戻すかも知れない．従来，数論が何か本質的な手段を見出すときには，その手段そのものが，数論との結び附きによって大きく発展するのが常であった．将来も又常にそうであろう．そして我々は，この様な新発表の胎動を感じ始めているのである．

(5 号，1956 年 5 月，署名数論グループ運営委員会)

数学の歩み　第4巻

代数的整数論における ζ-函数

これは今年の春の学会での特別講演の要旨である．詳細はいずれ論文として発表されるし，共立社の講座の中の「近代的整数論」にも関係ある章が加えられる予定なので，ここでは主として考え方の筋道を説明して見たい．

1. L-函数と類体論

ζ 函数や L-函数は，普通，解析数論に現れ，素数，素イデヤルの分布を調べるのに使われる．念のために定義を思い出して見る．有限次代数体 k の L-函数とは

$$L(s,\chi)=\sum_{\mathfrak{a}}\frac{\chi(\mathfrak{a})}{\mathrm{N}(\mathfrak{a})^s}=\prod_{\mathfrak{p}}\left(1-\frac{\chi(\mathfrak{p})}{\mathrm{N}(\mathfrak{p})^s}\right)^{-1}$$

で定る（s の）解析函数である．但し χ は k のイデヤルの合同類群（mod $\mathfrak{f}$）の指標（＝1次の表現），$\mathrm{N}(\mathfrak{a})$ は $\mathfrak{a}$ の絶対ノルム（＝剰余環の元の個数），そして $\mathfrak{a}$, $\mathfrak{p}$ はそれぞれ，k の整イデヤル，素イデヤルで $\mathfrak{f}$ と素なものを動く．ここで χ が恒等的に1（即ち単位指標）のとき，特に ζ-函数と言って $\zeta_k(s)$ と書く．

この級数・積は s の実部 >1 の所でしか収斂しないが，全 s-平面に解析接続出来，而も $L(s,\chi)/L(1-s,\bar{\chi})$ がガンマ函数と指数函数とで表される．この関係は函数等式と呼ばれ L-函数に対し基本的に重要である．

解析数論は Dirichlet に始まる．彼は有理数体の L-函数を考え，等差数列の中に素数が無限にあることを証明した．

$\{a+nd\}$ $(n=0,\pm1,\pm2,\cdots)$ なる数列は，mod d での合同類を作るから，L-函数が出て来るわけである．ζ-函数を応用したのは Riemann で，素数定理が目標であった．それを代数体に拡張したのは Dedekind である．

所でこの様な函数から得られる命題は，本質的に，イデヤル類とノルムとだけに関するもので，例えば，あるイデヤル類に，$\mathrm{N}(\mathfrak{p})\leq M$ なる素イデヤル $\mathfrak{p}$ がどの位あるかという問題は解けても，それ以上の精密さは望めない．有理数体ならこれで十分だが，n 次の代数体では n 個のパラメーターのあることが望ましい．

ここに目を着けたのが Hecke で，量指標というものを考え出した．即ち k の数には n 個の共軛数があるが，その中実数は絶対値と符号，共軛複素数の組 $\alpha,\bar{\alpha}$ は絶対値と偏角とで決るから，本質的に n 個のパラメーターが対応している．これ等のパラメーター（共軛数の絶対値と偏角）を適当に組み合せて，k のイデヤル群の指標となる様にしたものが量指標で，独立なものが $n-1$ 個ある．これとノルムとを合せて，k のイデヤルに対する n 個のパラメーターが得られたことになる．これを使えば，虚2次体の素数で，複素平面上の与えられた角領域に入るものの密度という種類の問題も解ける．

所でこれ等の函数は，別の立場からも把えられる．それは代数的整数論の立場で，その基本課題は，ある代数体の素イデヤルが，有限拡大体の中でどの様に分解するかということである．ζ-函数の定義には，素数 p の k での分解の様子が反映されていることから，この理論との関連が生じる．これが Artin の観点であった．

ここで Artin の L-函数を説明しよう．上に書いた（普通の）L-函数の定義式から

$$\log L(x,\chi)=\sum_{\mathfrak{p}}\sum_{m}\frac{\chi(\mathfrak{p}^m)}{m\,\mathrm{N}(\mathfrak{p})^{ms}}$$

となるが，今度はこの χ として，$\mathfrak{p}$ の分解の様子を表すものを考える．

一般に，K を k のガロア拡大体，そのガロア群を G とし，$\mathfrak{P}$ を，K における $\mathfrak{p}$ の素因子の一つ，$\mathfrak{P}$ は k 上 f 次とする．G は $K \bmod \mathfrak{P}$ の剰余体のガロア群（$k \bmod \mathfrak{p}$ の体上の）を induce するが，後者は $\xi \to \xi^q$ ($q = \mathrm{N}\mathfrak{p}$) なる automorphism により生成される f 次の巡回群である．この automorphism を induce する G の元 $\sigma_{\mathfrak{P}}$ を $\mathfrak{P}$ のフロベニウス置換というが，これは $\mathfrak{p}$ の分解の様子を精密に表したものと云える．$\mathfrak{p}$ が K で不分岐なら $\sigma_{\mathfrak{P}}$ は $\mathfrak{P}$ により一意に決る．又一般に同じ $\mathfrak{p}$ に属する $\mathfrak{P}, \mathfrak{P}'$ のフロベニウス置換は，G の内部自己同型で移り得る．そこで χ を，G の行列による表現の trace（所謂群指標）とすれば $\chi(\sigma_{\mathfrak{P}}{}^m)$ は $\mathfrak{p}$ だけで決るから $\chi(\mathfrak{p}^m)$ と書いて差支えない（$\mathfrak{p}$ が分岐するときは $\sigma_{\mathfrak{P}}{}^m$ に対する平均値を取る）．この χ を使って

$$\log L(s,\chi) = \sum_{\mathfrak{p}} \sum_{m} \frac{\chi(\mathfrak{p}^m)}{m(\mathrm{N}\mathfrak{p})^{ms}}$$

により定義される $L(s,\chi)$ を Artin の L-函数というのだが，これは K/k でのイデヤルの分解の様子を指示していると見られる．

特に G がアーベル群ならば，$\sigma_{\mathfrak{P}}$ は $\mathfrak{p}$ だけで決るから $\sigma_{\mathfrak{p}}$ と書いて良い．このとき K は k のアーベル拡大で，類体論が使える．その基本定理の一つに K で完全に分解する $\mathfrak{p}$ は，ある合同式を満す $\mathfrak{p}$ で，これ等の合同式による合同イデヤル類群が G と同型になるというのがある（又 K で分岐する $\mathfrak{p}$ は，合同式の法を割る $\mathfrak{p}$ である）．ここで Artin は，$\chi(\mathfrak{p}) = \chi(\sigma_{\mathfrak{p}})$ がこのイデヤル類群の指標になること，即ち $\sigma_{\mathfrak{p}}$ が $\mathfrak{p}$ の属するイデヤル類だけで決ることを証明した（相互法則）．（G がアーベル群だから，分岐する $\mathfrak{p}$ に対する平均値 $\chi(\mathfrak{p}) = 0$ となることに注意．）云いかえれば，今の場合 Artin の L-函数は，始めに述べた普通の L-函数に等しい．これが，類体論の基本定理の L-函数の言葉による表現で，これにより L-函数と代数的整数論とが結び付くのである．

話は変るが，類体論は大変難かしかったので，多くの人々がその簡単化につとめた．その過程で局所類体論，即ち k の $\mathfrak{p}$ 進拡大体 $k_{\mathfrak{p}}$ の上の類体論が生れた．これは主として Hasse-Chevalley によるが，それによれば，$k_{\mathfrak{p}}$ の最大アーベル拡大体の $k_{\mathfrak{p}}$ 上のガロア群は，$k_{\mathfrak{p}}$ の乗法群の，適当な位相による completion と同型である（この位相は $\mathfrak{p}$ 進単数群 $U_{\mathfrak{p}}$ 上では $\mathfrak{p}$-進位相と一致する）．

その後，Chevalley は，多分このことに示唆されて，イデールを考えた．イデール群とは，k のすべての素イデヤル及び無限素点に対する $k_{\mathfrak{p}}$ の直積の適当な部分群 J で，$J = J_0 \times J_\infty$ と分解される．ここで J_0 は素イデヤル $\mathfrak{p}$ に対するもので，位相的には完全に非連結，一方 J_∞ は無限素点 $\mathfrak{p}$ によるもので，何個かの実数体，複素数体の乗法群の直積と同型である．ここで J は k の乗法群 k^* を含んでいるが，それによる商群 $C = J/k^*$ を k のイデール類群と云う．これを使って類体論を考えるのだが，k の最大アーベル拡大体の k 上のガロア群 $\mathfrak{G}$ と C との関係を考えると，第一に位相が違う．$\mathfrak{G}$ は有限群の極限だから完全に非連結だが，C には，J_∞ に由来する連結成分 D がある．所で $\mathfrak{p}$ が素イデヤルなら $k_{\mathfrak{p}}$ の最大アーベル拡大のガロア群（$=\mathfrak{G}$ における，$\mathfrak{p}$ の分解群）と $k_{\mathfrak{p}}$ に上記の関係があるが，$\mathfrak{p}$ が無限素点のときは，$k_{\mathfrak{p}}$ は実数体又は複素数体だから，その最大アーベル拡大体は $k_{\mathfrak{p}}$ 上 1 次又は 2 次で，この次数は，$k_{\mathfrak{p}}{}^*$ における連結成分の次数と等しい．そこで商群 $C' = C/D$ が $\mathfrak{G}$ と同型になることが予想されるが，事実そうであることが証明されている．これが，イデールの言葉による，類体論の基本定理である．

無限次拡大を扱うときには，イデヤル群で

は話が済まないことに注意して欲しい．何故なら $\mathfrak{G}$ は連続の濃度を持つのに，イデヤルは可付番個しかないから．

この Chevalley の論文は 1940 年だが，イデールが非常に本質的な概念であることが，戦後段々に判って来た．例えば，Hecke の量指標が，イデール類群 C の指標と 1 対 1 に対応すること，又 L-函数の函数等式が J の上のフーリエ解析を使って証明されることが岩沢と Tate により独立に証明された．イデールでは，J_∞ の所に k の共軛数が並んでいるし，J_0 の方はイデヤル分解を指定すると見ることも出来るから，後から考えれば，この様な関連も不思議ではない．又函数等式の方は，Hecke，岩沢，Tate とも，Poisson の和公式を使って証明するので，違う所は，古典的なフーリエ解析のそれであるか，locally compact アーベル群 J の上のそれであるかという点である．

そこで単なる云いかえに過ぎないと云う批評も出て来る．尤もこの様な批評は，色々な場合に繰返されるが，大低は，新しいことの嫌いな年寄りか怠け者の云うことで，的を外れていることが多い．たとえ単なる云いかえに過ぎないとしても，それにより，本質的で一般的な概念に結びつき得るならば，それはより豊かな方法を約束し，より広い展望と，より深い認識とを可能にするわけで，やはり重要な step であると云わなければならない．勿論，単に形式をもて遊ぶだけで，この様な批評も成程とうなづける様な論文も少くはないのだが．

とにかくも，量指標は，イデール類群 C の指標として説明されたが，尚これでは不十分である．ガロア群と同型になるのは，C でなく $C'=C/D$ だから．そこで，普通の指標に対する $\chi(\mathfrak{p})=\chi(\sigma_\mathfrak{p})$ なる関係と類似の関係は成立たないか．即ち Artin の相互法則と直接に結びついて，量指標を C' の指標により説明できないか，という問題が生れる．

他方，特別な量指標は，abelian variety の ζ-函数で説明されるが（月報 3 巻 1 号，33 頁参照），このことの最初の証明からは，両者の関連の必然性が読み取りにくい．そこで，量指標のより本質的な説明が望ましいのである．

2. 量指標と $\mathfrak{q}$-進表現

ここで問題を少し堀り下げて見る．C' とガロア群 $\mathfrak{G}$ との同型対応を与える Artin の相互法則は次の通り：$\tilde{\pi}_\mathfrak{p}$ を，$\mathfrak{p}$ 成分が丁度 $\mathfrak{p}$ で割れる（$k_\mathfrak{p}$ の）数で，他の成分がすべて 1 であるイデールとすれば，$\tilde{\pi}_\mathfrak{p}$ の C' への像が，$\mathfrak{p}$ のフロベニウス置換 $\sigma_\mathfrak{p}$ に対応する．この $\tilde{\pi}_\mathfrak{p}$ は $\mathfrak{p}$-進単数群 $U_\mathfrak{p}$ を除いて定り，$U_\mathfrak{p}$ の像が $\mathfrak{p}$ の分岐群に対応している．そして殆んどすべての $\mathfrak{p}$ に対する $\tilde{\pi}_\mathfrak{p}$ の全体の生成する群の，C' への像は，C' で稠密だから，上の対応により $C'\longleftrightarrow\mathfrak{G}$ の同型が定まるわけである．

所で，量指標 χ から C の指標を作るには次の様にする．χ が，イデヤル $\mathfrak{f}$ を法として定義されているとき，$\mathfrak{f}$ と素な $\mathfrak{p}$ に対する対応 $\tilde{\pi}_\mathfrak{p}\to\chi(\mathfrak{p})$ は C の位相で連続になり，$\tilde{\pi}_\mathfrak{p}$ の稠密性から，C の（連続な）指標に一意に拡張出来る．

$\tilde{\pi}_\mathfrak{p}\to\chi(\mathfrak{p})$ なる対応は C' の位相では連続にならないことに注意．C' は完全に非連結であるのに，$\chi(\mathfrak{p})$ なる値の群（＝絶対値 1 の複素数の群）の位相は連結だから．連続性のためには，連結成分 D で割る以前の C が必要とされるので，C は，上の様な対応がすべて連続になるための最小の群である．

従って，量指標とガロア群とを直接に結びつけるために，表現の値の群として完全に非連結な群を選ぶことが考えられる．これまで使って来た，絶対値 1 の複素数の群は複素数の乗法群の最大コンパクト部分群で，又複素数体は代数体の無限素点での completion である．そこで，代数体の有限素点（＝素イデヤル) $\mathfrak{q}$ での completion $k_{\mathfrak{q}}$ (=$\mathfrak{q}$-進拡大体) の，最大コンパクト部分群，即ち $\mathfrak{q}$-進単数群を考えたらどうなるか？　この群の位相は，確かに完全に非連結である．

困ったことに，量指標の値 $\chi(\mathfrak{p})$ が常にこの様な群に含まれるとは限らない．そこで，$\alpha\equiv 1 \bmod \mathfrak{f}$ なる単項イデヤル (α) に対し

$$(*)\quad \chi(\alpha)=[\prod_i \sigma_i(\alpha)^{n_i}]/[N(\alpha)]^{\nu}$$

という形に書ける χ だけを考える．ここで $\sigma_i(\alpha)$ は α の共軛数，n_i は適当な正整数，又分母の ν は $|\chi(\alpha)|=1$ なる様に定めてある．このとき明かに，$\chi(\mathfrak{p})$ はすべて，一つの有限次代数体 K に含まれている．そこで K の素イデヤル $\mathfrak{q}$ を一つ定める．$\mathfrak{p}$ が $\mathfrak{f}\cdot N(\mathfrak{q})$ と素なら，$\chi(\mathfrak{p})$ は $\mathfrak{p}$-進単数群 $U_{\mathfrak{q}}$ に含まれ，前と同様に，$\tilde{\pi}_{\mathfrak{p}}\to\chi(\mathfrak{p})$ は，$C\to U_{\mathfrak{q}}$ の連続な準同型に一意に拡張出来る．今度は $U_{\mathfrak{q}}$ が完全に非連結だから，この準同型は連結成分 D 上で 1 となり，実は $C'=C/D$ から $U_{\mathfrak{q}}$ への表現になる．

即ち χ から，C' の表現が得られた．これは，昨年シムポジウムで Weil の話したことの一つである．(数学，7巻4号，205頁参照)

所で，これでは未だ量指標を説明したことにならない．なぜなら $\mathfrak{G}\cong C'$ から，$U_{\mathfrak{q}}$ への表現は沢山あって，そのすべてが量指標に対応するわけではないから，どの様な表現が量指標に対応するかは，改めて決定されなければならない．

そのためには $\chi(\mathfrak{p})$ の代りに，$\psi(\mathfrak{p})=\chi(\mathfrak{p})\cdot(N\mathfrak{p})^{\nu}$ を考えた方が都合が良い．定義 (*) から，$\psi(\mathfrak{p})$ は K の整数で，k のイデヤル $\mathfrak{p}$ の共軛の積として表され，$|\psi(\mathfrak{p})|=(N\mathfrak{p})^{\nu}$ となる．この ψ により与えられる表現 $\psi_{\mathfrak{q}}$: $\mathfrak{G}\to C'\to U_{\mathfrak{q}}$ の kernel は $\mathfrak{G}$ の閉部分群だから，それには (k の最大アーベル拡大体の) 部分体 $k(\mathfrak{q})$ が対応する．この $\psi_{\mathfrak{q}}$ は $\sigma_{\mathfrak{p}}\to\tilde{\pi}_{\mathfrak{p}}\to\psi(\mathfrak{p})$ により決るから，$\sigma_{\mathfrak{p}}$ の像はすべて同一，つまり $\mathfrak{p}$ のフロベニウス置換はすべて，$k(\mathfrak{q})$ の上では同じ作用を持つ．これは $\mathfrak{p}$ が $k(\mathfrak{q})$ で不分岐ということに外ならない．又この表現 $\psi_{\mathfrak{q}}$ を，K のすべての素イデヤル $\mathfrak{q}$ で考えれば，$\mathfrak{q},\mathfrak{q}'$ が $N(\mathfrak{p})$ と素のときには $\psi_{\mathfrak{q}}(\sigma_{\mathfrak{p}})=\psi_{\mathfrak{q}'}(\sigma_{\mathfrak{p}})\ (=\psi(\mathfrak{p}))$ は $\mathfrak{q},\mathfrak{q}'$ にはよらないことに注意する．

所がこれ等の性質が逆に，量指標を特性づけるのである．正確に云えば次の通り．$\mathfrak{G}$ を今迄通り，k の最大アーベル拡大体の k 上のガロア群，K をある有限次代数体とする．K の殆んどすべての一次素イデヤル $\mathfrak{q}$ に対し，$\mathfrak{G}$ から $U_{\mathfrak{q}}$ への表現 (=連続な準同型) $\psi_{\mathfrak{q}}$ が与えられているとし，$\psi_{\mathfrak{q}}$ の kernel に対応する体を $k(\mathfrak{q})$ とするとき，次の条件が成り立つとする．

1) $\mathfrak{q}$ によらない，一定のイデヤル (k の) $\mathfrak{f}$ があり，$\mathfrak{f}\cdot N\mathfrak{q}$ と素な $\mathfrak{p}$ はすべて $k(\mathfrak{q})$ で不分岐 (これより $k(\mathfrak{q})$ の k 上のガロア群における $\sigma_{\mathfrak{p}}$ は一意に決り，従って $\psi_{\mathfrak{q}}(\sigma_{\mathfrak{p}})$ は $\mathfrak{p}$ と $\mathfrak{q}$ だけで決る)．

2) 1) における $\mathfrak{p},\mathfrak{q}$ に対し，$\psi_{\mathfrak{q}}(\sigma_{\mathfrak{p}})$ は K の整数で，$\mathfrak{q}$ によらない．即ち，$\mathfrak{q},\mathfrak{q}'$ が $N\mathfrak{p}$ と素なら，

$$\psi_{\mathfrak{q}}(\sigma_{\mathfrak{p}})=\psi_{\mathfrak{q}'}(\sigma_{\mathfrak{p}}).$$

(故にこれを $\psi(\mathfrak{p})$ と書くことが出来る．

3) $\psi(\mathfrak{p})$ のすべての共軛数の絶対値は $(N\mathfrak{p})^{\nu}$ に等しい．ここで ν は $\mathfrak{p}$ によらぬ有理数．(これと，$\psi(\mathfrak{p})$ が整数であることより $\psi(\mathfrak{p})$ を割る素イデヤルは $N\mathfrak{p}$ の約数だけである．)

4) $\psi(\mathfrak{p})$ は k の共軛体のイデヤルの積とし表される．

(4) は本質的ではない．k を少し拡大すれば常に満されるから．)

上に注意した通り，$\psi_{\mathfrak{q}}$ が，(*) なる型の量指標から得られるときは，1)～4) が成立つが，逆に 1)～4) を満す system $\{\psi_{\mathfrak{q}}\}$ は常にある量指標から，上の様にして作られることが証明される．そのためには $\sigma_{\mathfrak{p}}\to\psi(\mathfrak{p})$

が，K の $\mathfrak{q}$-進位相で連続なことから，4) における，$\psi(\mathfrak{p})$ のイデヤル分解の型が $\mathfrak{p}$ によらず決ること，そして $\psi(\alpha)/[\mathrm{N}(\alpha)]^{\nu}$ が (*) の型に書けることを示せば良いわけで，基本的にはかなり簡単な考えに基いている．但し技術的には多少面倒になる．

とにかくもこれによって，(*) なる型の量指標が，ガロア群 $\mathfrak{G}$ の $\mathfrak{q}$-進表現の system $\{\psi_{\mathfrak{q}}\}$ の中で特性づけられた．これがこの小論の基本定理である．

序でに云っておくと，上に現れた $k^{(\mathfrak{q})}$ の（すべての素イデヤル $\mathfrak{q}$ に対する）合併体 $\bigcup_{\mathfrak{q}} k(\mathfrak{q})$ を，表現 $\{\psi_{\mathfrak{q}}\}$，或いは対応する量指標 χ に対応する体という．この体におけるフロベニウス置換 $\sigma_{\mathfrak{p}}$ の性質は $\psi_{\mathfrak{q}}(\sigma_{\mathfrak{p}})$ から，従って $\chi(\mathfrak{p})$ で決るから，この体での分解法則は χ により定り，非常に目立つ性質を持っている．後で話す様に，これは特別な場合には虚数乗法による類体の構成と関連している．

3. Abelian Variety の ζ-函数

この様に話して来ると，すべてが極めて自然である様に思われるが，以上は後からつけた理屈であって，始めからこの様に考えたわけではない．前にも云った様に，特別な形の量指標は abelian variety の ζ-函数によって説明されるが，問題はここから始まる．

ここで始めに，ζ-函数の定義に戻って見る．代数体 k の素イデヤルは，k の数 α に，mod $\mathfrak{p}$ での剰余類を対応させることにより，k から有限体への準同型を与え，この有限体の元の個数が $\mathrm{N}\mathfrak{p}$ であった（ここで，又以下でも，体の間の準同型というときは，常に，値として ∞ なる記号も含めるものと約束する．今の場合は，分母が $\mathfrak{p}$ で割れる数は ∞ なる値を取るわけである)．

そこで V を特異点のない代数多様体とし，V の函数体から有限体への準同型をすべて考え，これ等の有限体の元の個数をそれぞれ q_i $(i=1,2\cdots)$ として，少し大雑把に云って，

$$\log\zeta_V(s)=\sum_i\sum_m\frac{1}{m q_i^{ms}}$$

により，V の ζ-函数 $\zeta_V(s)$ を定めれば，代数体の場合の類似になっているといえる．ここで V が q 個の元から成る有限体上定義されているときは，この準同型は V の点 P により与えられる．即ち V 上の函数 f に P における値 $f(P)$ を対応させれば準同型になる．そこで点 P に対して上の和を作り少し計算すれば，

$$\log\zeta_V(s)=\sum_m\frac{\nu(m)}{m q^{ms}}$$

となる．これが正確な定義である．但し $\nu(m)$ は，定義体の m 次の拡大体（q^m 個の点から成る有限体）上 rational な点の個数である．

一方 V が代数体 k 上定義されているときには，V の函数体から有限体への準同型は2度に分けることが出来る．つまり $\mathfrak{p}$ を k の素イデヤルとし，V を modulo $\mathfrak{p}$ で考えた多様体を $V_{\mathfrak{p}}$ と書けば，$V_{\mathfrak{p}}$ は $\mathfrak{p}$ の剰余体なる有限体上定義されている．そこで問題の準同型は

V の函数体 → $V_{\mathfrak{p}}$ の函数体 → 有限体

と分れる．この中第二段階には，上に定義された $\zeta_{V_{\mathfrak{p}}}(s)$ が対応する．そして V の ζ-函数 $\zeta_V(s)$ は，

$$\log\zeta_V(s)=\sum\log\zeta_{V_{\mathfrak{p}}}(s)=\log(\Pi\,\zeta_{V_{\mathfrak{p}}}(s))$$

により定義される．ここで mod $\mathfrak{p}$ の reduction がうまく行かない $\mathfrak{p}$ は除く．この $\zeta_V(s)$ は，特別な場合に，Hasse が始めて考えたので，Hasse の ζ-函数と呼ばれる．一方有限体の場合の $\zeta_{V_{\mathfrak{p}}}(s)$ を考えたのは Artin だが，この方は合同 ζ-函数と呼ばれている．

合同 ζ-函数については，一般的にわかっているのは，V が1次元（即ち代数曲線）のときだけだが，特別な V に対しては，実際にその形を求めることが出来る．Hasse の函数については，それが s の有理型函数になっ

て，函数等式を満すだろうというのが Hasse の予想だが，非常に特別な場合にしか証明されていない．割合に良くわかっているのは abelian variety のときで，それを以下に説明しよう．

一般に，n 次元の abelian variety A の上で，r 倍して 0 になる点のなす群は r 次の巡回群 $2n$ 個の直積に分解される．但し r は定義体 k の標数と素とする．所で l を素数とすれば，有理数体の l-進拡大体 Q_l の加法群を l-進整数の群 Z_l で割った群 Q_l/Z_l の中で，l^m 倍して 0 になる元は丁度 l^m 個ある．そこで A の上で l の何乗倍かして 0 になる点全体のなす群 $\mathfrak{g}(l, A)$ は，Q_l/Z_l の $2n$ 個の直積と同型になることがわかる．

今 λ を A の endomorphism（＝虚数乗法）又 σ を，k の代数的閉包 $\overline{k}$ の，k 上の automorphism とすれば λ, σ はそれぞれ $\mathfrak{g}(l, A)$ の変換を引き起すから，上の同型対応により，Q_l/Z_l の $2n$ 個の直積の変換として，$2n$ 次の l-進整係数行列 $M_l(\lambda)$，$M_l(\sigma)$ により表現される．特に $\sigma \to M_l(\sigma)$ は，位相群としての連続表現になる．又 λ に対しては，$M_l(\lambda)$ の特性多項式が有理整係数で，而も l によらないことがわかっている．又 A が代数体 k 上定義されているとき，l と $\mathfrak{p}$ とが素ならば，$\mathfrak{g}(l, A)$ の点は，modulo $\mathfrak{p}$ により，$A\mathfrak{p}$ 上の対応する群 $\mathfrak{g}(l, A\mathfrak{p})$ の上に一対一に移り，従って $M_l(\lambda), M_l(\sigma)$ は，A で考えたものと，$A\mathfrak{p}$ でのそれとが等しくなることも注意しておこう．

さて，代数体 k 上定義された A の ζ-函数は，A を mod $\mathfrak{p}$ で考えた $A\mathfrak{p}$ の ζ-函数から計算されるが，後者を求めるには，$A\mathfrak{p}$ の有理点の個数 $\nu(m)$ がわかればよい．$A\mathfrak{p}$ は，k mod $\mathfrak{p}$ の剰余体，即ち $q=\mathrm{N}\mathfrak{p}$ 個の元から成る有限体上定義されているから，$A\mathfrak{p}$ の点 a が，この定義体の m 次の拡大体上 rational であるということは，それが $\xi \to \xi^{q^m}$ なる automorphism で不変なことと同等である．一方この automorphism は $A\mathfrak{p}$ の endomorphism を引き起す．それを $\pi_{\mathfrak{p}}{}^m$ と書けば，この a に対する条件は，それが $\pi_{\mathfrak{p}}{}^m$ で不変なこと，或いは

$$(\pi_{\mathfrak{p}}{}^m - \varepsilon)a = 0$$

が成り立つこととなる（ε は恒等置換）．所がこの様な点は，丁度 l-進表現 $M_l(\pi_{\mathfrak{p}}{}^m - \varepsilon)$ の行列式（それは l によらぬ有理整数）だけあることがわかっている．

即ち $\nu(m) = \det(M_l(\pi_{\mathfrak{p}})^m - E)$，$E$ は単位行列．これを定義の式に代入すれば $\zeta_V(s)$ が出来る．

一方 $\pi_{\mathfrak{p}}$ を induce した automorphism $\xi \to \xi^q$ は $\mathfrak{P}$ のフロベニウス置換 $\sigma_{\mathfrak{P}}$ を定めるときのそれと同じである（$\mathfrak{P}$ は $\mathfrak{p}$ を割る素イデヤル）．従って上に書いておいた注意から

$$M_l(\pi_{\mathfrak{p}}) = M_l(\sigma_{\mathfrak{P}}),$$

或いは

$$M_l(\pi_{\mathfrak{p}}{}^m) = M_l(\sigma_{\mathfrak{P}}{}^m).$$

即ち A の ζ-函数は $M_l(\sigma_{\mathfrak{P}}{}^m)$ から計算される．

又，表現 $\sigma \to M_l(\sigma)$ の kernel は，$\mathfrak{g}(l, A)$ の点を動かさない automorphism から成るから，それに対応する体は，k に，$\mathfrak{g}(l, A)$ の点の座標を付け加へて生じる体 $k^{(l)}$ である．所が $\mathfrak{p}$ と l とが素なら，今見た通り $M_l(\sigma_{\mathfrak{P}}) = M_l(\pi_{\mathfrak{p}})$ で，右辺は $\mathfrak{P}$ だけで決る（同じ $\mathfrak{p}$ に対する別の $\mathfrak{P}'$ を取れば，$\mathfrak{g}(l, A)$ と $\mathfrak{g}(l, A_{\mathfrak{p}})$ との同型対応が変って $M_l(\pi_{\mathfrak{p}})$ も変換される）．即ち $k^{(l)}$ における $\mathfrak{P}$ のフロベニウス置換は一意に決り，$\mathfrak{p}$ は $k^{(l)}$ で不分岐であることになる．そして A の ζ-函数はこの $\sigma_{\mathfrak{P}}$ の性質を表わすわけで，つまり $k^{(l)}$ におけるイデヤルの分解法則を反映していると云える．これが abelian variety の ζ-函数と，代数的整数論との結びつきの生れる根拠である．

4. Hasse の予想と q-進表現

以上が一般論であるが，ここで，A が十

分多くの虚数乗法を持っている場合を考えて見よう．正確にいえば，A の虚数乗法の環が，$2n$ 次の，可換な semi-simple algebra を含んでいるとする．このときは $A_{\mathfrak{p}}$ の endomorphism $\pi_{\mathfrak{p}}$ に対応する A の虚数乗法があって，今云った algebra に含まれる．だからその表現が，従って又 $M_l(\sigma_{\mathfrak{P}})$ も，同時に対角形に変換される．$\sigma_{\mathfrak{P}}$ はガロア群の中で dense だから，$k^{(l)}$ の k 上のガロア群は可換，即ち $k^{(l)}$ が k のアーベル拡大であることがわかる，だから $\sigma_{\mathfrak{P}}$ は $\mathfrak{p}$ だけで決り，$\sigma_{\mathfrak{p}}$ と書いて差支えない．更にこの $M_l(\sigma_{\mathfrak{p}})$ の，第 i 番目の対角元は，各 i に対し，ある定った代数体 K_i の整数で，勿論 l にはよらない．そこで $\sigma_{\mathfrak{p}}$ にこの対角元 $\psi_i(\mathfrak{p})$ を対応させれば，これは，$k^{(l)}$ のガロア群から，K の $\mathfrak{q}$–進単数群 $U_{\mathfrak{q}}$ への連続な準同型 $\psi_{\mathfrak{q}}$ に一意的に拡張できる．但し $\mathfrak{q}$ は l を割る K の素イデヤルである．これを，k の最大アーベル拡大の k 上のガロア群 $\mathfrak{G}$ の表現と考えれば，その system $\{\psi_{\mathfrak{q}}\}$ は，§2 における条件 1)〜4) を満す．1) の不分岐性は §3 の終りで述べた通り．但し $\mathfrak{f}$ は，mod $\mathfrak{p}$ での reduction がうまく行かない $\mathfrak{p}$（有限個）の積とする．2) も上に云った通り．3) は，$\psi_i(\mathfrak{p})$ が

$$M_l(\sigma_{\mathfrak{p}})=M_l(\pi_{\mathfrak{p}})$$

の固有値であることから出る．即ち，合同 ζ–函数に対する Riemann 予想より，$M_l(\pi_{\mathfrak{p}})$ の固有値はすべて絶対値 $(\mathrm{N}\mathfrak{p})^{1/2}$ であるから．又 k を少し拡大すれば 4) が成立つ．

従って基本定理から $\{\psi_{\mathfrak{q}}\}$ が，ある量指標から得られることが云える．$\zeta_A(s)$ を実際に $M_l(\sigma_{\mathfrak{p}})$ の固有値 $\psi_j(\mathfrak{p})$ で書いて見れば，このことは $\zeta_A(s)$ が，量指標の L–函数によって表されることを意味している．かくてこの場合に Hasse の予想が，更めてより本質的な立場から証明されたのである．

実は，2. で述べた $\{\psi_{\mathfrak{q}}\}$ の満すべき性質 1)〜4) は，今の様にして abelian variety から作られた表現の性質を基礎にして考え出されたものなので，基本定理もこの Hasse の予想を証明するために定式化されたのである．だが一度証明されてしまえば，この基本定理は abelian variety の理論から独立な位置を占め，それと独立に理由つけられることは，§1，§2 で見た通りで，数学はこの様にして発展して行くのである．

又，§2 の終りに云った意味で，この system $\{\psi_{\mathfrak{q}}\}$ に対応する体は $\bigcup_l k^{(l)}$，即ち k に A の位数有限な点すべての座標を付け加えた体，云いかえれば A の虚数乗法により生ずる体である．従って，虚数乗法による体における奇妙な分解法則は，実は量指標により統制されていたことになる．

ここでもう一度，A の ζ–函数の定義に戻って見よう．A の函数体から有限体への準同型は，別な形で2段階に分けることも出来る．即ち，a を，A の位数有限な点とすれば，$f\to f(a)$ により，A の函数体から代数体 $k(a)$ への準同型が得られる．$k(a)$ から有限体への準同型は勿論素イデアルにより定まる．所がこの様に分けると，同一の準同型が2度以上数えられることがある．これは，a の位数と素でないイデアル $\mathfrak{p}$ に対しては，a は mod $\mathfrak{p}$ で，何個か重なって同じ点に写されることに基く．そこでこの第2段階に対応する函数，即ち $k(a)$ の ζ–函数から，この余分な因子を除いたものを $\zeta'_{k(a)}(s)$ とすれば，

$$\zeta_A(s)=\prod_a \zeta'_{k(a)}(s)$$

となることがわかった．但し右辺の a は，A の位数有限な点すべてを動く．

ここで，A が十分多くの虚数乗法を持っていれば，左辺は量指標の L–函数により表されるから，この式は大雑把に云えば，量指標の L–函数の有限積が，普通の ζ–函数の無限積で表わされることを意味する．

所でこの関係も又 alelian variety の理論か

ら独立に論ずることが出来，一般に (*) なる形の量指標に対して同じ結果が得られる．但し $k(a)$ の代りとして，適当な方法で定る有限次拡大体の列を取る．この関係は Artin の L-函数と普通の L-函数との間の関係と似たものとみることも出来る．実際，それは量指標に対応する無限次アーベル拡大における分解法則の表現なのである．

然し，$\zeta_A(s)$ の積表示は，全く一般的な A に対し成立つのだから，今の関係を更に一般化することが考えられる．実際，代数的閉体 $\bar{k}$ の k 上のガロア群の l-進表現系 $\{R_l(\sigma_{\mathfrak{p}})\}$ が A に対する $\{M_l(\sigma)\}$ と同様な性質を持っているとき，それに対し適当に ζ-函数を定義すれば，それを，代数体の ζ-函数の無限積として表すことができる．

ここで一つ問題が生れる．アーベル拡大の場合に，$\{\psi_{\mathfrak{q}}\}$ が量指標に対応したのと同じく，この様な性質を持つ表現の system $\{R_l(\sigma)\}$ は，何かある複素係数行列による表現 $R(\sigma)$ に対応しないか，つまり，フロベニウス置換 $\sigma_{\mathfrak{P}}$ に対し，$R(\sigma_{\mathfrak{P}})=R_l(\sigma_{\mathfrak{p}})$ となる $R(\sigma)$ は存在しないか？ 位相を考えに入れれば，この表現はガロア群の表現ではあり得ないことがわかる．そこで，C' の代りにイデール類群 C が現れたのと同様に，ここでも，ガロア群を適当に拡大した群が必要になろう．この群とはどんなものか？ 又この群の上でフーリエ解析を使って，$\{R_l\}$ に対応する ζ-函数の函数等式が証明されないか？ これは一般の abelian variety の ζ-函数に対する Hasse の予想と関係している．然しこの様な ζ-函数は，無限次非アベール拡大における分解法則を表現していると見られるから，この問題は非常に難しい様に思われる．実際まだ少しも手がつけられていないが，何等かの形でうまく行けば，非アベール拡大の整数論への手掛りも生れるであろう．

もう一つ問題がある．今まで議論して来た量指標はすべて (*) の形のものであったが，これは，共軛数の偏角に由来するもので，従って，多い場合で量指標全体の半分，少い場合，例えば総実な体 k に対しては，一つもない．そこで，共軛数の絶対値に由来する量指標，特に総実な体の量指標に対しては，全く別な説明が必要となる．

この両者は，実際色々な点で違うので，例えば，虚２次体の量指標の L-函数に対応するフーリエ級数はモヂュラー函数になるのに対し，実２次体のそれに対するフーリエ級数は複素解析函数にはならない．これは函数等式に現れるガンマ函数の中の変数の形が違うからである．この問題も，解けて見れば至極当然な形になるかも知れないが，abelian variety の ζ-函数の様な補助的な段階が一つもないので，やはり手がついていない．

参考文献

解析数論一般．ζ, L 函数の理論については，末綱，解析的整数論（岩波）が便利である．量指標についてはHecke, Eine neue Art von Zetafunktionen und die Beziehung zur Verteilung der Primzahlen II (Math. Zeitschr. 6)．岩沢－Tate の理論は，刊行されていないが，その中 Tate の講義録が手に入る様になると思う．代数的整数論，類体論では，高木，代数的整数論（岩波）が良いが，少し古い．一番新しいのは Chevalley, Class field theory (Nagoya University, 1953～1954) だが，判りやすいとは云えない．又 Weil, Sur la théorie du corps de classe (Journal of the Math. Soc. of Japan 3) はL 函数の理論の発展をも含んでいて，非常に面白い．局所類体論では Artin の講義録, Algebraic Numbers and Algebraic Functions (Princeton) が良いと思う．1940 年の Chevalley の論文を読むときは，イデール群の位相の入れ方が，現在の理論とは少し違っていることに注意．

ここに使われている abelian variety の理論の基本的なことは，Weil, Variétés abeliennes et courbes algébriques (Actualité, Paris)にある．合同 ζ-函数については Weil, Sur les courbes algébriques et les varietes qui s'en déduisent (Actualité, Paris) が面白い．Hasse の ζ-函数には文献はない．Hasse は個人的に伝えたのである．modulo. $\mathfrak{p}$ での reduction は，Shimura, Reduction of algebraic varieties with respect to a discrete valuation of the basic field (Amer. J. Math. 77) が基本的である．但し以上はすべて，Weil, Foundations of algebraic geomety (New York) を読まないと読めない．

ここで扱われた問題に関しては，谷山，虚数乗法と合同 ζ-函数（月報，3巻1号），谷山，代数幾何学と整数論（月報，3巻2号），及び数学7巻4号（国際数学会議特集号）の Deuring, Weil, 志村，谷山等の記事，非公式討論会の記事参照．詳しい文献はこれ等に付いているのを見られたい．又近く出る Symposium の Proceedings 参照．共立社の講座の「近代的整数論」の中にも，まとめて書かれる筈．

以上，主要な文献の二，三を挙げたが，勿論，これ以外に好学の士のための良書，論文が数多くないというつもりではない．

付　記

Hasse の ζ-函数について，Hecke の Operator T の理論，Eichler の理論（月報2巻2号，同4号の谷山の記事参照）と関連して，最近志村氏が，楕円函数の理論と楕円モヂュラー函数論とを関係させることによって，非常に面白い理論を作っている．始めの予定ではこれについても紹介する筈だったが，予定の枚数も尽きかけたし，第一，志村氏自身に書いていただく方が良いと思うので，割愛することにした．詳細は多分次号に報告されると思う．

（1号，1956年7月）

数学の歩み　第5巻

巻頭に寄せて

我々新編集委員会が今後一年間この雑誌の編集に当ることになった．時間の制約のため，この第一号では大体従来の方針を引継いだのであるが，次号以後，この雑誌の性格は次の様であることが望ましいと我々は考える．読者諸賢の理解と協力とをお願いする．

先ず，去る6月の運営委員会の決議によって，この雑誌がSSSの機関誌となったことを報告しよう．SSSの目標とする所は，「日本における数学の全面的な発展を民主的におし進める」ことである．そのためには様々な活動が必要であろう．事実，数学界の現状にあっては，広汎な組織活動と効果的な啓蒙活動が，またそのための雑誌の存在が必要であると思われる．然しながらこれは，日本数学会の手によって，また学会の機関誌の改善によってなされるべきことであって，SSSが，またこの雑誌が，その（必然的に不完全な）代役を演ずることは，たとえ可能であるにせよ極めて望ましくないことであると我々は判断する．SSSの目標を実現するために何よりも重要なことは，各自が，それぞれの場において，学生として，研究者として，教育者として，乃至は応用に当る者として，自からを高め，生長し，発展することであり，しかもそのための努力が，相互の協力により，協同の場においてなされることであると我々は考える．自分自身生長しつつあるのでない者が，他の人々の生長を助け得るであろうか．どのような組織活動，どのような啓蒙活動もそれが，活動する人自身の問題の解決，彼自身の力の伸張と結びつくのでない限り，或いは片手間の奉仕として実効を挙げることなく終るか，さもなければその活動家の生活を破綻に至らせるかであろう．不幸にもこの様な例の幾つかは我々の周囲にも見出されるのである．

互いに協力して才能をきたえ，可能性を伸ばし，大きく生長するための真剣な努力がこの雑誌に反映することを我々は希望する．この雑誌の記事が，その片手間に読者への奉仕として書かれるものではなくて，このような努力の過程そのものから生れ出るものであることを望むのである．記事を書くことにより問題が更に広い視野において把えられ，また効果的な解決法が予見されるようになることも期待できるであろう．また，同じ問題を考え，或いは同様な立場にある人々からの反響も起るであろう．このようにして，各自の数学人としての生活の中心から生れる問題の解決に役立ち，またそれについての交流の討論の場となることこそ，この雑誌の果すべき主要な任務であると我々は思うのである．その問題は特殊であるかも知れない．多くの人々の理解を期待できる性質のものではないかも知れない．然しながら，何人かの人々が真剣に立ち向った問題についての記録は，内容的に理解され得ない場合でさえも，その生々しさ，その真剣さの故に，多くの人々の同様な努力を力づけ，また形の異った他の方面にまでも，直接或いは間接の示唆を与えるに至るであろう．

専門の研究の上で，勉強の仕方の点で，教育制度の面で，教育それ自体について，或いは生活条件，就職等の問題について，数学的であれ非数学的であれ，その努力と関心とを集中する対象を持たない人があろうか．その様な努力が，協同の場で，各サークルにおいて，活発に，恒常的に行われるようになること，しかもその過程において，その努力の記録，問題の提示等が，自発的に編集部に寄せられる様になることを我々は切望する．その形式，内容は問わない．この雑誌の，これまでのスタイルにはとらわれないで欲しい．我々が読者に提供しようと欲するものは，整った形ではない．真剣な声なのである．

（1号，1957年7月，署名編集委員会）

少数精鋭主義について

少数精鋭主義とは，大学を純粋な研究機関として運営するという原則である．それは更に，次の様な考えと結びついている．即ち，人間の才能は不変な，固定したものであって，しかもそれを見分けることは容易であるというのである．曽って L 氏が問題にし（月報3巻1号36頁），又倉田氏が批判した（数学の歩み4巻3号102頁）のは，まさしくこの様な原則なのである．尤もL氏及び倉田氏は，更に数学の発展の過程について，少数精鋭主義の抱く“誤った考え”について語り，A. Weil の“共鳴箱の説”を批難する．即ち数学の発展が本質的に少数の優れた人々の手によってなされ，二流以下の数学者は共鳴箱であるに過ぎないという説は誤りであるというのであるが，筆者はこの批難は取らない．第一に，Weil の説は 19 世紀の数学については正しいと思われるし（文末の註参照），現代の数学については歴史の審判を俟つのみである．第二に，少数精鋭主義者は必ずしも Weil の説を信奉しているわけではなく，大低は共鳴箱を尊重しているし，彼等自身共鳴箱であることも少くない．だから共鳴箱云々は差当り現実の問題とはならないのであって，問題は共鳴さえも出来ない人達なのであると思われるからである．

その他少数精鋭主義について論じられたことは少くない．或る人は，自分が尊重されないのは少数精鋭主義のせいであるとして憤慨する．不愉快な現象をすべて少数精鋭主義のせいにするのは甚だ景気のよいことではあるが，結局，一切の悪を共産主義の存在に帰して快哉を叫ぶ右翼の心理と軌を一にするものであって，もとより筆者の取る所ではない．又，“五流数学者の精鋭なんて実際バカバカしい話”といいながら，“少数精鋭主義は必要悪”であるという所に落ちつき，“孫子の代まで数学はやらせたくない”とする考え（月報3巻2号72頁）も，ここでは問題にしない．それは，いささか性格の違う問題であると思われるから．現実に存在し，しかも多くの点で問題とされる少数精鋭主義は，この小論の始めに規定した通りのものであると筆者は考える．

大学は研究機関であると同時に教育機関でもある．これは自明なことであるが，大学を純粋な研究機関として運営しようとする原則は相当根強いものである．この原則に従えば例えば次の様な考えも許容される．即ち，大学において研究者を養成することさえも必要ではない．またそれに所属する人が日本にいようと外国にいようと，研究さえ続けていれば問題ではない．これ程極端でない意見としては，大学では将来研究者になる人だけを養成すればよい．又，研究者になるのはかなり少数でよい．現に適当な少数で結構研究がスムースに進んでいる．だから，自然に放任して，頭角をあらわす人だけを問題にすればよい．勿論大学院にはその様な人だけを入れれば良い等々．L 氏が繰返し批判したのはまさにこの様な考えなのであって，それは大学を純粋な研究機関であるとする所から生れるものなのである．

人によると，理学部の数学教室は研究機関であり，教育学部のそれは教育機関であると考えるが，この考えも誤りである．両者とも同時に研究機関でありまた同時に教育機関である．そしてこの二つの道の比重も，それ程違うべきではない．勿論，“教育”の内容が異るのは当然である．ここでは理学部における教育を主として問題にする．

数学は応用されることの少い学問である．現代の数学が自然認識からも，技術的な応用からも一応切り離されて発展しているという説をL氏は攻撃するが，これは事実なのだから仕方がない．勿論，例えば数理統計学の様に，直接に応用される部門も存在する．この様な部門については独立したコースが設けられて，或る程度の人数の実際家を毎年社会に送り出すようになることが望ましい．この点で，一部の関係者の勢力争いのためとか聞くが，統計専門のコースが未だに大学院に設けられないのは甚だ奇怪なことである．然しこの様な部門を別にすれば，理学部の数学科で教育される学生の大部分は，直接に数学を応用する職に就くのではない．彼等は，或いは小・中・高校の教師に，或いは大学の教養課程の教師に，或いは理学部の数学科の教師になる．いずれにせよ，数学を教えるという職に就く．そして大抵は，数学を応用する人々のために，自然科学に，工業技術に，或いは経済学や管理技術に数学を応用する人々のために，それ程尖端的ではない数学を教えるのである．

では，数学科の学生に対して，何よりも先ず教育技術を教えるべきなのであろうか．そもそも教育技術とは何であろうか．

或る人の説によれば，教育とは要するに学生を上手に“まるめこむ”ことであるという．つまりもっともらしいことを云って，それをうまく学生に信じ込ませてしまうことで，だから大道易者の方が，その辺の教師よりも遙かによく数学を教えることができると彼はいうのである．この様に極端な説は滅多に聞かれるものではないが，これと似た様な考え方をする人は少くない様に思われる．つまり教育技術とは“まるめこむ”術だというのである．定理や公式などを覚え込まして，決った型の練習問題を解ける様にすればよいというので，問題の解き方を，易者の託宣よろしく，のたまうのである．所で教わる人々にとっては，数学を使える様になることが必要なので定理や公式と同時に，数学的な物の考え方，数学的な“何物か”を身につけることが望ましいのである．数学を，生き生きとした感覚でとらえる様になることが望ましいのである．彼等にそれを得させるためには，教える人自身，その様なものを身につけていなければならない．つまり彼が教えることを形式的に理解しているだけでなくて，“数学というもの”がわかっていなければならない．それを生き生きとした感じでとらえていなければならないのである．（易者に数学が教えられないゆえんである）．つまり，数学を受動的に呑み込むのではなく，数学に対して積極的な態度を持つことが望ましい．第一，応用する人は数学を能動的に把えざるを得ないのであるから．

従って，教育技術を教えるか否かということは枝葉の問題なのであって，大学の数学科における“教育”の主眼は，この様な態度を身につけさせる，つまり数学というものを，生き生きと把えさせることに置かるべきであり，これは学生が将来研究者（兼教育者）になるときでも，教育者になるときでも，等しく重要なことである．

所で，この様な態度を身につけることは，単に講義を聞き，本を読むだけでは困難である．それだけでは理解が上すべりし，形式的になって，“本当の所”が中々把握できない．この様な目的のためには，勉強したことを自分なりの方法で再編成して見，更に新しいことを考える．少くとも考えようと努力すること，即ち一口に云って“研究”と呼ばれることが最も役に立つ．勿論始めは，余り程度の高いことは出来ないであろう．然し例えば，２年間マスター・コースで勉強して，修士論文を書くということは，将来研究者になるのでない人に取っても，極めて有意義なことなのである．

勿論この様な場合の研究の題目は極めて特殊なものであって，恐らく彼が将来教える様になるかも知れない数学とは甚だ縁の遠いものであるかも知れない．然し一事に通ずれば万事に通ずということもある．特殊な題目においてであるにせよ，本格的に数学と取組んだ人は，数学全体の性格を，より良く把握出来るようになるのである．

この様に見て来るとき，大学の数学科における“研究”と“教育”とは本質的には矛盾しないことがわかる．更に研究者養成と教育者養成との，二つの異った態度があるのではなく，数学を本質的に，主体的にわからせるという，一つの態度があるに過ぎないこともわかる．そして，上に書いた様に指導して行く時，研究に適した人は研究者になるであろうし，そうでない人は研究者にならないであろう．然しこの後の場合でも，彼の努力は無駄ではないのである．更に，本格的な研究者と，何もしない人との間には様々な段階の人がいる．余り結果は出なくても研究の態勢を続けている人から，かなりの程度の研究者までに至る．この様な人の存在は，数学の発展にとって大して意味がないという人もいるかも知れないし，この様な人こそ数学の発展を支えているのだという人もいるであろうが，これは結局歴史の審判に委ねられるべき問題である．何れにせよ，この様な人々の層が厚く広くなることは，数学の発展のためにかなり有意義である様に思われるのだが，それは別にしても，このことは，日本の数学教育のために極めて望ましいことであるのは，以上見て来たことから明かであると思う．

結局，数学科における教育で，本格的な研究者になる見込のある者とない者とを分けて，前者しか問題にしないという方針は，極めて望ましくないのであり，この点で筆者はＬ氏の説を支持するのである．

以上の議論の中で，実は或る一つの問題に意識的に触れずにおいた．それは人間の才能乃至能力に関する問題である．

数学に向いている人と向かない人がいることは明瞭であり，又前者の中でも，数学の研究に適した人とそうでない人とがいることも明かである．従って研究に適した人だけに目を付けて養成しようという態度が可能になるのであり，又研究の能力もないのに大学院に入っても始まらないと考える学生も出てくるわけである．この様な考え方や態度は共に，人間の才能が不変であるという前提に基いている．それは“生れつき”のものであり，今更どうにもならないというのである．少し目先のきく人はそれに意志の力を付け加える．才能があっても遊んでばかりいれば何も出来ないのは明かだからであるが，意志の力も結局はその人固有のものであって，中々変るものではない．だから，出来ない人に研究を強いるのは無駄であるばかりでなく本人の不幸でもある，ということになる．

この考えは誤りである．“出来ない者が出来る様になる”ことは，事実屢々起るのである．――こう云うとき，多くの反対論が出ることを筆者は知っている．或る人が出来るか出来ないかを見分けることが出来ると自慢する人は少くないし，又，出来ると思っていた人が大きく伸び，ダメだと見られていた人が結局ダメだったという例は，出来ない者が出来る様になる例よりも遙かに多いであろう．とすれば，余り多くない偶然を頼りにする外ないということになるかも知れない．

所が，この“偶然”を詳しく検討して行くと，多少異った結論が生れて来るのである．

数学に限らず，すべて学問，技術或いは芸術的技能等の修得にあっては，質的に異る幾つかの段階があり，その一つの段階に到達した人は，それから先ある程度までは割合に骨を折らずに発展して行けるのに対し，その段階に到達しなかった人は，一寸怠けると元の木阿弥になってしまい，それまでの努力が水泡に帰するものなのである．この性質は，この段階に固有のものであって，どのような手段乃至事情によってその段階に達したかには余り関係がない．才能により，努力により，良い環境により，或いは幸福な偶然により，等．様々の原因によるとしても，とにかく或る一定のレベルに達した人は，そのレベルを維持し，高めることが容易になし得るのである．この間の事情は語学の修得の例などを見れば極めて明らかである．ここでは最初の段階は，一通り初等文法を覚えるか否かである．散慢な勉強をしていたために，何年かかっても一冊の文法書を読み終えることのできなかった人を知っているが，その様な努力の時間の総計がどの様に多くなろうとも結果から言えばこの人にとっては，何も勉強しなかったのと同一である．この際彼は語学の才能がなかったわけではなく，努力をしなかったわけでもなく，ただ勉強の仕方を誤ったのである．

数学において，かなり程度の高い一つの段階は例えば，次の様なものである．即ち，何かある問題意識を持つか否かということであって，問題意識に従って論文を読む時は，必要な知識を体系的に，しかもかなり確実に身につけることが出来るのに反し，そうでない論文の読み方から，研究に必要な知識を得ようとすれば，比較にならぬ程多くの努力を要することも稀ではない．又，何か自分で考える際，一つの問題意識に従って考えるのと，漫然と思いをめぐらすこととの違いは言うまでもないであろう．又もう一つの段階は，自分の才能を訓練し，発展させる機会をもてる様になるか否かということである．どの様な才能も，訓練なしに大きくなることは殆んどない．又それ程高い才能でなくとも，訓練によりかなりの程度にまできたえ上げることが出来るのである．現実には，多くの才能を持った人程，この様な段階に達しやすいという事情のために，“才能のある人が伸び，ない人が伸びない”という現実が起るのであり，又，“出来ない人が出来る様になること”や才能もあり勉強もしていてもウダツの上らぬ人がいるのも，この様な事情によるのである．初等的な学習から本格的な研究に至るまでに存在するこの様な段階をすべて数え上げることはしない．実際筆者も，そのすべてを良く知っているわけではない．ここで強調したいのは，その様な段階が存在し，それに達するか否かが重大な結果を生み得るということである．

しかも，この様な段階を経て進んで行くことの出来る期間，またその間に自分の才能をきたえて行くことの出来る期間は割合に短い．年を取ってからでは遅いのであって，或る年令に達すれば大体そこで固定してしまうのである．頭が固くなってしまってからでは遅いのである．

多くの人は“眼力”を誇ることが出来るのもこの様な事情に基く．つまり誰かが或る段階に達したか否かを見分けるのは容易であり，しかもその段階に達した人はその後割合に発展するからであり，又或る年令以後はそれが大体固定するからである．

以上の様な考えが，数学科における教育に反映する様になることが望ましい．この時，“出来ない人に研究を強制するのは本人の不幸である”という説の論拠は，かなり薄くなる．それに，本格的な研究者になることは一般的にいってあまり容易なことではないが，数学の論文を書くことはかなり容易なことだからである．自分で論文を書いたことのある人は誰でもこの説に同意するであろう．そして，論文を書く様に指導することによって，その学生に，一つの段階を踏み越えさせる様にすることが望ましいのである．尤もこれはマスターコースの後半における教育であって，その以前に多くのなすべきことがあるのは勿論であるが，一つの例として挙げたのである．

数学科に於ける教育の問題であると同時にこれは，数学科の学生の勉強の仕方の問題でもある．即ち，一般的に云って，多くの時間をかけて努力すれば良いというだけではいけないのであって，より合理的な，科学的な方法が求めらるべきである．第一に，そもそも何らかの方法があることが望ましいということを確認すべきである．この点につき，経験者は自分の経験について語るべきであり，又多くの人が協同して検討すべきであろう．この“方法”は，初等的な学習から本格的な研究に至るまでの各段階において様々であり，異った段階にあっては必ずしも同一ではない．

或る段階においては正しかった方法を後の段階にそのままもち込むことによって失敗する人も少くないことは注意すべきである．

少数精鋭主義に対する反対は，一方において“勉強と生存の権利”を主張すると同時に他方この様にして自己の才能と可能性とを伸ばして行く努力に結びつくならば，真に望ましい結実を得るであろう．

（註）　筆者は，19世紀後半に ϑ-函数について，百余人の人により書かれた四百余りの論文のリストを持っているが，この数多くの論文の中，ϑ-函数の理論の発展を直接或いは間接に助けたと思われるものは，余り多くはないのである．しかも，ϑ-函数の理論そのものについても，そのかなり多くの部分は現代の数学から見て余り意味がない様に思われる．例えば Prym は ϑ-函数の理論に少なからず貢献しているし，当時は相当に評価されていたことは間違いないと思われるが，今日では彼の業蹟どころか，その名を知る人さえ少い．

（1号，1957年7月）

整 数 論 展 望

表題は大げさだが，勿論整数論の全分野にわたって展開するつもりではない．主として代数幾何学と関連する部門での最近の話題の幾つかを取り上げ，その指し示す方向を尋ねて見たいと思うのである．この範囲に話を限っても，井草によるネロン多様体の研究，Weil によるトレリの定理の証明等，当然紹介すべきものの幾つかを，種々の事情により割愛せねばならなかった．従って本稿は整数論展望の第一部として読んでいただき，いずれ日を更めて第二部を書くことにしたいと思う．

我々の問題にしている分野で，現在興味の中心を占めているのは，様々な種類のζ函数であり，その幾つかに対応するモヂュラー函数（一又は多変数）であると思われる．又，方法的には，微分の概念の再検討が要求されているのではないかと思われる．従来専ら利用されて来た第一種微分の他に，第二種，第三種の微分の代数的な理論が要求され，更に現在解析的にしか定義され得ない微分，高階の微分や調和微分などの代数的な説明，或いはこれ等に代るべき何等かの代数的な概念の構成が要求されているのではないかと思われるのである．或いは，或る型の微分方程式の理論の“代数化”も望ましいかも知れない．要するに，多くの人々が漠然とした予感を持ちながら，明確な形では誰も提出し得ない何者かがある．それが現れるとき一つの革新が起る様な何者かがあるのである．我々は恐らく，この革新の前夜に居るのであろう．

1.　群多様体と微分

代数多様体とアーベル多様体との関連の根源的な形は，一変数函数体のリーマン面の上の第一種積分の週期によるヤコービ多様体の構成であり，又函数体の因子類群とヤコービ多様体の点の加群との同型対応を与えるアーベル・ヤコービの定理である．高次元の多様体に対しては，第一種積分から作られるアルバネーズ多様体と，因子類群を代表するピカール多様体とは一般に同じでないが，それ以外の点では一変数の場合と同様であり，この関連は任意標数の体の上の多様体にまで拡張されている．そして，この関連を様々な方向に拡張することが，現在の一つの問題である．

ピカール多様体の理論は現在完成している．そして，Chow による最近の研究［1］により，ピカール多様体とアルバネーズ多様体との間に成立つ“双対性”について著しい成果が挙げられているという（論文未発表）．一方アルバネーズ多様体が，多変数函数体の数論に対して如何に有効であるかを Lang は明かにした（［4］，［5］）．即ちピカール多様体が因子類と対応する様に，アルバネーズ多様体は0次元のサイクルの同等類を定めるのであるが，数体の不分岐な類体論に対し，絶対イデヤル類が持つのと殆んど同じ意味を，この同等類は，多様体の不分岐アーベル拡大に対して有し，定義体が有限体の場合には普通の意味の類体論が成立つのである．又この類群の指標により多様体の L-函数が定義され，数体の場合と同じ形式の理論が成立つ．更に，アルバネーズ多様体の代りに，可換な（完備でない）群多様体への写像を考えることにより，数体の Strahlklasse に対応する分類が得られ，分岐する場合の類体論を作り得ることを彼は指摘している．

所でこの Lang の注意は，少し前から続けられていたRosenlichtの研究（[7]，[8]）に負うている．Rosenlichtは代数曲線の場合に，“一般ヤコービ多様体”を考えているのであるが，これは一方では第二種，第三種微分の理論であり，他方ではStrahlklasseに対応する分類を与えるのである．そして最近森川は（[6]）Langとは独立に，この一般ヤコービ多様体を用いて，代数曲線の上の，分岐する場合の類体論を作っている．所で高次元の多様体に対しては，一般ヤコービ多様体の概念はやはり二つに分れて，因子類に対応する一般ピカール多様体と，第二種，第三種微分に対応する一般アルバネーズ多様体となるべきである．前者については問題の因子類を形式的に定義することは出来るが，それを幾何学的にアーベル多様体により表現することは簡単ではない．後者については，どの様な定義が適当であるかもわかっていない．Rosenlichtによる最近の結果（[10]）が，望ましだけの精密さを持っていないので，それを普遍写像の性質により定義するには今の所多少難点がある．むしろ，類体論を成立させるように定義することを考えた方が良いのかも知れない．

望ましい定義の最低の条件は，可換な群多様体の一般アルバネーズ多様体はもとの多様体自身であるということである．これらが定義された場合，やはり双対性が成立つか否かを検討することは興味があるであろう．尤も可換な群多様体の自己準同型環において，ロザッチの自己反同型やカステルヌオーヴォのレンマを考えることは余り有用でないと思われる根拠もあるのであるが．

以上の様な問題を考えるためにも，又それ自身の興味のためにも，可換な群多様体の一般論を作ることは有意義である．上に述べた双対性もその一つであるが，先づ標数0の場合に，解析的な方法で取扱うことが有効であろう．アーベル多様体に対するテータ函数と同様な役割を果す函数がこの場合に存在するであろうか？ Severiの本（[13]）は参考になると思うが日本には来ていないらしく，見る機会がない．任意の標数の場合，一般論の段階（Rosenlicht［9］における様な）において未解決の問題は，標数 $p \neq 0$ のアファイン空間に，同型でない群構造がどれ位入るかという問題である．trivialでない例としては，長さ有限のウイット・ベクトル（mod p）の群があるが，この群はアファイン群の中でどの様に特性附けられるであろうか．この見地から，ウイット・ベクトルの新しい定義が得られるかも知れない．

第二種，第三種微分の理論は，モヂュラー函数の理論に応用できる筈である．－2次元のアイゼンスタイン級数がその様な微分に対応するから．

代数多様体とアーベル多様体との関連においても，残された大きな問題がある．中間次元のサイクルの分類がそれである．現在，有理同等の概念があるが，これは（群多様体による）幾何学的な表現を持たない．そして恐らく本質的に持ち得ないという難点がある．むしろ，アーベル多様体或いは可換な群多様体で表現される同等の概念を直接に考えるべきであろう．そのためには，アルバネーズ多様体の性質を徹底的に研究することが必要かも知れない．例えば，ある多様体の，十分多い個数の対称積から，そのアルバネーズ多様体への写像に対し，正則なcross-sectionが存在するであろうか．それは曲線の場合にしか，わかっていない．0次元のサイクルに対しアルバネーズ多様体による分類と一致する様な最も狭いEquivalence Theoryが存在するか，又存在したとすればそれは因子に対し一次同等と一致するであろうか？ 等々．

然しながら，以下に見る通り，この様な理論が整数論に対して持つであろう有効性についてはかなり疑問の余地がある．勿論，以下とは全く別の関連の中でそれが有効に利用される可能性は相当大きいと思われるのであるが．

2. 合同 ζ-函数

有限体の上で定義された代数曲線の ζ 函数は，数体のそれの形式的な類似とした定義され，それに対しリーマン予想の類似が成立つか否かが興味の中心であった．然しこの予想が成立つことが証明されて見ると，数体の整数論に現われる幾つかの量の絶対値の評価に対しこの予想が有効に利用されること，或る場合にはその問題の本質的な部分をなしていることが次第に明かとなり，合同 ζ-函数は単なる形式的な類似物以上の重要性を持つことが認められたのである．

さて，一般に有限体の上で定義された，完備な，特異点のない代数多様体の ζ 函数（合同 ζ 函数）については，有名なヴェイユの予想（[16]）がある．この予想は位相幾何学におけるレフシェッツの不動点定理と，或る意味で同等である．より詳しくいえば次の通り：ある写像の不動点の個数を，様々な次元のホモロジー群による写像の表現の跡（trace）により計算する方法を与えるのがレフシェッツの定理であるが，標数 $p\neq 0$ の多様体に対し，有理係数又は，或る標数0の環を係数とする各次元のコホモロジー群を適当に定義して，それによる有理写像の表現の跡により，レフシェッツの定理と同じ形で不動点の個数が計算できるというのが，ヴェイユの予見の含蓄する意味である（[18]）．標数 p の多様体に対し，“適当な”標数0の環を見出すことは難題に属する．所で，中間次元のサイクルの，或る類別が，アーベル多様体により代表されるとすれば，有理写像はサイクル類の写像を引き起すから，対応するアーベル多様体の準同型を引き起し，従って l-進整係数の行列で表現される．事実代数曲線に対しては，この行列表現から不動点の個数が計算できてヴェイユの予想が成立つのであるし，又一般の多様体に対しても，1次元，$2n-1$ 次元のコホモロジーによる表現の代りに，この行列表現が取れることが予想されている（n は多様体の次元）．これ等は，ピカール或いはアルバネーズ多様体から得られる表現であって，標数0の解析的な場合には，一般に奇数次元のホモロジーは，“高次ヤコービ多様体”（Weil [17]）により表現されるのであるから，この高次ヤコービ多様体を代表的に定義出来れば，標数 p の場合にもそれで十分な筈である．所が，中間次元のサイクル類を表わすアーベル多様体が出来たとするとき，その次元は一般には，対応するヤコービ多様体の次元より低くなる．従ってサイクル類では不十分だということになる．事実，具体的に計算できる例について考えてみると，中間次元のサイクル類から生ずると思われる項と同次に，それからは生じ得ない項が，ζ-函数に現れるのである．

さて，奇数次元のホモロジーは ζ-函数の零点に，偶数次元のそれは極に対応する．従って上記の方法が成功したとしても，ζ-函数の極については何もわからない．寧ろ，始めから，全然別な環を係数とするコホモロジーを考えた方が良いのであろう．Serre による最近の理論［12］は，その一つの試みであって，ウイット・ベクトルの環を係数とするコホモロジーを考えるのである．この理論は，代数曲線の p-ベキ次の拡大の問題などに極めて有効であるが，まだ試みの範囲を越えていない．高次ヤコービ多様体は調和積分の週期により定義されるのであるから，解析的な調和微分の概念が何等かの形で代数的に定義され，標数 p の多様体に対し標数0の“調和微分”が自然な方法で定義されれば，それは非常に有意義であろう．然しこれは空想乃至妄想に属する事柄である．

数学界には，“コホモロジー恐怖症”ともいうべき傾向があって，コの字のつくものは何でも毛嫌いする人もいるが，この傾向が健全なものであるか否かは，ヴェイユの予想がどの様な方法で解決されるかにより判定できるであろう．

以上の様な考え方とは別に，Lang [5] は，いわゆるネロン多様体の方法を使って問題を代数曲線の場合に帰着し，変数の絶対値が非常に小さい所での合同 ζ-函数の行動を研究している．この方法により，1次元のホモロジーがアルバネーズ多様体により代表される（即ち ζ-函数の絶対値最小の零点がアルバネーズ多様体の ζ-函数のそれと，重複度を含めて一致する）という予想は証明できるかも知れない．これだけでも証明できればかなり応用範囲は広いと思われるが，この方法でそれ以上の結果を得ることは無理であろう．

3. Hecke の結果の拡張

モヂュラー函数についての Hecke の様々な業績の中で，特に彼が力を注いだと思われるものの一つに，或る型の函数等式を満すヂリクレ級数の中における或る種の代数体の ζ-函数，L-函数の特性附けの問題がある．ラマヌジャンの数を係数とするヂリクレ級数は，この様にして特性附けられるものの一つであって，Hecke の目的の一つは明らかに，このラマヌジャン級数の理論を一般化することにあった．これが有名な“ヘッケの作用素”の理論であり，その後他の型の函数等式を満すヂリクレ級数に対して，Maass により，様々な場合に拡張された．この拡張はいささか“形式的”なものであって，或る型の偏微分方程式の理論とも関連し，その数論的意義をうかがい知ることはかなり困難である．以下に述べる Selberg の理論と同様な枠の中で論ぜらるべきものかもしれない．

所が最近，Eichler ([2])，次いで志村([14])により，ヘッケの作用素の理論は，モヂュラー函数体のハッセの ζ-函数，-2 次元のモヂュラー形式の展開係数の評価の問題に応用された．特に後者は，後に述べるラマヌジャンの予想との関連において注目すべきものである．さて，志村の取り扱いにおいては，楕円曲線の系列によって解析的な問題が“幾何学化”されている．ここで，この様な幾何学化を，より広い範囲のモジュラー函数に対して考えることは非常に大切な問題である．さし当り，ヒルベルトのモジュラー函数を，一つの総実な体を虚数乗法として持つアーベル多様体の系列を考えることにより，代数的に取り扱えないであろうか？　これは，総実な体の量指標の L 函数の問題と密接な関係がある様に思われる．総虚な量指標の L 函数は，或る意味で算術的な特性附けがなされているが（谷山 [15]），この特性附けも，上記の関連の中で新たな光の中で照し出されるかも知れない．一方，“自然”に考えれば，多変数の ζ 函数に導かれるのであるが，これはどんな意味を持つであろうか？　以下に述べる Selberg の理論をここに適用して得られる多変数の ζ-函数とは，どの様な関連にあるであろうか？　この様な関連を追求して行くことにより，多変数の ζ-函数の一般的な性格と，数論に対する意義とを解明する手がかりが得られるかも知れない．我々はまだこの未開の原野の開拓に必要な道具を十分に手に持っているとはいえないのであるが，それにも拘らずこの問題に手をつけることが必要であり，問題の重要性は，“悪戦苦斗”に十分に価するものであると考えられる．

Hecke の業蹟は，まだ色々あり，いづれも大切な問題を含んでいる様に思われるが，その詳論は他の機会にゆずりたいと思う．

4. Selberg の理論

今世紀になってから発展した多くの現代解析学の思想と手法は，再び解析的整数論に応用されて，いくつかの発展の緒をひらきつつあるように思われる．Selberg の最近の仕事([11]) は，そのような傾向の諸労作のうちの最初の決定打といってもよいもので，Siegel, Maass 等のドイツ学派の巨大な計算の累積，Hecke が全生涯をかけて建設した解析的数論の一手法，E. Cartan の対称空間を中心とする微分幾何についての諸知識，Bargmann, Gelfand, Neumark, Harish Chandra などの，群のユニタリ表現‥‥etc. がここに出会うのである．

Selberg の仕事から暗示をうけて，どれだけのものが今後構成されるかは，まだ見通せないけれども，彼が explicite に表明したものの内重要と思われるものを並べて見る．

（1） 非可換群のばあいの Fourier 変換の正しい拡張をあたえたこと，そしてそれに対して，Poisson の和公式を拡張したこと：古典的な Fourier 変換論，とくにその中の Poisson 和公式が，解析的整数論に果した役割は大きい．Θ 公式や，円の内の格子点の問題はその例である．Fourier 変換論は実数直線の上でばかりでなく，一般の Abel 群（locally compact）の場合に拡張され，ある場合には Poisson 和公式が成立する．それが非可換群の場合に拡張されたのである．そのことは二次形式の整数論，algebra の ζ-函数等に意味を持つかも知れない．

（2） Riemann 予想における一つの試み：彼はフックス群に関係してきまる一つの Dirichlet 級数を構成し，それに対してリーマン予想が殆んど成立することを示した．すなわちその函数の non-trivial な零点は実数軸上の区間［0.1］にあり得る有限個の例外をのぞいた外は皆 $\Re s=1/2$ の上にあるのである．その事の証明の核心は，楕円型偏微分方程式の固有値が負であることである．このことは，リーマン予想の成功した他の例の，congruence ζ-函数の場合には，証明の根拠が Castelnuovo の Lemma $\sigma(\xi\xi')>0$——すなわち二次形式 σ が正値であること——であったこと，及びある種の位相群の上での distribution が正値になるという命題が通常のリーマン予想と同値な命題であるという A. Weil の注意とくらべて見て興味がふかい．

それでは，本来の ζ-函数に対して，楕円型微分方程式や，二次形式 σ に対応する構造は何であるか？　そのような疑問に我々のモウ想をたくましくするのは止めよう．徒労におわるのは目に見えている．まだ材料はそろっていないのである．

しかし，Selberg の ζ-函数をもっと一般なる不連続群に対して拡張しておくのは，意味があるように思われる．

（3） 二次形式に対して多変数の ζ-函数を導入したこと．今まで二次形式に対応した ζ-函数が Epstein や Siegel などによりあつかわれたが，それらは一変数のものであった．しかし Selberg の方法では多変数のものが自然に出てくる．そしてそれらはある函数等式を充す．たとえば三変数の二次形式のばあいには二変数の函数 $\xi_r(s,s')$ が出て来て，これは置換 $(s,s)\to(s+s'-1/2)$, $1-s')$，$(s,s')\to(1-s,s+s'-1/2)$，$(s,s')\to s,3/2-s-s')$，$(s,s')\to(1-s',1-s)$ によって不変である（この置換の全体は S_3 に同型な群である）．これらの函数の研究はまだ行われていないが，行きづまっていた二次形式の数論に活路を与えるものであることは認められよう．

（4） Kronecker の類数公式の拡張．modular 群の働く上半平面を base とし，torus を fibre とするある fibrespace のような物を考えることにより modular form の概念を modular 群のある種の invariant として定式化することができる．それを用いて，その空間に (1) でのべた一般論を適用して，modular form に働く Hecke の作用素 T_p の trace を計算した．このことから二変数の二次形式の類数の間に成立つ，いくつかの関係式がえられる．又 Ramanujan の数 $\tau(n)$ を二次形式の類数を用いて explicit に表わすこともできる．-2 次元の modular form は函数体の微分と見なせるから，その場合には代数曲線の全知識がフルに使えて，類数公式は割に容易にえられる．residue 公式を用いるか又は不動点に関する Lefschetz の定理を用いて，不動点の個数を考えればよい．又は Eichler-Shimura の合同式 $T_p \equiv \pi + \eta_p \pi'$ (mod p) を用いて数えてもよい．これが Kronecker の類公式で，Selberg はそれを $-k(k>2)$ 次元の modular form に拡張したのである．

しかし Selberg の方法はあまりに“解析的”である．すなわち，彼の論法の中には，T_p が algebraic curve の correspondence であるという事実は explicit には現れて来ない．そのため，彼の方法では T_p のもっとくわしい性質，たとえばその整数論的性質を出すことは難しいと思われる．

5. Ramanujan の予想

さて，Ramanujan の予想 $|\tau(p)| \leqq 2p^{11/2}$ は，そのような T_p の整数論的問題の一つである．それは -12 次元の Stufe 1 の cusp form に働く T_p の固有値の絶対値が $2p^{11/2}$ より小さいということであって，これは合同 ζ-函数に対する Riemann 予想に類似した問題であることを Petersson が注意した．実際 Eichler, Shimura は，-2 次元 Stufe N の cusp form に働く T_p の固有値が，$2\sqrt{p}$ より小さいことを示したのであるが，その方法は modular 函数体を表わす algebraic curve を modulo p で reduction して，問題を congruence ζ-函数のリーマン予想に帰着させたのであった．

このように考えると，Ramanujan の予想のような問題が解かれるためには，modular 函数をあらわす curve が reduction されねばならない．そしてその際，高次元の modular form が一緒に reduction されねばならないことが肯けよう．modular form が reduction されるためには，それが何か代数幾何的に意味のある何者かで表現されねばならないが，そのような試みは未だ何処にも現れていない（-2 次元の場合，それは微分そのものであった）．それは何か fibre-bundle のようなものであろうか？――Selberg の仕事にあらわれる空間は fibre space のようなものであった．――又は或種の微分方程式の議論が，そろそろ代数化されねばならぬ時期が来ているのであろうか？　とにかく，今までの代数幾何学の道具，第一種微分と abelian variety だけでは片のつかぬ問題であるらしい．

このような問題に入るとき，Selberg の仕事が，――そのままではすぐ役に立たぬのは前にのべた通りであるが――，何かの指針を与えてはくれるであろう．

しかし，このような問題に関する限り，Eichler の最近の仕事（[3]）の方が我々に身近であるように思われる．

Eichler の仕事の目標も Hecke の作用素 T_p の trace の決定である（結果は勿論 Selberg の結果と一致している）．しかしその方法は Selberg の方法とは違って，もっと代数的である．その詳細をのべる余裕はないけれども，modular form の間に適当な内積を考えたり，kernel function 的なものを構成している点で，Selberg の議論に似ているし，しかもそれらのものの構成法が代数的であるという点で，我々にいくらかの望みを与えるのである．

だが Eichler の方法といえども，modular form とは何か？　という問に対して解答を与えているわけではない．――modular form は modular form としてとりあつかっているのである．――彼の方法は半代数的である．

要するに Ramanujan の問題を扱うためには我々はまだあと一つの“何物か”に遭遇しなければならないように思われる．

それが何物であるかは，今後の数学の発展が教えてくれるだろう．

Selberg の仕事にもどって，彼の論文の最後に，ある代数的な関係を持つ積分作用素の固有空間の次数の間に，何か代数的整数論的なカンタンな関係のあることを示しているが，その事実の意味，暗示するものについて我々は語る言葉を持たない．或いはつまらないことかも知れない．

この他，整数論の問題を離れれば，Selbergの仕事は尚多くの問題を我々に投げかけている．たとえば拡張されたFourier変換の逆変換はどのようにして得られるか？（或いはすでに出来ていることかも知れない）．

参考文献

[1] Chow, W.L., Abstract theory of the Picard and Albanese varieties: to appear.

[2] Eichler, M. Quaternäre Quadratische Formen und die Riemannsche Vermutung für die Kongrenz-zeta-Funktionen (Arkiv der Math. 5. (1954) 355-366)

[3] —,Eine Verallgemeinerung der Abel'schen Integrale (Math. Zeit. 67 (1957), 267-298)

[4] Lang, S., Unramified class field theory over function fields in several variables, (Ann. Math. 64 (1956) 285-325)

[5] —, Sur les serie L d'une variété algébrique (Bull, Soc. Math, France, 84 (1956) 385-407).

[6] Morikawa, H. On generalized jacobian varieties and separable abelian extension of function fields of one variable, to appear.

[7] Rosenlicht, M. Equivalence relations on algebraic curves (Ann, Math. 56 (1952) 169-191)

[8] —, Generalized jacobian varieties (Ann. Math. 59 (1954) 505-530)

[9] —, Some basic theorems on algebraic groups. (Aner. J. Math. 78 (1956) 401-443)

[10] —, A universal covering property of generalized jacobian varieties (Ann. Math. 66 (1957), 80-88)

[11] Selberg, A. Harmonic analysis and discontinuous groups in weakly symmetric Riemannian spaces with applications to Dirichlet series (Proc. Int. Colloq. on zea-functions at Tata Institute, Bombey, 1956)

[12] Serre, J-P. Sur la topologie des variétés algébriques en caractéristique p (Symposium on Algebraic Topology, at Mexico, 1956 (mimeographed note)

[13] Severi, F. Funzioni Quasi Abeliane, 1947

[14] Shimura, G. La fonction zéta du corps des fonctions modulaires elliptiques. (Comptes Rendus, 244 (1957), 2127-2130)

[15] Taniyama, Y. L-functions of number fields and zetafunctions on abelian varieties (Journ. Math. Soc. Japan, 9 (1957) 330-366)

[16] Weil, A. Number of solutions in finite fields. (Bull. Amer. Math. Soc. 55 (1949), 497-508).

[17] —, On Picard varieties. (Amer. J. Math. 74 (1952), 865-894)

[18] —, Abstract versus classical algebraic geometry (Proc. International Congress of Math. 1954, Amsterdam, Vol. III. 550-558).

尚，本誌，月報——数学の歩み．に掲載された志村・谷山の幾つかの記事も参照されたい．又本号の，Rosenlichtの論文の飜訳も見られたい．その他，[6]，[12]，[14]，については．「代数幾何学とその応用」（赤倉セミナー報告，1957年，数学振興会発行）に詳しい紹介がある．虚数乗法とζ-函数，モヂュラー函数等との関連については，志村．谷山「近代的整数論」（共立社現代数講座）に書いてある．

（久賀道郎と共著，3号，1957年12月）

数理科学研究所設立の問題について

——全国の研究者各位に訴える——

最近我が国では核物理学研究所，基礎物理学研究所，物性論研究所など基礎科学の全国共同利用研究所が相ついで設立されております．数学部門からも昨年来数理科学研究所設立の準備が進められ，来る4月7日にこの問題について第一回の公聴会が開かれようとしています．数学の総合研究所の設立という点だけでも注目を引くに足るのでありますが，特にその趣旨にうたわれている全国共同利用と云う性格に着目する時，もしこの研究所が適切に運営されるならば，我国の数学の今後の発展にとって極めて大きな意義を有するものと考えるのであります．我々はこのような見地から研究所の意義と，その望ましい運営について検討し明らかにすることに努力して参りました．以下がその概要であります．我々としては，設立される研究所がこのような形で実現されることを強く望むものであります．

然しながら研究所に期待される大きな役割も，それがどのような経過を経て設立されるかに深く関係している事に注意しない訳にはいきません．それは，研究所が如何なる理念と制度とをもって設立されるかによると共にこれらがそれを利用すべき主体たる全国の研究者の間の大きな関心事となり，その意義と望ましいあり方について十分の検討が行われ，その結実として実現を見るか否かに大きくかかっていると思うのであります．

このような見地から，全国の数学研究者各位がこの研究所の設立と運営に関して活発に検討され，積極的に意見を発表されて，この研究所が真に共同利用の研究所として成果をあげていくように作りあげていかれることを強く訴えるものであります．

研究所はなぜ望まれるか

現在，数学の研究は主として全国に散在する大学の教室乃至講座の単位に編成された研究体制の下で行われております．然しながら研究の実際のあり方について云えば，これらの単位の内で行われているのでなく，同じ問題に興味をもつ同地域内の研究者間の日常的な協力は勿論，各地の研究者が直接接触してそのテーマについて討論し，idea を交換すること，一定期間の共同研究など各種の交流によって最も効果的にすすめられることが知られており，そのように行われている例もあります．所で問題は，このような研究体制と，研究の実際の在り方とのくい違いによって，多くの不都合な事態が生じていることであります．講座制が研究に関する限り事実上廃止されている例は多いのでありますが，難点の主な部分はいまのままでは全く解釈できず，何等かのこれに代る新しい研究体制が望まれているのであります．そしてこの事は，日本の数学界の次の様な状況から云って特に必要が痛感されるのであります．

今日，日本の数学は高い水準に達したと云われ，事実若干のすぐれた人々によって大きな成果が収められており，一方相当数の研究者層が形成されてもいるのであります．然しながらひと度これらの研究の指導的中心，あるいは流れの中心に着目するならば，それは依然として海外に仰いでおり，我国に指導的な伝統が形成されているとは申せません．

また国内において研究上の実質的な交流の仕組は極めて未発達であります．例えば数学者の大部分が接触の機会としている日本数学会の年会を中心とした交流が実質的なものであると考える人は稀でありましょう．行われたとしても，極めて個々に偶然的に行われねばなりません．これに代るシンポジューム，研究グループによる交流など多少試みられているとしても，学界全体としての努力を伴わないならば大きな解決は全く望み得ないのが実状であります．まして研究を中心とした人事の交流などは不可能と云ってよいでありましょう．各地の研究者が互に意を尽して研究テーマについて意見を交換し，自信をもってそれを追求するというような状態からは甚だ遠いのであります．このような事態は上述の伝統の欠除と相まって我々における研究の発展にとって甚だ不利な条件となっております．いわゆる数学者の海外流出の原因が経済条件だけにあるのではないと云われる所以であります．このような行き詰りを打破して交流の道を開き，伝統の形成を計るためには，現在の研究のあり方に最も適した仕組，即ち研究体制を採用することが不可欠と思われるのであります．これについて論ずるためには連絡交流の果す役割についてやや詳しく反省する必要がありましょう．

先に屢々“接触して討論する”事について言及したのでありますが，もしも現在，各人の研究が一人一人全く散発的乃至偶発的な動機に従い極めて狭い範囲の問題を興味の対象とし，それだけに着目して行われているならば――例えば何等かの偶然的動機に従って選んだ個々の論文を少しずつ modify するような研究が一般的であるなら，各人の興味の共通部分はごく狭いのでありますから，この交流にさほどの意義や必要も感じられないのであります．誰でもが必要とする基礎的事項の勉強会とか，極く限られた範囲の共同研究などがせいぜい必要なものでありましょう．所が，現在或程度着実な乃至興味ある研究を意図するならば上述の方法では困難であって相当の広さを持った題目を系統的に追求して行くことが必要なのであります．このことは孤立して研究する場合でも勿論，何の変りもありません．

所でこのような方法で研究が行われている場合には共通の問題に興味をもつ研究者と比較的多く互に接触して研究の方向について討論したり idea を交換すること，更に一定期間の共同研究を行うことが，original な研究を行う時にも，それを目ざして現在までの研究に follow する段階にあっても極めて有効なことは明らかであります．世界的に研究者の増大が一般化している今日，この事は益々痛感されるようになっております．近来流動研究員制度や研究グループの必要が強調されている背景には，このような事情の意識的乃至無意識的な認識があるのであります．

殊に日本に於ては，かって学問の輸入に国をあげて努め，どの研究者の研究も大体は直輸入であった時代，次に大多数の人々が各地毎に教室中心，或いは全く孤立して主として外国の研究の modify につとめ，若干のすぐれた人達が外国の伝統の上に大きな成果をあげるようになった時代を経て，次第に研究者も増加し，その follow もはるかに系統的となり，一般的水準も向上して次の段階への移行が求められはじめていると云うことが出来ましょう．全国交流を基礎にした研究体制の問題と，伝統の形成の問題との間にはこのような内的な連関があるのであって，いわばひとつの問題とも申せましょう．勿論この事は数学に限らないのであって，本文の冒頭に掲げた各種の共同利用研究所設立の動機はまさにそこにあったのであります．このような例にならって新らしい研究体制をうみ出し，その中心としての機能を果す所に，そしてそのことによって我国の数学研究の真の意味での指導的中心を形成していく可能性を与える所に，我々は数理科学研究所設立の最大の意義を見出すのであります．従ってこの研究所の機能が，研究所本位に若干の俊秀を養成して研究の発展を計ることも， 単なる“全国交流”の中心事務局の如きものに終ることも，或いはこの二つを機械的につなぎ合せたものに止ることも，全く望ましくないと考えるのであります．

今少し具体的に研究所の役割を考えるために，研究所を中心にした“体制”について述べる必要がありましょう．それは次のようなものであってほしいのであります．即ち，各地の研究者の間に各自の興味に従って，共通のテーマを中心とした研究グループなどの形で実質的な交流がはかられること，これらを背景に全国の研究者の合議により必要に応じていくつかのテーマを選び，その指導的中心を共同利用研究所に置いて多様な接触交流をはかる．内容的にくわしく云えば，研究所の研究が系統的，活発であることによって，グループ全体の研究に刺戟をあたえてその中心のひとつとなる．symposium，財政的な裏づけを持った一定期間の滞在，事務機構を利用する communication などがこのために利用される．勿論研究の進行に応じてメンバーは移動する．従って所の人事は研究上の要請を第一義として上記の合議によって流動する．但し次のことはつけ加える必要がありましょう．ここに研究グループとは狭く固定したものを意味せず，必要に応じて流動する．個々の人は自からの興味に従って自由に選んだ広いわくの中で自由かつ自主的に研究する．またここにいう“指導”とは“実質的にそのような役割を果す”ことを意味するのであって，研究テーマを強制することでないのはもちろんであります．

このような体制の一環として次のような行事を行うことが望まれます．然しながら研究所の役割はこれにとどまらず，上述の機能を全面的に実現することにあるのは勿論であります．

a）テーマ別に一定期間の全面的なシンポジュームを開いて専門家間に研究方面の見通しについての討論等を含む接触協力の機会を作る．

b）共同研究の場を提供するほか，研究上の必要に基いて全面的な人事交流をはかる．

c）指導的研究者を外国からまねき，共同研究が行われるようにする．

d）国際的な学術交流のセンターのひとつとして全国の研究者に事務的な便宜を提供する．

望ましい運営の原則

研究所に期待される大きな役割を果すには，それに応じた運営方式が必要とされるでありましょう．そのためには，運営の理念と適切な制度とが不可欠であります．

運営の理念とは，研究所の果すべき役割を明確に認識して，すべての運営をその方向に向けて行うことであります．このような明確な認識に基いて始めて，様々な現実の制約の中から，より有意義な，より理想的な研究所を作り上げる方向を見出し，それを実現させて行くことができるのでありまして，これは個々の当事者の見識や心構えの中に解消することのできないものであります．

さて，理念の実現を保証するものは，それに応じた制度であります．もしも適切な制度を欠くならば，事の成否はその時々の当時者の見識と手腕とのみによることになり，我々は偶然の幸運を期待する外はなくなるでありましょう．従って，運営の原則は制度の形で実現されなければならないのであります．

望ましい原則の第一は，研究所が，全国の研究者の手によって直接に運営されることであります．これは，前に述べた研究所の役割から当然のことでありまして，詳述する要はありません．このために，全国の研究者から直接選挙で選ばれた運営機構が研究所の運営に当るという制度を作るべきであります．

第二の原則は，研究所の運営が，研究を第一義として行われることであります．これも当然すぎる程当然なことでありますが，その実現のためには，因襲にとらわれない革新的な制度が必要となります．先ず，研究所の人事，予算等は，研究計画に基いて計画的に行われなければなりません．更に，大きな研究計画の下で，各研究者間の自由な接触，交流協力が進められなければなりません．これを制度的に保証するために，先ず講座制を撤廃し，また教授・助手等の職階に基く身分的従属関係や差別を打破することが必要であります．職階に基く差別等は，研究者の研究能力や，識見，指導力と何等関係ないからであります．さらに同じ理由によって，研究計画の立案，研究方針の検討，また，それに応じて人事，予算等一切の運営は，各研究者の平等な権利に基く会議によって行われる必要があります．最も適切な計画，最も正しい結論を保証するものは権利の平等の上に立つ自由な討論に外ならないからであります．これを要するに，研究を第一義とする運営は，民主的な制度によって最も良く保証されるのであります.

研究所の及ぼす影響

始めに述べました通り，研究所の問題は全国的な研究体制の問題と切り離すことはできません．従って，研究所の理想的な運営のためには，学界全体，すなわち各研究室や数学会の運営方式が改善されることが必要になると思われます．たとえば“研究を中心とした”人事交流が各教室間でも盛んに行われることが必要となりましょう．然しながら，このことを逆に申せば，先ず第一に研究所の運営が理想に近い形で行われ，全国の研究者がその運営に参加し，または研究所を中心とする研究に参加して行くならば，それによって逆に各研究室や学会の運営が改革されて行くことが期待できるということになります．実際大きな見通しと実質的な交流，共同とに基く研究方式は，各地域の研究者の研究を刺戟し，促進するでありましょうし，また，研究所の民主的な運営は，各研究室の運営を改革し民主化するための一つのモデルとしての役割を果すでありましょう．そしてこの様な改革が必要とされていることは，現在では誰の目にも明らかとなっていると思われるのであります.

さらに特筆すべきは，日本数学会に対する影響であります．現在，学会の合議機関は実質的な役割を持たず，従って会員の関心を失って形式化しております．このことは，評議員会の選挙を，ソビエトの選挙と同じ単一名簿方式によって行われているにも拘らず，それに対して積極的に異議を申し立てる人が一人もいないことからも明らかであります．この形式化はまた，これらの合議機関を支える分科会や支部の実質的な基盤の脆弱なことによって定常化され，逆に，この基盤の脆弱さは，合議機関の形式化によって促進されている状態であります．この意味で，研究所を中心とする実質的な研究グループが全国に結成されること，また，実際に権力を持つ民主的な運営機構が全国から選出されて研究所の運営に当ることは，学会の現状に革新的な影響を与えるでありましょう.

学会の運営を問題にするとき，研究費の配分方式に注目しないわけには参りません．本来，文部省の科学研究費は，研究テーマを中心としたグループすなわち研究班に対して，実質的な研究活動の支えとして支払われるものでありますが，数学の場合そのような活動には利用されず，単なる金の配分に終っているのが現状であります．然しながら，このような資金は，たとえその額は僅かであっても，あるいはテーマ別のシンポジューム開催の費用として，あるいは短期間の“武者修業”の費用の補助として実質的な研究の促進に役立てることもできるのであります（本誌参考資料の項の，素粒子論グループの例参照)．またこの様に利用することにより，実質的な研究班が形成され，研究費の利用の細目に対しても多くの人々が関心をひき，その決定が民主的なシステムに従って行われるようになることが期待されるのであります.

この小さな例から見ましても明かな通り，明確な理念と積極的な意欲さえあれば今すぐにでも実行できることで，放置されたままになっていることも少なくないと思われます．またその様に実行して行くことによって，全国的な，民主的な基盤が徐々に成長して行くでありましょう．研究所の問題について我々が始めに述べた提案も，実質的な基盤を云々するよりも，先ず理念と意欲とによって解決されることが少なくないと信ずるのであります.

二，三の論点

これまで述べてきましたことは，いわゆる“応用”“純粋”数学について何等の区別もあるべきものでないと考えるのでありますが特にこの面について付け加えるならば，その研究が真に理論的にふかいものを目指して活発に行われることを希望したいのであります．このような研究こそ，応用面に関し我国において全く他所に期待し難い貴重なものであると同時にいわゆる“応用”“純粋”にわたる多方面の研究者の自主的協力によって特に稔り多い成果が期待される方向であって，我国の数学の健全な伝統を形成するという観点からも非常に意義深いと思われるからであります．尚，設立準備委員会による趣意書でのべられている共同研究の方法は[注1]，このために全く適切なものと考え，賛成するものであります．

さてしばしば体制と制度に言及したのでありますが，これらの運営上の難点として，現在，例えば基礎物理学研究所に対する素粒子論グループのような完備した実体の存在しないことは当然あげられるでありましょう．この例では，日常的な学問的交流に基き，民主的に選出された委員会により研究計画の討論からその実施に及ぶ一切の運営が行われているのであって前述の体制を考えるにあたってひとつのモデルとした所であります．物性論研究所には物性論グループが存在しています．然しながら，やや詳しくみるならば[注2]，これらも決して最初からそのようなものではなかったことを知るのであります．前者について云えば，戦前よりの討論の伝統が一部少数の上級研究者間にあったとは云え，グループとしての発足は講演予稿集の発刊が行われたこと，就職，人事問題の話し合いの会などを機会とする戦後の事にぞくし，研究内容をふかく討議する学問的交流のグループとしての性格を確立したのは基礎物理学研究所設立後の国際会議開催前後の時期にあると云われ，そののち基研を支えとして漸次充実完備してきたのであります．注目すべきは，このようにひとつひとつを取り上げれば或る程度偶然とも云える重要な機会を最大限に活用して，それぞれ一時期を画する進歩の機縁として発展して来たことであります．

物性論グループの場合は我々にとって更に教訓的であります．即ち，それが極く小さな萠芽の状態からごく短期間に飛躍的に現在に至ったのは，前記基研に物性部門が（素粒子方面の主唱によって）設けられ，俄かに必要となった研究所運営上の事情に基いて急速に素粒子論グループのシステムを導入するという，極めて外的な事情を転機としている事であります．しかも物性論は，どちらかと云えば個々の具体的問題の追求を特徴として居ると云われており，単一のグループとして運営する上に予想される多少の難点にも不拘，これを巧みに克服して来た事情は数理研の場合のよい参考になろうと思われるのであります．

このように見る時，数学に於ても，積極的に伸長せしめる意図をもってみるならば，多方面にその萠芽を認めることができるのであります．尚若干の難点を見るとしても，適切な準備と時間とをかすならば十分の可能性が存在するのであります．むしろ現在ない故にこそ，理念はそれを目標とするに足るだけ高いものであり，制度はそれを満すに足るだけ広いものであって，両々相俟ってその成長を促進させる必要性が痛感されるのであります．出発点においてその枠を狭め，あらかじめその可能性を摘みとるような在り方は永く禍根を残すものと云えましょう．望まれる方向に向っての積極的な意志の重要性を強調したいのであります．

以上考察したことは，いわば主として数学界内部の努力によって解決さるべき体制の問題に限られております．然しながら我国の数

注 1）「数理学研究所設立に関する公聴会」（日本学数会）参照
注 2）「数学の歩み」5巻4号「参考資料（素粒子論グループと基礎物理学研究所について)」参照

学の在り方を考える時，これは勿論その一部にすぎません．日常話題になる研究ポストの不足や経済条件から来る困難など山積していることは明らかであり，その早急な解決など到底のぞめないのが実状であります．これらに目をふさぐことは誤りでありましょう．

然しながら上述した問題は，一応それとは独立にも解決できる部分であり，それによってもたらされる所は，我国の数学の将来にとって計り知れないものがあると考えるのであります．この機会に，これらの問題，特に数理科学研究所のあり方が，広く活発に検討せられ真に望ましい方向に向って実現されるよう，全国の研究者各位に重ねて訴える次第であります．

（署名新数学人集団，4号，1958年4月）

米国留学について二つの意見

最近，A大学大学院ドクター・コースのB氏が，アメリカのC大学に留学するについて，C大学にいる日本人数学者D氏と打合せた．また滞米中のE氏からも意見が寄せられた．それは次の様なものである．ちなみにB氏は，日本にいれば来年3月に学位を取る予定であるが，C大学の research assistant の fellowship に apply するつもりであった．

D氏の意見

貰える金額は，授業料を除いて年に1700ドル，独身でも相当つらく（日本よりいくらかましな程度），夫婦では不可能，しかし大体これが標準で，良い方である．別に，月に10ドル位の，余り時間を喰わないアルバイトの口も学校からもらえる．

この fellowship は，マスターの学位を取った位の人にはすすめられるが，B氏の様に相当上級に進んだ人にはすすめられない．アメリカでドクターの学位（Ph. D）に近い連中は大体2000ドル（10カ月）位を最低に取っている．悪い待遇でもB氏が行くとなると，そのことが，数年間アメリカにいる間にB氏にとって不利なことになる可能性が十分ある．また，それから後でC大学に行く日本人をも，出来るだけ少い待遇で呼ぼうという風潮が起る心配がある．

B氏の立場から考えて一番良いのは，日本でドクターの学位をとり，その上で，自分の業績を表面に出して "Post Doctor fellowship" というのに apply することである．そのときC大学でもよいが，一番良いのはプリストンである．

E氏の意見

日本で学位を取り相当の仕事をしてから出かける方が有利であるが，それは生活という点からで，数学の研究という立場からいえば問題は別である．ただし研究の方も，人によって，日本にいる方が良い人，外国に行く方が良い人，どちらでも同じ人と，いろいろあり，一概には云えない．

生活は，月150ドルでは楽でないことは明かで，大体住居費50ドル，食費90ドル，小づかい10ドル位になる．留学など，この頃は当り前のことの様に思っている様だが，やはりチャンスということもある．

アメリカにいる若い連中は，日本で業績を上げてから，良い待遇でよばれなくちゃという考えだが，自分はそうは思わない．尤も日本で教授という人が悪い待遇で甘んじてやって来るのは同意しかねる．大体老人では余り留学の効果が期待されないのに．

D氏は，日本人数学者の，一種のダンピン

グを憂えているのであるが，数学第一の E 氏にはそれが余りピンと来ないようである．A 大学の教授にもピンと来ないようであった．

一方 D 氏は，F 大学にある，より良い条件の地位に，B 氏を推せんして呉れたのである．

（4号，1958年4月，無署名）

数学の歩み　第6巻

大学院の問題について

東大の（数学科）大学院学生の何人かが一夜を飲み明した折，ふとしたことからなぐり合いをしたとかしないとか，マスター・コースの学生が，修士論文を書く見通しが立たずに消耗してしまうとか，あるいは，フルブライト法によるアメリカ留学生の試験を，落ちても落ちても受け続ける人が少なくないとか，こんな話ばかり耳に入って来ると，新制度の大学院というものは余程住み心地の悪い所に違いないと考えないわけには行かない．しかも，底抜けの善意と極端な空想力とを持たない限り，この考えに対する反証を現実に見出すことは余り容易でないと思われる．

そんな住み心地の悪い所に何故入るのかと聞くのは野暮である．研究者たらんとする人は，アメリカにでも行かない限り，他に行く所がないのだから．しかも，マスター・コースには，研究者志望の学生だけが入るのではない．精深な学識と，研究者，教授者たるべき能力あるいは実社会において指導的役割を果すために要する“能力を養う”ことが修士課程の目標とされているのだから，多少とも数学を身につけて社会に出ようとする人が修士課程に入るのは極めて当然なのである．実際学部２年間の専門課程だけで数学がわかったというのは，余程の秀才か，さもなければ，自分が数学がわからないことさえ気付かない程の大人物かのどちらかであろう．マスター・コース２年間を終えて始めて“数学を勉強した”という気になれたという人が大部分なのである．所が，他に行く所がないにも拘らず，大学院，特に博士課程の入学志願者は年々減少しつつある．志願者は全部入学させろと言って教授を“吊し上げ”た頃から見ると隔世の感なきを得ない．

では何故住み心地が悪いのか，どこに問題があるのか？　事情は修士課程と博士課程とではかなり異っている様に思われる．

マスター・コースに入りたての学生は余り物を知っていない（例外もあるが）．そこで専門分野の基礎知識の修得から始めなければならないのであるが，これに１年から，２年全部かかることもある．この期間はいわば“学習”の段階である．2 年間に 30 単位に当る講義を聞かなければならないので，この“学習の段階”という性格は一層強められる．講義は勿論，個々人に取ってはその専門課目以外のものも幾つか聞かねばならないわけであるが，最近の学生は講義を聞くことが好きであり，出席率はかなり良い．所で問題は，この“学習”が，学部学生の頃の学習と余り変りばえがしないということである．つまり，一種の“数学的教養”として，与えられた素材を受動的に受け入れるに止まることが多い．例えばある人は書物ばかり読んで論文は読まない．論文を読むには色々な基礎的な知識が必要なので，自分にはまだ無理だというのである．そして一冊の書物を読み終ると今度は隣接分野の他の教科書を取り上げる．この様にして知識は段々増えるが，書物というのは本来出来上ってしまった体系を一つの閉じた世界の中に整理したものだから，知識は増えても数学の動いて行く姿は一向につかめない．瞬くまに時はすぎ２年目の秋ともなれば修士論文のことをいやでも考えざるを得くなるのであるが，こんな調子では修士論文とか研究とかは，何時になっても遠いカスミの彼方にある．そこで消耗してしまったり，あるいは卒業を一年延そうかということになったりするが，卒業を延ばせば奨学金が打ち切られる．人によっては絶体絶命になることもある．

問題の所在は極めて明白である．修士論文を書かせるということは，研究，少くとも研究のための試みを要求されるということであり，事実この要求を果すことによって始めて数学を主体的に身につけ，数学をやった気になれるのである．所で出来合いの料理を呑み込む様な，学部の時と同様な学習の段階から直ちに研究の段階に移行することは，天才でもない限り余り容易ではない．両者の中間の過渡的な段階を通して研究の段階の入口にまで学生を導くことが，マスター・コースにおける“指導”の主要目的でなければならない．所が，“出来合い丸呑み”式の勉強は気分的に割合に楽であるのに，そこから次の段階に進むことには或る程度気分の上での抵抗があるので，放っておくと何時になってもこの段階から抜けようとしない人も居るかも知れない．しかも面白い講義は色々あり，書物はいくらでも手に入る．食べるべき料理には事欠かないわけである．講義の出席率が良いことは，学生の好学心の現れと手放しで喜ぶわけにも行かない様である．

放任して置いてはいけない学生に対して適当な助言を与え勉強の方向を指示し，時には修士論文のテーマを与えてその方向に勉強を集中させることが，指導教官の果すべき主要な役割とならなければならない．例えば書物ばかり読んでいてはいけないと言っても，無方針な論文乱読は一層弊害がある．学生が専攻しようとする分野の現状とその進歩の方向とを大雑把に指示し，そこに到るまでに読むべき論文のリストを与え，読み方について注意する位は最低限なさるべきであろう．そして，出来れば，学生が研修テーマを自分で見出せる様に導くことが望ましい．所が，この様な“指導”が必要であるということが実感として理解されていないのではないかと思われる節がある．大学院の先生方が勉強されて来た頃とは時代が違うのである．専門知識は殆んど何も持たずに修士課程に入り，アルバイトと講義とに追われながら基礎知識を詰め込み次の年の終りにはもう研究成果が果げなければならないというのである以上，相当能率的な勉強が必要である，ということが実感されていないのではないか．成程，もう少し時間的に余裕があれば，学習の段階から研究の段階への移行が極めてスムースに行われ，本人はそれを自覚することさえ無いかも知れない．然し極めて短時間（旧制に直せば卒業して１年後或は半年後位いに当る）にこの移行を行うには，両者が質的に異る段階であることを注意して，前者から後者への移行が意識的・能率的に行われる必要があるのである．

“だから，修士論文の質は問題にしない．何でも書けば好いんですよ”という人もあるかも知れない．実際，修士論文の審査基準はかなり甘いのが通例の様である．然しそれでは学生の自尊心が承知しないらしい．多くの学生はこの様な説明を単なる気休めとしてしか受取らず，決して心からは納得しない．だから相変らず消耗を続けるということになる．それに，能率よくやれば一応まとまったものが書ける様になるにも拘らず，“何でも好いですよ”と云うことは，その学生に対して親切なことではなく，研究者の養成，ひいては日本数学の進歩という面から見てもマイナスの作用をしか持ち得ないのである．

ドクター・コースの学生は，同時に研究者であり学生であるという二重性に悩まされる．

勿論学生といってもピンからキリまであるのは当然であるが，ドクター・コースの２年３年生ともなれば大体は既に立派な研究者であり，数学教室における研究活動の有力な支え手の一人となっているのである．所が我が国では学生は研究者として扱わないという伝統的な風習があるらしく，彼等はあらゆる面で，研究者としての待遇を受けることができない.

それも，彼等は研究者としては駆け出しなのだから仕方がないではないかというのなら（感心しないまでも）一応話の筋は通る．所が，研究者としてはやはり駆け出しである筈の助手は，立派に研究者としての待遇を受けているのである．だから問題は研究者としてどうこうというのでなく，国家公務員として給料を貰う身分であるか，学生として授業料を払う身分であるかによって区別されるのだと考えざるを得ない.

例を東京大学の数学教室に取ろう．ここでは，助手は 3，4 人で一つの部屋を持ち，各自とにかくも自分の机を持っているのに対し大学院学生は（修士博士両課程とも）自分の机所か自分達の部屋一つ持たない．従って，講義，ゼミナールの時以外に学校に出かけても居る場所もない．時間を浪費するまいと思えば，なるべく学校に寄り付かず，学校に来てもゼミや講義が終ればサッサと帰る方が良いということになる．だから同じクラスであり，お互いに怠けているわけではないのに半年位相手の顔を見ないということもある．実際に研究を進めている先輩と合って話をしたり，或いはその研究の進め方を実際に見たりする機会が少ないということが，修士課程の項で述べた，学習から研究への移行の困難さの原因の一つになっていることは明かである．さて，では学校に出かけなければ勉強できるかというと，住居の周囲が騒々しくて，家でも落ち付けないという人も少なくない．でも学校に出かけるよりはマシだというのであろうが，能率が落ちることは否定できない．大学院学生が受ける唯一の研究上の便宜は図書雑誌の借し出しであるが，それも助手と比べて差別待遇されている状態である．学会に出席するための旅費もほとんど貰えず，辛うじて雀の涙程の科学的研究費の配分にあずかることでも出来れば有難いという有様である．これらは勿論，予算や施設等の制約があってどうにもならないという面もあるが，単にそれだけではないということは，例えば大学院学生のために旅費を調達したり，部屋を一つ取ってやろうとしたりする努力が教室として真剣になされているという話を聞いたことはついぞ一度もないことからもわかる．個々の事柄としての便不便はもとよりであるが，“研究者として待遇されていない”という気分の与える精神的な影響も決して過小評価し得ないことである.

この様に，彼等は研究者としての待遇を殆んど受けていないのであるが，しからば“学生”として正当に待遇されているか？ 例えばその研究を指導され，あるいは学位論文を書くための色々な援助，指導を指導教官から受けているか？ 勿論否である．なぜなら彼等はすでに一人前の研究者なのであるからそれを指導するなど口はばたいことであるし，まして学位論文の指導など本質的に不可能なことである．‥‥というわけで，彼等は学生であるが故に研究者としての待遇を受けられず，研究者であるが故に学生としての指導を受けられない．勿論，このことの一つの原因が，予算，教官定員の不足にあることは明かである．周知の通り大学院を作るとき，文部官僚の詭弁にゴマ化されて，予算，教官定員等は据え置きのままとなったのであるからあちこちに矛盾が起るのは当然である．しかし，その矛盾の多くが学生の方にシワ寄せられ，しかもそれが早急に解決される見込もなく，また誰一人それを解決するために骨を折って呉れている人もなく，寧ろ学生に低い待遇を納得させることによって現状を維持する精神的基盤を作ろうという気分さえ無いことも無いとあっては，事は単なる予算の不足の問題ではなく，これだけでクサってしまうに十分である.

しかし，これは序の口であって，次には学位論文と就職の問題が控えている．学位の問題についてはいずれ詳しく論ずる人があると思うので，ここでは簡単に触れるに止めておきたい．学位論文の基準については従来いろいろな意見があり，今年理学博士を出さなかった大学もある様である．“学位のことに余りこだわる必要はない”という人がいるが，それは旧制の感覚で考えているからで，卒業して5年の後に取るべきものと決められている学位を取れないことは，一つのマイナスのレッテルを貼られることを意味し，就職などに直ちに差支えるであろう．勿論名の通った大学ではそんなことはないだろうが，小さい大学や私立大学等で，数学科以外の人が大部分を占める教授会を納得させるのに，学位のあるなしが無関係であるとは言いきれないであろう．外国留学する際に学位のある無しが如何に大きく影響するかは良く聞く所である．またいずれ，或る地位に就職する際学位のあることが必要条件となることが予想される．以上は“学位”という形式の果す役割に関することであるが，これを別としても，或る程度必要以上の論文を一定の期間内に書けるか否かということは，自分の能力に対する一つの客観的な試金石となるものであるから，何れにせよ学位の問題はかなり憂欝な問題として目の前にブラ下っているのである．それに，彼等は“一人前の研究者”なのだから，論文作成に当って何等かの指導乃至援助を指導教官から期待することが出来ないことは，既に触れた通りである．

では，学位さえ取れば万事オーケーかというと，世の中はそれ程甘くは出来ていない．数学科ではないが，今年理学博士を取りながら適当な就職口の無かった人が東大にも何人かいる．第一，その人に如何に能力があろうとも，たまたまその時に適当なポストに欠員があるので無い限りお話にならないわけである．余り気の進まない所に就職するか，それも出来ずに完全失業という可能性は常に存するのである．

そんなわけだから，コースの途中でどこか適当な所から助手の口でも提供されればそれに飛びつくのは当然である．第一に助手は研究者として待遇され，大学院学生より有利な条件で研究を進められ，しかもその地位は法律によって永久に保証されている，給料は奨学金よりよい上にボーナスもあり，年に一度の昇給もある．更に学部卒業後5年目に学位論文を書かなくてもその能力を疑われることはない．多少雑用をしなければならないのが難点だが，大学院学生でも結構なんのかのと雑用を押し付けられているのだから大して違いはない．‥‥このことの結果，ドクター・コースは潜在失業者乃至は“産業予備軍”の溜り場と化し，その学生はいわば大部屋のワンサであって，どこかから口がかかって来るのを待っているという状態におかれる．違う所といえば彼等は自分達の大部屋さえも持っていないということ位のものである．だから，5年間も大学院に“ゴロゴロ”していて，学位を取ったなんて人は余程の“阿呆”であり，途中で就職をして学位をとらずに悠々自適している人に比べて，数学的能力がかなり劣ると見られる様な雰囲気が醸し出されるのである．

そもそも修士，博士の課程を作り，年限を定め，取るべき単位数を指定し学位についての規則まで作ったのは，入学した学生の大部分は中途で就職せずに課程を修了し，修了した人の大部分が学位を取れることを予想しての上であろう．それが上の様な状態になったのは，制度の中にどこか根本的な欠陥が含れていたか，または運用が根本的に誤まっていたかであるに違いない（因みに，東大数学教室では，ドクター・コースの第1回入学者7人，修了者3人，学位を取った者1人，第2回は9人入学したが修了者は 0~4 人と予想されている）．現在では，新制度を採用したことの意義が失われ，その欠陥だけが大写しにされている状態であろう．制度の欠陥について一つ云えることは，入学者の大部分が課程を修了して学位を取ることは“Post-Doctoriel fellowship”の制度がない限り不可能に近いということである．なぜなら一度にそれだけの人数の新博士に適当な就職口が見つかるであろうとは到底考えられないから，この制度がないため，本来なら学位を取り Post-Doctoriel fellowship を貰って研究を続けている人々が，順次になし崩し的に就職していくという状態が，ドクター・コースにおいて（つまり Pre-Doctoriel コースで）起り，新しい大学院制度の持ち得たであろう意義を完全に失なわせているのである．そして迷惑するのは，いつもながら先ず学生である．

以上に加えて経済的な問題がある．僅かばかりでも奨学金が貰えればよいが貰えない人も多い（全国，全学科の大学院の平均で，実人員の 62.2％が奨学金を貰っている）．たとえ貰えたとしても，アルバイトで不足を補わなければならない．そのため勉強の時間は分断されるが，その時間の中で高度の集中と持続とを要する数学の研究を続けてゆくのはそれこそ生やさしいことではない．大学院学生の間に重苦しい空気，非人間的な雰囲気がただよい，時として爆発することがあるとしても，敢えて不思議ではないであろう．

この様な問題について多くの人々に真剣に考えていただくために，ここでは敢えて解決策を提示しないことにする．ただそれについて一言いわしてもらえば，どうせアメリカの制度をまねするものなら，その外形だけでなく，細い運用の面まで徹底的にまねしないと破綻するのではないか，例えば奨学金制度，指導教官の指導方法，Post-Doctoriel fellowship 等々．しかしながら今は，始めから一切を再検討すべき時であろう．このまま放置すれば，その中に，ド・ゴールでもヒットラーでもこれよりはましだと考える人物が出て来ないとも限るまい．

以上，多少筆の滑った所もあるかもしれないが，事の本質はそらしていないと思う．敢えて識者の高覧に供する次第である．

（1号，1958年7月）

応用数学小委員会への質問状

——数理研の問題について——

1

a　研究連絡委員会，応用数学小委員会は，数学会会員に対しどの様な責任を持っていると考えるか．またその責任を果すためにどの様な具体的手段を考えているか（研連は数学会会員の選挙で選ばれるのであるから，その選挙母胎たる会員に対し何等かの責任を持つのは当然であるが，それを具体的にはどう考えているか）．

b　研究所について，調査委員会，設置委員会等を設けることが予想されているが，これ等の委員会の構成法・運営の仕方，現在の委員会がそれ等に移行する時期，方法はどうなっているか．

c　これまで，種々の具体的な質問に対して「まだ決っていない」という返答が屢々くり返されたが，具体的なやり方が決っていない場合，広く一般会員から意見を聞くことを考えているか．考えているとすればその方法について．

d　これまで，シンポジウムその他の機会に多くの意見が述べられたし，これからも述べられることと思うが，それ等の意見はどの様に処理されているか．

より詳しくいえば

i）単に聞き放しにして，取り入れないことにしているかどうか．

ii）聞き放しにしないとすれば，何を取り入れ入れないかをその理由と共に責任を以て報告すべきではないか．

2

a　全国共同利用方式についての具体案を持っているか．

b　或る特定の個人が，研究所を中心とする全国交流に参加しようとするとき，どの様な具体的手続を取ればよいか，また参加を希望しても拒否されることがあるか，あるとすれば参加の資格は何か．

c　科学研究費の別枠を使った共同研究は，研究所を中心とする全国交流の一つの原型となると思われるが，これについて

i）どの様な機関（乃至組織）が，この予算を取扱うのか．

ii）この別枠の利用は，研究所の設立にとってどの様な意義を持つと考えるか．

iii)）具体的にはどの様に使うのか．

研究班の組織方法・研究題目の決定法・共同研究の具体的な進め方・その時期等について．

3

a　研究所の組織，運営について，具体的な案を持っているか．

b　具体案を持っている場合にはそれを，持っていない場合には幾つかの試案を作って，一般の批判，検討を求めるつもりはないか，あるとすれば具体的にはどの様に行うか．

c　研究題目の選定法．人事の決定法の要綱は定まっているか．定まっていなければ何時，誰がきめるのか，これ等についても試案を提示して一般の検討を求めるつもりはないか．

4

a　具体的な研究題目について議論したことがあるか，もしあるとすれば，その結果どの様なことが決ったか，また，その際の少数意見としてはどの様なものがあったか．

b　発表された講座題目に変更があったか，もしこの規模で実現しない場合には，何から設けて行くかについて議論したことがあるか．もしあるとすればその結果どう決ったか．

c　このa，bは非常に大切なことと思うがこれについて議論されたことがあれば，その議事の経過，細目を公表し，一般の批判と，意見とを求めるつもりはないか．

もし議論されたことがなければ，広く一般から意見を求めるつもりはないか．

また意見が寄せられた場合に，その意見をどの様に処理するか．

d　この様な，すべてにわたる講座を持つ研究所が出来た場合，将来，数学の応用を主要目的とする研究所，いわゆる応用数学をやる研究所，純粋数学をやる研究所等を作ることが妨げられると思うか否か，また上の様な内容の研究所を将来作ることが将来必要になると思うか．

以上について項目別に，具体的に答えていただきたい．

（2号，1958年10月，共著）

ヒルベルトのモヂュラー函数について

—— 1 ——

ヒルベルト・モヂュラー函数の歴史については，以前，この雑誌に少し書いたこともあるので繰り返さない．ヒルベルトがこの函数を考えたのは，専ら虚数乗法による類体の構成に応用する目論見からであるが，我々が現在この函数を取上げるのは，主としてヘッケの作用素の理論の立場からである．尤もこの二つの立場は，その見かけ程に違うものではない．このことについては，楕円モヂュラー函数についての志村氏の理論をお読みになれば納得されるであろう（本誌４巻２号掲載．また巻末の参考文献参照）．すなわち楕円モヂュラー函数を，楕円曲線の不変量として把えることによって，類体の構成も，ヘッケの作用素の理論も，共に導き出されるのである．ヒルベルト・モジュラー函数に対して我々の取る立場も当然この立場であって，今度は，或る型の“虚数乗法”を持つアーベル多様体の系列を考えることになるのである．この見地からするとヒルベルト・モヂュラー函数の研究は志村氏により進められて居り，筆者もまた，志村氏とは多少違う方向に向ってこの問題に手をつけている．以下に筆者の考えの大要を紹介したいと思うのであるが，まだ未完成の状態なので，差当って本稿が一種の“中間報告”的な性格のものになることは止むを得ない．不充分な点は，いずれ後の記事で補うことにしたいと思う．

1

ヒルベルト・モヂュラー函数の定義は簡単である．それは，総実な有限次代数体 k の各々に対して定義される．k を n 次とし，k の数 α の共役を $\alpha^{(1)},\cdots\cdots,\alpha^{(n)}$ と書くとき，複素 n 次元空間の点 $\tau=(\tau_1,\cdots\cdots,\tau_n)$ に対して

$$\tau'=(\tau_1',\cdots,\tau_n'),\quad \tau_i'=\frac{\alpha^{(i)}\tau_i+\beta^{(i)}}{\gamma^{(i)}\tau_i+\delta^{(i)}}\quad (i=1,\cdots,n)$$

を対応させる変換を簡単のため

$$\tau\to\tau'=\frac{\alpha\tau+\beta}{\gamma\tau+\delta}$$

と書くことにする．ただし $\alpha,\cdots,\gamma$ は k の整数で $\alpha\delta-\beta\gamma=1$ となるものとする．この変換によって複素 n 次元空間の“上半平面” T（すなわち各 τ_i の虚数部>0 なる τ の集合）は T 自身に写される．上の様な変換全体の作る群 Γ がヒルベルトのモヂュラー群である．この Γ に対して，普通の型の変換公式を満し[1]，或る種の正則性の条件を満す函数（定義域は T）を，ヒルベルトのモヂュラー形式というのである．k が有理数体のときには，普通の楕円モヂュラー形式となるわけである．

我々の目的は，この様な形式に対して，ヘッケの理論を拡張することである．より詳しく云えば，

i）ヒルベルト・モヂュラー形式 $F(\tau)$ に対し，函数等式を満すヂリクレ級数（の系列）を対応させる．

ii）k の各整イデヤル $\mathfrak{a}$ に対し，ヘッケの作用素 $T_{\mathfrak{a}}$ を定義し，$F(\tau)$ がすべての $\mathfrak{a}$ に対する固有函数になることと，対応するヂリクレ級数がオイラー積に分解することが同

注 1）すなわち，Γ の変換に対し，

$$\frac{1}{N(\gamma\tau+\delta)^g}F\left(\frac{\alpha\tau+\beta}{\gamma\tau+\delta}\right)=F(\tau)$$

の形の変換公式，但し $N(\gamma\tau+\delta)=\prod_i(\gamma^{(i)}\tau^{(i)}+\delta^{(i)})$.

（このとき $F(\tau)$ を $-g$ 次元の形式という．$g=0$ のとき，ヒルベルト・モヂュラー函数というのである．）

値な条件になることを証明する.

さらに問題は発展して

iii) $T_{\mathfrak{a}}$ に対し跡公式 (Trace formula) を導く.

iv) ヒルベルトのモヂュラー函数の体に対し, その標準的な生成元を求め, 類体の構成を, この生成元の特殊値によって行う.

さて, i), ii) は, 既に一応解決されている. すなわち de Bruijn は, k の類数が 1 に等しい場合にヘッケの作用素を定義し, Herrmann は, 任意の k に対して同様な理論を作ったのである (本文の終りの文献表参照).

ところで, この様な問題においては, 大きく云って一つの難点があり得る. その一つは多変数函数を扱うことから起るものであるが, この難点は Maass の研究により (十分とは云えぬまでも) 一応除かれたと考えてよい. 第二の難点は k の数論的な構造の複雑さによるものである. たとえば, 楕円モヂュラー形式の場合, ヘッケの作用素は, すべての整数 m に対し定義され, それは, 整係数, 行列式 m の 2 次正方行列の, ユニモジュラー群による右代表系から定義されるものであった. 一般の代数体 k に対しては, 従って, イデヤル $\mathfrak{a}$ に対し, "行列式が $\mathfrak{a}$ に等しい" 様な, k の整数係数の行列を考えることが必要になる. de Bruijn が, 類数 1 の場合にしか成功しなかったのはこのためであるが, Herrmann はこの難点を避けるため, ヘッケの "イデヤル数 (ideale Zahl)" を利用し, すべてのイデヤルを "単項化" するのである. このため彼の論文は甚だ難解なものとなり, また "Stufe" のある場合への拡張は試みられていない. また, この様な扱い方では, アーベル多様体との関係は少しも明確でなく, またそこで用いられる概念に対する "自然な" 説明も得られないのである (というよりすべてが, 甚だ技巧的なのである).

2

さて, 類数が 1 より大きいという難点を除くだけならば, イデヤル数を用いるよりは, イデールを使う方が簡明であるとは, 誰しも考える所であろう, 筆者も最初この考えから出発したのである. より詳しくいえば次の通り.

Herrmann が考えたように, k の各イデヤル数に対し一つずつ, ヒルベルト・モヂュラー形式を対応させ, これ等は "上半平面" だけでなく, 変数の虚数部が 0 でないような領域で定義されているとする. そのとき, k のイデール群 J の上に, 適当な函数 $\vartheta(\tilde{\mathfrak{a}})$ ($\tilde{\mathfrak{a}}\in J$) を定義すれば, この様な函数の組は, 次の形に表される: (ϑ は適当な条件を満す (J 上の) 函数である):

$$\sum_{\mu\in k^*}\vartheta(\mu\tilde{x})=f(\tilde{x})$$

ただし, $\tilde{x}$ の無限成分 $(x_1,\cdots,x_n)$ に対し, $\tau=(\sqrt{-1}x_1,\cdots,\sqrt{-1}x_n)$ とおいて $\tilde{x}$ の有限成分 $\tilde{x}_0$ を固定して $f(\tilde{x})$ を τ の函数と考え, それを複素 n 次元空間に解析接続するのである. こうするとき, 異った有限成分 $\tilde{x}_0$ に対して異った函数が得られるわけであるが, その中で本質的に違うものが丁度 k の類数だけあり, それが始めのモヂュラー形式に対応するのである. 所でこの ϑ に対して, 岩沢–Tate 流のフーリエ解析を行えば, 始めのモヂュラー形式に対応するフーリエ級数が得られる. すなわち, f が上の形であるから, ポアソンの公式によって J/k の指標群の上での和, すなわち k の量指標の上での和

$$\sum_{\lambda}\zeta(s,\lambda)\qquad(\lambda\text{ は }k\text{ の量指標})$$

が得られ, 各 $\zeta(s,\lambda)$ が, 函数等式を満すヂリクレ級数になる. 最後に $f(\tilde{x})=f(\tilde{x}_0,\tau)$ と書くとき, Hecke の作用素 $T_{\mathfrak{a}}$ は,

$$f(\tilde{x}_0,\tau)T_{\mathfrak{a}}=\sum_{\mathfrak{b}\in\mathfrak{a}}\sum_{\substack{\zeta\in\mathfrak{b}^{-1}\\ \zeta \bmod \frac{\mathfrak{a}}{\mathfrak{b}^2}}}f\left(\frac{\mathfrak{a}}{\mathfrak{b}^2}\tilde{x}_0,\tau+\zeta\right)$$

の形で得られ, これが必要な性質を満すこと

が容易に示される.

この様にして，Herrmann の理論が非常に簡単に再構成できたのである.

この方法は，要するに，イデアル数の代りにイデールを使って，イデヤルを単項化することに基くのであるが，その際イデール群の上のフーリエ解析が使えることから，i) の，ヂリクレ級数との対応が非常に自然に行われる所に特徴がある．然しながら，ヘッケの作用素の定義は，Herrmann のそれより以上に技巧的となり，またアーベル多様体との関係も明らかでない．更に，玉河氏からの注意もあり，もう少し見方を深めることにしたのである.

3

ヘッケの作用素 T_m は，一方では行列式が m に等しい整係数の行列の完全代表系により定義されるものであり，他方それは，アーベル多様体を，位数 m のすべての部分群で割った商多様体の全体を考えることになるのであった．従って我々の場合には，イデヤル $\mathfrak{a}$ に対する作用素 $T_{\mathfrak{a}}$ を定義するのに，先ず，（イデールでなく）附値ベクトルを係数とする2次の正方行列を考え，一方では行列式が $\mathfrak{a}$ の整係数行列の代表系を取れば良いわけであるが，他方それが，アーベル多様体の"位数 $\mathfrak{a}$"の部分群による商多様体に対応することを明かにしなければならない．この様な考え方に従って，附値ベクトル係数の行列とアーベル多様体との対応ということを先ず考えたのである.

少し詳しく説明しよう.

k の附値ベクトル（すなわちアデール）を係数として行列式がイデールになる様な，2次の正方行列の作る群を V_2 とする．V_2 の行列 $\widetilde{D}$ の有限成分を $\widetilde{D}_0$，無限成分を $\widetilde{D}_\infty$ と書く．$\widetilde{D}_0$ に対し，$k+k$ の部分加群 $\mathfrak{m}$ を

$$\mathfrak{m}=\{(\alpha,\beta)\in k+k\,|\,(\alpha,\beta)\widetilde{D}_0=(\widetilde{a},\widetilde{b})\ \text{が整の附値ベクトル}\}$$

によって定義する．また $\widetilde{D}_\infty$ は，

$$\begin{pmatrix} a_i\,b_i \\ c_i\,d_i \end{pmatrix}\quad (a_i,\cdots d_i\ \text{は実数},\ a_id_i-b_ic_i\neq 0$$

なる形の行列の $i=1,\cdots,n$ に対する列であるが，これに対し

$$u_1=(a_1\sqrt{-1}+b_1,a_2\sqrt{-1}+b_2,\cdots$$
$$\cdots,a_n\sqrt{-1}+b_n),$$
$$u_2=(c_1\sqrt{-1}+d_1,c_2\sqrt{-1}+d_2,\cdots$$
$$\cdots,c_n\sqrt{-1}+d_n)$$

とおけば，u_1,u_2 は，複素 n 次元空間 $\mathfrak{D}^n$ の点となる．そこで，

$$D=\{\alpha u_1+\beta u_2\,|\,(\alpha,\beta)\in\mathfrak{m}\}$$

とおけば，D は $\mathfrak{D}^n$ の，階数 $2n$ の discrete な格子であり，商空間 $A=\mathfrak{D}^n/D$ はアーベル多様体となる．更にこの A の乗法子環は，k の全整数環に同型な部分環を含む．逆に A の乗法子環が，k の全整数環を含むような，n 次元のアーベル多様体はすべて，ある $\widetilde{D}\in V_2$ から上記の様にして得られることがわかる．従ってこの様なアーベル多様体の系列が上記の対応によって，V_2 により Parametrize されたわけである.

この対応によれば，A の，位数有限な点は，すべて $\gamma u_1+\delta u_2(\gamma,\delta\in k)$ の形に表され，そのとき，$(\gamma,\delta)\widetilde{D}_0=(\widetilde{c},\widetilde{d})$ とおけば，この c,d が，点 $\gamma u_1+\delta u_2$ の"l 進座標"の，すべての l に対する組を表していることになるのである（正確には，k の整イデヤル l 全体に対する"l 進座標"の組を表す）．従って，k の整イデヤル $\mathfrak{a}$ に対して，A の"位数 $\mathfrak{a}$ の"部分群という概念も明確に定義することが出来，同時に，それで A を割った商多様体を作ることが，A に対応する $D=(D_0,D_\infty)$ に対し行列式のイデヤルが，$\mathfrak{a}$ になる様な整ベクトル係数の或る行列 M を掛けることに対応することも明かである．すなわち，ヘッケの作用素に対する二通りの見方が，上記の対応を通して統一されるわけである.

この様にして，ヘッケの作用素に対し，最も自然で数論的に扱い易く，しかも，アーベ

ル多様体との関連も明確な定義が得られたわけである，そして，A の偏極を自然な方法で定義するとき，偏極多様体として同型な A に対応する行列 D が，V_2 の double coset を作ることがわかる．この double coset の上で定数値を取り，D_∞ について解析的な函数[2] が，丁度ヒルベルト・モヂュラー函数になるのである．また，A の等分点を不変にするという性質によって，“Stufe” のある函数も，全く同様に取り扱うことでできる．

さらに，k のイデール $\tilde{a}$ に対し，行列

$$\begin{pmatrix}\tilde{a} & 0\\ 0 & 1\end{pmatrix} \text{ は } V_2 \text{ に属するから．}$$

V_2 は，k のイデール群をこの形で含んでいるわけである．そこで，ヒルベルト・モジュラー函数（或は形式）を，V_2 の函数と考え，それを，V_2 に含まれるイデール群に制限して，それに対して，2における様にしてフーリエ解析をやれば，ヂリクレ級数との対応が得られるのである．

4

以上で文句ない様であるが，実際には更に色々問題がある．第一に，V_2 で Parametrize されるアーベル多様体に対し，“自然な” 偏極をどう定義するかが問題である．勿論，筆者は一つの方法を考えたわけであるが，それが最も “自然” であり，従って最も有効であるか否かは多分に疑わしい．もう少し良く考える必要があるかも知れない．

更により本質的な問題がある．折角行列群 V_2 を持ち出しても，フーリエ解析は，もとのようにイデール群でやるというのでは，新しい方法の効用も大分割引きされるわけである．更に，問題 iii）すなわちヘッケ作用素の trace formula を導くためにも，行列群 V_2 自身の上でフーリエ解析をやることが最も望ましいと思われるのであるが，V_2 は勿論非可換群であるから，その上のフーリエ解析なるものについてはまだまとまった理論が出来ていない．ただ，以前に本誌で紹介した様に（5巻3号）Selberg が，弱対称リーマン空間に対してフーリエ解析を行い，ポアソンの公式に対応する和公式を導いていることが注目される．我々の場合には単に局所コンパクトな群であるが，これに対して，この Selberg の理論と同様なフーリエ解析の理論を作ることが出来，和公式を証明できれば，ヘッケの作用素に対する trace formula も勿論求められるであろう．この辺が，差当って次の攻撃目標とされるのである．

次に，楕円モヂュラー函数の場合と同様に Stufe のあるヒルベルト・モヂュラー函数体のハッセの ζ-函数を問題にすることもできるが，これは，合同 ζ-函数の理論が（高次元の多様体に対し）出来ていないため，今の所余り見込はない．

また，以上の様な取扱い方は，勿論，ジーゲルのモヂュラー函数に対しても行うことも出来る．この場合には，有理数体の附値ベクトルを係数とする $2n$ 次の行列群を考えることになるわけである．ただし，この様に扱うことがどれだけ有意義であるかはこの場合には甚だ疑わしい．なぜならば，第一に，$2n$ 次の行列群に対して Selberg の様なフーリエ解析論を作ることは恐らく非常に困難であると思われるからであり，第二に，ジーゲルのモヂュラー函数に対しては，ヒルベルトのそれの様な基礎に取る体の数論的な構造の複雑さに基く困難さが余りないので，わざわざ附値ベクトルやイデールを持ち出す必然性がどれだけあるか疑わしいからである．勿論筆者は，ジーゲルの函数に対してこの様な扱い方が無意味であると云い切る根拠は何も持ち合

2）$D_\infty=\left\{\begin{pmatrix}y_i, z_i\\ v_i, w_i\end{pmatrix}\middle| i=1,\cdots,n\right\}$ とするとき．$\tau_i=\dfrac{y_i\sqrt{-1}+z_i}{v_i\sqrt{-1}+w_i}$ とおき，$\tau=(\tau_1,\cdots,\tau_n)$ を変数と見れば，V_2 の上の函数は（有限成分を固定して考えるとき），複素 n 変数 $\tau=(\tau_1,\cdots,\tau_n)$ の函数となる．また，“解析的” である上に，正確には，∞ での，或る種の正則性が要求される．

せていない．何れにしても，手を附けて少し考えてみれば，どの程度意味があるか否かハッキリするであろう．ただ筆者は今の所やってみる気がしないというだけである．

参考文献

（本文中に直接引用されたものだけを挙げた）．

de Bruijn, V., G.: Over Modulaire vormen von meer veran derlijken. Amsterdam (1943).

Herrmann, O.: Über Hilbertsche Modulfunktionen und die Dirichletschen Reihen mit Eulerscher Produktentwicklung. (Math. Ann. 127 (1954) 357-400)

Shimura, G.: Correspondances modulaires et les fonction ζ de courbes algébriques (J. Math. Soc. Japan, 10 (1958) 1～28)

（2号，1958年10月）

(自 己 紹 介)

1950 年浦和高等学校卒業，53 年東京大学理学部数学科卒業，両方とも旧制の最後で，追い立てられる様にして卒業した．浦高時代に読んだ「近世数学史談」（高木貞治著）が病みつきで，数学に非常に興味を感じる様になったが，本格的に始めたのは大学に入ってからである．東大の数学科にはもともと整数論の強い伝統がある上に，当時本郷で代数を講義されていた菅原正夫先生の強い影響などもあって，整数論を専攻する様になった．

現在では専ら，楕円函数，アーベル函数などの理論を応用して整数論の問題を解決することに力を注いでいるが，これなど「近世数学史談」の影響が尾を引いているものと云えよう．整数論には，壮大な構成に加えて，非常に深い，神秘的とさえ云える面もあり，これを専攻したことは幸運であったと思っている．著書「近代的整数論」（志村五郎と共著）共立社現代数学講座第 8 巻 1957 年．

（駒場，1958 年 3 月）

科学読売

「零 (ゼロ) の 発 見」*

（吉田洋一著　岩波新書）

“0 の発見” というと，不思議に思う人もいるかも知れません．「1，2，3‥‥という数がある以上，0 があるのも当り前だろう．第一，さもなければ 10 とか 108 とかいう数を書くことも出来ないし，足し算や掛け算も出来ないではないか」といわれそうです．

ところが “十” とか “百八” とかという数を 10，108 と書いて，紙の上で計算するようになったのは，実は割合に近ごろのことなのです．このような書き方，計算法が使われるようになった歴史，それが「0 の発見」の歴史なのです．

また，幾何を勉強している人は，直線は点が集まって出来ているが，点には位置だけがあって長さはない，ということを教わったことでしょう．そのとき，こんな疑問が浮んだことはないでしょうか．「長さのない点がどんなに集まっても，長さのある直線になるはずがないではないか．点の長さ 0 をいくら加えても，やはり 0 だから」

○

実際昔の人々は，点は非常に小さいけれども，やはり大きさがあるもので，直線はそのように小さい点が数多く（しかし有限個）つながり合っているものだ，と考えていました．しかし，こう考えると，どうしてもぐあいの悪いことがいろいろ起って来たので，多くの人々が長い間苦心したあげく，ついに “大きさのない点” という考え方に行き着いたのです．

このほか，皆さんが学校で，何でもないことのように教わっているいろいろなことも，それが考え出され，使われるようになるまでには，長い歴史と多くの努力とが必要でした．このような歴史を，わかりやすく説明しているのが，この「零の発見」という本なのです．

この本は，「零の発見」と「直線を切る」との二つの部分にわかれています．この第一

* “名著紹介” シリーズの一つとしてかゝれたものである（編者注）

部では，数の書き方，計算法の歴史が，また第二部では，直線や数についての考え方の移り変りの歴史が書かれています．ここでは，第一部「零の発見」を，少しくわしく紹介しましょう．

第一部　零の発見

ナポレオンのロシア遠征のとき，ロシア軍に捕えられたポンスレという数学者が，ロシアのソロバンをもち帰り，フランスで大変珍しがられたという話から，この本は始まっています．ところが，このときから300年ほど前までは，つまり 15，16 世紀ごろまでは，ヨーロッパでも，計算といえばいつもソロバンを使っていたので，そんなに珍しいものではありませんでした．しかも，そのころのソロバンは日本のソロバンと違い，板の上に珠を並べたり取ったりして計算するので，ちょっとした計算にも大変手間がかかるめんどうなものでした．だから，それを使ってうまく計算の出来る人はそんなに多くなかったのです．では，なぜそんなめんどうなソロバンをいつまでも使っていたのでしょうか？

僕たちがいま使っている 1，2，3‥‥という数字を使って数を書きしるす方法は，インドで生れました．そこでこのような書き方を「インドの記数法」といいます．二千七百五十二は，2752 と書きますが，ここで一番右の２は二を，次の５は五十を，次の７は七百を，一番左の２は二千を表わします．つまり，数字は，左から何番目に書かれているかによって，一の位，十の位，百の位，千の位を表わしているので，この書き方はまた「位取り記数法」ともいいます．この「位取り記数法」を使うから，百八十，百八などを表すのに，180，108 と，どうしても０という数字が必要になるのです．これは当り前みたいですが，日本語で百八十，百八と書く書き方は「位取り記数法」ではなく，また０という字も必要ないことに注意してください．皆さんが使いなれているこの「位取り記数法」こそは，インド人の発明であって，それが世界にひろまるまで，ほかのどの国でも発明されなかったものなのです．

○

では，昔はどんな記数法が使われていたのでしょうか．例えば，中世の終りごろまでヨーロッパで使われていた「ローマ記数法」では，

1	**5**	**10**	**50**	**100**	**500**	**1000**
I	**V**	**X**	**L**	**C**	**D**	**(1)**

という数字を並べて数を表わします．例えば

2 7 5 9

(1)(1) DCCLIX

1 8 0　　　**1 0 8**

CLXXX　　　**CVIII**

この書き方でも，“0”を表わす数字は必要ないことがわかるでしょう．このローマ数字は，今でも時計の文字盤などによく使われているので，ごぞんじの皆さんも多いと思います．時計の文字盤は 12 までですから，どんな記数法を使っても大して違いはありませんが「ローマ記数法」で大きな数字を表わそうとすると，大変長くなって読みづらくなります．おまけに１けた増えるごとに，二つずつ新しい数字を使わなければならないので，とても天文学的な数を表わすことは出来ません．例えば，二百三十万を表わすには，(((**1**)))という数字を二十三個並べなければなりません．これにくらべると「インド記数法」では０から９までの 10 個の数字を並べるだけでどんな大きい数でも表わせるのですから，大変便利なわけです．

ところで「インド記数法」の便利な点は，これだけではありません．これを使えば，数字を紙に書いて，紙の上で計算できます．（これを筆算といいます）．次の例を見てください．

```
   2759          (1)(1)DCCLIX
×   108        ×         CVIII
---------      ---------------
  22072                  ?
  0000
2759
---------
297972
```

「ローマ記数法」は「位取り記数」ではないので，紙に書いても，計算は出来ないのです．これで，「インド記数法」が使われるようになった15, 6世紀ごろまで，ヨーロッパでも，計算はみなソロバンに頼らなければならなかったわけがわかったでしょう．このソロバンが不便だといっても，足し算，引算，掛算はまだよいのですが，割算となると大変です．日本のような，割算の“九々”がないため，引き算を何度もくり返さなければ，割算は出来ませんでした．例えば，745÷57 を計算するには，745 をソロバンに並べ，それから 57 を何度も引いて行きます．13 回目に余りが4となることから，745÷57=13 余り 4と答が出るのです．数が少し大きくなると，その手間がどんなに大変なものになるかがわかるでしょう．計算のうまくできる人が少なかったというのも無理ありません．

インドでは昔から，ソロバンは使わず，板の上に砂をまいて，その上に数字を書いて筆算していました．上に書いた計算を見てもわかるように，そうすると，ある数に0を掛けたり，加えたりする計算が必要になって来ます．こんなところから，

$$a\times 0=0,\quad a+0=a,\quad a-0=a$$

という計算規則は，インドではずっと以前から知られていました．このように，0を加えたり掛けたりすることが考えられるようになって始めて“0”も数の仲間入りをすることになったのです．だから，“0”という“数字”だけでなく“0”という“数”も，インドで生れたといえるでしょう．

○

それでは，こんな便利な，インドの記数法，計算法は，どのようにしてヨーロッパにひろまったのでしょうか？まずそれは，8世紀の末ごろにアラビアに伝わりました．その当時のアラビアは，ヨーロッパよりも文明が進んでいたので，パレスチナの聖地に巡礼する人々によって，また少し後には十字軍の遠征などによって，アラビアの文明は盛んに，ヨーロッパにとり入れられました．「インド記数法」もこのようにして，アラビアからヨーロッパに紹介されました．今日 1，2，3‥‥という数字をアラビア数字というのはそのためです．しかしこの記数法が，広く使われるようになるまでには，長い時間がかかりました．

第一に，印刷術のなかったころのことですから，手から手へと書き写されて行くうちに，アラビア数字の形が少しずつ変ってきて，時によると間違いの起ることもありました．それに，そのころは紙といえば，パピルスとか，けものの皮をなめして作ったものとかしかなく，どちらも大変値段の高いものでした．だから紙を使って計算するのはもったいないので，計算はもっぱらソロバンですまし，数字は，その結果を書きしるすだけそれもそんなに大きいけたの数字ではありませんでした．これでは，特に「インド記数法」を使う必要もなく，それまでの「ローマ記数法」で何とか間に合っていたのです．

そのうちに印刷術が発見されて，アラビア数字の形もいま使われているようなものにきまり，中国の製紙法がヨーロッパに伝わって，安い紙が手に入るようになりました．これが大体 15,6 世紀ころです．ここで初めて，紙が計算のために使われるようになり，インドの記数法や計算術が，広く使われるようになったのです．それに，そのころは，いわゆるルネッサンスのころで，商業や航海，科学などが急速に発達し，大きい数の計算をする必要が起り，またこのころに，インド筆算法による掛算，割算の仕方が，いま使われている形のものに完成し，などと，いろいろな原因

が重なり合って，「インド記数法」と「計算術」は，間もなく，ヨーロッパのどこででも使わるれまでにひろまりました．逆にいえば，このような計算法が伝わっていなかったならば，ルネッサンスのころに，自然科学があれほど急速に発展しなかったとも考えられます．

とにかく，16世紀の終りになって「位取り記数法」による小数の書き方が完成されました．分数はずっと昔のエジプトにもありましたが，小数が発明されたのは，このように最近のことなのです．実際，小数は「位取り記数法」によって初めて考えられるのですから．

さて，普通の有限小数はすべて分数で表わせます．例えば，

$$0.25 = {}^{25}/_{100} = {}^{1}/_{4}$$

ところが逆に，分数を小数で表わそうとすると，必ずしも有限小数だけでは間にあいません．${}^{1}/_{3}$ を小数で表わせば 0.33333‥‥と，3が無限につづく無限小数となります．それでは，すべての無限小数はみな分数を表わすのでしょうか？　いいえ，皆さんもごぞんじのように，円周率：$\pi=3.14159\cdots\cdots$はどんな分数でも表わせません．0.333333‥‥は，1を3で割った答をどこまでも書きならべたものですが，分数で表わせない無限小数3.14159‥‥は一体どんな意味をもっているのでしょうか．

高等学校で解析Ⅱを勉強した人は，無限小数が，収束する無限級数の和を表わすことを教わったことと思います．この本では，そのようなことが解析Ⅱを知らない人にもわかるように，ていねいに説明してあります．ここで分数を表わさない無限小数は，何か新しい数を表わすと考え，それをひとまとめに無理数と呼ぶのです．つまり，円周率πも無理数だというわけです．このように「位取り記数法」から出発すれば，自然に，無理数という考えに行き着くのです．

位取り記数法によって計算が便利になったといっても，天体観測が精密になり，大きい数の掛算，割算が必要になると，やはり計算はめんどうです．そこで，掛算，割算を足し算，引き算になおす対数計算が生れ，それを使って，計算尺が発明されました．この対数と計算尺の原理の説明で，今まで紹介して来た「零の発見」は終ります．

第二部　直線を切る

第二部「直線を切る」では，ギリシアの初期の，数と直線とについての考え方が，無理数の発見によって非常な混乱におちいったことからはじめて，解析学の基礎になる直線とか連続とかいう考え方が育って来た筋道が書いてあります．第一部よりは少しむずかしくなりますが，ピタゴラスの定理ぐらいを知っていれば，中学生の皆さんでも楽に読めると思います．

この本ではいつも，いきなり数学の話を始めず，エジプトやギリシアの文明についての説明から始めて，いつの間にか数学の話に移り，その間にもいろいろなエピソードが入っているので，数学がそれほど好きでないような人でも，面白く読むことが出来るでしょう．またこのような書き方によって，数学が，その時代時代の文明や物の考え方，また自然科学などの発達から独立しているものでなく，おたがいにいろいろ影響をおよぼし合いながら発達して来たことが，よくわかることと思います．これがこの本の大きな特長といえるでしょう．

また皆さんがこれまで当り前のこととして教わり，または何の不思議もないと考えていた多くのことが決して当り前でもなく，またいろいろ疑問の余地もあるものだということがわかるでしょう．目の前にある数や直線，空間などの簡単なものでも，決して当り前だと考えてすましてしまわない態度，それが現代数学の態度なのです．

（科学読売7巻12号，1955年12月）

「科　学　と　方　法」*

（H. ポアンカレ著，吉田洋一訳）

19世紀の終りから20世紀の初めにかけて，数学と理論物理学とは，ともに激しい動揺の中から新しい方向，新しい理論を生み出そうとしていました．この本に収められた幾つかの科学的評論は，そのような動きの中で書かれたものです．もちろん約50年後の今日では，その内容のあるものはすでに過去のものとなっていますが，その場合でさえもポアンカレが問題を取り扱った態度，方法はなお十分に興味があります．特に彼は当時議論の中心であった相対性原理，数理論理主義などに対し，現在から見ても割合に妥当といえる見解を述べています．

だがここでは，科学の方法そのものについて，また数学上の発見の心理的な契機について，彼の語るところを紹介しましょう．この部分にはポアンカレの思想が最もハッキリ現われていて，現在なおその新鮮さを失っていません．（ポアンカレの科学哲学的評論を集めたものは，この本のほかに「科学と仮説」「科学と価値」「晩年の思想」の3書があり，みんな非常に興味深いものです）．

真実の選択

まずポアンカレは「科学のための科学」とは不合理な概念であるとするトルストイの言葉を引用します．「われわれは一切の事実を知りつくすことはできない，従って“科学の対象とすべき事実”を選択しなければならないが，この選択に際し，好奇心のおもむくままにまかせて差支えないだろうか．むしろ実益を，特に道徳的要求を標準とする方がよいのではないか．地球上に何匹テントウムシがいるかを計算するよりも，さらに価値ある仕事がないであろうか」． しかし科学の進歩を支えて来たのは常にただ科学のためにのみ研究をつづけた多くの狂熱家でした．彼らなくしては科学は存在せず，従って実益も得られなかったでしょう．この狂熱家たちは後の世の人々のために思考の労力を節約してくれたのに反し，目先の実益だけを追った人々は後の世に何も残しませんでした．科学の法則は普遍的であればあるほど，思考の労力の節約に役立ち，貴重なものになるのです．

これが事実の選択をなすべき基準です．つまり最も興味ある事実とは普遍的な事実，すなわち繰りかえして起る機会の多い事実です．それは結局単純な事実，あるいは単純らしく見える事実にほかなりません．そしてこれこそはまた，科学者が本能的に採用して来た方法なのです．

単純な事実は，例えば天文学の中に見出されます．ここでは天体の間の距離が大きいため，各天体は一つの点と考えることができ，その性質の違いは消え失せるからです．またすべての物体を構成する要素が最も単純で，また最も興味ある研究対象であることはいうまでもありません．

何度も繰りかえされる事実，すなわち規則的な事実の研究からはじめるのがよいわけですが，いったん規則ができあがってしまったうえは，例外的な事実を研究することが重要になって来ます．例えば天文学上の事実，地質学的過去など，日常の規則の当てはまらないような事実の研究によって，日常の事実そのものも，より深く理解できるようになることも少なくありません．

しかしさらに大切なのは，見たところバラ

* 名著紹介シリーズの 1 つとしてかゝれたものである（編者注）

バラで何の間係もないようなものの間に隠れた相似を見出すことです．これにより規則は次第にその範囲を広めるのです．例えばエネルギーという言葉により，力学，熱学，電磁気学などの多くの事実が一つの法則にまとめられたことを考えてください．

科学者はテントウムシの数をかぞえはしません．その数は複雑に，気まぐれに変化するのに対し，科学者が興味をもつのは法則的な事実だけだからです．

だがこれは問題の一面に過ぎません．科学者は何よりも自然の美しさのために，すなわち宇宙の調和，その単純さ壮大さなど，知性のみのつかみ得る内面的な美のためにこそ，苦しい研究に身をささげるのです．ここで思考の経済の原則は真の実益を生み出すだけでなく，このような美の源ともなることが見られます．一体この一致はどこから来るのでしょうか？

ポアンカレは，知的美を愛する心こそ知性に力を与え，人類の真の利益を生み出すものであることを強調するのです．

このような事情は数学において特に著しく認められます．

数学では人間の知性がより純粋に働くからです．数学者は特に方法と結果の優美ということを強調します．そして異った部分の調和や対称，あるいは孤立した事実に秩序をもたらし統一を与えるもの，従って細目も全体をも同時に明りょうに見通させるものほど優美の感を起させるのです．一方このように見通しがよくなれば，それと似た対象との類似がさらによく目につき，一般化も可能となり，従って産出力もまた豊かになります．

例えば，長い計算の後に単純で目ざましい結果に到達したとき，その結果の重要な特長は計算をする前から予想できたはずだということを示せないうちは数学者は満足しません．なぜなら，前もって結果を予想させる推理は，半ば直観的であり，短く，一目の下に全体を見通せるものなので，類似の問題に対し繰りかえし役に立ちますが，単なる計算は問題ごとに全部やりなおさねばならないからです．

このように真に価値ある結果を得るには機械的な計算だけでは十分ではないことがわかります．また優美の感と応用の広さとが，同じ源から出ていることが見られるのです．

数学上の発見

数学が論理の法則以外の何物をも必要としないのならば，全然数学に向かない人，数学を理解できない人が余りにも多いのはなぜであろうか？　まずポアンカレはこのように質問します．もちんすべての人に数学上の発見ができたり，証明を全部記憶できたりするわけにはいかないのは当然ですが，万人に共通な論理だけに基く証明がなぜ理解されないのでしょうか．さらに驚くべきことは，数学者自身でさえも，その推論の途中でしばしば誤りに陥るということです．

この誤りの原因としては，証明が長い時には，最初の部分の記憶がおぼろげになり，それを後の推論に使うとき，少し意味を取り違えてしまうということが考えられます．もしそうであるなら，数学の才能とは確実な記憶力と非凡な注意力であることになるでしょう．つまり多くの指し手を記憶し，誤りなく使うことのできる将棋の名人が最も優れた数学者であり，またすぐれた数学者は将棋の名人でなければなりません．

もちろんそんなはずはありません．実際，数学の証明は単に推論の式を並べたものではなく，その並べ方には一定の順序があって，この順序の方が，個々の推論式よりも遙かに大切なのです．そしてこの順序についての直覚をもち，一目の下にその全体が見通せるならば，個々の推論の細かい点まで記憶しておくことは，それほど必要ではないのです．

数学上の発見のできるのは，記憶力，注意力のすぐれた人ではなくて，このような直覚，すなわちかくれた数学上の調和と関係とを見とおす直観をもっている人なのです．ここで数学上の発見というのは，単に数学的事実の新しい組合せを作ることではなくて，有用な組合せ，一般法則に導くような組合せを作ることです．発見とはこの意味で選択することです．

次にポアンカレは自分自身の経験を語ります．彼は後にフックス函数と名づけられた函数を研究していました．２週間の間，机に向かってはたくさんの組合せを作りましたが，うまくいきませんでした．ところがコーヒーのために眠れなかったある夜，多くの考えがわき起り，たがいに衝突し合いましたが，その中の２つが密着して１つの考えにまとまるように思われました．朝までに彼は１つの目ざましい結果を得，あとは起きてからその証明を書き上げる仕事が残っているだけでした．さらに少し研究をつづけた後，彼は地質旅行に参加して，数学上の研究のことはしばらく忘れていました．そのようなある日，乗合馬車の階段に足をふれた瞬間，突然にフックス函数と非ユークリッド幾何学との関係についての考えが浮びました．この考えは確信に満ちたもので，事実，後で１人になってから，ゆっくり確かめることができました．その後も似たようなことが何度かあって，この研究は完成したのです．

このような経験に共通なことは，第一に突然天啓が下るように考えが開けることですが，これはそれまで長い間無意識的活動が行われていたことを示すものです．例の眠れなかった夜には，この無意識活動がある程度知覚されていたわけです．もちろんこの無意識活動は，しばらくつづけられた意識的な研究に刺激されて，はじめて起されたものです．それにしても，この無意識的，潜在的な自我は意識的な自我よりもすぐれていると考えるべきでしょうか．むしろ，無意識の中に数多くの組合せが作られ，その中で興味ある組合せだけがわれわれの意識に浮んでくるのだと考える方が正しいでしょう．

意識的になり得るのは，われわれの感受性，数学者の審美的感情に最も強く訴えるものだけです．この優美という特質は，各部分が調和的に配列され，その細部に徹しながらしかも全体を一目の中に見通せる所から生れるものであって，従ってまた数学的法則へと導くものであります．有用な事実とはまた法則的な事実のことですから，この有用と優美との一致した特質をもつものだけが，数学者の感受性を魅することができるといえます．

時としては，数学者に突然現われる啓示が誤りであるとわかることもあります．しかしその時でも，その啓示は，もしそれが正しかったとすれば，われわれの数学的審美感を大いに喜ばせたに違いないというようなものであることが多いのです．このことは，以上の考えを裏書きするものともいえましょう．

数学的感受性をもたない多くの人に数学上の発見ができないのも，また計算機械がどのように発達しても人間の思考に代ることができないのも，このゆえなのです．

科学至上主義，科学のための科学これがポアンカレーの一貫した立場です．原子力の現在に彼を生れかえらせたならば，彼はやはり同じ立場を主張しつづけたでしょうか？

（科学読売 8巻6号，1956年6月）

数学月報*

1955 年国際数学会議

日本に行くと教授 (Professor) が変じて予言者 (Prophet) になる, 先年来日した或る数学者が, 帰国後こういったそうである.

数学の発展に対する見通しや, 色々な問題についての予想などをしばしば質問されるからだという. 今度の数学会議には, プリンストン大学のアルチン教授, コロンビヤ大学のシュヴァレー教授, シカゴ大学のヴェイュ教授など, 10 人の予言者を海外から迎えたわけである.

実際, 数学は物理学などと違い, 時の中心問題というものもなく, 研究の対象自身, 研究者の創造にまつ所も少くないのだから, 将来の発展, 研究の方向など, 学生が真剣に質問するのも当然である. また数学には, 有名な予想が数多くあって, それを解決しようとする努力が発展の原動力となることも多い. 例えばフェルマーの問題, つまり

$$x^l + y^l = z^l \qquad (l \geqq 3)$$

という方程式が, $x=y=z=0$ 以外の整数解を持たないという予想の研究が, 現代の代数的整数論の基礎を築いたのである. だから, 良い教授は同時にすぐれた予言者でなければならぬともいえよう.

今度の会議でも, 単に研究の成果が発表されただけでなく, 色々な予想, 問題も提出され, また非公式の討論会などは, まさしく予言者の集いの観があったのである.

この会議は正式には, "代数的整数論についての国際的シムポジウム" と言い, 国際数学連合 (I. M. U) の主催で毎年開かれるシムポジウムの一つで, これが日本で開かれたのは, 高木先生の類体論以来, 代数的整数論の優れた研究者を多く産み出した, わが国の真価が認められたものといえよう. 会議の主題は, もちろん整数論であるが, それがさらに三つに分かれる. 1). **類体論とその拡張**. 2). **代数系の理論と整数論との関係**. 3). **代数幾何学と整数論との関係**. 現代整数論は一つの転機に立っているので, このシムポジウムが, 新しい発展の一里塚となることが期待されていたのであるが, まずこれらの部門について簡単に説明してみよう.

1920 年に高木先生の類体論が発表されてから, この理論の整理, 簡単化と, それから派生する問題とが, 代数的整数論の中心問題であった. この面では, 単に理論を書きかえるという以上の重要な研究が数多く生れたのだが, 30 余年過ぎた現在, 類体論は徹底的に整理研究しつくされ, 後にはその拡張という問題だけが残されているのである. ところでこの拡張という問題は, 非常にむずかしく, 多くの数学者の努力にもかかわらず, まだその第一歩が踏み出されたともいえない状態である. 何か新しい発展の方向が見出されなければならない. まさしく一つの転機ではないか.

しかし類体論は重要な副産物を産み出した. 抽象的な代数系の理論はその一つであって, 類体論を新しい方法で構成し直す時に必要とされるものである. この面ではまだ多くの発展の余地があるのだが, 整数論への応用という立場が忘れられ, 小手先だけの細工になりがちなので, 口の悪い連中は, "あれは整数論でなく代数" だという. いずれにせよこの方面からは, 整数論の大きな発展は, もはや望めないように思われる.

シムポジウム第1日 (9月9日), 第2日

* 小山書店発行新初等数学講座月報

(10 日) は，東京の第一生命会議室で開かれた．第 1 日は類体論とその拡張とに関するもので，将来の発展のために必要と思われる結果とともに，興味ある予想も幾つか発表された．しかし第 2 日，代数系に関する議論が始まると，会場をエスケイプし，控室で別な問題を討論しているフランスの学者の姿も見られた．

10 日の午後にロマンス・カーで一同日光に行き，会場は金谷ホテルの舞踏室に移る．11 日は日曜なので会議はなく，日光付近の観光に過ごされた．ついで12日，13日には，代数幾何と整数論との関係が中心題目となった．

ここで，代数幾何学というのは，代数方程式で定義される曲線，曲面 (例えば双曲線，楕円面など) の理論である．それが整数論とどんな関係があるのかと不思議に思われる方もあるかと思うが，例えばフェルマーの問題は，$x^l+y^l=1$ で定義される平面曲線上に，座標が有理数である点が存在するか否かという問題にほかならないことを考えていただきたい．

このような関係の中で，特に重要なのは虚数乗法論とよばれている理論で，これが議論された 12 日は，シムポジウムの山とでもいうべき日であった．虚数乗法論の特別な場合は，古く 19 世紀に研究され，現代の類体論の母胎となったものだが，それを一般の場合に拡張することは，長い間懸案の問題であったのである，ところがこの会議ではからずも，シカゴ大学のヴェイュ教授と，東京大学の 2 人の若い研究者とが，この問題に対し，それぞれ独立に，深い研究を発表し，この 3 人の結果を総合すれば，ほとんど完全な解決が得られることがわかるという，きわめて興味ある事態が起った．

そしてこの理論はさらに，解析的整数論とも深いつながりを持っていることが明らかとなり，今後の発展に対する一つの道標が打ち建てられたわけで，シムポジウムの大きな成果の一つといえよう．この日は夜おそくまで，非公式の討論が続けられたので，翌 13 日には，ねむそうな顔をした出席者も少くなかった．この 13 日には純粋の代数幾何学が問題にされたが，発表された研究の間にはあまり密接なつながりは見出されないように思われた．

全体を通じ，会議は終始緊張した空気に包まれていた．特にアルチン教授は底力のある声で鋭い質問をくりかえし，発表者もそのために，たじたじとなるという場合も見られた．

今度の会議で特に目立つものは，発表者の年齢の若いことである．全部で 19 人の発表者の中，約 3 分の 1 が 20 代で，35 歳以上の人は半数以下であり，特に日本側の研究者は大部分 20 代であった．大体，数学という学問は，若い中でなければできないといわれているくらいだから，このこともそれほど，不思議ではないかもしれないが，特に日本でこのことが目立つのは，40 歳前後の中堅で，経済的事情のために，アメリカの大学の教授になっている人が多いということにもよると思う．だが中堅の指導者のいないことがかえって若い人々の気持を刺激したことも見逃せない．例えば東京では，新数学人集団 (略称 S. S. S.) という，若手研究者及び学生の会が生まれ，自分たちだけの力で勉強を続けていこうと努力してきた．その成果が今度のシムポジウムにもあらわれ，日本側発表者のうち約半数までが，この S. S. S. に属する人々なのである．

とにかくも若い人々が多いために，会議の空気はきわめて活潑であった．来日した大予言者たちもなかなか気が若く，ヴェイュ教授などは会場で猫の鳴きまねをするなど，相当な茶目振りを発揮していた．ところでフランスでは，約 20 年程前，(当時の) 若い数学者の集団ができ，ブルバキという名まえで，フランス数学の改革に力をつくしてきたのである．今度来日したヴェイュ，シュヴァレー，セール教授などはこのブルバキの仲間なので，S. S. S. の運動に大いに同情的であった．ことに日光では S. S. S. とブルバキとは大いに語り合い，大いに暴れ廻って，会場外での収穫もきわめて多かったと思われる，

このような活溌さは，13 日，全部の日程を終った夜のサヨナラ・パーティで，その頂点に達した．まず，シムポジウムを題にしての詩のコンテストが行われ，3 篇の応募があった．各篇に予言者または予言ということばが出てきたのはいうまでもない．ついで余興に入ったが，ブルバキー派のフランスの歌，またドイツ人のアルチン，ブラウアー，ドイリング教授たちのローレライの合唱など，おそらく前代未聞のことではないかと思う．S. S. S. を中心にする日本の若手は，コンテストに応募したシムポジウムの歌を合唱した．参考までにその全文をご紹介しよう．（これは広瀬中佐の歌の替え歌である）

驚くアルチン脅ゆるドイリング
暴れて廻るヴェイユの故に
黒板貫くシュヴァレーの叫び
定義はいずこ定理はいずや

室内隈なくにらめる三たび
問えど答えずただせど知らず
ヴェイユは次第に闘志に溢れ
げきぜついよいよあたりにしげし

今はと予言にかくれるヴェイユ
飛び来る問にたちまちつまり
国際学会効果のうすき
予言者ヴェイユとその名残れど

こうして最後に，全員による螢の光の合唱のうちに，シムポジウムの幕は静かにおろされたのである．（数学　月報 4，1955 年 9 月）

自信ある仲間

Quelques　Questions

最初に，僕がQ君の立場に同意することを書いておくのも無駄ではないでせう．然しQ君の所説は，御覧の通り，曖昧であり，不充分であり，更に余り感心できない部分もあります．つまり此れは，Q君の夢を，乃至は“非合理的思考”を，どうにか他人にも了解できないこともない形で，表現したものなのでせう．表現形式について，揚げ足を取るつもりはありませんが，本質的と思はれる二，三の点について疑問を述べ，Q君の御教示を仰ぎたいと思います．

第一にコトバの機能を考へて見ませう．御説の通りコトバは communication の手段として生れ，現在でもそれがコトバの主要な役割であるには違ひありません．だが生れ出たコトバは独り歩きする様になり，それ以上のものとなってしまひました．と云うことの意味を以下に述べて見ましよう．

Q君の云はれる様に“思考”する人間を見出さうとしたら，松沢病院にでも行くか，思ひ切ってポリネシアあたりまで出掛ける必要があるでしよう．人間にとって最も重要なものは，他の人間達であるとしても，霞を喰ひながら清談してばかり居るわけには行きません．人間は自然を対手にし，それを征服し，遂には“人工の自然”つまり様々な機械を作りあげました．そして毎日，それらのものを相手に過さなければならないのです．‥‥と云うのは，人間の思考が“合理性”の枠を要求されるのは，単に，社会との communication に於てだけではないと云ふことです．人間の思考が，天然の又は人工の自然に対して有効である為には，それは合理的でなければならないのです．

人間の対手にする自然が，極めて簡単なものであった時代には，此の“合理性”はいはば本能的な形で“非合理的”な（Q君の云う意味で）思考の中に表われることが出来ました（例えば，チンパンヂーが椅子を積み重ねて吊されたバナナを取る場面を想像して下さい)．然しながら，複雑極りない機械を作り，又は動かすとき，人に納得させることのみを目標として合理的な表現を取る様な思考で間に合ふかどうか，一つQ君に実験して見て欲しいと思ひます．

他の人々に伝へることを全く予想しない場合にも，高度の合理性が要求されることがわかるでせう．それなら，この合理性は如何にして得られるのでせうか？　実際に僕達は，そんな場合，如何に思考するでしようか？

コトバを使はずに，更には，普通のコトバの延長として数式乃至記号を使はずに，すますことは不可能であることがわかるでせう．と云ふより此の場合“思考”は，頭の中に於けるコトバの連続として現はれる筈です．Q君の云ふ直観は，この言葉の流れを一定の方向に進め，或いは進めないための役割を果すのです．此の場合，コトバは，合理性を保証しないとしても，そのための必要条件となるのです‥‥‥ここにコトバの第二の役割が見出されます‥‥つまり“過程としての思考”の中にも，コトバが現われるのです．

以上のことは，数学にそのまま当てはまります．Q君の云ふ“思考”によって，簡単な問題一つでも，解くことが出来るか否か，これは実験をせずとも明らかでせう．ポアンカレだってただ馬車に乗っただけであの輝しい業績を生み出したのではありません．

要するに，我々が問題にする程度の思考はすべて（直観，又は非合理的飛躍により推進されるとはいえ）コトバに依り行はれるのです．従って，そのコトバに要求されるのは単に“communication の機能を果す”ことだけではなく，そのコトバに依り行われた思考が有効であることでもあるのです．

一つ，南の海に乗り出してみませう．そこには「ポポロ族はおうむである」と話す人々が住んでいます．その他，彼等の使うコトバは，僕達を驚かせるでせう．然し，彼等の社会の中で，そのコトバは，立派に communication の手段としての役割を果しているのです．――それは，いはば，呪術的なものを背景とするコトバです．そのコトバに依る思考は，彼等の程度の“文明”を維持するは充分でせうが僕達が，そんなコトバを使ひ始めたとき，現在の文明が維持されるか否かを真面目になって考へる人は居ないでせう．

logique の対象となるコトバが，僕達のコトバであるべきか，ポポロ族のコトバであるべきか，コトバを単に communication の手段と考へるとき，それを決定することが出来るでせうか？

各々のコトバに対して，それぞれの logique があるといはれるかもしれません．でもやはり問題は残ります．我々は何故此のコトバを使ひ，あのコトバを使はないのか？更に“形式的体系”を問題にするとき，その唯一性の根拠を，どこに求めるか？ Q君が今問題にしようとしている体系が，何時の日にか，無意味なものとして捨てられることがないと，どうして保証するのでせうか？

勿論，僕達が考へるとき，論理的に完全な statement に依るとは限りません．寧ろ必要あらば完全になし得るといふ予想の下に省略された文章に依ることが多いのですが，それは，日常の会話とても同じことで，要するに，完全なコトバなるものが予想され，その基礎となっている点が重要なのです．

従って，“考え方の consistency を保証する規則”は一時的に社会から切り離された個人にとっても，その人が何か，自分以外の“もの”を問題にする限り，意味を持ち得るのです．此のことは，発表する意図もなしにガウスのなした老大な計算を考へて見れば明らかでせう．

Q君が，此の，コトバの，第二の機能を無視したのは，或る意味で賢明なことでした．何故なら，それは，絶望的に困難な問題を提出するかもしれないのですから，だがそれが無視され得るためには，先づ第一に，僕の疑問が答へられなければならない筈です．（他の疑問を提出する人もあるでしよう．）

ここで“判断”の問題を考へてみませう．このように考へるとき，判断は“declarative sentence”から，如何にして区別され得るでせうか．強ひていへば，判断とは，思考の中に生起した sentence（又はそれが表現される際に sentence として完成さるべき，コトバの組合せ）に対して“yes”又は“no”という機能だといへませう．だが真面目になって「これはこうであり，あれはああなる」と考へ，又はいっている時に一々それに対し，“yes”又は“no”と註釈をつける人があるでせうか？ たとえ気紛れな人が居たとしても，それを，特に“判断”として区別する意味がどこにあるでせうか？ この点僕は，Q君の意見を支持します．

要するに，“明析な形での論理学の対象”を“コトバ”として規定すること，此れは非常に結構なことです．ただ残念ながら，Q君はコトバの一つの大きな機能を，無視してしまったのです．さうして，精神病者乃至は未開人の思考に対する分析で，数学的思考の分析をすりかえることにより，此の無視をもっともらしくみせることに“成功”したのです．所が，僕は，見聞の狭いせゐでせうが松沢病院から偉い数学者が出た話も聞かなければ，ミクロネシヤに数学の一流派が栄えたことを書いた本をみたこともないのです．

（自信ある仲間，1952年6月）

無 題

余り気乗りもしませんが，此のノートを，余り長く僕の手許に置いておくわけにも行かないので，何か知ら書いて見ませう．

ヴェリテ君，君の文章，面白く拝見しました．①然し，此れを読んで，「諸兄」が，ムキになったり，弁解したりすると思っているのでしたら，君は非常に自惚が強くて，純真だと云ふことになります．

勿論，②僕は自分自身に対して相当に，時には必要以上に，批判的です．さうして，他の人の批判が，僕の問題としている点に触れるなら，或は，僕の心の底まで動かすなら，その人は鋭い観察家であり，さもなければ皮相な観察家だと云ふわけです．③皮相な観察家の多いのに時々ウンザリすることもあるのですが，それにしても，或る人が，どんな観察家であるかと云ふことは，僕にとっては，云はば，どうでも良いことであり，従って，僕は他の人の僕に対する批判には余り興味を持っていません．④それに実際の話，今まで，自分の考へている以上のことを，他の人から教はったことはありません，問題はそんな所にあるんじゃないのです．

⑤僕は自分から積極的に活動するのでない限り何も出来ない性格なのです．⑥引きまはされたり，人の後をついて行ったり，自分のものでない言葉を口にしたりすることは大嫌ひ，と云ふより事実上不可能なのです（此のことを不思議に思ふ人があるかも知れません．僕は，⑦non-essential と思はれることは，すべて人に任せてしまひますから‥‥それは，エネルギー節約の為です）このことを最初に強く感じたのは，出（いで）さんの選挙の時でした．⑧どう云はれて見ても，僕が，乏しいエネルギーをさいて，選挙運動する必要性は認められませんでした．⑨それにも拘らず，それ以前の僕の行動から当然何かするものと期待されているのを知って，少し不愉快になりました．同じ様なことはまだありますが，⑩一つの行動を共にすることは他の行動を共にすることではなく，ましてや思想を共にすることを意味しません．勿論，それ等のことについて議論しませんでした．しかも意味のない議論はしない方がましです．そして，何もしないことによって僕の考へを表明したのです．それまでに，話を聞き，或ひはして，一応さうしなければならないと頭で考へて，さうした，と云ふことが，何度かあったのですが，その結果，僕は，自分の頭の中だけでの判断は，信頼出来ないことを知ったのです．⑪人間の頭なんて，なまじ論理を知っているだけに，論理的な仮装に，容易に欺かれるものです．さうならないためには，鋭い批判的精神を必要とします．そしてそれは，既に頭だけの問題ではないのです．——頭だけに信頼出来ないとき，感覚に信頼する外ありません．さうして，感覚は，容易に欺くことは出来ません．（勿論，感覚のみに，とは云ひません）．⑫そして僕は，自分で，さうする必要があると考へ，そして又さう感じたことの外は，しないことにしたのです．外からの考へに従って行動すること，——それは既に精神的に，非常な無理をしなければ不可能ですが，無理をして，さうしたとしても，‥‥その結果が成功であった時でさへ‥‥後になってからしなければ良かったと思ふ位が関の山です．

⑬例へば，僕は，学生大会には，常に出なければならないと考へています．然し，最近の学生大会のバカバカしさを見て御覧なさい．そして，僕がそれに出ることに，どれだけの意味があるかを考へるとき，僕は，どうしてか出たくないと感じないわけには行きません．

此の様な喰ひ違ひを，大低の人は，適当に妥協させて，頭の中の立派な考へは，そのまま保存して，行動の方は，「適当に」すませておくのです．⑭然し，それが嫌な人は．飽くまでも自己を強制して，考へた通りにやるか，考へを変へるかの外に道はありません．⑮前者の結果は既に見た通りです．人間の精神は，永い間の強制に堪え得るものではありません．と云って，考へを変へたと表明すれば，「適当に」やっている人々から（此等の人々は，そのことを自覚していないのが普通です）激しい非難を浴びるでせう．

大部ゴタゴタしてしまひましたが要するに，無理は永く続かないと云ふことです．⑯僕の考がブルヂョア的であると云ふのなら，それも結構です．又，他の人が僕を「赤」だと云っても，一向に差支へありません．⑰その他，「非常識」であらうと，「常識的過ぎ」ようと，「コケシ人形」であれ「風船玉」であれ，そんなレッテルには，何の興味も持ちません．人間は自己を強制することは出来ません．現在の自己を進歩させることは出来ますが，全く違ふ自己を持つことは出来ません．⑱生活革命ですら，真に自己の内からの要求によるものでない限り，線香花火的なものに終るでせう．自己のブルヂョア根性を払拭しようとしている，或はしていると自称する，人々の，高潔な心性は，僕もそれを認めるに吝かではありませんが，僕としては，個々の場合に，その必要を心から認めるのでない限り，そんな馬鹿げたことはしない積りです．

序でに，此の機会に明かにしておきますが，⑲僕は決して，ロシアを讃美してはいません．アメリカに対する考へは御承知でせう．⑳「祖国日本」に対しても，何の愛着も持っていません．総じて僕は自分の属する団体を愛したことはありません．㉑それは常に，僕の自由を制約するものとしか考へられませんでした．㉒自分の属しない団体に至っては尚更です．㉓そして「同志」なんて言葉は僕の辞書にはありません．㉔勿論僕はどんな人々の「仲間」でもありません．（と云って人間を軽蔑しているわけでもありません．）

・・・・一つの国を絶対的に正しいものとして信頼したり，又は絶対的に悪いものと考へたりすることは，余りにも馬鹿げたことの様に思はれます．勿論資本主義社会の機構とその動き方，社会主義社会の理想と云ふ様なことは知っていますし，みんな人々は必ずそれを引き合ひに出すものですが，此処に一つの言葉のトリックがあることに注意して下さい．さうして，たとへ，非常に純粋な理想を持った国家が現はれたとしても，現在の様な，（或は多少変ったとしても）国際情勢の下では，それは非常に歪められずには居られないと云ふことを考へて下さい．・・・・僕がロシヤを弁護するのは，余りに馬鹿げた非難を聞くからであり，アメリカを非難するのは，非難する人が余り少な過ぎると思はれるからに外なりません．――そんなことは，皆当り前のことで，今更云ふまでもないことばかりだと云はれさうですから，此の辺でとめて置きませう．

ヴェリテ君の文章を読み直して見て気附いたのですが，㉕9頁にある「知る」と「感ずる」と云ふ言葉は，僕の云ふ「考へる」「感ずる」とは，恐らく非常に違った意味を持っているのでせうから，両者を混同したり，類比したりしない様に注意して下さい．総じて僕の書いたものは，ヴェリテ君のものとは，独立したものとして読まるべきであり，その間の関連を，余り密接なものと考へないで下さい．㉖例へば，此れは，ヴェリテ君に対する僕の「アポロジー」ではありません（僕はその必要を認めませんから）．・・・・そしてヴェリテ君は，僕の此の文章に，余り満足されないだろうと思はれますが，それも仕方がないでせう．僕は人が良いから，外の人を余り失望させたり，不満に思はせたりしたくないのですが，「自信ある」彼のことですから，此れを読んでも失望したり憤慨したりしないだらうと思って，割合に自由に，ペンを取って見たのです．

㉗此のノートが発生してから，色々な事件が起きています．それ等のことについても，一応触れて見たいと思ったのですが，問題が別になりますし，此の調子で書いていったらキリがない‥‥それに，「もう時間がない！」‥‥真面目な話，僕の健康は，今直に，床に就くことを僕に命じているので，此の辺で一応切り上げます，「又後の機会に」‥‥なんて約束はしない方が安全でせう．今度書く時があったとしても，その時書きたいと思ったことしか書かないでせうから．

最後に，此の様な冗舌を，飽きずに（或はアクビしながらでも）読んで下さった方々に，感謝します．　（自信ある仲間，1952 年）

或る映画を見てある人がどう考へるかと云ふことについて

――一つの実例――

此れは或る試みの出発点をなすスケッチに過ぎない．勿論出発したものが目的地に到達するとは限らず，スケッチはスケッチのままに終ることも多い．此れを書くのは，それ以上のことを考へるためでもあり，又それ以上のことを書かない為でもある．

「欲望と云ふ名の電車」此れは暗く重苦しい．然し人の心を惹きつける．演技の巧さ，その他は抜きにしてその原因を考へて見たい．

主人公が精神錯乱に陥ったと思はれる最初の時に，此んな意味の独り言を云ふ「‥‥私は，此んなに知性が高く，美しい心を持っているのに，あんな職工などに，自分を安売りしようとしたなんて‥‥」彼女が実際に知性が高く，心が美しいか否かは問題でない．重要なのは，彼女が常に，心の底に，かうした考へを抱いて居り，現実と妥協し，それに頭を下げて，可能な限りの幸福を得ようとする，もう一方のより現実的な自分に反撥していたことである．その可能性が失はれるとき，‥‥彼女は此れ以上の現実への屈服を肯んじなくなるのは当然である．つまり現実を無視することにより，それに反抗する．‥‥それが精神錯乱に外ならない．

僕は此の言葉に，非常な興味を感じた．‥‥同じ様な気持は，大抵の女が結婚しようとする際に抱くものであらう．然し此れはそれ以上のものである．

機械文明の発達した社会，その典型としてのアメリカを考へて見給へ．個人の「成功」の機会は昔に比べれば，非常に少くなっている．然も，現実は常に流動して行き，それに従ひ，何とか生活を保って行くためには，精神の強度の緊張，自己の心性の犠牲，心理エネルギーの過度の使用を強要される．人間関係は益々非人間的となり，その中にあって自己の道を切り拓いて行くために，以上のことは益々必要となる．‥‥然しまだそれは，人間の堪え得る限界までは達しない．

だがそれは Normal な状態に於てである．恐慌は定期的に訪れる．その他，種々な原因による社会の変動，それ等はそれまでの努力の結晶を一瞬にして無にしてしまふ．

現実を受け入れ，或はむしろ，現実に屈服して，多くの犠牲と努力とを払って築き上げた，或は上げつつあった，或はさうしようとしていたものが，破壊されてしまふとき，然もその原因が，自分には責任のない，社会の変動によるものであるとき，‥‥或は，破壊されないにしても，絶へずその危険に戦いているとき‥‥人間がどう感ずるか，——

彼は自己を犠牲にし，現実に屈服して，或る程度の現実的な利益，幸福を得ようとする此の「取引き」が割に合はないものであると感ずる．特に多少精神構造の弱い者に取って，現実に順応することは，既に非常な苦痛である．その結果が，その報酬が此れでは‥‥‥彼等は最早，そんな「取引き」は御破算にしてしまはうと感ずる．そして空想の世界の中に自分だけの幻想の世界に閉ぢ籠り，精神病院に送られる．

‥‥‥‥‥‥‥‥‥《余白がない！ Fermat》

（自信ある仲間，1952 年）

四　行　詩

——Q君に——

君を夢中にさせるもの，それは
君の持っていないもの
折目正しい「素晴しい」ズボン，ただ
穴のあることは御存知ない．

（自信ある仲間，1952 年）

屋　上　屋　を　架　す*

（ノートの「左半分」の「真空状態」を埋めるために）

註　訳「無題」

①

V：勿論自惚が強くて純真です． 尚，cf. “CARTE FRANC” のはじめ．

L：非常に結構ですね．

②

V：必要以上に批判的！ とは何ですか？ Je ne sais pas！

L：Naturelment, vous n'en savez point.

③

V：どんな観察家であってもよいならウンザリするなんていうのはおかしいね．

L：蚤がどんな動物であっても，構はないとしても，余り刺されれば，いやになるでせう．

④

V：自分で考へている以上のことを人に教わったことがない．天才ダナ君ハ（地球上にいない天才！）

L：さもなければ，君達が間抜けと云ふわけさ．（地球上にも居る間抜！）

⑤

V：勿論，勿論．

L**:

⑥

V：共通の弱点！

L：「弱点は美点である」（アイマイ弁証法より）

⑦

* これは上掲の「無題」に対してV氏が加えた反論に，谷山自身が再び反論を加えたものである．原文では頁の右半分に書かれたV氏の文章の左に余白があり，L氏（谷山）がその余白を埋めてある．こゝでは編集の都合上V氏の言葉の下にL氏のものを記すことにした．

** 答なし　（編者注）

V：trivial といわなかったことのみが良心的であるところか？

L：事実上，trivial でないことも多いのでね.

⑧

V：君なんかに選挙運動をやれといった奴がいたのか？

L：恐らく君より非常識な人だったのでせう.

⑨

V：こう云った方がよい，‥‥こっちの云い分「当然何か期待されていると期待されている面をみたので少し不愉快でした！」

L：多分さうでせう.

⑩

V：当り前

L：勿論

⑪

V：なまじ論理を知っていると思っている丈おかしいね！

L：自分が知らないからって他人も知らないと思っちゃいけません.

⑫

V：

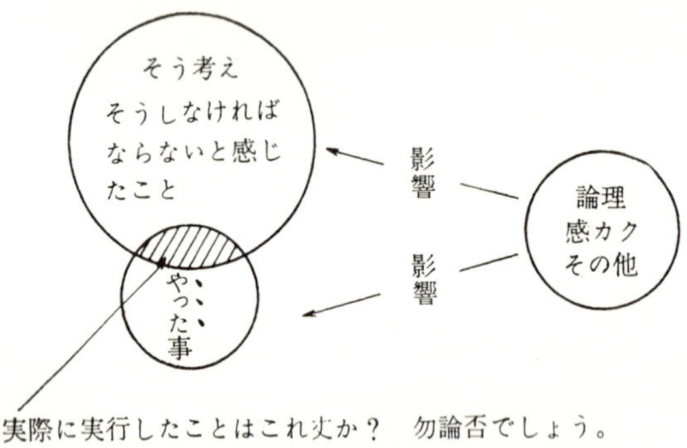

実際に実行したことはこれ丈か？　勿論否でしょう。

L：例へばQ君から「此れは何だ」と聞かれてそれを見もせずに返事する様なことでも，僕のしたことなのですから勿論それだけじゃありません．大低の人に対しては，その程度で済ませるのです「ヤア，シバラク！」

⑬

V：君達の（L氏をふくめて）学生大会ではなかったのか？　学生大会に君の建設的な異見はシャツト・アウトなのですか！　満足でないものを押しつけられて精神的抵抗を感じつづけている哀れなインテリの姿！

L：此れは一例に過ぎない．それにいつでも，そんな問題に対しても，「建設的」意見を持っているとは限りませんからね.

⑭

V：退嬰的会社主義！

L：それとこれとどんな関係があるのですか.

⑮

V：行動の前に結果を陳列してみせている手品師！

L：それを「先見の明」と云ふわけさ.

⑯

V：「赤」若しくは「ブルヂョワ」好きな方を宣伝したらどうです.

L：残念なことにどっちも好きじゃないのでね.

⑰

V：レッテルというのは中味に関係なくもない場合に注意！

L：勿論それを考慮に入れての話です．さもなければ，始めから問題になりゃしない.

⑱

V：「生活革命，云々」勿論，だが，そんな理想的な場合が起り得ると妄信している人の方がコッケイ！

L：だから多くの人がコッケイに見えるんです.

⑲

V：勿論，Vとても同じこと.

L：勿論，ね.

V：自分の住んでいる社会には愛着を持ってほしいね.勿論,「祖国日本」などとゼスチュアしなくてもよいが，個人的な愛着と同じように，この愛着は自分自身への真の愛着に近づくことである.

尚ロシアといわないで，ソビエトといった方がより適切である.

L：「愛着」が自分や他人を拘束するのを恐れるのでね．その上それに，値しないものに「愛着」することは出来ないでせう．更に「『真の』自我を実現するため」特攻機に乗った人も居るんですからね.

尚 U.S.A. をアメリカと云ふ様に U.C.C.P. をロシヤと云っても変ではないでせう？

㉑

V：ヒガイ妄想狂

L：此の言葉の正しい意味を知っていますか？

㉒

V：自分の属しない団体と君の間に mapping でもあるのか？

L：ないから，さう云ふのです.

㉓

V：多分その辞書は数千年前に絶版になったのじゃないですか．〈out of print !〉

L：恐らくさうでせう．僕も見たことがないのですから.

㉔

V：俺は誰ともちがうんだ！ということが自分の価値を高める如くに，この楽章は Andante ——

その後は当り前！というに吝かでなし.

L：僕はわざわざ自分の価値を高める必要を認めません．それに誰とも違ふなんて，余り気持の良いものぢゃありません．（さう思ひませんか？）

㉕

V：カテゴリスト的反撥！ 独立か！

L：反撥じゃなくて老婆心と云ふものさ，言葉の内在的意味に因はれちゃいけません.

㉖

V：Vチャンに対するLの「トポロジイ」といった方が適切.

「むきになって，反撥しないことは決して自慢でもなければ，良くもない．君達はI君を見習え．彼がムキになるのをカラかう前に，勿論L君はどういう立場か知らんがね！

L：君にとって良い気持のしないことは，良くないことである，と云ふわけでせう.

一つI君でも見ならって君や Ts. 老先生とケンカでもしませうか？ 正しく「純情二重奏」とでも云ふ所でせうね.

㉗

V：「発生」という言葉が気に入った．Lが文学的才能の片鱗を示したことは拍手するに足る．何かS学の論文集なんかのおしまいみたいだね．網らしなくてもいいのだよ！さよなら，さよなら．・・・・

L：此処では何も云はないのが礼儀に適ったことでせう.

（天野君著，文部省発行「漢文ニヨル道徳ト礼儀ノ教育」763 頁を見よ）

◦ 断片的にあげ足とりみたいですが，まとまった異見は他の機会又はこれからのいろいろな論説，創作なりで表明します.

Lよ，がんばれ！ワルク思ウナ.

◦ 僕のは「あげ手とり」位の所でせう．此の行のLを $\pi/4$ だけ回転して，僕の最後の言葉とします.

（自信ある仲間，1952 年）

正雄さんの話

昔，ある灰色の街に，大きい石の建物が立っていました．鉄の扉には，錠が固くおりていて，住んでいる人もない様に見えました．或日，正雄が，友人と一緒にその前に立っていますと，品の良い，中年の女の人が来て，二人に，ついていらっしゃいと云って，家の前に立ちますと，扉はひとりでに開きました．そして，誰もいない，大きな，寒々とした部屋を幾つか通り抜けると，裏の玄関に出，その外に出ると，……此の暗い街に，此んな所があると，誰が想像出来るでせう，広がった砂浜の先は海が静かに波打ち，太陽は，まぶしいばかりに輝いています．人っ子一人いない砂浜には，ブランコの様な，ギョチンの様なものが，ただ一つ，ぽつんと立っていました．明さと，強い色彩，そして不気味なばかりの静けさに，正雄は，少し気分が悪くなって，へなへなと腰を下してしまひました．

その日，どうして家に帰ったのか，正雄は，良くおぼえていません．然し，その時のことは，強く頭に焼きつけられて，度々思い出すのでした．でもそれから，あの，灰色の街を通っても，鉄の扉の固くとざされた，大きな石の建物は，見当りませんでした．正雄は，何度も何度も，歩き廻りました．あれは夢だったのでせうか？　でもそんな筈はない．そう思ひながら，暗くなるまで探しては，失望して帰って来るのでした．

所が，或る日，正雄が，又，二三の友人と一緒に，その街を歩いていると，あの，大きな石の建物が，昔の様に，立っているのでした．それを見たとき，正雄はどんな気持だったでせう．扉は，押すと自然に開きました．そして，人のいない，大きな，寒々とした部屋の幾つか，それも昔のままでした．いつかの，女の人がいるかと思ひましたが，誰もいる様子はありませんでした．一人の友人が，一番奥の部屋まで行って見ようと云ひ出しましたが，正雄は首を横に振りました．そして又，裏玄関に出，扉を開けました．所が，………いつかの海は，どこにもありませんでした．低くたれ込めた暗い空の下には，一面の雪景色がひろがっていました．そして，その雪の間から，青々とした草の芽が，あちこちに覗いていました．正雄は，一人の友人と，雪の中を歩き廻りました（他の友人は，どこかへ行ってしまっていました．）水の無い谷に沿って下り，又丘に上りました．笹の葉が首を出している雪の道を，その友人と一緒に，どこまでも歩いて行きたいと，正雄は思ひました．所が気がついて見ると，いつの間にか，又，あの裏玄関の前に戻って来ているのでした．

そこには，誰もいないと思っていたのに，一人の，下男らしい男が，ぼんやり立っているのでした．そして，正雄たちを見ると，次の様な話をして呉れました．

その家は，昔から，此んなに人気のない，淋しい家ではなかったのです．いつか，正雄達を，連れて入った女の人は，此の家の奥さんなのですが，一人，女の子があり，御主人は，特に，その女の子を可愛がっていました．そのほか，奥さんの，二人の姉妹もその家に住んでいました（でも此の三人の姉妹は，お互に，あまり仲が良くない様に見えたといふことです）．

暖くなり，雪が溶けると，今正雄達が歩き廻った谷の岸に，一むらの百合が芽を出し，透き通る様な，輝く様な黄色い花をつけるのでした（そんな百合の花を，正雄は見たことがないので，本当かしらと，不思議に思ひましたが，黙って聞いていました）．そして，その女の子は，此の百合の花が，特別に好きなので，良く，その傍に来ては遊んでいました．

所が，或る年の夏の，蒸暑い午後，その女の子が，此れと云ふ病気もせずに，突然死んでしまったのです．御主人は，勿論，非常にがっかりしました．……そしてそれから，此の家が，此んなに淋しくなってしまったのです．此の広い家に，いつも，誰かいるのかどうか，それさえも，此の下男は，良く知らないのだそうです．

でも毎年，暖くなり，雪が溶けて，あの黄色い百合の花が開くと，御主人は，それを切って，女の子のお墓に捧げるのです．そして，その前に，その百合の花束を抱えて，永いこと，淋しそうに，谷に沿ひ，又丘に上っては，歩き廻るのでした．……その下男が，御主人を見るのは，一年の中，唯，その時だけなのでした．……

そんな話を，いつまで聞いていたのか，正雄は，良く覚えていません．気がついて見ると，鉄の扉に錠が固く下りているのでした．又此の建物が，消えて失くなってしまふのかしらんと，心残りに思ひながら，正雄は，暗い，灰色の街を歩いて行きました．

その場所を，見失はない様に，今度こそ，忘れてしまはない様に，そこを大きな字で，手帳に書きとめておこうと決心したのでした．正雄さんが，大きな字を書く様になったのは，そのときからなのです．

mai, 26, 1953. Y. T.

駒場の四季*

正　夫　さ　ん　の　話

その頃正夫さんの家は海の近くにありました．海岸は切り立った崖になっていて，波打際に出るには，ゆるい傾斜のある長いトンネルを下りて行かねばなりませんでした．そのトンネルを出た所に平な広い岩があり，お天気の良い日には皆そこに集って，魚釣りなどして遊ぶのでした．

ある晴れ渡った満月の夜，正夫さんは一人でそのトンネルに入って行きました．滑りやすい道を進んで行って，出口から月の光が差し込み，波がキラキラ輝くのが見えて来た所で，壁に一つの横穴があるのに気付きました．入って見るとそれはまた一つのトンネルで，ジメジメした道がどこまでもどこまでも続いていました．やっと出た所は，月に白く光る広い広い砂浜で，舟着場などもあり，少し先には森が暗くかすんで軽いざわめきが聞えて来ました．近くの丘に登って見ると，どうでしょう，その森の中には芝生の張りつめた広場があり，大勢の人々が楽しそうに歌ったり，踊ったり，ギターやアコーデオンをひいたりしているのが見えました．正夫さんはいきなり丘をかけ下りて森の中に飛び込んで行きましたが，森には道もなく，さんざん歩いたあげく出た所はもとの砂浜でした．すっかり疲れて正夫さんは，またあのトンネルを通って家に帰りました．

正夫さんはこのことを誰にも話しませんでしたが，何としてもあの広場に行って見たく

* 東京大学教養学部理科1類8組（1956 年入学）クラス雑誌

てたまりませんでした．然し，いつもトンネルの中は暗く，あの横穴はいくら探しても見つからないのです．所が，ある晴れ渡った満月の夜，仲良しのミエ子さんと一緒に入って行きますと，月の光は出口から差し込み，波はキラキラ輝いて，あの横穴が暗く開いているではありませんか．二人はその長い長いトンネルを歩いて行きました．所がどうでしょう，広い砂浜のある海はひどい嵐で，風は激しく吹きつけ，波はすさまじくさかまいているのでした．あたりは暗く，いつかの森も丘も見わけられませんでしたが，良く良く見廻すと，この嵐の海に，小舟が一艘浮び，灯りを持った男がしきりに海の中をのぞきこんでいるのでした．段々近寄って来るのをよく見ると，乱れた髪を風になびかせ，狂ほしい目附きで何か県命にさがしている様でした．恐ろしくなった二人が，逃げ帰ろうとすると，後ろに白髪のおじいさんが立っていました．そして，舟の男のことなど気にとめる様子もなく，「二人は仲良しだから御褒美を上げましょう．大事に持っているんだよ．」と，美しい飾り紐のついた金のメタルを一つづつ呉れるのでした．二人はお礼をいい，舟の男のことを聞こうとしている間に，おじいさんは消えた様にいなくなってしまいました，が，結局，少し先の漁師の小屋で，親切な漁師から詳しい話を聞くことが出来ました．それはこんな話です．

この近くの森の中に，森の谷間と呼ばれている広場があって，晴れ渡った満月の夜には，辺り一帯の若い恋人達が集り，踊ったり歌ったりして夜を明かすのでした．新しく森に来る恋人達には，森の精が，美しい飾り紐のついた金のメタルを呉れるのでした．（「そのメタルならわたし達もさっき貰ったわ．」とミエ子さんが口をはさみましたが，正夫さんは，シッシッと留めました．）今沖にいるあの男も，そんな夜は，恋人と一緒にその広場で楽しく過していたのです．所がある嵐の夜，この沖で難破した船がありました．若者達は皆海に漕ぎ出し，激しい波や風にもまれながら乗っていた人々を一人残らず救い上げたのですが，その最中に，あの男は金のメタルを海に落してしまったのです．助けられた人々は皆に厚くお礼を云って陸伝いに帰って行きましたし，あの男も，金メタルを落したことを別に気にもとめず，相変らず恋人と仲良くやっていました．所が次の満月の夜，二人して森の谷間に行こうとすると，森の精が二人そろってメタルを持っているのでない人は駄目だといって，どうしても入れて呉れないのです．仕方なしに二人は戻って来て，あんな広場に行かなくたって良いさと互に慰め合いました．然し三ケ月経ち四ケ月経つうちに，空が晴れ渡り満月の輝く夜になると，女は，淋しい顔をして一晩中森の方を見詰めている様になって来たのです．男の慰めの言葉も耳に入らない様子でした．そこで男は，森の精の所に相談に行きました．もともとあのメタルを落したのは悪気があってのことではない．溺れかけた人を助けている中に間違って落してしまったのだから，もう一度新しいのを貰えないだろうかと持ちかけて見たさうです．所が森の精の答は意外でした．あの嵐の夜に助けに行った人は大勢いるが，メタルを落したのはお前だけだ．それは，他の人々はメタルを大切にしていつも首にぶらさげているのに，お前はぞんざいにポケットの中などに押し込んでおくからだ．あの晩に落さなかったとしても，いつかは落してしまったに違いない．そんなやつにもう一度メタルをやるわけには行かないと，剣もほろろの調子だったということです．これを聞いた時から女は次第に男に冷く当る様になりました．しまいには，あなたがメタルを落したのは，もう私を嫌いになったからだなどと駄々をきこね出す仕末でした．そして，空が冷く澄んで霜の一杯下りたある冬の朝に，女は一人で遠くに旅立ってしまいました．それ切り戻って来ませんが，きっとどこかでメタルを持った男でもつかまえたのでしょう．一方置き去りにされた男の方は，嵐の晩になるといつも沖に舟を出しては，落したメタルを探して歩くのです．ホラ，あんな風に，と漁師の指さす方を見ると，さっきの舟はもうかなり沖に出たらしく，灯りだけがボウッとかすんで見えるのでした．正夫さんとミエ子さんは，さっきのメタルを落しては大変と，さっそく首に掛けて，大切そうに両手でおさえました．

さて次の朝，目をさました正夫さんが真先に胸に手を当てて見ると，どうでせう，昨夜しっかりと首に掛けた筈のメタルが，あとかたもなく消えているではありませんか．青くなって蒲団をめくったり，あちこち探し廻りましたが，どうしても見つからないのです．ションボリとしてミエ子さんの所に行き，メタルを落してしまった，もうあの森にも広場にも行けない，嵐の晩には海へ出てメタルをさがし廻らなければならないと云うと，ミエ子さんは，「メタルだの森の広場だの嵐の晩だのって，一体何のお話？，昨夜はお月様がとても明かるかったから，きっと悪い夢でも見たんでしよう」と，笑って取り合いませんでした．

その後，晴れ渡った満月の夜，月の光がトンネルの出口から差し込み，波がキラキラ輝くのが見える時でも，正夫さんは，あのトンネルの壁の横穴をどうしても見付けることが出来ませんでした．僕がメタルを落したとき，ミエ子があんなことを云って冷やかしたものだから，きっと森の精を怒らせてしまったに違いない．だからもう横穴がみつからなくなってしまったのだ．正夫さんはついこの間までそう考えていました．

（駒場の四季，1957 年 11 月）

講演記録

現代数学の性格について

現在，数学で何をやっているかということを説明するのはむずかしい．何の学問でももちろん何をやっているかを本当に説明するのはむずかしいわけですが，たとえば物理学でいえば，原子核なら原子核をやっている，といわれれば，もちろんだれも原子核というものを見たことはないわけですが，何をやっているか大体わかったような気になる．原子核のどういう問題をやっているのかはわからなくても，考えている対象は大体見当がつく．同じようにして数学でこういうものをやっているといっても，それが実際何のことなのかさっぱり見当がつかない．数学の対象自身を説明するのはむずかしい．だから現代数学とはどういうものか，話をするのは非常にむずかしいのです．数学の対象というものは，ときによって変ってくる．始めは数とか図形とかいうものが対象だった．そのころは数学で何をやっているか，誰にもよくわかったわけです．ところが対象が高度になって来ると，数と数との関係，函数とか，図形の運動とかを対象にする．現在ではそれを更に抽象化している．というように，数学の対象というものはときによってどんどん変って行く．変って行くというのは，今までの対象を考えなくなったという意味ではなくて，対象がだんだん高度になっていく，概念化され，抽象化された形の中に今までの対象も組み込まれて行く，それだけ視野も広まり理論も強力になるのですが，対象そのものはだんだんわかりずらくなってくる．そして現在何をやっているのか普通の人にはさっぱりわからないということになるのです．だから例をあげて，それがどういうものかということを簡単な場合に説明しようとすると，たとえば円を廻転させるとか何とか，非常につまらないことになる．子供だましみたいなことになってしまうのです．

そこで一般的に現代数学ということを話すことはやめにして，一つの問題について，実際に数学というものはこういうふうに抽象化されるのだということを話してみることにします．抽象化する場合に‘ただ単におもしろ半分に抽象化するのじゃないことはもちろんで，数学にはいろいろな分野がある．そのいくつかの分野の間に共通な性質があるとき，もう一つ高い概念を作り上げてこのような共通な性質を一つにまとめる．そこに抽象化ということが起るわけです．そういうやり方によって，多くの方法で今までばらばらに考えられていたことを一まとめにすることができるし，また単純化されるわけですから，方法自身もさらに深められる．その深められた方法を始めの分野に使うことができる．そうやって数学はだんだんと発展していくのです．

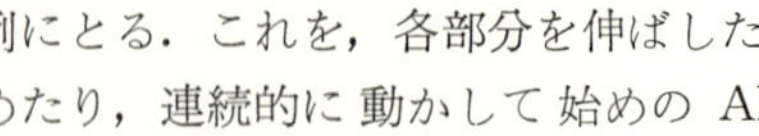

図 1

一つの例として不動点定理というものについて説明します．これはどういうものか，一番簡単な例でいうと，図1のようにゴムひもを例にとる．これを，各部分を伸ばしたりちぢめたり，連続的に動かして始めの AB に重ねる．一般的にいえば動かす前と後とで，各部分はズレるわけですが，前と後とで変らない点が出てくる．動かした後でも両端はやはり両端の位置にあるわけですから，A の端が A の端に，B の端が B の端に重なるとすれば，少くとも A, B 二つの点は動かない．一方ひっくり返して，A を B に，B を A に重ねると，途中のどこかに動かない点がでてくる．たとえば伸ばしもちぢめもしないでひっくり返せば真中の点が動かないことになる．このように，連続的に変形して重ねるときに動かない点があることがある．この動かない点を不動点といいます（図 2）．不動点がいくつあるかは変形の仕方で変ってきますが，この線分の場合には，どんなふうに変形して重ねても少くとも一つは動かない点がでてくる，ということはすぐわかります．これは円板でも同じです．たとえば円板を廻転させれば，中心が不動点になる．ところが円周にするとそうでなくなる．円周をたとえば中心のまわりに 90 度廻転させると，やはり始めの円周に重なりますが，このとき円周の上の点は全部一様に動きます（図 3）．

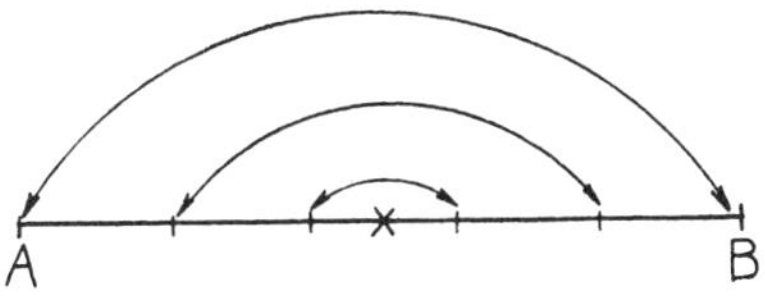

図　2

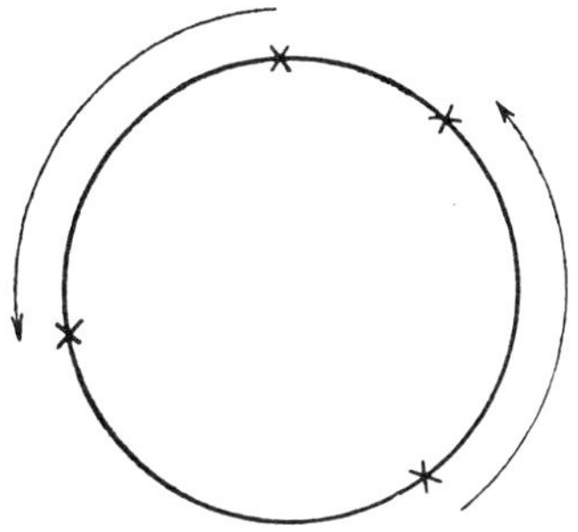

図　3

つまり不動点は一つもなくて，みんな動いてしまう．このように，どういうふうに動かしても不動点のでてくる図形と，動かし方によっては不動点がでてこない図形とがある．この二つの図形では，ほかのいろいろな性質も違ってくる．不動点がいくつあるかは計算できます．このように，不動点があるかないかとか，またはその数の計算の仕方とかをきめる定理を不動点定理といいます．これだけいうと，まったく子供だましみたいな話になる．しかし，これと同じ考え方はいろいろな場合に使うことができるのです．その一つの例として不動点定理を微分方程式に応用することができます．

微分方程式では，始めはその方程式を解いて答がどうなるかということを考えたわけです．ところが，まあ計算して解ければよいのですけれども，解けない方程式が多い．その解けないというのも，数学者が頭が悪いから解けないのではなくて，その解が，始めから簡単な初等函数では表せないようなものなのです．そこで，解けない方程式は考えないということにすれば簡単ですけれども，それではすまされない．物理などに出てくる微分方程式でも，解けないものが多いのです．だから最後には，どういうふうにして解くか，つまり答を式で表すにはどうするか，ということは考えないで，その答の性質がどのようなものであるかを考えるようになる．答を式で表さなくても大体の性質がわかることが多い．たとえば微分方程式の答をグラフで表すと，図 4 のようないろいろな場合が考えられる．つまり x が無限大になるときは，答のグラフが無限大になるとか，一定の値に近づくとか，振動して一定の値には近づかないとか，このような性質をしらべるには，必ずしも答が式で表されている必要はない．微分方程式が解けなくても，また答を計算して出さなくても，方程式の形から適当な方法で考え

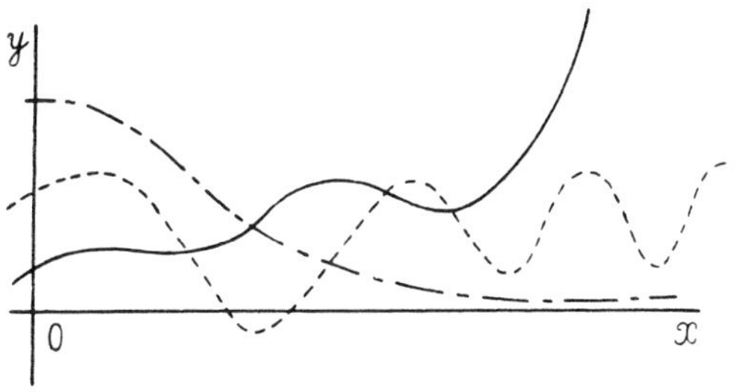

図　4

ればわかることが多い．そしてこういう定性的な性質も，非常に役に立つものなのです．こういうことで，微分方程式の答を計算して出さなくても，答の性質がこうなるとか，ああなるとか，大体のことはわかる．そうして今度は，ただ計算して答を出すのではなくて，微分方程式の解そのものを対象にして考えることになる，その解の性質を考える．そんなふうに変ってくるのです．このような問題を考えるとき，この微分方程式の解で，これこれの性質を持つものがあれば，その解は更にもう一つの性質も持っている，という形の議論が多い．たとえば，グラフが原点を通る解があれば，その解はxが無限大になるとき0に収束する，という形のものです．ところがこれだけでは，そのような性質を持つ解がそもそもあるかないか，ということは全然わからない，ただ，もしあればこうなるというだけです．そこでどんな条件のときに，ある性質を持つ解があるか，あるいはないか，ということを考えることが問題になる．

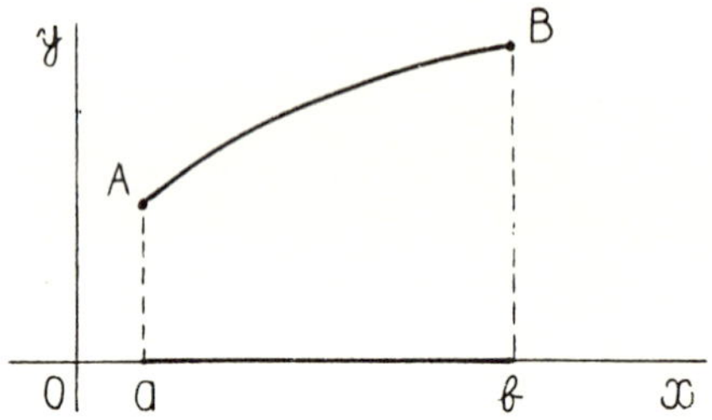

図　5

このような形の定理を存在定理といいます．つまり解の存在を保証する定理です．存在定理にはいろいろ大切な役割りがありますが，少くとも，答を計算しないで解の性質を考えようとするときにはなくてはならないもので，微分方程式の理論の中で一つの中心的な問題になってきているわけです．一番簡単な例でいえば，ある方程式の解で，そのグラフが，図5のように，二つの点 A, B を通るものがあるかどうか，こういう問題がありま す．これは境界値問題といわれています．つまり，考えているxの範囲の境界の a, b 点で，解の値を指定して，そのような値を持つ解があるかないかを考えるという意味です．

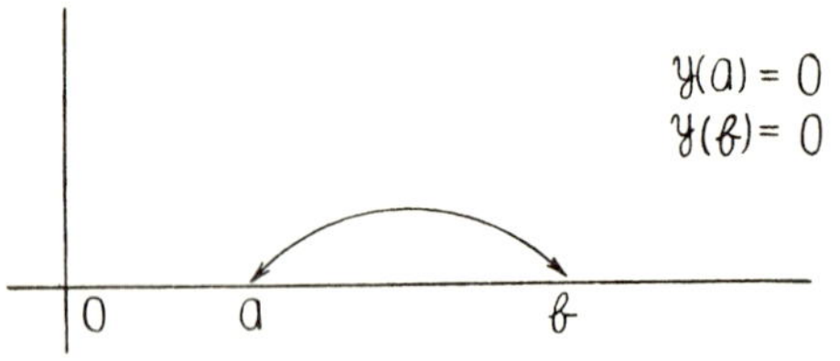

図　6

こういう問題を考えるときに，前の不動点定理というものを応用することができる．簡単な問題として次のような方程式を考えてみます．

$$(1)\qquad \frac{d^2y}{dx^2}=F\left(x, y, \frac{dy}{dx}\right)$$

この方程式の答で，aにおいてゼロになり，bにおいてもゼロになる，つまり図6の点 a, bを通るような答があるかどうか，あるいはa, b がどのような範囲にあるときにそういう答が存在するか，境界値問題の典型的な例としてこういう問題を考える．この問題と不動点定理と，ちょっと見た場合に，何も関係はなさそうに見える．しかしこれは次のようにして関係づけられるのです．この方程式を変換して見ると，次のように書き表すことができます．つまり，

$$\begin{cases} y(x)=-\displaystyle\int_a^b H(x,s)F(s,y(s),z(s)\ \ )ds \\ z(x)=-\displaystyle\int_a^b \frac{\partial H(x,s)}{\partial x}F(s,y(s),z(s))ds \end{cases}$$

この方程式を満すyが始めの方程式の解になり，zはdy/dxになります．ここでHはaとbとできまる函数ですが，その具体的な形は差し当り書く必要はない．ですから，始めの方程式の境界値問題を解くということは，この連立方程式を解くということになります．ここで注意すべきことは，y, z が左辺

にも右辺にもある．だからこう変形しても，ちょっと見た所やはり手のつけようがない問題に見えます．しかしこういう考え方をする．y と z が両辺にあるからむずかしくなる．そこで

$$(2)\begin{cases} -\int_a^b H(x,s)F(s,y,z)ds=u(x) \\ -\int_a^b \frac{\partial H(x,s)}{\partial x}F(s,y,z)ds=v(x) \end{cases}$$

とおくと，問題は

$$(3)\begin{cases} y(x)=u(x) \\ z(x)=v(x) \end{cases}$$

となります．こうすると，y, z という函数をきめれば上の式 (2) から u, v という函数がきまるわけです．そしてこの u, v が始めの y, z に等しいような y, z が答になる．

ここで，y, z をある大きな枠の中に入れて考える．つまり，a と b との間で連続な函数の全体をひとまとめにしてそれを一つの空間と考える．函数を空間の点と考えるわけです．この空間に遠いとか近いとかいう関係，つまり距離を入れる．どういう距離を入れるかというと，二つの函数があった場合に，そのグラフがお互いに近い所にあるときは近いと考え，グラフが離れているときには遠いと考える (図 7)．こうして，函数の間に遠い近いの関係が入り，この関係を適当に式で表して，函数の作る空間に距離が入る．こういう空間を函数空間といいます．空間といってもこれは目で見るわけにはいかない．しかし数学者は目で見ることのできないような空間も考えるのです．その次に，二つの函数を並べたものを一つの点とするような空間を考える．そしてこの空間にも前のようにして距離を入れます．そのとき，今考えている函数 y, z の組 (y, z) はこの空間の点になる．そして，y, z から (2) の式できまる函数 (u, v) もこの空間の点になります．そこで，(y, z) に対してこの (u, v) を対応させる．

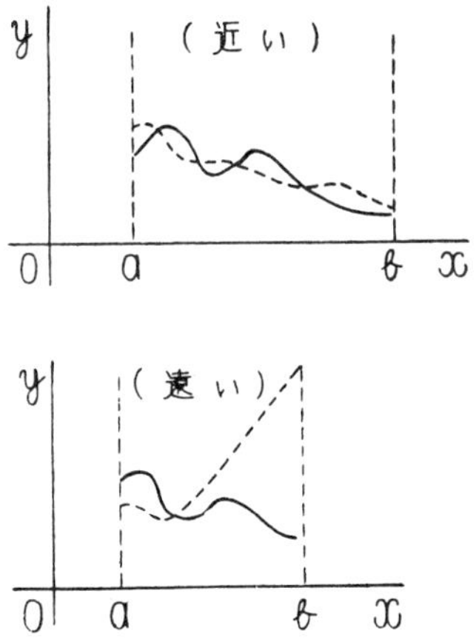

図　7

$$(y, z) \to (u, v)$$

この対応は，空間の中の運動になる．つまり一つの点 (y, z) が (u, v) に移るような運動になります．この運動は連続的である．つまり，前にいった意味で互いに近い二つの函数は，やはり互いに近い函数に移ります．これはすぐわかる．こう考えるときに問題は (u, v) が (y, z) に等しいような点，つまりこの運動で動かない点，不動点を求めるということになります．こういうふうにして，始めの微分方程式の答で a, b を通るものがあるかないかということは，この連続的な変換の不動点があるかないかという問題になるわけで，さっきの不動点定理と同じ考え方の問題になります．このように，不動点定理は微分方程式の問題にも出てくる．しかし空間の構造がちがうわけです．だから前の定理をそのまま使うわけにはいかない．しかし，ほとんど同じようなやり方でこの場合にも不動点定理を証明することができるのです．

こういうふうにして，不動点定理を使って微分方程式の理論を考えたのは 1930 年代のことで，日本の数学者の福原さんや南雲さんなどもその研究の一つの中心になっていたのです．もちろん世界的に，そのほかたくさんの人が研究しています．

さて，このように，普通の図形と函数空間との間にはいろいろ似た性質がある．両方と

もその点の間には，遠い近いという関係がきめられているし，また不動点定理というものを考えることができる．そこで，函数空間だとか，円や線分だとかに共通な性質をとり出して，一般的な概念を作り，今までの空間や図形はその特別な場合になってしまうということが考えられる．つまり抽象化するわけです．この抽象化したもので不動点定理が考えられれば，それは図形にも函数空間にも両方に使えるということになります．どのように抽象化するのかというと，余計なものは全部解消してしまって，遠い近いという関係だけを残す．つまり，二つの点の間に遠い近いという関係だけしかないような空間を考える．このような空間を位相空間といいます．今までの空間は全部位相空間の中に含まれてしまいます．そしてこの位相空間で不動点定理を考えることができる．こんなふうに，数学の対象，数学の問題というものは段々一般化されていくのです．ところでこの位相空間というものをちゃんと説明するのはむずかしい．もちろんちゃんとした定義があるわけですけれども，その定義を書いて見ても，それを読んだだけでは何のことかさっぱりわからないでしょう．しかし，こういうふうにして出てきたものだと説明すれば，大へんわかりやすくなります．さて，位相空間というものを作って見ると，今のような問題に使えるだけでなく，解析学のいろいろなほかの問題もこのような考え方に従って解くことができるのです．

ここで不動点定理というものをもう一度考えてみる．今まではすべて遠い近いという関係が基本的だったわけで，これがなければ連続ということも考えられない．ところが，そうじゃなくて，これをもう少し広い範囲で考える必要が起きてくる．それはどういう場合かというと整数論で次のようなことが問題になっている．たとえば，

$$x^n+y^n=z^n$$

という方程式の解で，x, y, z が整数になるものを求めるという問題がある．これはむずかしい問題ですが，簡単なものではこういう問題がある．

$F(x, y)=0$，(または $G(x, y, z)=0$，等々) という方程式を考える．これを実数の範囲で解くのではおもしろくないわけですけれども，別な範囲で考える．たとえば整数の範囲で考える．しかしここでは，有限の範囲で考える．つまり有限体というものの中で考える．この有限体というものの説明は面倒ですから省略しますけれども，有限個のものの間に，普通の数と同じように，加，減，乗，除がみんなできる．そういうものです．この有限体の中で，上の方程式の答がいくつあるかということを考える．x，y は有限個の値しか取れないわけですから答も有限個ですが，この答がいくつあるかを考える．そして，段々大きい有限体を取っていった場合に，この答の個数も段々増えて行くわけですが，どのような割合で増えて行くかを考える．これも非常にむずかしい問題です．ちょっと見ると余り意味のない問題のように見えますが，これは，ゼータ函数というものに関係しているので，整数論の非常に深い所に関係してくるのです．そしてこの問題を考えるために，いろいろな数学の理論が使われる．そしてこの問題に使われることによってそういう理論そのものが発展していく．そういう意味で数学の発展に非常に役に立つ問題の一つです．この問題を考える場合にやはり不動点定理が出てくる．それはどういう形で出てくるか．x，y の範囲を非常に大きく取れば，問題の方程式はこの大きい範囲の中で無限に多くの答を持っている．この答になる (x, y) に適当な変換をやる．そのとき，この変換で動かない x，y が，始めに考えている有限体の中に入るということがわかるのです．そこで有限体の中での答の個数は，この変換の不動点の個数と同じになる．そして有限体を段々大きく

して行けば変換は段々複雑になって，不動点の個数も段々増えて行くのです．そこでこの不動点の個数が，一つの公式で計算できれば，この問題は解けるわけです．

こんなふうにしてやはり不動点ということが問題になるのです．所で，普通の図形，または位相空間のときでも，適当に条件をつければ，不動点の個数を計算する公式は求まっています．しかしこの公式をそっくりそのままこの場合に使うわけにはいかない．どうして使えないかというと，有限体ですから，有限個の点しかない，だから遠いとか近いとかいうことを考えても意味がないわけです．つまりこの場合には遠い近いという関係は考えられない．しかも，どうしてもさっきの不動点定理というものを考える必要がある．所が不動点定理というのは連続的な運動を問題にするのですから，遠い近いという関係は本質的なものなのです．そこでどうするかというと，今度は遠い近いという関係はやめにしてしまって，空間の別な性質に目をつける．つまり不動点の個数を計算する公式に出て来る因子に目をつけて，その意味を考えて見る．そうすると，この因子は，必ずしも遠い近いという関係がなくても考えられる対象に関係しているのです．そこでこの対象と同じ意味のものを今の場合にも考えればそれを使って不動点の個数を計算できるだろうと考えられるわけです．このように，抽象の仕方はいろいろあって，問題に応じていろいろ考えるのです．ところで，今の問題では，この方法はまだうまく行っていない．色々な人が考えているのですが，まだできていないのです．もちろん方程式の形が特別な場合には，こんなふうに一般的に考えなくてもできる．とくに，二変数の場合，つまり $F(x,y)=0$ という形の式のときにはできている．然し三変数，

$$G(x,y,z)=0$$

となると，もうわからない，という状態です．

ここでもう一度，一般的な話に戻って，現代の数学の性格というものを考えて見ます．さっきも云ったように，現代の数学では，いろいろなものの間に共通な性質を抽象して一般的な概念を作る．そしてその一般的なものについていろいろな定理を証明していく．たとえば遠い近いという関係を取り出して位相空間という概念を作り，それに対して不動点定理を証明する．そのとき位相空間というのは具体的にどういうものであるか．そういうことは考えない．それは函数の空間であるかもしれないし，円や線分であるかもしれないし，またほかのものであるかもしれない．そんなことは問題にしないで，位相空間を定義する性質だけに基づいて考えるのです．つまりあるものがこれこれの性質を持つとすればそれについてこれこれの定理が成り立つ，という形になる．この性質のとり方をいろいろに変えればそれに応じていろいろ変った対象が考えられることになり，いろいろ変った定理が出てくる．つまり，

$$If\cdots,\ \text{then}\cdots.$$

という形が現代数学で普通の形です．この $If\cdots$，という部分を公理といいます．つまり位相空間では，遠い近いという関係を厳密に数学的に書き表したものが公理となっているわけです．この公理を満すものは何であっても，then 以下の定理が成り立つ．所でこの公理，つまり抽象的な概念を定義する性質は，普通の場合にはいろいろな具体的な対象から取り出したものですが，極端にいえば，何を持って来ても差し支えない．論理的に矛盾していなければどんなことを考えても数学になる，という議論も成り立つのです．ところが実際に，思いつきの性質を持ってきて公理を作って，こういう性質があればこういう定理が成り立つ，そういっても誰からも問題にされない．だから，形式的には，If の次につまり公理として何を持ってきてもよい．し

かし実際にはそうはいかない．公理としてどういうものを取れば一番有効な理論ができるか．たとえばさっき云った通り遠い近いという性質から非常に有効な理論が出て来たというようなものです．そのために勝手なものではだめで，これがいいだろう，という性質に目をつけなければならない．それからまた，抽象化し公理を作るだけでは意味がない．それを具体的な問題に応用して行かなければならない．位相空間の理論が発達したらそれをまた微分方程式の理論に応用する，そしてこの応用によって更に高度の位相空間の理論が発展する．というように，抽象，一般化と，応用とは，お互いに作用し合いながら前進して行く．ここで応用というのは微分方程式や応用数学だけではない．さっき云った通り，整数論にも応用されて，それによって位相空間の理論やその他の抽象化された数学が更に深められ，進歩して行くのです．このように，広い意味での応用ということが，ある理論，ある公理が良いものであるか，思いつきだけのつまらないものであるかを見分ける一つの目安になる．だいたい，数学の理論というのは勝手に作られていくのではありません．何をやるか，大体の目安がある．たとえば応用数学なら，実際に応用して有効に使えるかどうか，ということが目安になる．ところで，応用されない数学，たとえば整数論などでは，美的感覚とでもいうものが一つの目安になる．言葉がちょっと適当でないかも知れませんが，何かそんなものがある．これは深い理論だとか，壮大理論だとか，美しい理論だとか，そんな感覚を誰でも持っている，この感覚に従って理論を進めていく．また，今まである数学の分野を統一する，それらの間の関係を説明する，たとえば位相空間によっていろいろな分野の問題がひとまとめに考えられる，そういう理論もよい理論である．こんなふうに，数学をやるときの目安はさまざまですが，実は見かけほどさまざまではない．数学者の頭の中では，このようないろいろな基準が，大体一つの感覚にまとめられていて，これはよい，これはつまらない，ということが直覚的にわかる．また，数学の発展の歴史を見てもそういうふうになっている．たとえば，美的感覚から見てすばらしいが，応用には関係のないように見えた理論が，思いがけない方法で応用されるとか，逆に応用上有効だが数学的にはごたごたしてあいまいだった理論が，手を加えられて美的感覚から見てもすぐれたものに発展するとか，そのとき同時に他の分野の理論をも統一するような概念が生れるとか，そういう例はいろいろある．ただ，人により，また場合によって，応用が強調されるとか，統一性が強調されるとか，美的感覚が強調されるとか，いろいろなニュアンスが出てくるのです．このような歴史の教訓と現実の状態から，数学者の中には一つの信念とでもいうものができている．それは，「数学は一つだ」ということ，つまり，いろいろな分野に分れているように見えても，数学というものは一つの統一されたものであって全体としてひとまとまりになって発展していくものだということです．分化と統一とを繰り返しながら，全体として伸びて行く．だから，たとえば，微分方程式を発展させようとして，そればかりやっていても，位相空間をやらなければ，ある所で行きづまる，ということになる．数学の応用を考えても，応用に役立つことばかりやればいいと考えてそればかりやっていると，数学の発展が片わになり，行きづまり，貧困になる．一つのものを解明するために，いろいろな分野が使われることはよくあるし，逆に一つのものを深く考えることによって，いろいろな分野が刺激を受けて発展することも多い．いつでも全体的な見方に立たなければならない．数学全体を高めることを目標にしなければならないのです．

現代数学とはそのようなものなのです．だ

から，ちょっと見て何をやっているかわからないからとか，実用になりそうもないからといってある部門を軽視することはいけない．実際，少しちゃんとした国ではどこでも純粋数学と応用数学と，両方に力を入れてやっている．アメリカもソビエトもみなそうです．

とにかくも，こんなふうにして現代数学というものはだんだんと発展して行くのであります．

(1957年11月第8回駒場祭における講演，雑誌「駒場」掲載，1958 年 3 月)

第三部

書　　簡

松本皓一　へ

1946 年 5 月 16 日
埼玉県　騎西　より　(はがき)—原文縦書き

御手紙有難う．

僕もとうとう休み一杯御無沙汰してしまって．

四月の三日に発熱して，（二月の過労が悪かったやうです.）今に至るまで，一寸も良くならず，今だに病床に呻吟して居る．右の肺尖が悪い様です．そんなわけで当分学校へも出られまいと思って居ります．今度の休みぐらゐつまらない休はなかった．試験で無理したのもいけなかったのでせう．

君も盲腸にかかった相ですが，もう良くなられたのでせう．

近藤は，例の病気が再発して一ケ月余り病床に伏して居るとか云って来ましたが，彼も暫く学校へ出られない相です．

病気の話しばかりですが，でも僕のは，そんなに悪くはなってゐない様ですから，その中良くなると思ひます．近藤はどんな工合か，手紙を見ただけでははっきりしません．では，おからだをお大事に．さやうなら．

平岡仙之助　へ

1947 年 3 月 16 日
埼玉県　騎西　より　(封書)—原文縦書き

君から屢々お手紙をいただきながら，すっかり御無沙汰してしまった僕の怠慢をお許し下さい．僕は今，此の癒し難い怠慢のために苦しめられて居ります．僕の病気は又近頃，悪くなって閉口して居ります．でも，もうそろそろ良くなるつもりです．今度こそは完全になほしてしまはうと思って居ます．そして六月頃には学校に行ける様にと.（四月から出るのは一寸無理ですから）

もう試験も終り，君も家でのんきに暮して居る事でせうが，（もしそうでなかったら失礼！）　それにしても，去年のことが思ひ出されます．あれからもう一年になるのですね．長い長い一年，それで居て，いつの間にか過ぎ去ってしまった様な此の一年間に，僕は，自分自身が全く変ってしまった様な気がします．どう変ったのかは，はっきり解らないし，進歩か退歩かも，判然としないけれども，何かしら僕の中に変化が起きた事は事実です．それは，終戦の頃から芽ぐんで居たのかもしれませんが，此の一年間の，暗い環境の中で，歪んだ成長をしたのではないかと心配して居ります．索索然とし漠漠然とした療養生活の中に，正しく強いものが育つなどとは考へられませんからね．

然し何かしら変化が起きたのは事実です．尤もさうでなくても一向構ひませんし，寧ろさうでない方が良いのかも知れませんが．

病床から眺めると，春の暖かい光に，草々の緑が，すきとほる様です，そして時折吹き込んで来る風は，明るい春の気分を伝へます．此んな時に寝て居るなんて馬鹿々々しいのですが．

　　春は再び来つれども
　　などて我が春　来らざる‥‥

何かの本で此んな書き出しの詩を読んだ様に覚えて居ます．それは又，僕自身の気持でもありませう．ただ僕は今，精神が分裂して居るので，その様な，突きつめた，センチメンタルな気分に浸り切れないのです．精神の分裂は不愉快です．そして病気にも良くないのです．先づ此の方を癒すのが先決問題でせう．今何を読んで居りますか．僕は又チェホフに引かれて居ましたが，僕の元気を失はせるので止めてしまいました．今は何も読んで居りません．しばらく，読書は中止して，気楽に暮さうと思って居ります．

好い気になって勝手なことばかり書き並べてしまいましたが，お許し下さい．もう此の辺で止める事にします．

三月十六日，夜．乱筆乱文多謝．

追伸，新井君に会ったら，よろしく伝へて下さい．

平岡仙之助　へ

1947 年 9 月 6 日

埼玉県　騎西　より　(封書)—原文縦書き

先日は，わざわざ見舞に来て下さり，その上結構なお見舞までいただいて，有難う．お陰様で，僕は，今月の三日に退院した．丁度，一ケ月，病院に居たわけだ．此れからしばらく家で静養して，六月の末頃から学校に出ようと思ふ．実際の処，家で療養するのは，.退屈で，馬鹿馬鹿しいものだが，無理は出来ない．学校に出ても，当分は，何もせずに，ぶらぶらして居なくてはならないだろう．さう考へると憂鬱だが，学校へ出られるまでに良くなっただけでも，良かったと思ふ．君が来て数日してから新井が来た．そして君達二人とも，大して変って居ない様に思はれた．或は大いに変って居るのだが，あの場合に，その変った処を発揮出来なかっただけなのかも知れない．一年半の間には，それだけ生長するものだから，それに，ああ云ふ場合には，人の個性が強く表はれて，内面的なものは，表面に出ないものだから．

併し人間と云ふものは，さう簡単に変るものではない．その思想に変遷はあっても，人間の性格と云ふものは，中々変るものではない様だ．僕は一年半，世間と絶縁して，唯新聞に依って，世の推移を知って来た．だが，病院に入って，色々な人に接して見ると，新聞に依って築き上げて居た僕の幻想は，簡単に崩れてしまった．新聞に表はれるのは，日本人の思想である．実際に接して知るのは，人々の性格である．僕も日本人の思想が変ったのは疑はないが，その性格は，少しも変って居ない．又さう簡単に変るはずのものでもない．

今日は此れで失礼する．又僕が学校へ出る様になったら，又ゆっくり話しをしよう．

さやうなら

平岡仙之助　へ

1947 年 9 月 21 日

埼玉県　騎西　より　(はがき，速達)—原文縦書き

その後如何ですか．

試験は何時から始まるか，判ったら至急お知らせ下さい．水害で鉄道事情が悪い為当分学校へ出られません．（勿論僕の家の辺は水害はありませんでした．試験の時は出ます．）

頸の手術の痕は中々良く成りません．

綿貫は未だ休んで居りますか，心配して居ります．新井に出した手紙が「受取人さがし当らず」で戻って来ました．住所が変ったのですか，新井にもよろしく伝へて下さい．

以上お願ひまで．

松本皓一　へ

1948 年 4 月 14 日

浦高　武原寮　より　(はがき)　原文縦書き

永らく御無沙汰しました．その後もお元気の事と思ひます．

また遂に本格的な春になってしまひました．外を歩くと美しい若葉，色とりどりの花々，華麗な，或は可愛らしい，が心を楽しませて呉れます．然し，その春の歓びの中に，心から飛び込んで行く事の出来ない僕なのです．もう，そんな風になってから，何年に成るで

せう，それよりも，暗い部屋で，静かに勉強して居る時に，しみじみとした喜びを感ずる事が多いのです．勿論，外の緑色の明るさ，風の爽やかさは僕を誘惑せずには置きません．然し，僕はいつも失望して帰って来るのです．何故此んな事を書いたのでせう．

明日は新入生の入寮式です．又明日から新しい生活が始まります．かうして，武原寮には，過去二十年間に，どれだけの多くの夢が花咲き，埋もれて行った事でせう．夢は人間を欺く物かも知れません．然し僕の様に，普通の意味での夢と憧憬とを持たない生活は淋しいものです．

僕と君とは，とうとう親しく交際する暇がありませんでした．そしてもう君は卒業してしまはれましたが，然し僕はやはり君と永くつき会って行きたいと思ひます．友人の非常に少い僕の事ですから．

此の一年間は君に取って，暗い陰鬱な一年間かも知れません．然し，真剣に生きるとき，その暗さの中から何物かが生れて来ると信じます．御健闘を祈ります．

松本皓一　へ

1949 年 1 月 1 日
浦高　武原　寮より　(封書)—原文縦書き

先づ何よりも，「新年おめでたう」と云はねばなりません．

御無沙汰のお詫びは後廻しにして，新しい一年が，君にとって，君のより良い生活への為の一年である様に，そして結局，君にとって，より良い一年である様に．

御元気の事と思ひます．そして，勉強に精出して居られる事を，

僕は相変らず，相変らず，何時になっても相変らずで，相変るのは，病気をする時位なものです．

人気の無い寮に正月を迎へて，陰気なお天気の為に，およそ正月らしく無い正月，でも，此れが，此処数年間で，僕がまともに迎へる事の出来た最初の正月です．

そんな訳で，僕としては，大しておめでたいわけでも無く，さりとてつまらなくもなく，部屋を掃除して，プラトンの「シンポジオン」と，チェホフの「かもめ」を読んで，いささか妙な気分に成って居る所です．

日常生活の細々した事に気を取られ，それ等をうまくやって行かうと考へて居ると，もっと大切なこと，もっと重要なことを考へる暇が無くなってしまって，習慣の安易な繁雑さの中に自己を失い，月並な，卑俗な生活に埋れる様になる．かって頭の中に存在した些細な事が，批判されずに，そのまま居据り，存在権を主張する．かうして，まともに考へたら考へ得ない様な考へを，人は持つに至るのです．細かい事に損をすまい，うまくやらうと考へて，大事な生活を失ふに至る．僕にはそれが恐ろしい．かうして何かの機会に，自己の喪失に，転落に気附いて，愕然とする．自分は今まで生きて居たのではない，頭に蓋をかぶせられて，何物かに，引きずり廻されて居ただけだと感ずるのです．

大切な事と，さうで無い事とを区別しなければならない，さうして常に批判的で無ければ，‥‥さう考へるのは容易する事なのですが‥‥．

此れは年頭の感ではなくて，年末の思ひ，さうして，その原因となった他の事情も手伝って，僕は眠り少き一夜を明した．それが昨年．昨夜はぐっすり眠って，此れが僕のお書き初め．

かう云ふ事は此の辺で切り上げて，いろいろお話ししたい事もありますが，僕は書くのが下手だから，（それに字は一層下手！）而も手紙と云ふものは，時には便利な事もありますが，何と云っても，紙の切れ端しに過ぎませんから．

もしお暇だったら，（正月早々から勉強しては毒ですよ．） 遊びにいらっしゃい．僕は

いつも寮に居ますから．（午後は外へ出掛ける事もありますが）部屋は北二の十六室，下の一番端です．何もありませんが，ゆっくり話す事は出来ませう．一週間程前，綿貫が遊びに来ました．従って当然，今度は君の来る番ですよ．休みは十六日までです，（その頃僕は家へ帰るかも知れません．）待って居ます．

環境の力は恐ろしい，何故なら人は環境に働きかけるものだから，そして働きかける事に依って，その人自身が変へられて行くのだから，勿論受動的な影響もあります．かうして，人は過去から切り離されて行く，過ぎ去った一切が夢の様に思はれて来る．さうしてその夢を通してのみ，過去の人々と接触すると，現在生きて居る人をも，ある種のヴェールを透してしか見る事が出来なくなる．さうしてその影も，深い靄の中に，段々と薄れて行く，人間の弱さ，はかなさ．

生きなければならない．常に前進しなければ．そこに傾向的な同一が生じませう．

だが過去は消え去らない．何物も，空しく過ぎ去る事は無い．

その時の気分に従って，人はああも考へ，かうも考へる．

今日は此んな事しか書けません．何か知ら僕を落ち着かせないものがある．

此れは，手紙ではなくて，まるで独り言の様です．

春・夏・秋・冬．いつもそれぞれに美しい．それなのに，今は，夏を，春を，はっきり思ひ出すことが出来ない．

Amo ergo sum．おかしいですか？

勉強しないで居ると，妙な事ばかり考へる．

勝手な事ばかり書いて失礼しました．

何よりもお身体を大切に．

さやうなら．

一月一日夜

松本皓一　へ

1949 年 4 月 15 日

浦高　武原寮　より　(はがき)—原文縦書き

合格おめでたう．心からお祝ひします．

もうそろそろ学校も始まる頃かと思ひますが，どうして居られますか．

暖い春と共に，君の暗い生活も，終った訳ですね．本当に，僕としても嬉しく思ひます．此の上も，力強く進まれる様に．

今度，西三の三室に引越しました．北寮は閉鎖です．一度，君の家へ遊びに行かうと思って居たのですが，色々な事情で行けなくなりましたから，一度，寮へ遊びにいらっしゃい．今度は，右の六畳は，僕一人ですからゆっくり出来るでせう．

僕は近頃，少し変調で，どうにも成りません．今年も，春が徒らに過ぎて行く，と云ふより，何か，生活の支へを失ってしまった様な，気がします．毎日が無意義です．何をしたら良いのか．どうしたら良いのか，皆目解りません．試験勉強は，馬鹿らしくて，する気に成れません．本を読めば，やるせない気持に成るばかりです．

さなうなら．

松本皓一　へ

1949 年 6 月 4 日

浦高　武原寮　より　(はがき)—原文縦書き

暫く御無沙汰しました．相変らず御元気な事と思ひます．

幾度か寮に来られたさうですが，その度に僕が居なくて会へなかったのは残念でした．一日に，宮川にさそはれて，（君も来ると云ふ事でしたので）東大の映画会へ行く予定でしたが，此れも他に用事が出来たため，行けませんでした．と云ふのは，午後に拡大自治委員会があったためです．二日には生徒大会でスト決議．今日は二十四時間スト．行動隊として東大へ行きましたが文学部へ寄るひまはありませんでした．それに，君が居るかどうかも解らないし．

此処数日，会議会議で，それに今日は寮へ帰って来たのが八時．もうクタクタです，然し，今度だけは，どうしてもやり通さなければならないと思ひます．問題が余りに大き過ぎるのです．単なる個々の事件を超えた或ものがあるのです．僕は理論的に，マルキシズムに賛成出来ませんが，行為的にはマルキスト達に同調して居ます．多少の相違を云ふよりは，何としても此処で，戦ひ抜かなければならないのです．日本の将来を左右する大きな問題ではないかと思ひます．幾分興奮して居ますので，妙な手紙に成ってしまひました．

谷山清司　へ

1951 年 2 月 10 日
上野桜木町　より　（はがき）

谷中デ，ズボンヲ買ッタカラ来テ下サイトノコトデス．

試験ハ 28/2　1/8 及ビ 9/3．ソノ近クノ日ニ来ルノハ，ナルベク避ケテ下サイ．

他ニ，別ニ変ッタコトモアリマセン．モシアッタトシテモ，ソレニ気ガ附カナイカ，又ハ変ッタコトトシテ感ジナイノカモシレマセン．

谷山清司　へ

1951 年 4 月 3 日
上野桜木町　より　（はがき）

御手紙拝見．

何モ変ッタコトハアリマセン．

暇ガナクテ困ッテイマス．

住所ハ 7 月頃マデ変ラナイデセウ．尤モ，近クニ良イ下宿ガ見ツカレバ別デス．

本ハ，名前ヲ知ラセテ呉レレバオ送リシマス．金ガナイタメ，新シイ本ハアマリ買ヒマセン．

16 日カラ講義ガ始マリマス．午後ニ講義ガアルノハ週ニ 1 日ト 1/4 ダケ．割合ニ自由ニ勉強デキルデセウ．デハ又

及川広太郎　へ

1951 年 7 月 16 日（消印）
上野桜木町　より　（はがき）

ズット前ノ約束ニヨリ，土曜日ニ（ツマリ昨日）君ノ函数論ノノートヲ借リルコトニナッテイマシタノデ，学校デ（アマリ当テニシナイデ）待ッテイタノデスガ，遂ニ君ガ現ハレナカッタノデ，オ手紙スル次第デス．

何時借シテ呉レルノカオ返事下サイ！

（日時ガハッキリシテイレバ，オソクテモ結構デス）学校ニ来ルノガ面倒デシタラ，君ノ家マデ借リニ行ッテモ結構デス．ソノ際ハ地図ヲ教ヘテ下サイ．（勿論日時ヲ指定シテ下サイ．）

休ミニナッテモ，オ天気モ，従ッテ気分モハッキリセズ，期待シタ程勉強モデキマセン．整数論ヤ，van der Waerden ノ代数幾何等．ソノ他，大シタコトモアリマセン．

極ク近イ将来ニ引越スノデ（ソシテ引越先キノ所番地ヲ聞キハグッタノデ）御返事ハ理学部数学教室宛ニシテ下サイ．

及川広太郎　へ

1951 年 7 月 21 日
大塚坂下町　より　（はがき）

アレカラ三浦ノ家へ行キソレカラ，柴岡ノ家へ行キマシタガ，柴岡ガ，ヤッテ呉レルト云ヒマスノデ，君ノ家ノピアノノ，偉大ナル

騒音ニ敬意ヲ表シテ，柴岡ニヤッテモラフコトニシマシタカラ，悪シカラズ．

……輪講ハ結局流レテシマヒマシタ……サウト判ッテイレバ急グノヂャナカッタ！

タマニハ数学カラモ解放サレタイト思ヒマスガ，目ノ前ニ，シナケレバナラナイコトガ余リニモ積重ッテイテ，ドウニモナリマセン．結局，重要ナコトハ，ソンナニ数多クナイト思フノデスガ，ソレヲ見分ケルノガ大変デス．……余リ多クノモノニ，手ヲ着ケスギタ様ナ気ガシマス．

谷山清司　へ

1951 年 8 月 6 日
埼玉県　騎西　より　（はがき）

9 日ノ 9·55 ノ準急デ，ソチラニ行クコトニシマス．出来タラ迎ヘニ来テ下サイ．（万一予定ガ狂ッタトキハ電報シマス）

風呂ニ入ルノハ閉口デスカラ，温泉ハヨシテ，他ノ所（遠足ニ行ク様ナ所デモ）ノ方ガ良イト思ヒマス．無ケレバ無クテモ結構デス．

尚，アチコチデ，林檎ガ欲シイサウデス．金ハ少シ持ッテ行キマス．

以上．

及川広太郎　へ

1951 年 9 月 4 日
大塚坂下町　より　（はがき）

今日出テ来マシタ．

函数論ノ Notes ハ直接森本ニ渡シテ良イデセウネ．……講義ヲ聞クヨリ，君ノ Notes ヲ読ンダ方ガ，良クワカル様デス．

8 月一杯，ボンヤリシテイタ訳デハナイノニ，何ト云フコトナシニ過ギテシマヒマシタ．何ダカ時間ノ計算ヲゴマカサレタミタイナ気ガシマス．

ソレデモ VAN DER WAERDEN ノ代数幾何ヲ大部読ミマシタ．WEIL ノ本デ，直観的ニ摑ミ難カッタ概念モ，此ノ本デハッキリシテ，思ッタヨリ有効デシタ．

今日ノ夕刊ヲ，見テゴランナサイ．又政令 325 号！

谷山清司　へ

1952 年 2 月 28 日 午後
大塚坂下町　より　（封書）

同封の申告書を出さなければなりませんので，適当に記入捺印して返送して下さい．今年になって，今日始めて区役所へ行ったら渡されたので，期限には間に合ひませんが，成るべく早く返送して下さい．試験は明日と 3 月 12 日，あとは 4 月，暫く忙しいと思ひます．少し身体の調子が悪く，困っています．……大したことはないと思ひますが．

その他，別に変ったこともありません．良い方に変ったことは当分，起きそうにもありません．

例の「東大事件」で，大部騒々しいけれど，情勢の悪化するきっかけにならなければ良いがと思っています．此のことには，色々考へさせられます．

松本皓一　へ

1953 年 月 1 日
大塚坂下町　より　（はがき）

新年おめでたう．

又 1 年経った．時と共に，段々過去の印象が薄れて行き，「何年，何ケ月」前と云ふ言葉は無意味になって，すべては，不定の時の「過去」と云ふ，漠然としたものの中に，沈んでしまふ様な気がします．

然し，その「時」を越えて，生きているものもあるに違ひありません．

何時か又お会ひしませう．話しはその折に，御多幸を祈ります．

及川広太郎　へ

1953 年 1 月 1 日
大塚坂下町　より　(はがき)

新年おめでたう．

今年が，君に取って，良い年である様に，そして君のポアンカレーが君を見捨てない様に．

成田正雄　へ

1953 年 1 月 1 日
大塚坂下町　より　(はがき)

新年おめでたう．

もう一年君と議論出来ると思ふと，……少くとも，君の，詭弁的名論を聞けると思ふと，少し嬉しくなります．

杉浦光夫　へ

1953 年 7 月 20 日 頃
埼玉県　騎西　より　(封書)

御手紙有難う．君は今何処に居るのかと思っていたのですが，未だ東京に居たのですね．

思いがけない暑さに見舞はれたため，少し熱を出してしまひ，勉強も思ふにまかせません．24 日には，是非行きたいと思いますが，それも暑さ次第，更に僕の健康次第（熱と云っても別に大したことでもないのです）余り期待しないでいて下さい．――行くとすれば，遅刻せずに行きます．

modular 函数は，当分一休みです．abelian variety の，虚数乗法，つまりその endomorphism の ring の構造を，Lefschetz や Weil に従ってまとめて見ようと思っているのです．出来たら，涼しくなってから，始から講義しようかと思っています．数論グループの人には参考になるでせう．classic をまとめるのは，余り楽ではありませんから．此処で faisceau が，一つの基本的な役割を演ずるのが，わかるでせう．

実際，代数幾何，特に abelian variety の理論は，不思議な領域です．第一に，それは，classic な函数論に密着しています．そして，又純代数的に取扱うことも可能で，そうすれば整数論との類似及び関連が特に目立ちます．又それは compact analytic manifold の一つとして，topological な，又，新しい解析学による取扱ひが可能になり，そこで classic な理論の一部が更新されるのが見られます．然も，此の 2 つの，対立した取扱ひは，丁度此の部門で交っているだけで，全く別な領域に連っているのです．

そして虚数乗法は，少くとも最初は，此の両方の面を必要とする様です．虚数乗法，つまり correspondence は，第一種積分の周期が 0 となる analytic cycle により定められます（faisceau は，此の証明に現れるのです）．而も，その全体の作る ring は，或る代数数体の Hauptordnung 又はその直和と同型になる（とは限らないのですが，主要なものについては，多分さうなるらしいと思はれます）そして，その体の上のアーベル体がそれにより構成される，その深い原因はどこにあるのでせうか．――

勿論，幾つかの原因が，それを説明している様に見えます．第一に，アーベル函数体の不分岐アーベル拡大は，すべて Teilung により得られること，そしてそれが，剰余類体に移ることに依って定義体のアーベル拡大を induce すること，又 endomorphism の ring の構造は，確かに，類体論と同じ根拠に基いているに違ひありません．然し，それで一応尤もらしい説明は出来るとしても，尚隠れた，深い原因が存在しないとは云へないでせう．

幾つかの都合の良い実例について，elliptic な場合の理論を拡張することは，それ程難かしくないかも知れません．井草さんの取扱った場合，‥‥1 の 5 乗根の体‥‥は，極めて例外的な場合で，Hecke の Dissertation でも除外されています．何故彼が実例としてそれを取上げたのか，一寸了解に苦しむのです．）然し，すべての場合に，虚数乗法より得られる結論が何を意味するかを考えることも無意味ではないでせう．elliptic な場合には一つになっていたものが，一般には幾つかの方向に分れます．そして問題は，アーベル体の構成に限らないかも知れないのです．

同時に，その過程の中で，Siegel の modular 函数を，より自由に使ひこなし得る様になるでせう．それが，elliptic modular function と同じ様に自由に使へる様になれば，整数論は，多くの分野に於て，かなり発展し得るでせう．Hilbert の modular 函数が実２次体数論に対し，多くの貢献をしているのと同様に．

大部勝手なことを書きました．そして，大部空想的なことを．････本当はもっと夢想的な事を書くつもりだったのですが，ペンが勝手に動いてしまって，････此れも熱のせいでせう．

具体的な群の表現と云っても，････僕は良く知りませんし，････まあ，一般的な理想論でも並べて見ませう．

第一に，結果が出ない程困難な計算を必要としないこと．そして，余り一般的な群では，一般論の研究と大差ないでせうから，何か他の structure と関連していて，その方の理論を採用すれば，割合に容易に表現を求め得るかも知れないと云ふ様なもの．

第二に，慾を云へば，その表現を求めること自体が，何か他の理論に対し有意義である様なもの．

重要なのは，結果を得ることであって，それが如何にして証明されたか，或はそもそも，それが何等かの方法で証明されたか，否か，と云うことではありません．一度結果がわかれば，それを一般化し得る様な方法で証明し直すことは，それ程困難ではないでせう．

第三に，unitary 表現が，その群に対し，特に essential な意味を持っている様なもの，つまり，有限次表現だけで済む様な群を考えてもつまらないと云ふわけです．

此の意味で，Lorenz 群なんか，最も適当なのでせうが．････

僕は余り具体的な群を知らないので．

まあ誰でも考へるのは，classical group でせう．それも，non-compact と云うと，余りありませんね．

リー群でない，non-compact，non-abel な locally compact group の例としては，例へば Weil の群がありますが，此の構造は大体わかっていますし，････然し，ユニタリ表現がそれに対しどんな意味を持つか．

もう大部遅くなったので，健康上の理由から，この辺で止めにして置きませう．

Symplectic group については････Siegel のものは，その discontinuous subgroup と基本領域が主ですし，その他余り文献も知りませんが．････構造と云っても，勿論 general linear group の subgroup ですから，余り変な構造でもないでせう．ただ，此の群で興味があるのは，普通の一次変換群と同様に，matrix space の“単位円”の変換群に移すことが出来ること，更に，“単位円”のそれ自身の上への１対１ analytic な変換群は，丁度此の群と同じになることでせう．従って，symplectic group を考へる時は，それを常に変換群を考へることが必要で，その時，変換される空間の“幾何学”は，割合によく解っていますから，都合よいでせう．此の，“単位円”の変換群として表はすことが，Cayley 変換（直交群の場合の）と同様に有用であるか否か，その辺のことは，余り考へたことはないのですが．

もし 24 日に僕が行かなかったら，僕の家に遊びに来ませんか．Hilbert の講演の原文を，清水さんに渡さなければならないので，上書しようと思っていたのですが，始めに書いた様な事情ですから，何もおもてなしは出来ませんが，少くとも僕が居ますよ．道順は，上信越線又は高崎線で，鴻巣（コーノス）で下車，停留所の前の道から出る，騎西（キサイ），加須（カゾ）行きのバスに乗り，騎西町で下車，谷山医院と聞けばすぐわかります．又は東武線（浅草から出る）に乗り加須で下車騎西鴻巣行きのバスに乗って来ることも出来ます．

いづれにしてもお会ひした折に．

清水達雄　へ

1953 年 8 月 19 日
埼玉県　騎西　より　（はがき）

御無沙汰しています．

Hilbert のパリ講演のこと，種々な都合で，もう数日かかります．それから原本を，図書室にでも預けて置きませう．その時は又御知らせします．

一応，序論（約 12 ページ）と，問題 VII—XVIII まで訳すつもりです．問題は，問題の部分の約 1/3 の分量です．enumerative geometry や 7 次方程式の所などは，良くわかりません．

序論は，大変興味深いものです．あの当時の Hilbert の抱負が良くうかがはれます．

全体に，変分法を特に強調しているのは，今日の我々から見れば一寸意外ですが，変分法が Weierstrass により，最初のそして目ざましい近代化を受けたばかりの時であり，又それが厳密さや Axiomatik を愛好する彼の好みにも合ったためなのでせう．20 世紀前半に，基礎論が経験した，そして今も経験している絶望的状態を知らなかった当時の彼は，極端なそして明快な楽観論を持っています．又後に彼自身手を附け，20 世紀の解析学に，あれ程大きな影響を及ぼした積分方程式について，何も語っていないのは，いささか不思議に思はれます．　　　では又．

中村得之　へ

1953 年 8 月 19 日
埼玉県　騎西　より　（はがき）

御無沙汰しています

Hilbert のパリ講演のこと，僕は序論と，問題 VII–XVIII とを訳す予定です（問題は此れで問題の部分の約 1/3 です）まだ終らず，もう数日掛りさうです．

問題には，見慣れない概念や言葉が多くて困ります．僕のやる部分にも，僕の良く知らないことが多いのですが，他の人よりは，少しは余計に知っているだろうと思はれるものを当って見たのです．

又暑くなってしまひました．君は，又真黒になっているかも知れませんが，僕は一日中閉ぢこもった切り，何しろこう暑くては，頭も身体も，運かす気には成れません．

ではいづれ，涼しくなってから．

清水達雄　へ

1953 年 8 月 25 日
小石川局（出）

大変オソクナリマシタガ，Congrès ノ記事（Hilbert ノ，パリ講演ノ出テイル），今日図書室ニ返シテオキマシタ．

以上．取敢エズ．

成田正雄　へ

1953 年 10 月 25 日
大塚坂下町　より　（はがき）

昨日は，折角いらして下さったのに，居なくて失礼しました．文理大での輪講がなくなったと聞いて，君が帰り途に，僕の下宿へ寄るんぢゃないかなと思ったのですが，学校で，杉浦とおしゃべりしている中に，ついおそくなってしまったので……．僕は 28 日の水曜日に帰るつもりです．それまでも，何やかとあって，もうお会ひする機会もないかと思ひますが，暇を見て，家の方に，遊びにいらっしゃい（その前に一寸，ハガキででもいつ来るか知らせて下さると好都合です．勿論，僕の都合の悪い日など，ありませんが）

僕は夢を気にしませんし，憶えておくことにもしていませんが，それにも拘らず，僕の記憶に残る夢があれば，それは非常に面白いものであるからに違ひないと思っています．その面白さが，一見わからないとしても．

以上，取敢へず．

追伸．学会に行くとき，君の都合がよかったら，一緒に行かうと杉浦が，云っています．そして，一緒に，山菅の下宿に泊らうと．28 日の，朝の汽車で行くさうですが，彼と連絡して御覧なさい．一人旅は退屈ですからね．

再伸．僕の机の引出しやノートなど，かきまはさなかったでせうね．

杉浦光夫　へ

1953 年 11 月 16 日
埼玉県　騎西　より

先づ要件から．

例の，Hecke の追悼文の，意味不明の個所につき，君の友人の独文学者が何と云ったか，至急知らせて下さい．原稿は，(多分) 今週末上京する際お渡しします．

こちらに来てから，何となく落ちつかず，恢復も思はしくなく，いささか参っています．でも悲観しているわけではありません．良くなることは確実ですから．唯，そのために払はねばならぬ犠牲が，不当に大き過ぎる様に思はれるのです．

清水さんが，書評で 2 等を取ったさうですね．1 等を取れるのではないかと思っていたのですが．…． とにかくも，S.S.S. の財政も，一息つけて，喜ばしいと思います．

こちらに来てから，色々本を読みました．今「詩と真実」を読んでいますが，余り僕の興味を引かないせいか，中々読み終りません．「渋江抽斉」は読みましたが，僕に興味を起させたのは，鷗外の文体だけでした．古書勘考も，又考証家も，僕にとって全く無縁ですから，……君の愛読書に対し，此んなことを云ふのは甚だ失礼なことですね．

数学は全然やりません．何もかも忘れてしまひさうです．君のユニタリ表現は，どうなりましたか？　半月位の中に，それ程進展するわけもないでせうが，でも本当の所，どれ位見込があるのですか？　新しい理論が出来るには，何か一つ，本質的なキーポイント或いは強力なテコが必要ですが，それが見つかりさうですか？　これは重要な問題です．では又．

成田正雄　へ

1953 年 11 月 22 日
埼玉県　騎西　より（封書）

お手紙有難う．

僕に「お変り」が無かったら，それこそ大変です．前より多少良くなりましたが，まだ，そんなに思はしくありません．

昨日，一寸学校へ行ったのですが，遅くなって，君達はセミナリーに行ってしまっていました．終るまで待っていると，帰れなくなるので，又お会ひする機会を失ってしまひました．

こちらに来てから，すっかり意気銷沈してしまひ，景気の良くないこと夥しいものがあります．色々小説を読みましたが，それも，余りスピードを出せない．君が 1 日で読む分を，1週間かかって読む程度です．最近，始めて，「詩と真実」を読みました．此れについては，「恐れいった」と云ふ外はありません．あの頃のドイツ文学や，宗教について，余り知らないため，良くわからず，興味も持てない箇所も多くありますが，‥‥あの，憎らしい程の落ち着きには，実際，恐れ入らざるを得ません．その他，又あとでゆっくり書きませう．

実際，こちらに来てから，急に寒くなったので，驚いているのです．去年はもっと暖かかった様な気がしたのですが．然し，世界を夏と冬との「二大陣営」に別ければ，僕は，いつもと同じ様に，ここでも，君の反対の陣営に属する光栄を担ふものです．いつでも僕は，秋に身体をこわし，春に良くなるのですからね．僕の身体と精神が，如何に夏に堪えられないかは，此れでもわかるでせう．5月から9月始めまで，僕の精神は，憂欝な無気力状態に沈み，秋から冬にかけては，安定を失って，無意味な焦燥と憧憬とに満ちる．然しとにかくも，不安定こそは，創造と進歩との源泉ですからね．

とは云ふものの，近頃は，年を取ったせいか，冬の寒さには閉口です．でも，やはり暑いよりはましです．

病気になると，良く夢を見るものです．昔は，それを一々覚えていて，思い出しては，夢に見た生活をなつかしんだものですが，最近は，大低は，夢を見たと云うことしか覚えていません．でも，今日の夢は少し覚えています．石垣やその仲間，それから警官隊が出て来て，大騒ぎが起る．大変景気の良い夢です．残念ながら，君は出て来ませんでした．(勿論，出て来ないに決ってますね)．

暫く数学をやらずにいると，高等学校に居た頃の気分が多少戻って来る様です．その頃は勿論，その気分に浸って，好い気になっていたのですが，今ではさうは行きません．

今，何を勉強していますか？　昨日掲示板を見たら，大部セミナリーをやっている様ですね．でも，率直に云って，algebraic variety の上の arithmetic ばかりをやっているのは，感心しません．それだけでは，一歩も進めないでせう．此の理論が，何に基いているか，そして何を目標にしているのかを，もう一度考へて下さい．そして特に，我々が何を目標にすべきかを．なんて大きなことを云って，僕も良く知らないのですが，それ等のことがわかるためにも，代数幾何や，代数函数論の，一般的，或は特殊な理論についての，相当の知識が必要でせう．(例へば，“Geometry of numbers” の理論の意味は，あのリプリントを読んだだけではわからない様に)．

もう寝るべき時間です (何時だと思ひます？　10時ですよ．馬鹿馬鹿しい！)

Minkowski は，“正規の” 手続は取らずに，図書室に返して下さい．但し僕の名前で借りていたと断って，念を押して，さもないと，僕の借出票がそのままになっていて，後になって又催促されるなんてことになりかねませんからね．

僕の方も景気が悪くて，甚だまとまらない手紙になってしまひました．どこもかしこも不景気ですから，僕達だけ景気の良い筈もないでせう，まあせいぜい，景気の良い夢でも見ませう．

来月半頃，写真を取りに出かけて，少しゆっくりしているつもりです．その時にはお会ひ出来るでせう．

では又，
お手紙下さい，こちらにいると，手紙でも

貫はないと，やり切れません．

宮原克美　へ

1953 年 11 月 22 日

埼玉県　騎西　より（封書）

大部御無沙汰しました．

先週の日曜日に，杉浦が来てくれた時，一寸話を聞きました．そしてすぐにお便りしようと思ったのですが，一寸延ばした所，勉強を少し始めて，その方に気を取られたので，大部延びてしまひました．序でながら，身体の方も大部のびてしまひ，一寸弱っています．勉強は，自分で考へ出すと，ズルズルといくらでも続いてしまふので，度を過し勝ちです．それに，今考へている超楕円函数といふのが，いはば泥沼みたいなもので，うっかり引込まれると，足搔きがとれなくなる，········
そんなわけで，考へるのは一応打切りにして，ささやかな結果で我慢することにしました．

杉浦から，Loomis の “Abstract Harmonic Analysis,, と Artin のものを受取りました．どうも有難う．清算は，今度（多分来週）お会ひしたときにします．Loomis は一寸読んで見ました．他人の書いた本を読むのは，少し退屈な代りに，度を過す心配はありません．·········· Bourbaki は，僕の持っているのは，Ensemble (Livre I), Livre II の Algèbre は Modules sur les anneaux principaux, Groupes et corps ordonnés のある分冊まで，Livre III. Topologie générale は全部，Livre IV は，始めの 2 分冊，Livre V Intégration は始めの一冊，Livre VI Espaces vectoriels topologiques は始めの一分冊（多分），以上です．その後のが出たら取っておいて下さい．その他，どんなものが出ましたか？

Loomis は，始めの三章は，かなり要領のよい，しかし完全に無味乾燥な要約，第四章は，吉田耕作 “位相解析” にも書いてあることですが，扱ひ方がより体系的である様な気がします．然し，何だか少しゴタゴタしている．五章以下はまだ見ていません．

所で，杉浦から一寸，僕の予想以上に，何か不愉快なことが起ったことを聞きました．尤も，君に取って，かなり不愉快であったとしても，要するにそれは，本質的に重要な意味は持っていないのでせう．S.S.S. を脱会するのは止めなさい．（もう脱会しようとは思っていないでせうが）S.S.S. に於て，誰誰が中心であるか又はあるべきかと考へるのは馬鹿げています．様々な考えと要求を持った人々が集り，その結果幾つかの行動がとられる．その個々の行動の中心となる人々はいるでせう．だからと云って，その幾人かとの意見の衝突乃至感情的対立は，S.S.S. の方針と矛盾することにはなりません．（会の統制なんてつまらぬことです．S.S.S. は K.P. ではないのですからね）君は君の思ったことを，······但し一般の討論を経て····実行できるし又するべきです．尤も，詳しいことは僕にはわからない····つまり，一体今度の衝突乃至衝突らしきものが，単なる感情的対立により起ったのか，意見の相違によるものか，そしてその中心となる問題点はどこなのか．
········尤も此んなことはどうでも良いのかも知れません，もう衝突もおさまり，平衡に達していることでせうから．

そして，「進適」の問題も，その後も順調に進んでいることと思ひます．

それはさうとして，君の挙げた a)—g) etc. 一人で此れだけ背負っているのは，いくら君でも大部大変ですね．殊に，身体の調子には呉々も気をつけて下さい．君の様に，少しセンチメンタルで，案外気の弱い所のある人間は，殊に気をつけなければいけない．少し消耗したときに，決定的な打撃の起るのを警戒し避ける様にすれば，余りひどいことにはならないのですが．

「深刻な生活劇」については，後でゆっくりお話をうかがひませう．

数学と絵とについては，一番重要な問題ですから，「尚一層の御努力を」此れはキマリ文句の中で一番便利なものです．

そんなわけで，色々な気苦労を放り出して，少し休養したらいかがでせうか．……長い間静養するのは，退屈でウンザリするものですが．

いずれ暇を見て「見舞」にいらっしゃい．尤も，来週中頃一寸上京する予定ですから，そのときお話する機会はあると思ひます．

とにかくもお元気で．

月報はまだ出ませんか．又少し何か書いても良いのですが，今度は遠慮しておきませう．

宮原克美　へ

1953 年 11 月 ? 日
埼玉県　騎西　より　(封書)

お手紙有難う．

身体の調子は，‥‥勿論１カ月前よりは良くなりました．然しまだ前途遼遠です．

所で，昨日一寸学校へ行ったのですが，君は居ませんでしたね．色々君に云うことがある様な気がするのですが，何から始めましょうか．

先ず S.S.S. の諸君は最近，非常に張切っているそうですね．会計の方も一息つけて，とにかく，非常に結構でした．進適についての，アサヒへの君の投書及びその反駁を読みました．そこで，B大のK君の反論に対する僕の感想を一寸書いて見ましょう．

人間の身長又は体重の分布が，正規分布になることは「これまでの膨大な研究結果」によって明かでしょうから，身長なり体重なりにより知能を測定することは「此の意味で‥‥測定が適正に行われたことになる」従って進適を止めて身長の大きいものを入学させることが，どうして行われないのか，「わたくしには理解できない」と云った調子です．―知能を「適当なモノサシ」で測定する云々という，「モノサシ」とは何でしょうか．もしそんなモノサシがあるなら，それによる測定と進適の結果との相関関係を調べるべきでしょう．そのモノサシが即ち進適に外ならないと云うのなら，又何をか云わんやで，K君の云うことは，何の意味もありません．どちらにしても，あの投書の中に，「科学的に実証された資料」が利用されている形跡はないことになります．彼の云うことは要するに，正しい測定のための必要条件（正規分布になること！）が満されたと云うに過ぎず，それだけなら，身長の測定によっても満されているのだから，それから，「適正」さを云々するとは，「科学の実質論理」と云うものは，何と向う見ずな，大胆な，そして都合の良いものでしょうか．彼が，学会とか専門誌上での討論を望んでいるのなら，勿論喜んで応ずべきでしょう．いよいよ面白くなるでしょう．僕が居なくて，君に「助言と勧告」を与えられないのが残念ですが……S.S.S. の諸君は皆有能の士ですから，頑張って下さい．（君の投書は時宜を得たものでした．今，あちこちで進適が問題になっているときですから，急ピッチで追撃すべきでしょう）

整数論グループのこと，今後も色々と知らせて下さい．Weil の Formula と云うのは合同 ζ-函数の積が，Größencharakter の L 函数の積になると云うものなのでしょうか．君の文面では，一寸推測し兼ねますが．

それから，それから，それから……後，何を云う筈だったか？

君の絵がアトリエに載ったこと．とにかくもオメデトウと云いましょう．尤も「既成作家として扱われ」るなんてウヌボレるのはまだ一寸早い様に思いますが，まあ君のことだから，それでもケンソンしているつもりなのかも知れませんね．どちらにしても，もうそろそろ，自信と自覚とを持って仕事を始めても良い頃でしよう．大した才能はなくても，他人から，何か相当のことをするだろうと期待されていると考え，又何かする様に強制される様な位置に置かれたために，或る程度のことをなしとげると云う人も少くないのですから，君の様に才能のある（と少くとも自分では考えている）人は，自覚することにより大いに伸び得るものです．そうすれば，もう少し落ち着きと重味とが出て来ますよ．大分生意気なことを並べ立てましたが，まあ誰も悪くは思わないでしよう．

その他色々のことを書く筈でしたが，忘れてしまいました．思い出したら，又書きましよう．

銀林君の，何とかと云う，11 月 8 日の会はどんな工合でしたか．君は勿論出席して，主観的並びに客観的に楽しみ，宮原氏の観察を下したことでしようね．いつかお聞きしたいと思います．

時々お手紙下さい．僕は意気銷沈しています．昨日，帰りの汽車の中で，前に坐っていた2人の青年の会話を聞いて，非常に面白く思いました．中位の「教育のある」人々で，気取りがなくて，正直で気が小さくて，人が良くて，虫が良くて，尤もらしい意見を持っていて，而も違う意見の人と衝突もせず，おまけに余り頼りにならない．と云った人々です．

では又，お元気で，……

杉浦光夫　へ

1953 年 12 月 7 日
埼玉県　騎西　より　(はがき)

先日はどうも有難う．今度は，少しゆっくりしていられる様にして，又いらっしゃい．京都行きの件，こちらに残ることになると，又，指導して呉れる人のいないことで多少困るでせうが，勉強が進むにつれて，指導者の必要度は段々少なくなることでせうから，寧ろ同好の士を募って，やって行けば，その方が良いかも知れません．米田さんを中心にするトポロジーのグループだって，余り頼りになる指導者が居たわけではありませんからね．持って来ていただいた参考文献のおかげで，少し勉強が進みました．気をつけないと，度を過し勝ちになるおそれもありますが．

そして，始めの予想とは，一寸違った結果が出たので，一寸まごついています．然し，例へば，$y^2=1-x^5$ の函数体（標数 p）の Hasse-Witt の不変量の Rang が，1の5乗根の体での p の素因子分解の型により決ることは確かです．

君の Neumann と Dixmier との関係の話，後で詳しくお聞きしませう．さもないと一向にわかりません．

宮原克美　へ

1953 年 12 月（日付不明）
埼玉県　騎西　より　(封書)

その後お変りないことと思ひます．此の間上京した折は，色々と忙しくて，君と余りゆっくりお話する暇も元気もありませんでしたし（序でに，お話することもなかったのかも知れません）何と云ふことなしにペンを取った次第．年賀状を書いてしまったし，重複する様ですが，それは，云はば忘年会の後に新年宴会がある様なものでせう．

年末ともなれば，忘年会とか，借金取りとか，中々と賑やかで良いものですが，余り賑やか過ぎて，主観的に飲み過ぎ，客観的に借金を取られすぎて，消耗していらっしゃるのではないかと思っています．まあそれも良いでせう，……年に一度だから，せいぜいビタミンとメチオニンでも飲んで，朝寝することですな．

帰って来てから，余り勉強もしません．健

康状態は，少しづつ良くなっているとでも云ふのでせう．此れから先がむずかしい所なのですが．例の，“Weil の Formula” と云ふのを，少し考へています．問題の本質がどこにあるのかを理解するのがむづかしい．符号の決定が重要であると云ふこともあり得るのでせう．

過ぎ行く1年を振り返って，年末の感想を．と思って見ても，一向にそんなものは浮んで来ない．飲み過ぎて裏返しになった胃袋に，青い顔をして，借金取りにでもせめられなければ，年末の感慨なんて湧かないものなのでせう．そこで，そんなことはあきらめて，何かしら考へて見ませう．試験の答案じゃあるまいし，白紙を君の所へ送るのも，一寸変な様な気がしますから．

年の暮と云ふものは，色々なことがあるものですね．君が，例のハンガリー人と絶交するとかいきまいて，来年から何かに専念すると宣言したのは，去年，いや，たしか一昨年の今頃でしたね．････（去年のクリスマスに何とか云っていたから）．その前の年には，君の親友に関し，何かゴタゴタがありましたね．･･････････････その後どうなったか知りませんが，何かに専念するのは，余り望ましくないと悟ったと見えて，君は，相変らず，何にも“かぢり附か”なかった様ですね．それに，君は淋しがり屋だから，他のもの或はことを忘れて一つのもの或はことにかじり附くなんて，始めから出来ない相談だったのでせう．その間にも時は流れて行き，君の“自信ある仲間”は，それぞれの考へに従ひ，世の中の成り行きに応じて，勝手な道を歩いて行ったので，そこに一つの断層が出来て来た様に感じられたのではないでせうか．君の好きな“人間的結び附き”も，人々が年を取るにつけて，そして世の中が移って行くにつれて，その形が変って行く．そして多くの人々は，自分自身のことに忙しい様になる．（最近結婚した誰かみたいに）．然し時々は過ぎ去った親密な感情を懐しんで，大いに飲んで，当時を思ひ起さう．････････と云ふわけで，あちこちで忘年会が盛大に行はれる，･･････君の“仲間”も，その中に，そんな風になるのではないでせうか．すべての人が，他の人々から取り残され，だからせめて，何か一つのもの，一つのことに専念して，自分の廻りに，そこで安心できる様な境地を作り上げようとする，そんなものなのでせう．

そんなわけで，年の暮には色々なことがあるのでせう．今年はどんなことがあったか知りませんが（こちらは相変らず･･････何もなし），とにかくも，ビタミンとグロン酸でも飲んで，ゆっくり朝寝をすることですな．

良い年を迎へる方は，年賀状に書いてしまったから，今年を良く送られる様に．過ぎ行く年に対して，サヨナラを云ふだけにしておきませう．

暇があったら遊びにいらっしゃい．

成田正雄　へ

1954 年 1 月 1 日
埼玉県 騎西 より （はがき）

明けましておめでたう
そして
“ヤア，シバラク”

去年一年は，何だかひどく散慢に過ぎてしまった様です．きっとお天気のせゐだったのでせう．

今年も又，大きい鉄の扉のある灰色の建物が，僕の前に現れるのを待つことにしませう．

大して“おめでたく”もない新年に思うこと．

（今度は，一寸も失礼な言葉はないでせう）

山崎圭次郎 へ

1954 年 1 月 1 日
埼玉県 騎西 より (はがき)

明けましておめでたう.

今何を考へていらっしゃいますか. 最近, すべてが行き詰りです.

数論グループも, 今年は, 大きく伸びることを期待しています.

杉浦光夫 へ

1954 年 1 月 1 日
埼玉県 騎西 より (はがき)

新年おめでとう.

新年第一日の朝寝を楽しんで居られることと思ひ, 御同慶の到りです.

昨年は, 色々つまらないこと. 面白いことがあり, 大部お世話になりました. 今年もよろしく.

今年は何が起ることか. そして僕達が, どの位進み得るか, 甚だ心許ない様です.

志村五郎 へ

1954 年 1 月 23 日
埼玉県 騎西 より (はがき)

御手紙拝見致しました. Math. Ann. 124 は, 今, 下宿に置いてありますので, 今度上京した際にお返し致します. 今月末か来月始め頃になると思ひますが, それまでお待ち願へるでせうか. Deuring の論文は, 類体論を使はずに, 虚数乗法から, 出来るだけの結論を出すことを目的としたもので, 最後に, "現在の Reduktion の理論の不満足な状態のために" 云々, と書いてあります. 今度 Reduktion の, 有力な理論を作られたそうですが, それをお使ひになったら, もう少し先まで行けるのではないでせうか.

僕の, Reduktion についての結果は, 余り大したものでなく, 大体 Deuring の方針に従ったものですが, abelian variety を目的として, non-singular variety が, 殆んどすべての $\mathfrak{p}$ に対し, non-singular variety に行くことの証明が眼目です. それは, Weil の, algebraic variety の上の arithmetic を使ふので, すぐ云へます. 又 $\overline{f(P)}=\bar{f}(\bar{P})$ (−は Reduktion した結果, P は point). つまり, 0 や ∞ でない値は, 一般に, mod $\mathfrak{p}$ でも, 0 や ∞ にならぬこと. などです. 少し整理して, 何かに書かうと思っていますが, 今の所発表の予定はありません.

いつか, 今度お作りになった理論の大要を御教示願ひたいと思ひます.

御返事まで.

杉浦光夫 へ

1954 年 1 月 25 日
埼玉県 騎西 より (はがき)

御手紙有難う.

今月始めから, 少し調子が悪く, 大部消耗しました. 今, 段々良く成っていますので, 多分, 今月末頃上京するつもりです.

やはり, 二学期一杯静養しなければいけない様です.

最近, 余り何も読みません.

暇つぶしに, 少し考へて見たのですが, Weil の Thèse の, アーベル函数に関する部分の証明が, 代数幾何も何も使はずに出来る様です. それには, 簡単な Lemma を 2 つ使い, 第三種積分の, 変数とパラメーターの変換法則に対応する内容を, 代数一算術的に, 証明するのです. 超楕円函数体の "Zweiteilung" については, classical な結果に対応する定理 (分岐点までの積分の表す半週期の Charakteristik が互いに azygetisch なこと) が, すぐに云へます.

いずれお会いした折に.

清水達雄　へ

1954 年 4 月 13 日
埼玉県　騎西　より（手紙）

前略

先日は，僕の下宿のこと，お世話様でした．やはりああ云ふ所は，僕には向かないと思ひましたし，それに約束した日曜日には，引越しのための荷物整理に取りまぎれて，木村さん（と云ひましたね？）のお宅に伺はなかったのですが，先方で，待っていらしたのではないかと思ふと少し気掛りになりますし，清水さんにも御迷惑をかけたことになってしまったのではないかと，心配になります．とにかくもおくればせながら，御心配下さったことに，御礼申し上げます．

ところで，Hilbert の飜訳のこと，序論は，も少し日本語らしく訳し直しましたが，さうすると表現はどうしても少し長たらしくなります．難かしいものです．早い方が良ければ，郵便でお送りします．フランス訳への飜訳者 L. Langel による 2 つの註（一つは此の飜訳の原典（Gött Nachr. に出たもの）について，一つは Corps du cercle (Kreiskörper) について．‥‥此の頃は，円体はまだ new face だったのでせう）は省略しました．註をつけるとしたら，此処に現われる言葉，概念について，又扱はれる問題などについて簡単に説明した方が親切でせうが，適当に短くしないと尨大になり，余り短くてはあってもなくても同じですから，訳註は，一応，一切つけないことにしました．もし，つけた方がよい様でしたら僕に書けるものは，少し書いても良いと思ひます．

手許に原稿用紙がないためあり合せの紙に線を引いて代用しました．

此の序文の中で，問題を特殊化することの重要性について述べてありますが，此の考へは Hecke，更に Siegel に受け継がれて，共に，その，多変数 modulor 函数論の研究の，一つの意義づけとなっています．Hecke は，“Analytische Funktionen und algebraische Zahlen, I. Teil” の Einleitung で，Siegel は “Symplectic geometry” の Introduction で，共に，複素多変数の一般函数論が満足な形に出来ていない一つの原因は，特殊函数の研究の不十分なことによると述べて，自分の研究は，その一助となり得るものだと云っています．（但し実際には，一般函数論がそれから受けた恩恵は，かなり少い様です．一般論はむしろ，Poincaré，Picard などの研究，それにアーベル函数論などを或程度 Vorbild としているのではないでせうか）．Hecke は，その modular 函数を，Hilbert から受けついだので，此の講演の頃には，Hilbert は既に，それを考へていたと思はれますから，特殊化の必要を述べた時には，そんなことも頭の中にあったのではないでせうか．

気のついたことを一寸書いて見ました．

今度の僕の下宿は下記の所です．S. S. S. の委員の人に話しておいて下さい．

世田谷区松原 4～1137　江島義之方

S. S. S. の，21 日の会合には多分欠席します．健康状態はまだそんなに良いとは云へませんから，

宮原克美　へ

1954 年 4 月 14 日
埼玉県　騎西　より（封書）

御無沙汰している中に，“ヤア，シバラク” と云ってもおかしくない位時が経ってしまひました．こちらに帰ってから，引越しの疲れやなにか，一度に出た様で，すっかり消耗してしまひ，つい御無沙汰する結果になってしまったのです．

所で，その引越しについては，君に，一方ならずお世話になりました．おくればせなが

ら，あらためてお礼を云ひます．更に，あらためて，何かお礼をしたいと思ひますが，今，少し財政困難なので，よく考へて，いづれ後程，と云ふことにしませう，‥‥君には，それまでも色々お世話になっていたので，いつかまとめて，と思っていたのですが．

いつもいつも廻り来る季節ごとに，何年か前の同じ頃を思ひ起すと云ふのは，精神が活動性を失って，後ろ向きになっているためで，いはば心理的エネルギーの頽廢によるものと云へませう．それに，君と違って，僕の，春の思ひ出は，生彩のない，憂鬱なものばかりなので，そんなものをせんさくするのは止めにして，少し外を眺めて見ませう．

20 世紀の悲劇は，人間が新鮮な感受性を失ってしまったことにあると誰かが云ひました．それは，恐らく余りにも強い刺戟が，余りにも屢々繰返されたことによるものでせう．水素爆弾が何度爆発しても，否，爆発すればする程，それに対する人々の心理的反応は強く表れません．みんな，二言三言口の中でボソボソ云って，肩をすくめてソッポを向いてしまふ．実際，どこの国でも，それに対して，政治を動かす程，世論が盛り上った所があるでせうか．勿論“良識ある”人々は，事の重大性を認識して，警告を発します．然し“良識”も，人間の心の底を，つまり感情的なものを動かさなければ，力とはなり得ません．所がその感情が，新鮮さを失ひ麻痺してしまっているのですから，どんな声だって，眠い耳にひびく様な子守唄の様なものです．此の麻痺から人間をたたき起したもの，それがナチズムであり，ファッシズムであったわけで，それは実はより完全なる催眠状態に外ならなかったのですが，とにかくも人々は，それによって，一種の新鮮な感じを取り戻したわけなので，ナチズムが，あんなにも根強い一つの原因は，そこにもあるのでせう．

とにかくも，我々は眠っています．そして外で起る物音を聞いて，夢見心地にブツブツ云っているにすぎないのです．我々，つまり人類全体が眠って居り，眠りながら水爆を爆発させ，眠りながらマグロを取り，眠りながら，スシ屋の前を素通りするのです．なんて云ふのは勿論誇張ですが，病気のために，感覚の新鮮さと活動性とを失ってしまった僕には，そんなことが特に強く感じられるのです．君や石垣の様に，或程度 vivid な感受性を持っている人間はうらやましい，此れは本当の話です．

そんなわけで，外を向いても愚痴になり，内を向いても気がふさぐ，春は良くない季節です．

但し，芸術家には，“暖い心臓”は不必要である．否邪魔にさえなると，別の誰かが云ひました．余りにもあたたかい心臓を持っている人は，歌はぬ詩人として，描かぬ画家として世を過すことになる．例へ彼等が筆を取っても，退屈な詩，売れない画をかくのが落ちだと．事の当否は知らず，傾聴に値する意見ではありませんか．但しその人も，芸術家は眠りながら筆を取り，眠りながらマグロを食べても良いとは云ひませんでした．否，飽くまでも目覚めた，冷静な眼と心とを要求したのです．何しろ時は第一次大戦前の，人類没落の萌芽期だったのですから．

そんな風に誰かの云っていることを聞き，つらつら世の成り行きを考へて，いささか感無量，なんて気にでもなればまだ良いのですが，‥‥‥まあ，世の中の成り行さはその辺にしておきませう．

もう少し元気になったら，君の絵について一寸書いて見たいのですが，‥‥‥僕の意見としては，まあ余り人の意見は聞かず，自分で，本当に，心の底からこうだと思う通りやるのが一番良い様ですね．自分を一番良く知っていのるは自分自身ですし，人は随分無責任なことも云ひますからね．此れは勿論，人の批判を無視せよと云ふのではなく，人の意見でも，自分の考えた絵画論にしても，時には実験して見てでも，本当に心の底から納得できない限り，それに従ふとろくなことはないと思ふのです．此れから，色々な人に色々なことを云はれる機会が多くなることと思ひますが，一寸老婆心から，大部口はばたいことを云ってしまひました．

21，2 日頃上京します．それから，お会ひする機会も，しばしばあることでせう．又その折に，余り疲れない程度に，お話ししませう．

勉強は何もしていません．例の Weil の講義の紹介，わかり易く，而も要領を得て，短かくまとめると云ふのは大部むつかしいことです．読む人の労力が少くなる様にしようとすると，それに反比例して，書く人の労力が多くなる様です．……

そんなわけで，毎日を家の中で送っています，今は，病気の爆発を防止することが，何よりも重要です．何しろ，春は良くない季節ですからね．

お礼の手紙のつもりで書きだしたのが，あちこち脱線して，大部取り取めなくなりましたが，此れもお天気のせい．とにかくもお元気で．

（僕の下宿の番地，お知らせしてあったでせうか？

世田谷区松原，4 の 1137　江島義之方です．

成田正雄　へ

1954 年 4 月 14 日

埼玉県　騎西　より　（封書）

その後，お変りないことと思ひます（と云っても，叱られないでせうね）．僕は，帰って来てから，それまでの疲れが出て消耗しました．20 日過ぎに上京するつもりです．

所で，新学年からのことですが，代数幾何をやる，或は少くともそれに関心を持って一通り勉強した人々は，東京にも，かなりいた筈なのですが，——例へば，玉河，松坂(和)，公田，銀林，の諸氏，又専問家には松阪(輝)，小泉両氏，その他，——いつになっても，一向にパットしない様ですし，小泉さんは入院してしまふし，前にあげた 4 人は，他のことに興味を移してしまった様ですし，と云ふわけで，又，今度卒業した人々の中，玉河さんの所で代数幾何をやった 3 人の中，2 人まで転向してしまふ始末，このままでは，東京の代数幾何も，立ち枯れてしまふおそれもあります．勿論立ち枯れてしまっても一向に差支へはないかも知れませが，僕達も，折角勉強して来たことですし，それに，格別，数学の此の部問が行き詰っているわけでもないのに，みすみす止めてしまふのもつまらない様な気がします．今まで，お茶の水又は文理大でやっていたセミナリーも，どれ程役に立ち得たか，一寸疑問にも思ひます．そんなわけで，4 月から，代数幾何をやる人達，差し当って，僕と君と，今度卒業した郡司君あたりで集って，もう少し突っ込んだ勉強なり討論なりを，始めから，やって行ったらどうかと思ふのです．今までも，君と，多少話しはしていましたが，それ等の話も，大体上すべりのことが多くて，お互ひに余り役に立たなかった様です．具体的なことは，一度，3 人で集って決めませう．正直の所，僕も，代数幾何について，それ程良く知っているわけではないので，此の辺で一度勉強しなほすのも悪くはないと思ひます．大体，代数幾何の色々な理論の現状，その相互間の位置，その役割についての評価，現在問題になっている事柄などにつき考へることと，もう一つは具体的な，或は半分位具体的な実例を研究して行くことと，それに応じて基礎理論を，たしかめ，勉強し直して行くこと，などをやったらどうかと思っています．要するに，研究能力を養ふこと，更には，隠された問題を見出して行く力を附けること，が目的です．

以上の点につき，君の御意見を伺ひたいと思ひます．

勿論，松阪さんや，その他の人々に相談したり，意見を聞いたりすることが必要になると思ひますし，出来れば，一緒になって，やってもらひたいのですが，それ等については追々考へることにして，小泉さんの病気の間だけでも，以上の様にやって行ったらどうかと思ふのです．（余りえらい人のいる前で，やさしいことをやるのも，一寸気がひけるかとも思ひますし）．

今度の移転先は，世田谷区松原 4～1137 江島義之方．

21 日頃まで，こちらにいます．君とはいつ会へるでせうか？

御返事下さい．（郡司君には僕から手紙を出しておきます）．

以上．要件のみ．

谷山義泰　へ

日付不明（1954 年6月～7月頃と推定）
世田ケ谷区　松原町　より（封書）

（一枚半省略，数学の質問の答）

此等のことは例へば高木貞治“解析概論”223 ページに書いてある．又少し高等な解析教程なら，どんな本でも，双曲線函数のことは書いてあり，科学者の常識である．

x の色々な値に対する $\sinh x$，$\cosh x$ の値については，表が出版されている．図書館で調べればわかることと思ふ．以上．

雨が止んだと思ったら急に暑くなり，いづれにしても閉口です．暇があったら，一度遊びに来るなり，但し，こちらの都合もあり，大抵下宿には居ないから，日曜日にでも，どこかで待ち合せて，どこかに遊びに行っても良いと思ひます．取敢へず．

大村佐登　へ

1954 年 7 月 1 日
世田ケ谷区　松原町　より（封書）

お手紙拝見致しました．――過分のおほめにあづかって恐縮です．係りの人に会ふ機会が中々無くて，少し遅れましたが，二三日前，入会の手続をしました．今後は，色々な催しについては，会の方から通知が行くことと思ひます．（尚，“新数学人集団”は，普通“S. S. S.”と略して居ます）．…… それ以前に通知を出してしまった二つの催しにつきお知らせ致します．

一つは今週土曜日（7 月 3 日），昨秋以来来日中の C. Chevalley 氏に，現在の数学の諸問題について，質問することになっています．（通訳つきで）．時間は，Chevalley 氏の講義終了後ですから，3 時頃になると思ひます．場所は東大理学部の数学教室，又は化学教室の 200 号室．…. 3 時頃に数学教室においでになってお聞きになれば，どこでやるかわかると思います．（数学教室の所在は御存知でせうね．理学部一号館（安田講堂の裏の，赤い三階建の建物）の三階，階段を上るとすぐに，小使室があります．）

第二は，来週水曜日（7 月 7 日）石膏会館で，一松　信氏（立教）の，多変数函数論の正則領域についての話があります．午後 5 時半－6 時頃から始まる予定….今学期の催しは，多分此れでおしまひです．

尚月報 5 号は，もしお出でになるのでしたらそのときお買ひになれば良いと思ひますし，さうでなければ，お送りしませう．

（此の間云ひ忘れましたが，会費は，主として通信費，会場借用費などに使はれるので，月報の会計とは別になって居ます．‥‥此れまでお送りした分の月報は，贈呈と云ふことになっています.)‥‥会費は，勿論，入会になった時からお払ひ下されば良いのです.)

以上取り敢へず，お知らせまで.

山崎圭次郎　へ

1954 年 7 月 17 日

世田谷区　松原町　より（はがき）

今朝，久賀君から速達をいただき，君が僕の下宿にいらっしゃるとのことでしたので，2時頃までお待ちして居りましたが，昨日締切りの筈の月報の原稿を，仕上げて渡さなければならぬ必要から，出掛けて，今，9時頃帰って来た所です．実は Riemann-Stahl は，昨日，図書室にあづけて来てしまったので，今手許にはないのです．又僕は明日，田舎に帰ります．昨日，僕の行く前に，学校に，久賀君から電話があったので，あの本は，君が取りに来るかも知れないからよろしくと，図書室に話してありますから，御面倒でも，本郷の方に行っていただけませんか．色々手違ひで御足労をお掛けしたことをおわび致します．

今度の月報の記事で，君の W-variety のことに軽く触れて，一寸批判してあります．悪しからず.

杉浦光夫　へ

1954 年 7 月 30 日

埼玉県　騎西　より（封書）

御手紙有難う．月報も受取りました．僕の，なぐり書きの原稿をもとにして校正をするのは大変だったでせう.

月報の内容をざっと見て見ましたが，やはりカルタンの業蹟が，読みごたえがある様です．数学的内容の，短かい記事が幾つかあっても良い様な気がします．今度は急いで作ったため，原稿を依頼したり集めたりする暇がなかったのでせうが，全国連絡会の記事にもある様に，数学の地図を作ること，各部門の歴史を作ること，それも，数学事典みたいに網羅的（と云ふことは，アクセントがないといふこと）でなく，又年代記的な歴史ではなくて，現在の中心課題にピントを合せて，それ等のものをまとめて行くこと，これを，名著解題と平行して，月報の恒久的な企画の一つとしたらどうでせうか．及川のリーマン面や君の第5問題の話など此の線に沿うものかも知れませんが.

最近一寸考へているのですが，数学の実質的部分は classic なものにあり，抽象的なものは，それを定式化し，解決するための方法にすぎないと云ふこと，此れは一寸変だと思ひませんか，classic 乃至具体的なものと，抽象的なものとを，此の様に機械的に分離することは，現在の数学を本質的に把握し，それを発展させるに適当な方法でせうか，数学とは無矛盾な公理系から論理的に導かれる体系であると云ふ考へ方，此れは，抽象数学を“輸入”した日本では，容易に根を張り得る考へ方ですが，此れに対するアンチ・テーゼとしては，以上の様な考へ方は，有意義ですが，やはり事の真相をとらへたものとは云ひ難いのではないでせうか.

数学的実体と云ふものが存在し得るとすれば，それは公理系により定義される抽象的な概念でもなく，又具体的に存在する，数，空間，物理現象，乃至それ等の関係，運動法則と云ふものでもない．常識的な様ですが，具体的な多くの異ったものが，一つの抽象的な概念の下に統一され，又多くの抽象的な概念が一つの具体的なものの中で関連する．此の二重の関係が，その本質を究明する鍵ではないでせうか.

例へば，合同 ζ-函数を考へませう．それは，抽象的な代数函数体に於る，Riemann の

ζ-函数の類似物であると考へれば，此の考へは，抽象的な把握です．又それは，代数数体のヂォファンタス合同式の解の性質を表すものであると考えれば具体的把握と云へませう．（此れが Gauss の立場です．）然し合同 ζ-函数の本質はもっと深い所にあるのであって，此の２つの把握は，それを，その各々の立場からとらへたものに過ぎない．そしてもっと深い立場に入れば，抽象的とか具体的とか一概に区別出来なくなる．尤も，合同 ζ-函数と云うのは，全く具体的なものだと云ふのなら話は別です．

では fibre bundle を考へませう．此れは，月報の記事によると，カルタンのRiemann 幾何から出たものである様ですが，確かにさうなのでせう．恐らく最初に此の概念を扱った Chern の論文では，大体その様なものとして，fidre bundle を把へていますから．然し此の概念は，非常に広い範囲で有効である，微分幾何，トポロジー，代数幾何だけでなく，例へば，多変数函数論の Cousin 第二問題が，本質的には projective line bundle の分類の問題であることは，此の間の一松さんの話にあった通りです．此れは，公理主義の一つの勝利とも見られますが，然し此の場合にも，公理主義の果した主要な役割は，整理することにあったので，例へば，岡さんの方法は，faisceau の概念の一つの基礎になったものですが，岡さんが，fibre bunde との類推からその様な方法に到達したのではない．とにかくも，此の時，２つの考へ方が可能である．「fibre bundle なる一つの概念が多くのものの基礎にあるのであるが，それを体系的に発展させるには，単なる抽象論では駄目で，それの表れている具体的な事実から，或る意味で帰納的に進んで行かなければならない．」「或る部門に於ける重要な事実は，fibre bundle なる概念によりうまく表現され又それにより，他の部門の同様な事実との関連が明かになり，此の抽象概念を使ふことにより，その具体的な問題を，見透しよく進めることが出来る」一何を目的とし，何を手段として考へるかの相違ですが，両方とも正しい考へとは云ひ難い様な気がします．具体的なものと抽象的なものとの交錯するその奥に，数学に於ける実体がある．大体，実体なるものは固定したものではなくて，時と共に移り変って行くものなので，例へば昔，計算の手段として考へられた複素数が，現在では実体と考へることに誰も異議はないでせう．そのとき，以前にはそれが実体であることがわからなかったのだと考へるよりも，18 世紀の数学では実体でなかったものが，19 世紀には実体となったと考へる方が自然でせう．大体，数学的実体なるものは存在しないと考へた方が良いか，さもなければ，実体であるか否かの判定法は，それから導かれる定理によるので，此れはいつか君の云っていた，「自然は簡単を好む」と云ふ原則を無条件に承認することで，例へば，有限単純群が，簡単な図式により分類されれば，それは数学的実体と云へるが，さうでなければ，それは一つの便宜的な概念にすぎないと云ふことです．此れは，多くの数学者が無意識的に取っている立場であって，要するに実体は存在しないと云ふのと大差ない．

少し混乱しましたが要するに一実体なんてものはどうでも良いのであって，数学の実質的部分は，興味ある定理の体系であり，或る体系に興味を与へるのは或は実際的な応用であり，或は具体的な，非実際的な種々の事実であり一或は抽象的な概念であり，そしてより多くは，此等のものの，様々な形での交錯である．そしてそれ等を通して，それを底から支へているのは，数学の統一性，法則性に対する無理由の信頼である，そして此の信頼が，何等かの事例により確かめられれば，我々は感激するのである．要するに数学とはそんなものなので，公理体系でもなければ，具体的なものの属性でもなく，その他色々なものでもない．こう考へるのは少し pragmatic

かも知れません．

所で，S.S'S. の財政困難のこと，余り名案も持って居りません．現代数学の紹介と云ふこと，非常に結構な考へだと思ひますが，それには，journalistic なセンスが大部必要で，相当案を練る方が良いでせう．調子を落さずに，而も一般の人にわかり易く書くと云ふのはむづかしいことです．暇なとき少し考へて見ることにしますが，僕の方の分野では，問題自体は大抵，わかり易い表現を持っていても，方法が甚だ難解なので，あまり良い例もない様です．とにかくも，公理主義的なことは，一寸読んでもすぐわかった様な気になれるのに反して，少し深いことを説明するのは大変です．君の挙げた例でも，リー群，ベッチ数，微分式，コホモロジーなどの概念の真の意味を説明するだけでも大変なのではないでせうか．形式的な定義だけなら簡単でせうが．尤も此の様な企画をやり遂げることにより，その対象及びその発展の歴史を本当に把握することが出来るでせう．つまり，此れまで当り前のことの様に使っていた概念の上にあぐらをかいた認識ではなくて，それ等の概念は，此の理論に本質的に，どんな形で，どんな根拠の上に，入り込んでいるかが，一番始めからわからなければ，わかり易く説明することは出来ないでせうから．

とにかくも，僕の方の分野でも，何か考へておきませう．

家に来てから今まで，つまらないことをゴタゴタ勉強している中に，いつか日が経ってしまひ，おまけに健康状態も少しあやしくなって来て，一寸弱っています．8月になってから，少し面白い勉強でもしようかと思っています．合同 ζ-函数の話は少し先にのばすことにしました．とてもひまがありません．

では又．

杉浦光夫　へ

1954 年 9 月 1 日
埼玉県　騎西　より（はがき）

御葉書拝見，談話会の相談の件，6日で結構です．

ここ二，三日涼しいものの，それまでの暑さで，いささか消耗しています．余り大した勉強も出来ません．代数というものは，勉強しているときは面白くても，あとですぐにつまらなくなります．代数的と云ふのは要するに Formalim に徹することで，而も Formalism だけでは無いのに，それだけで済むものですから，皮相な見方で満足することになり易い，代数が初学者に人気があり，論文も掃いて捨てる程出るのに，何となくつまらないのもそんな所にあるのでせう．昔の，計算ばかりしていた数学と同じ様なものなのでせう．真の代数学者とはどんな人を云ふのでせうか．

尚，次の両君に，月報6号送って下さい．
（代金は売掛にしておいて下さい）
新宿区余丁町 55　田代雄一
大田区雪ヶ谷町 886　大村佐登
では又．

成田正雄　へ

1954 年 9 月 1 日
埼玉県　騎西　より（はがき）

お互に御無沙汰しましたね．

此の夏はどんな風に過されましたか．まさか日曜学校が夏の学校になったりしたのではないのでせうね．

所で，例の Snapper の別刷，もうそろそろ読み終った頃と思ひますが，僕も一応目を通しておいた方が良いので，返してもらへませんか．9月6日の昼頃学校に行きますが，君の方がまだ始っていなかったら，そのとき持って来て下さい．

此の間から，仕方なしに Samuel の Thèse

読み始めました．local ring って，何か厄介なものですね．所で“Hilbert の Polynomial”について，その由来，主要結果，文献など，教へて下さい．君はいつか，何かで勉強していましたね．

ではいづれ近い中に．

成田正雄　へ

1954 年 9 月 17 日
埼玉県　騎西　より　(はがき)

Snapper，受取りました．君に手紙を書いても返事もないし，論文も来ず，どうしたのか，つまり，病気をして居られるのか，それとも流行の日本脳炎で死んでしまったのではないかと思っていた所でした．

とにかくも，まだ元気で出勤していらっしゃる由，非常に結構です．

Zariski の輪講は今度の水曜日，午前，君がやることになっていた筈ですね．

実を云ふと，最近，代数幾何には余り熱が無くなっているのです．実際，何となく退屈なものですね．それに，近頃，何かスコラ的に成って行く様な気配は無いでせうか．尤も君は，スコラは嫌ひではないかも知れません．

とにかくも又始めませう．その中面白くなって来るかも知れません．今度は君に，local ring のことを教はりたいのです．

以上，何となしに．

成田正雄　へ

1955 年 1 月 1 日
池袋　静山荘　より　(はがき)

新年おめでたう．

そして
ヤア，シバラク

何しろ，今年になってまだ一度も顔を合せないのですからね．

今年はどんな楽しみが期待できるでせう？

山崎圭次郎　へ

1955 年 1 月 1 日
池袋 静山荘　より　(はがき)

新年おめでたう

今年こそ，お互ひに，何か面白いことにぶつかりたいものですね．それはきっとすぐ近くにまで来ているに違ひありません．

それはさうとして，

僕が傍観者的興味を感じる所に，君が暗黙の強制を見るのは何故でせうね．……“真理”と云ふことに関して．

及川広太郎　へ

1955 年 1 月 11 日
池袋　静山荘　より　(はがき)

明けまして
おめでたう

整数論と函数論とが，数学の最高峰の中に結び合さった 19 世紀は，もう戻って来ないのでせうか．

志村五郎　へ

1955 年 7 月 18 日
池袋　静山荘　より　(封書)

もう京都からお帰りになられたことと思ひます．

今度は，不完全な，而も一部分誤った原稿で代読をお願ひして，大変失礼致してしまひました，とにかくも，お骨折りに，お礼申し上げます．

既にお気附きのことと思ひますが，あの原稿の，§3 の Lemma（基本的補助定理）は誤りです．証明だけでなく，Lemma そのものが成立ちません．誤りの原因は，canonical map．の一意性に，自己同型を忘れた所にあるので，さうすると，都合良く尤もらしい結果になるので，つい落してしまったのです．とにかく急いで考へましたので．

実は定理１自身，一般的には誤りです．今迄，Hecke の論文を読みちがへていたので（彼の記号で，群 $\mathfrak{S}\mathfrak{U}$ と $\mathfrak{S}\mathfrak{E}$ との混同），結果は少し形が違ふのです．

一般に，J の自己同型 ζ に対し，$E_l(\zeta^{-1}\Theta)=E_l(\Theta)M_l(\zeta'\zeta)$ となりますが，又，違ふ curve の Θ^* に対しては $E_l(\Theta)^{-1}E_l(\Theta^*)=M_l(\eta)$，$\eta$ は単数で real, total positive です．real なことは $\eta'=\eta$ より（$\pi\mathfrak{p}$ が R を生成する時），total positive なことは（$\eta=\delta'_{\Theta^*}$ ですから）$\sigma(\lambda'\delta_{\Theta^*}\lambda)=2\,d(\lambda,\ \Theta^*)\geqq 0\ \forall\,\lambda\in R$ から出るわけです．故に R_0 の，指数２の実部分体の，total-positive な単数群を，R_0 からの単数の Norm の群で割った群の位数を 2^ρ とすれば，此の J を jacobian に持つ，双有理同等でない C の個数は $\leq 2^\rho$ です（Θ と $\zeta\Theta$ とは同じ函数体を持ちますから）．Hecke の場合には実２次体の基本単数のノルムが $+1$ 又は -1 に従って２又は１，そこで此の様な C の不変量の対称函数を，k^* に添加した体での分解法則を考へる．此の体が，正に Hecke が扱っているものなので，此れなら，jacobian の同型から，すぐに分解法則が出て来るわけなのです．（勿論一次の $\mathfrak{p}$ に対し $\mathcal{A}(\tilde{J})=\tilde{R}$ となる体に対してしか云へません）

従って Teil-point の体の場合にも，此れに応じた修正が必要となります．不変量の体の reciprocity は，上のことから，$g\leq 3$ なら容易に出ると思はれます．$g>3$ だと，$\lambda_{\mathfrak{a}}J$ が jacobian になるための十分条件が，まだ求まっていません．

Hecke の論文で，Klassenzahl ungerade と云ふ条件は，彼が Klassengleichung を考へているから必要となるものの様です．Klassengleichung の性質に当ることも全部出したいのですが，まだよく考へていません．

以上の様なわけで，此れまで此のことに気附かなかったのは慚愧の至りです．誤った定理を証明しようとしていたのですから，どこかに誤った証明が入り込むわけです．先日の，証明の誤りの御指摘に，心から感謝致します．

以上，お礼とお詫びとお知らせまで．

白谷克巳　へ

1955 年 10 月 24 日

東大理学部数学教室　より　（封書）

お手紙有難う．先日いらっしゃった時は，こちらもかなり忙しかったため，ゆっくりお話も出来ず，大変失礼してしまひました．（９月）８日の夜，それからの予定を打合せしようと思っていましたが，うっかりしている中に過ぎてしまひ，それから連絡がとれず，（尤も９日に第一生命でお会ひ出来るかと思ってもいましたが），それからどうなさったかと思っていました．

それにしても，傍聴の件につき，僕達の方でももう少し積極的に運動すべきだったと思ひます．運動したとしてもうまく行ったかどうか勿論わかりませんが，結局こちらとしても何もしなかったことは非常に申し訳なく思って居ります．

日光に行かれたことは，お手紙で始めて知りました．それならばあちらでお話しする機会もあったわけなのですね．会場ではごたごたしていて暇がなかったかも知れないとしても，夜に，僕達の泊っている植物園にいらっしゃれば，割にゆっくりお話し出来たのではないかと思ひますが，何としても，一寸連絡していただければよかったのにと，残念に思ひます．

とにかくも，僕達の手落ちで，折角いらっしゃったのに，余りお構ひも出来ず，又余りお力にもなれなかったことを，お詫び致します．

所で，日光で会議を聞いてわからなかったと云ふことですが，それは，一般的に云って，わからないのが普通なので，正式のメンバーとして，すました顔をして聞いている人々も，大抵はわかっていないのですから，別に気になさることもないと思ひます．シンポジウムの意義は，会場外での個人的な話し合ひの方に，より多くあるのですから，然し，わからなくても，漠然と聞いていただけでも，結構何かの役に立つことが多いのではないかと思ひます．実際，どんなことが話題になっていたかと云ふことがわかるだけでも．

Weil と Artin とを読んでいらっしゃるのださうですね．Artin は，急いで読めばかなり速く読めると思ひますが，両方とも精読する価値があると思ひます．尤も此の間書いた文献のすべてが，精読に値すると云ふわけではありません．それにしても，無いものが多いのは困りますね．そちらに無い文献のリストを送っていただければ，こちらで簡単に整へることの出来るものはお送りしても良いと思ひます．

シムポジウムについては今度の月報に速報がのりましたが，大雑把なことはあれ位ではないかと思ひます．数学的な詳細については，数学会の雑誌「数学」に出ますから，もし興味があれば，その方を読めば良いと思ひますが，実を云ふと色々な話題について，今すぐにそんなに詳細なことを知っても仕方ないのではないかと思ひます．とにかくも，買っておけば，いつか読む必要が起きたときには読めると云ふわけです．

最近，Eichler の，Nancy での lecture note のリプリントが出来ました．少し代数函数論を勉強してからお読みになれば面白いと思ひます．非常に興味のあるアイディアを含んでいるものです．

いつも御返事が遅くなってしまって済みませんが，此れは僕の怠慢と気紛れによるもので，余り怒らないで下さい．又別にわからないことにぶつからなくても，時々お手紙下さい．僕に対しては遠慮は必要でもなく望ましくもありません．それに，わからないことを聞くだけでなく，そちらで考へていらっしゃること，又は此れからのプランその他についても，どんな簡単なことでもお知らせ下さい．その様なことを話し合ふことも，有意義だと思ひますから．

児玉君からもお便りをいただきました．末筆ながら，彼にもよろしくお伝へ下さい．

では又，そのうちに．さやうなら．

宮原克美　へ

1956 年 1 月 1 日
池袋　静山荘　より　(はがき)

新年　おめでたう

今年は，恒例の年賀状書きが何となく憶劫です，きっと原稿書きに追いまくられたためでせう

まあ余計なことは言はずに，とにかくオメデタイんだから，といふわけですな．

今年は，君の力作が出ることを期待します．といふ前に先ず，お説教，絵書きもたまには，数学者並の苦労をして見なさい，本当に．

では又，その中に．

成田正雄　へ

1956 年 1 月 1 日
池袋　静山荘　より　(はがき)

新年おめでたう．

何故おめでたいか？ですって，それはこっちにはちゃーんと理由があるんですから・・・・・・・・(以下忘れた!!)

ハテ？　前にもこんなことがありましたね

杉浦光夫　へ

1956 年 1 月 5 日
池袋　静山荘　より　(はがき)

とにかくも
新 年 お め で た う
今年は僕の方で驚いた，君から年賀状が来たので．

所で早速ながら，幾何学基礎論の話，あれは運営委員会の行き当りばったり主義ではなくて，チャンとした大方針に基いたものなのですが．基礎論がダメなら，射影幾何とか，その他，図に描ける幾何の話はどうでせうか．出来るだけ，専門以外の話をして貰ふといふのが，大方針の一つなのですが．……別に準備のため勉強する必要はありません．射影幾何は，この間講義したのでせう．….専門家向きの準備の必要なしといふのも大方針の一つ．では又

成田正雄　へ

1956 年 1 月 5 日
池袋　静山荘　より　(はがき)

先日は色々御馳走様，おまけに百人一首をボイコットしてどうも失礼，但し主謀者は僕に非ず．

所で，S. S. S. の例会の件，杉浦が幾何学基礎論の話を断って来たのですが，どうしますか，佐武氏もだめなので，他に人はいないと思ひますが，君が何か“図を描ける幾何,, の話をしませんか．或は誰か外に適当な人を探すか，何とかして下さい．期日が迫っていますからね．

来週月曜（9 日）午後，modular 函数のゼミのことで皆駒場に集りますが，良かったらそのときでも相談しませう．

末筆ながら，お母さんによろしく．

山崎圭次郎　へ

1956 年 1 月 5 日
池袋　静山荘　より　(はがき)

遅ればせながら
謹賀新年

昨年は，何か色々なことが多過ぎましたが，今年は少し落着けるでせう．

数論グループの新年会のこと，どうなりましたか．

宮原克美　へ

1956 年 1 月 7 日
池袋　静山荘　より　(封書)

拝復，久し振りに君から長い手紙を貰ひ，いささか昔が懐しくなりました．それにしても，今度の君の手紙が余り難解でないのは，確かに，少し文章がダレているせゐかも知れません．年は取りたくないものですな．

所で，僕が去年書きなぐった原稿の殆んどすべては，どっちみち一文にもならぬものばかり，“月報”とか“数学”とか，“売文”と云ふよりは一種の“売名行為”ですな．例へば宮原克美の如き，記憶喪失症にかかっている“現代人”の頭の中から，僕の名前が喪失されるのを防ぐためには，常に月報にでも，僕の名前が出ていなければならない，……内容を読む読まないは別問題，“ハハンあいつ，又わけのわからぬことを書いているな”と誰かが考えればそれで十分，……何だかうら淋しい世相ですな．

これもまた communication, カテゴリストはこの言葉の前に，どんな形容詞をつけますか？　extra communication, degenerate commun…., etc, etc…….

“人間のすべての仕事は分ってしまえば実につまらんものです．” 人はこれを “コロンブスの卵” と云ひます．

そして，結果から見てつまらん画を描くために，つまらん恋をするために，数学者並の苦労をするのはバカバカしいのかも知れません．何しろ合理主義の世の中，おまけにその合理主義が行き詰れば，地上数万呎とか，superism とか，便利な世の中ですな，実際！

つまらんことをつまらんと承知しながら，あきもせずに朝から晩まで，π とか $\mathfrak{p}$ とか ω とか書きなぐっている，それが数学者と云ふものです，正しく現代の錬金術ですな，“何故数学をやるか？”“美しいからさ！”確かに，simple にさう答へられれば，それで十分なのですが，一体誰が，そう答へられるでせうか？　成程数学は芸術かも知れませんが，余り芸術的な論文にお目にかかったことがないのは，僕の怠慢のせゐでせうか，雑誌の中から“美しい”数学を拾ひ出すのは，日展の中から生命に溢れた作品を探し出すのより困難です．成程 Riemann とか誰とか，たまにはそんな人も居ますが，誰もが Riemann になれるつもりで数学をやっているのでせうか．画ならば，下手でも何でも，一応審美感を満して呉れますし，物好きに買ふ人も居るかも知れませんが，つまらない論文など，Weil が云ふまでもなく，人間の暇潰しの役に立つに過ぎません，それも，銀座に行くだけの金の無い人間の暇潰しに．‥‥一体“美しさ”はどこに行ったのでせう．それとも，二流以下の数学者は，稀にしかない“美しい”数学を鑑賞していれば良いとでも云ふのでせうか？五流の画家が結構威張って画を書いている世の中なのに．

所で現代の錬金術師は何と云ふでせうか．彼はすべてのものにウンザリして，余生を暇潰しの中に送らうと決心しているのかも知れません．そして実際，錬金術位，完全な暇潰しは余りありません，誠に天才的な思ひつきですな．尤も本当に完全な暇潰しは自殺することですが，ノイローゼとか何とか，とかくうるさい世の中ですからな．尤もこの一事を見ても，世間の人が如何に暇潰しのネタに事欠いているかがわかるわけで，その点数学者は，“持てる者”ですな，‥‥暇潰しの材料を．だがもう一人の錬金術師は，もっとイカサマ師で，と云ふのは，もっと野心家で，色々なことを夢見ています．彼の夢は世の中の神秘を探究することですが，何しろ原子バクダンが，ありとある神秘を吹き飛ばしてしまったので，残す所としては存在と非存在の中間，即ち数学の世界だけです．ここでの彼の役割は，昔の錬金術師と同じことで，あらゆる神秘を，苦心の末に探り出しては，実はそれが神秘でないことを，即ち一般法則を，証明するのです．近代化学はこの様にして生れたし，未来の数学もこの様にして生れるでせう．それは，理論の書き直しや， Axiom の配列変へなどを事とする“デザイナー”などの手には及ばぬことです．

実際彼は神秘を求めているのでせうか？それなら神秘を探り出した所で立ち止るべきです．或は彼は神秘を無くさうとしているのでせうか？　それなら強いて神秘を探り出さうとせずに，最初から手を束ねてタバコでもふかしている方が賢明です．君の持っている様なマドロスパイプで．（尤もこれは，恋をしている人に，“幻滅を味はないために，中途で立止れ”と云ふ様なもので，効き目のない忠告ですな）

第三の錬金術師，それは奇蹟を行ふ者で，つまり鉛を金にすること，純粋数字から現象的な応用を汲み出さうとする山師です．昔の錬金術師が，それが出来ると宣伝して，バカな王様から金をせしめては実験を進めた様に，現代の錬金術師も，それが出来ると宣伝しては，ケチな政府から金をせしめて研究を続けようというわけで，まさしく思想を利用することですな．鉛が金にならなかった代りに，ウラニウムが鉛になったわけで，山師の云うことも長い目で見れば本当かも知れません．まあ王様とか政府とか，余り利口振らずに，適当にダマサれてケチな財布の紐をゆるめる方が功徳になる様です．どうせ汚職とか何かに使ってしまう金なんですからな．

所でまだ第四，第五の錬金術師もいるわけですが，“一度に３つ以上のことを云ふ勿れ”と“選挙当選虎の巻”に書いてありますから，後は此の次に．

そんなわけで結構色んな理由があるのですが，誰も本当のことを云ふのは気がひけると見へる．第一そんなことを云ったら給料を上げて貰へませんからな．だから“世の中に数学者がいるのは良いことだ，数学の栄えない国は滅びる”とワメき続けるのが一番良いのです．理由なんかいりません．何しろこれは教祖のミコトノリですからな．

大部脱線した．まだ云ひたいことがあったのです．

クロレラならずとも，優秀なものが次の世代に残るのは，数学でもその通り，ガロアを見給へ，リーマンを見給へ，そして彼等の当時に栄えていた下らない数学を見給へ，更に眼を転じて，現在栄えている下らない数学を見給へ，そこで君は忽念大悟するであろう，喝！

さて，縦のプロポーションの superist は，喫茶店で，プリンばかり食べていても，結構様々な変化を楽めるわけですが，残念なことに天才といふのは，プロポーションを失った人物のことなのです．つまり彼は一つの思想を信ずるだけで利用せず，犠牲にはおかまひなしに社会を改革し，人には迷惑をかけ，一つの思想を無制限に何にでも適用し，飯も喰わずに氷河に噛りつき，communication など，super の何乗が附くものでも，一切やらず，ただ同じことをワメき散らす，素人はバカにせずに愛するが同じ専門の天才共はすべて軽蔑する．誠に人間はお互ひを愛せないわけですな．そして結局この様な天才は．歴史に“書き込まれる”のではなくて，歴史を“作る”のです．一枚上手ですな，縦のプロポーションの天才よりも．論より証拠，君にはこの様な天才を説得することは出来ない，彼は“信は力なり”と云ふ格言を軽蔑しますが，それを実証しているのです．

所で，いつの世の中でも人のマネをしたがる徒輩はいるもので，彼等は，このタテのプロポーションの外れた天才のマネをしようとします．そんな連中が“信は力なり”とノタマって，カミカゼを期待したりするわけですが，その技神に入って来ると，少し前のドイツの独裁者達の様に，自分達の信じない思想を大衆に信じさせ，彼等すべてをこのプロポーションを外した天才に“仕立て上げ”て，その力を利用しようとする，誠に，縦のプロポーションを豊かに持ち，思想を信ぜずに利用する，素晴らしい天才ですな．

哀れなるはいつの世にも大衆なるかな，インテリからはお情けに愛され，縦のプロポーションの天才からはハメを外させれ，民主主義の政府からは税金を取られ，自由主義の政府からは検閲済みの映画ばかりを配給され，学校で物を教はれば，民主党から憂へられ，選挙の時はオダテられ，選挙が済めばあっちもこっちも強制測量で，アゲクの果が原子バクダン，アーメン！

とにかくも，真の天才は，プロポーションの存在しない所に特徴があるわけですから，失敗は彼等に附き者で，彼等の失敗が，その消滅を意味しない世の中になったとき，始めて，天才がその天才を発揮できる世の中になるのですが，それは夢の様な話．

それまでは，君や僕の様に，縦のプロポーションを楽しみながら，喫茶店でプリンやアイスクリームばかり食べている方が，天才的であるわけですな．

又脱線，要するに生活の犠牲の上に成立つ

天才と云ふものもあるので，誰も好んでそんな天才になるのではなく，彼等は，自己の天才から逃れることが出来ないのです．神よあわれみ給へ！

君の様に良く呑み屋に出入りする人間は，その片隅みで，飲んだくれたこの様な天才を見出すことも良くあるでせう，もし君が鼻がきくならば．

誠に super-super-communication は難しいかな！　多くの天才は，その手段を見出し得ずに，自分の天才を持て余すのです．

ここに至りてカゲの声の曰く，“お正月には，こんな下らんことは考えずに，地上数万呎の夢を夢見るに限る”と，噫！　これぞまさしく縦のプロポーションの声！

君の恋人によろしく．

Y. T.

A. Weil へ

1956 年 2 月 14 日
東大教養学部数学教室より（封書）

（欧文につき本文は 99 頁に組んである.）

杉浦光夫　へ

1956 年 8 月 3 日
池袋　静山荘　より（はがき）

お手紙有難う，先週火曜日にお会ひする筈でしたが，色々な都合で暇が無くなってしまったのです．所で Weil 特集のこと及川とも話しましたが，何故，今 Weil を取上げるのか納得できません．もう去年の秋から，月報でも何度も扱っているし，正直の所又 Weil について書くのは気が進まない，というのが駒場の二三の人々の感想です．寧ろシンポジウム特集をやったら如何．とにかく一度お話しませう．空いているのは5日か7日．7日の午後数学教室に行くことにします．或は5日のお昼頃（遅くとも1時迄に）僕の所に来て下さっても結構です．

君が出掛けてから東京は毎日物凄い暑さです．何も出来ずブラブラしています．

白谷克巳　へ

1955 年 6 月 16 日
東大理学部数学教室　より（封書）

お手紙拝見致しました．

差当り用件から片附けませう．

I. M. U. シムポジウムの傍聴は，人数は制限がありますが，出来る見込み（まだ確定はしませんが）です．申し込みは，本人から，書面でする方が望ましいと云ふことなので，「東京都文京区本富士町一．東京大学理学部内日本数学会」宛に，在学の学校名，身分（大学院学生とか何とか）を書いて，傍聴したい旨，ハガキででも申し込まれた方が良いでせう．なるべく急いだ方が良いと思ひます．尚人数に制限がある様なので，山本さんに，整数論を専攻している旨の推せんの言葉を，一寸書き添へて貰った方が有利でせう．勿論此れは，無くても差支へありません．

又シムポジウムの準備として開かれる“研究集会”は，7 月 15, 16, 17 日に京都で，7 月 22 日から仙台で，開かれます．京都のは代数幾何に関係のあるもの，仙台のは，コホモロジー系統のものです．僕は京都の方に行くつもりですが，健康状態が良くないため，確実ではありません．此れは自由に入って聞けるものと思ひます．

又来日する数学者に，講演，講義などを依頼する手筈も整って来ている様です．Weil は東京と京都に1ヶ月位づつ滞在する予定で，又 Serre は京都で講義，九州にも行く筈です．Artin, Deuring なども講義を引き受けて呉れたらしく，Brauer は，東京の外，北海道まで行ってもらふとかの話，その他，まだ細かいことは決ってないでせうが，シンポジウムに前後して，まあ，東京が主になるでせうが，各地方でも，色々面白い話が聞けるのではないかと思ひます．此れは公開ですから，その頃に東京に出ていらっしゃるのは，その点からも，面白いのではないかと思ひます．その外に，個人的につかまへて，色々話しを聞き，こちらの考へも話さうと云ふ計画もあるわけで，それも面白いことと思はれます．此れ等の点，細目まで決るには，8月末頃までかかるでせうが，詳しくわかったら，又お知らせします．

東京は，整数論は，かなり盛です．S. S. S. に整数論グループがあり，メンバーは不定ですが，今，シムポジウムの準備として，色々，多少泥縄的に，勉強しています．此れまでにあった話は，志村（五郎）さんの虚数乗法，久賀（道郎）さんの ζ 函数（イデール の上で扱ふもの），山崎（圭次郎）君の Faisceau（特に base space が分離公理を満さないとき――代数幾何や数論に出て来るもの），林田（侃）さんの，代数体と，代数函数体との類似など，今度から僕が，超楕円函数の場合の虚数乗法をやる予定で，以上からおわかりの様に，興味の中心は，代数体と代数函数体との関連と云ふ所にあります．又此の他，モヂュラー函数に興味を持っている人も多いのですが，シムポジウムの題目から外されているので，当分見送りです．勿論，此等若い人々の外に，先生方や，その他の人々も居り，セミナリーも，ありますが，それは後のことにしませう．

まだ色々，お話ししたいことは多いのですが，始めに書いた，傍聴の申し込みは急いだ方が良いので，取り敢へず此れで止めにして後の機会に廻しませう．

では又，池田さんによろしく．

（類体論を勉強されているのなら，Hasse の Berichte にも眼を通された方が良いと思ひます．リプリントがある筈）

中村得之　へ

1956 年 8 月 6 日
池袋　静山荘　より（はがき）

前略，

大部以前に，例の書類の件のお葉書を頂きましたが，その後，いつかお会ひする機会があると思って返事も出さずにいる中に，夏休みになってしまひました．もう君は千葉の方にずっといることと思ひますので，9月になってからでも見せて下さい．いつもの怠慢から，すべてが凄くルーズになってしまって済みません．

毎日飽きもせずに暑さが続きます．何も出来ないで，ブラブラしていたいと思ひながらそれも出来ないという所です．

お元気で．

谷山欣隆　へ

1956 年 10 月 17 日
池袋　静山荘　より（はがき）

御葉書拝見

「かき」は余り好きでないからいりません．くすりは適当に送って下さい．

本の方は仲々終らないので当分暇になりません．大体1ケ月や2ケ月で書けるものではないのです．

以上．

（此の間イカサマ師が洋服生地を売り附けに来て，うるさいので好い加減に一つ買ひましたが，物が悪くて何にも使えないので，和歌子にやらうと思ひます．値段もかなり安いから，洋裁の練習用にでも使う様に．失敗しても惜しくないものです.)

谷山欣隆　へ

1957 年 1 月 10 日
池袋　静山荘　より（はがき）

謹賀新年
最近大変忙しくて参っています.
そちらに年賀状行っていたら，回送して下さい．ボーナスを貰ったので，おばあさんに小遣ひを送るつもりですが，暇がなくて，その中送ります．和歌子，毛絲，何か変った色のもの欲しければ探してやります.
いづれ又

谷山義泰　へ

1957 年 1 月 10 日
池袋　静山荘　より（はがき）

謹賀新年.
今年は忙がしくて，年賀状は全部今日書いた．10 日おくれの正月.

山崎圭次郎　へ

1957 年 1 月 10 日
池袋　静山荘　より（はがき）

謹賀新年

今年もどうぞよろしく.
——せめて此のハガキが当る様に.

杉浦光夫・和子　へ

1957 年 1 月 10 日
池袋　静山荘　より（はがき）

謹賀新年

先日はどうもお邪魔様でした.
——年賀ハガキが余りさうなので流用した．悪しからず，洗濯機か切手でも当るかも知れない．今年は何かせわしなくて，今 30 枚位年賀状を書き上げた所，とんでもない新年だ.

松本晧一　へ

1957 年 1 月 10 日
池袋　静山荘　より（はがき）

謹賀新年
又 1 年たってしまひました．その後どうして居られますか．考へて見るともう随分お会ひしませんね.
10 年前，11 年前，何か夢の様です.

上野　正　へ

1957 年 6 月 13 日
池袋　静山荘　より（はがき）

我々の問題にしている学生層のことについて，市川君からかなり重大な疑問の点が提出され，それについて事前に相談する必要があると思ひますので，土曜日の午後 4 時頃に，ソバ屋においで下さい．市川，上村君も来ます.
尚渡辺君の住所，わかりませんので，出来たら君の方からも連絡して下さい．明日調べて，僕の方からもハガキを出すつもりですが,
以上取敢へず

上野　正　へ

1957 年 9 月 3 日
池袋　静山荘　より（はがき）

前略.
“歩み” 別刷りの校正を日曜日（15 日）にやる予定です．一人では大変ですから，出来れば手伝ってくれませんか．おひる頃三栄社に行く予定です（一応三栄社に電話して，出来ているかどうか確かめてから出掛けた方が安全です．番号は（35）4318
僕の受けた印象では，学生の動きを積極的

に援助しようという気運は，一般的に余り強くない様です．

新しい組織に移ったとき，常に情勢を検討して，“歩み”紙上に毎号報告する義務を負った委員会があることが望ましいと思います．これは合同会議乃至その一部分がやるべきでしょう．

谷山和歌子　へ

1957 年 10 月 30 日
池袋　静山荘　より　（はがき）

日暮里に行ったら頼まれたので洋服のキジを送る．(W巾) 1 ヤール 1250 円のもの 2.3 ヤールで 2875 円，裏地（S 巾) 1 ヤール 180 円のもの 2 ヤールで 260 円計 3235 円，それと送料で約 300 円不足．

緑色のキジは余りないし，選ぶのが難しい．明るい色だと安っぽくなるから，どうしても暗い色になる．青系統のものなどには，格子や何かでも，割合に安くて感じの良いものがあるが，緑と云われたから緑にした．無地だと中々引き立つものがないので，一寸変ったものにした．仕立てて見れば，かなり良くなると思う．（モノ は，3000 円程度ではこの位のもの（勿論純毛）．一寸良くなるとヤール 2000 円位する．）

一緒に紺の洋服地を送る（W巾 3 ヤール）．これはいつか押し売りが売りつけたもので，タテは毛だが横は人絹の安物で，3 ヤールで 1000 円位の値打ちしかないと思う．洋裁の練習用に，何でも作れば良い．見掛けは良いから背広位は出来る．

（キジは自分で出て来て探すのが最も良い．案内位はしてやる．）

以上．

杉浦光夫　へ

1958 年 1 月 2 日
池袋　静山荘　より　（はがき）

謹賀新年

（とはいうものの今年はサッパリ新年らしい気分もしません．暮の間中風邪で消耗していた上に，スミルノフに追い廻されている始末．）

今年もどうぞよろしく．

「数学の歩み」への御投稿もお忘れなく．

山崎圭次郎　へ

1958 年 1 月 8 日
池袋　静山荘　より　（はがき）

謹賀新年

今年もよろしくお願い致します．

スミルノフの翻訳に追われて賀状が大部おくれてしまいました．1 日 10 ページづつ訳すつもりでしたが，中々思う様には行きません．‥‥君は終りましたか？

北見悦子　へ

1958 年 1 月 8 日

おくればせながら
明けまして
おめでとうございます．

今年もどうぞよろしくお願い致します．

時の経つのは早いものですが．お会いする度に，いつも，段々大きく生長なさって行く跡が見られる様に思われます．僕は相変らず，停滞したままです．よいお年をお過しなさる様に．

駒場には何曜日に行かれるのですか．

谷山欣隆　へ

1958 年 2 月 15 日
池袋　静山荘　より（はがき）

この間，小包受け取りました．四角モチはもう食べてしまひました．カルシウム剤は大変結構です．

去年，和歌子に頼まれて洋服のキジを送りましたが，着いたとも着かないとも，気に入ったとも気に入らないとも，何とも云って来ませんがどうしたのでしようか．最近化繊布地の安物など多くあり，良ければまた送ります．

仙台から，たまには家に菓子でも送れと云って来ていますが，面倒なのでまだ送っていません．その中に暇になったら．

では又．

志村五郎　へ

1958 年 2 月 15 日
東大教養学部数学教室　より（封書）

大分御無沙汰しました．昨年末から健康状態が余り思はしくなかった上に，スミルノフに追いかけられて，何かしら落ちついた気分になれない日が続いたので，講義が終ってホッとしている所です．入試の採点の始まるまでしばらくの間，一休みできます．

先づ要件から，この間の校正，独断で二三ケ所訂正しました．

p. 5.　prop. 3　$\nu_i(p\delta_E)\geqq p$,　不等号の向き $\leqq$ となっていたのを逆に)．

p. 9.　prop. 10 の中の prop. 5 は prop. 9 に．

p. 16.　$d'(\widetilde{X}_p)=d'(\widetilde{X}_p)=p+1$　の第一項は何のつもりか良くわかりませんでしたので $d(\widetilde{X}_p)$ としておきましたが．

p. 17.　fonction canonique de Γ sur J, これは少し前には　application canonique de Γ dans J としてあったので，それに統一しておきました．

以上の外に自明なミスプリントが数ケ所．

また，pp. 2〜3（又ハ 1〜2）$x\to x^q$，$q=p^f$ という写像の説明の所で f は整数としてあって，それから何の説明もなしに，少し後で $x\to x^q$ は rational ˛map とありますが，途中に $f>0$ を入れておかないと具合悪いでせう．これは面倒なのでそのままにしておきましたが，どうしますか．

文献 8 の題目は，お送りしたゲラに書き込んであるようにしました．これも独断です．

さて，この［8］の件ですが，やうやくスミルノフが終るので，(あと少し，見直す必要あり），そろそろ書き始めるつもりです．また後で打合せしませう．

こちらは相変らず平穏無事です．平穏はとにかく，無事は感心しません．例の symmetric product の件，$V(n)$ の n が十分大きければ，Albanese への mapping による各点の逆像が irreducible なことはわかりました．有限体の場合に帰着させて，そこで rational point の個数を評価するのです．この評価はかなり trivial.　又 Albanese の mod $\mathfrak{p}$ での reduction が Albanese になる (almost everywhere で) これも同時に出ます．そして，V の ζ-函数の絶対値最小の 0 点が, Albanese のそれと，$V(n)\to A$ の，原点の逆像の var. のそれとからきまることがわかります．この逆像の ζ-函数に，あり得る絶対値最小の 0 点がないことがいえれば良いのですが，手がつきません．(V が surfaee のときいへれば，あとは帰納法でできますが)．　というより，サボリながらやっているので，‥‥元気がなく，中々続かないのです．n が十分大きくないときには，一点の逆像が irreducible でない例，また irreducible でもその Albanese が 0 にならない例があります．郡司君が，n が十分大きいとき，cross section $A\to V(n)$ が存在することを証明しました．これは多分 regular だらうというのですか，まだ不確実．

久賀氏は Hecke をやっています．彼から手紙が行くでせう．郡司君は，多分 Johns Hopkins Univ. に，この9月に，行くことになると思ひます．始め Northwestern Univ. に，というのが，途中で変ったのです．森川君も Johns Hopkins に行くさうです．何だか，皆居なくなってしまふみたい．

本郷では須藤氏がやめて文部省に行きました．後任は田村一郎氏(トポロジー)．又4月に福富氏がやめて日大に行きます．日大では数学科，物理学科などが新設されるので，辻さんや河口さんなどが中心になって数学教室を作るのださうです．福富氏の後任は佐藤氏．(多分志村さんと同級でせう．卒業して物理に行っていた人です)．佐藤氏は最近，analytic manifold の上の distribution (Schwartz の) の“新しい説明”を発見して，いろいろ面白い結果を得たということです．ただし噂で聞いたので，どんな説明なのかわかりません．

山崎君が結婚します．相手は久保田さん．これはもう御存知でせう．

数理科学研究所も段々具体化して来ました．SSS から色々申し入れをしていますが，時には難航します．研究所のあるべき姿について，御意見があったらお知らせ下さい．そちらの模様を知らせていただけるだけでも参考になると思ひます．

この間の御手紙の中の Hilbert modular etc. に関すること，もう少し詳しく説明していただければ有難いと思ひます（あの手紙，久賀氏が持っていってしまって，返してくれません)．

ではまた．遠からずお便りします．

佐武さんにもよろしく．(上林君が Northwestern Univ. に行くことになると思ひます．お伝へ下さい)．

Weil から1月始めに手紙をもらってまだ返事を書いていません．これもたしかにけしからん話ですが．

2月25日．午前4時半．

草場敏夫　へ

1958 年 3 月 6 日
池袋　静山荘　より（推定)（封書）

お手紙有難う．

病気は誤診だったとのこと，安心しました．でも，これからも身体には気をつけられるように．

2月 19 日附のお手紙に返事を書こうと思っているうち，忙しさにまぎれてつい日が経ってしまいました．もう試験も終ってせいせいしていることと思います．成績はどうであっても，余り気にする必要はないでしょう．

本郷に行ってからも退屈な講義があるかも知れませんが，少し前に比べて内容はかなり充実して来ています．そのせいか，学生が余り講義を聞きすぎるという批評もある位ですが，修業年限も短くなっていることですし，学部の間は良く聞いておいた方が良いでしょう．尤も，単位はかなり楽に取れますから，聞かなくてもその点は余り心配はいりませんが．

然し．何といっても，自分でやる勉強が結局一番役に立ちます．尤も，自分一人だけでやっていると，意味を取りちがえていたり，一応はわかっていても，あまり確実でなかったりということは良くあります．他の人に話す，或は説明することによって本当に自分のものになるという場合が多いのですから，何人かのグループを作って一緒に勉強して行くのが一番効果的です．このグループは同級生のそれとは限りませんが，三年の間は，上級生と一緒にやるのは一寸無理でしょう．（やさしいテキストを読む場合には可能ですが)．……それと同時に，練習問題をやったり，自分で実例について考えて見るのも有効で，或る段階では是非やらなければならないことです．

さて，秋月氏の調和積分論は，かなりむずかしい本で色々な予備知識が必要です．少くとも，位相幾何（topological groups にあるような“点集合論的”トポロジーではなく，algebraic topology と呼ばれているもの），manifold の理論，多変数の複素函数論の初歩，（勿論普通の複素函数論も必要）等，まだあるかも知れません．順序としては代数函数論から始めるべきで，普通の複素函数論→Riemann 面の理論→代数幾何→調和積分論と進むのが正統的です（その間に上の様な予備知識を勉強する），Complex analysis の Dirichlet problom など，かなりそれと関係して来る筈なのですが，放っておくのは余り感心しません．Riemann 面の理論としては，昔は Weyl の“Die Idee der Riemannsche Fläche”を読んだものですが，今では岩沢健吉“代数函数論”（岩波）か，Chevalley “Introduction to the theory of algebraic functions of one variable”が良いでしょう．（もっとも後者は内容が精緻すぎて，余り入門向きとはいえませんが）代数幾何は代数函数論をやってからにする方が賢明です．とにかくも，代数幾何をやる前には一応相談に来て下さい．かなり気をつけないと，変な所に迷い込んでしまうことが多い様ですから．

Chevalley の本は，読み出したのなら読み続けて行っても差支えないと思います．いずれにせよ，あの本の内容位は知っている必要があるでしょうが，どうも Chevalley の本は，どの本も，余り入門向きでないと思うのです．途中で退屈したり嫌になったりしたら（わからなくなることはないと思います），一応打切りにして，半年なり何なり時間をおいてからまた読み続けるのも良いと思います．かなり色々な部門に現れる内容が抽象してあるので，それ等の内容を知らないと，表面的な理解に止る恐れがありますから．

尤も Van der Waerden をⅡ巻の終りまで全部読んであれば，代数関係のそれ以上の勉強は，それ程急ぐ必要はないでしょう．むしろ他の方面をやる方が良いかも知れません．

差し当り読むべきものとして，次の様な plan はどうでしょうか．

Complex analysis の残りの部分を読む．

岩沢：“代数函数論”

Loomis：Harmonic analysis

この外，早い中に読んでおいた方が良いものとして

Bourbaki：Topologie générale 第1分冊（Ⅰ，Ⅱ章）……練習問題もやる．

代数的位相幾何関係の本（何が良いか，一寸すぐには思い出せませんが）

勉強して行く上に一つ注意．勉強が進んで行くにつれて，何度か，質的な飛躍がありますから，この調子で読んで行くと後何年たってどの位，などと考えるのは無意味です．また，初め読んだとき良くわからなかったり，上滑りしていたことも，他のことをやっている間に自然にわかって来ることが多いのですから，必要以上に拘らない様に．忘れることについても同様．必要なことは何度も出て来ますから，自然に記憶に残ることが多いものです．それに現在は準備段階ですから．…・本を読んで“理解する”だけでなく，自分で紙の上に証明を書いて見て“納得する”様な練習も必要です．知識の確実度が増します．要するに，自分で“動かして”見ること．

またお便り下さい．僕は二学期に本郷で講義をする予定ですから，さうなればお会いする機会も多くなると思います．

4月に Siegel と Eichler が来て，東京で何回か（英語で）講義をします．内容は難しいでしょうが一度位聞いて見るのも悪くないと思います．

ではまた．

3. 6.，1958

谷山欣隆 へ

1958 年 4 月 26 日
池袋 静山荘 より （はがき）

前略

今度の日曜には少し用事があるのでそちらに行けませんがよろしく．（いつも土曜の夕方（3時から）都立大学で講義があるので，行くとしても日曜に日帰りですが，今度はそれも一寸無理です）

4月から助教授になりました．多少月給が上る筈です．（どうせ多くはありませんが）

志村五郎 へ

1958 年 9 月 22 日
東大教養学部数学教室 より （封書）

自分でも驚く程御無沙汰してしまいました．本当に「シバラク」という所です．郡司が発ち斎藤も発ち，前にもましてすっかり淋しくなってしまいました．久賀さんと二人で気の抜けた様にしています．そのほか，余り変ったこともありません．また相撲が始まって，若乃花が優勝しそうだとか，東大が何年振りかで慶応に勝った（勿論野球の話）とか，かなり大きい台風が来て学期試験が一日延期になり，折角早起きしたのがフイになったとか，要するに毎年繰り返されるようなことばかり．台風のおかげでこちらはすっかり秋らしくなりました．蟲の声が盛んに聞えて来ます．何と云っても秋は一番良い季節です．然し早いもので，もう一年近くなるのですね．今頃はアメリカ行きの準備にお忙しいことと思います．

最近こちらで良く話題となることに，「日本は数学では田舎だ」ということがあります．こんなことは今更云うまでもないことですが，要するに，田舎だから，皆が外国に行くのも止むを得ないとか，或いは少くとも一度は外国に行かなければならないとか，といった様なことです．郡司からの手紙によれば，彼が考えていた様なことはアメリカで何とか云う人が 55 年頃からやっていて，もう paper が出る頃だということで，彼はガッカリしていました．森川君のやったことと Lang-Serre の類体論（分岐する場合の）とか，そう云ったことを指すのでしょう．……「田舎」である，ないというより，一種の雰囲気の問題，日本にいる日本人に共通している或る型の気質，それにセチ辛さ（気分的な人口過剰）といったことかも知れません．

こんなことを書いている中に思い出しましたが，この前お手紙したのは，たしか Siegel, Eichler が日本に来る前のことだったと思います．今頃こんなことを書くのもおかしな物ですが．思い出したついでに一寸．Eichler の講義では特に新しいこともなかったし，それ以外の話でも我々は Eichler を十分に利用したとはいえませんでした．（僕の英語の下手なせいもあって話が良く通じなかったこともある）．彼の様な人を“利用”するには，かなり永い間一緒にいなければならない様な気がします．（それが十分かどうかは別として）．あるいは，いくら永くつき合っても，彼の論文を読む以上の“利益”は得られないかも知れない．彼もまた“田舎”に居るのでしょう．

確かに Marburg は，というよりドイツ全体が，今では田舎でしょう，というのは，彼等は代数幾何を持っていないから．そして，田舎を特徴附けるのは，テクニック，或いは道具の不足乃至不備ということである様に思われます．これだけはどうにもならない．すべてを自分で考えるわけにも行かないでしょうし，テクニカルなことは論文に書かれないことも多く，書かれても一々探し出すのは，日常の会話から自然に耳に入って来るのとはケタ違いに面倒なことでしょう．この意味で，日本は確かに田舎です（reduction mod $\mathfrak{p}$ という様な道具があるが，使う人は殆んどいな

い）

大部つまらぬことを書きました．数学の話を少し．

最近，Hilbert modular 函数をイデールの上で扱うことをやっています．詳しくいうと次の通り．

k を totally real な体，V を k の valuation vector ring (adèle) J を k の idèle group. V_2 を，V を係数として行列式が J に入る様な2行2列の行列の作る環とする．$V_2 \ni \begin{pmatrix} \tilde{a} & \tilde{b} \\ \tilde{c} & \tilde{d} \end{pmatrix} = \tilde{\mathfrak{a}}$ の有限成分，無限成分を $\tilde{\mathfrak{a}}_0$，$\tilde{\mathfrak{a}}_\infty$ として，$\tilde{\mathfrak{a}}_0$ に対し，$k+k \ni (\alpha, \beta)$ で，$(\alpha, \beta)\tilde{\mathfrak{a}}_0$ の元が整の valuation vector になる様なもの全体の作る加群を $\mathfrak{m}$ とし，又

$$\tilde{\mathfrak{a}}_\infty = \left(\cdots, \begin{pmatrix} a_i & b_i \\ c_i & d_i \end{pmatrix}, \cdots\right) \left(\begin{matrix} a_i, \cdots, d_i \text{ は} \\ \text{実数}, i=1, \cdots, n \end{matrix}\right)$$

行列 $\tilde{\mathfrak{a}}_\infty$ の第 i 成分．

に対し，$u_1 = (\cdots, a_i\sqrt{-1}+b_i, \cdots)$，$u_2 = (\cdots, c_i\sqrt{-1}+d_i, \cdots)$ とおけば $\boldsymbol{C}^n$(complex n-space) の元 u_1, u_2 が定る．そのとき，$D=\{\alpha u_1+\beta u_2;\ (\alpha, \beta) \in \mathfrak{m}\}$ は $\boldsymbol{C}^n$ の格子で，$A=\boldsymbol{C}^n/D$ は abelian variety になる．こうして，$V_2 \ni \tilde{\mathfrak{a}}$ に対し，abel. var. を対応させる．A は勿論，$\mathcal{A}_0(A) \supset k$，より詳しく，$R$ を k の全整数環とすれば $\mathcal{A}(A) \supset R$，という性質を持ち，逆にこの性質 ($\mathcal{A}(A) \supset R$) を持つ abel. var. は，全部，この様にして得られる．R-同型な abel. var. に対応する $\tilde{\mathfrak{a}}$ は，V_2 の double coset を作る．詳しくいえば：U_2 ($\subset V_2$) を，有限成分 $\tilde{\mathfrak{a}}_0$ が integral である様な adèle を係数とし，$\tilde{\mathfrak{a}}_0$ の行列式が unit (イデアル1の idèle) であり，無限成分 $\tilde{\mathfrak{a}}_\infty$ の各成分 $\begin{pmatrix} a_i & b_i \\ c_i & d_i \end{pmatrix}$ が直交群×centre $\left(i, e. \begin{pmatrix} a_i & b_i \\ -b_i & a_i \end{pmatrix}\right)$ であるような行列の作るsubgroup，k_2 を，$\begin{pmatrix} \alpha & \beta \\ \gamma & \delta \end{pmatrix}$，$\alpha, \cdots, \delta \in k$，$\alpha\delta-\beta\gamma \neq 0$ なる行列の作る subgroup とするとき，R-同型なAに対応する $\tilde{\mathfrak{a}}$ は，丁度，double coset $k_2 \tilde{\mathfrak{a}} U_2$ を作る．また，A を適当な方法で一斉に偏極すれば，polarized var. として同型なものは，k_2 の代りに，k_2^+（つまり行列式が正の有理数となるものの作る群）とした double coset $k_2^+ \tilde{\mathfrak{a}} U$ を作る．ここで，$V_2 \ni \tilde{\mathfrak{a}}$ で，有限成分が一定 ($\tilde{\mathfrak{a}}_0$) なもの全体をとり，$\tilde{\mathfrak{a}}_\infty = \left(\cdots, \begin{pmatrix} a_i & b_i \\ c_i & d_i \end{pmatrix}, \cdots\right)$ に対し

$$\tau = (\tau_1, \cdots, \tau_n), \quad \tau_i = \frac{a_i\sqrt{-1}+b_i}{c_i\sqrt{-1}+d_i}$$

とおけば $\boldsymbol{C}^n$ の点が定まり，これを，上の double coset で分けると，丁度 Hilbert modular function の基本領域が得られる．($\tilde{\mathfrak{a}}_0 = \begin{pmatrix} 1 & 0 \\ 0 & 1 \end{pmatrix}$ のとき．一般の場合は，適当に変数変換すればそうなる．) これを V_2 全体の上で考えれば，本質的に相異る函数が丁度 h(=k の類数) 個出て来る．また，

$$J \ni \tilde{a} \rightarrow \begin{pmatrix} \tilde{a} & 0 \\ 0 & 1 \end{pmatrix} \in V_2 \text{ として } J \subset V_2 \text{ と考}$$

え，V_2 の函数を J に制限して，J の上で Fourier 解析をやれば，modular form と Dirichlet 級数との対応が自然につく．また，U_2 を少し小さくすれば，Stufe のある modular 函数（A の $\mathfrak{a}$ 分点を附け加えたもの）($\mathfrak{a}$ は k の整イデヤル) が得られ，対応する Dirichlet 級数は，“導手” が $\mathfrak{a}$ の約数となるものになることも自然にわかる．

Hecke の operator $T\mathfrak{p}$ は，V_2 の上で，行列式のイデヤルが $\mathfrak{p}$ の行列の代表系を掛けて（形式的に）加えることに対応し，これは勿論 abelian var. の上で，“order $\mathfrak{p}$” の subgronp で割ることによって決る modular correspondence に対応する，ということ．

全体に formulation の問題ですが，O. Herrmann の結果は全部，ずっと見通しよくなり，Dirichlet 級数との関係では，もっと色々なことがわかるのではないかと思っています．また，こう云った見地から，類体の構成

を idèle の上でやることも考えられるかも知れません．都合の悪いのは mod. $\mathfrak{p}$ の reduction と直接結びつかないこと．（結局 abel. var. を通してやることになる）．Siegel modular の場合にも同様な formulation が考えられますが，Fourier 解析がうまく行かない．本質的に non-abel になるのではないかと思われます．（Selberg が対応させている Dirichlet 級数はどんな意味を持つか？）・・・・ここ一と月ばかり，暑いので放り出したままですが，また考え始めるつもりです．

段々淋しくなるので，電蓄を買ってレコードを聞くことにしました．最近は Beethoven の第8ばかり聞いています．

ではまた．（アメリカに行ったら住所を知らせて下さい．）

ド・ゴールから始まって，レバノン，イラン，ヨルダン，最近の金門島と，騒がしい世の中，日本では今，教師の「勤務評定」というもので大騒ぎです．もう何年か経つ頃には，日本は益々住みずらい国になっているんじゃないかと心配です．

9 月 22 日　朝

以　上

遺　　書

昨日まで，自殺しようという明確な意志があったわけではない．ただ，最近僕がかなり疲れて居，また神経もかなり参っていることに気付いていた人は少なくないと思う．自殺の原因について，明確なことは自分でも良くわからないが，何かある特定の事件乃至事柄の結果ではない．ただ気分的に云えることは，将来に対する自信を失ったということ．僕の自殺が，或る程度の迷惑あるいは打撃となるような人も居るかも知れない．このことが，その将来に暗いかげを落すことにならないようにと，心から願うほかない．いずれにせよ．これが一種の背信行為であることは否定できないが，今までわがままを通して来たついでに，最後のわがままとして許してほしい．

本郷，駒場から借りた図書と研究室の鍵とは，この机の上においてある．レフェリーを頼まれた論文は電蓄の上にのせてある．（これは少し手を加えた方が良いと思うが，丁寧に読んだわけではないので意見は保留する．）

数学関係の書物，別刷，ノート等の処分は久賀道郎（目黒区駒場東京大学教養学部数学研究室）に一任する．（自分のものにするなり，寄附するなり古本屋に売り飛ばすなり，彼の自由である）．

電蓄，レコード，レコード・キャビネットは，（もしそれが彼女にとって迷惑でないならば）M. S. に進呈したい．ただし，レコードの中，バルトーク，ベルク，ブリッテンの三枚は斎藤正彦のものであるから別にする．）

これ以外のものの処分，あと始末は兄（谷山清司：荒川区日暮里 4~175）に一任する（面倒なことと思うが仕方ない）．

立机の左の引出しに富士銀行の貯金通帳が２冊入っている．（使用する印鑑は，同じ引出しの中のパス入れの中に入っている小さいもの）．都立大学の９月分の講師手当と，共立社からのスミルノフの翻訳料とは，まだ受取っていない．管理人に，今月分（つまり10月分）のガス代をまだ払ってない．それ以外に貸し借りはないと思う．

杉浦光夫，上野正等から何冊（或いは十何冊）か，本を借りている．適宜自分のものを持って行って欲しい．

講議の進行状況．

理Ⅰ，３組は，積分は全部終って，微分はテーラーの定理まで．（微分は，あと，その応用として極大極小，凸函数，近似計算，定積分の近似計算が残っている）．全部残っているのは級数と曲線論．

理Ⅰ，７組は，３組より一時間後れている．従って無理函数の積分（$\sqrt{2\text{次式}}$ を含むもの，二項積分），三角函数の積分，テーラーの定理（剰余項は積分で与える）も，まだやっていない．あとは３組と同じ．

理Ⅰ，６組は，ベクトル，行列，行列式，一次方程式は全部済んで（ただし内積，外積は，やっていない），n 次元アファイン空間の定義（Weyl 流）と，部分空間の平行性まで．残っているのは，部分空間のパラメーター表示，方程式表示，内積，外積，距離，角，平行体の体積，アファイン空間の座標変換（ベクトル空間のそれは済んでいる），直交変換，２次超曲面．

駒場のゼミナールは函数論で，線積分，複素線積分の定義と主要性質，正則函数の定義までしかやっていない．

大学院の講義は，清水君のノートを見て欲しい．

○追試験の問題を出してない．

いずれにせよ，駒場の方々にかなり御迷惑をお掛けすることになるのをお詫びしたい．

谷　山　　豊

第四部

新人紹介

Peter Roquette

(in München)

最近，Dissertationを出したばかり．Hasse, Deuringの流れを汲む．その業績には

1．Kongruenzzetaについての，Riemann予想の純算術的証明（Dissertation, 未発表）

2．Abel函数体，特にそのTeilungを純算術的に扱ったもの

3．Schurの予想およびArtinのL函数についての，R. Brauerの結果を，見通しよく導いたもの

などがある．論文の書き方は，極めて懇切丁寧．

（月報，1巻1号，1953年）

A. Weil「ゼータ函数の育成について」

今日のような蒸暑い日には，堅苦しい話より，'動物や植物' を '育成' でもするような話の方がよいであろう．ζ-函数についていえば，本質的な点は第一に ζ がギリシヤ語のアルファベットの一つであることで，第二にその変数が普通 s と書かれることである：

$$\zeta(s).$$

その歴史は Riemann にまでさかのぼる．しかし彼はこの動物の出生に対し責任あるわけではない．Euler が最初にこの動物を考えたのである．詳しく書けば：

$$\zeta(s)=\sum_{n=1}^{\infty}\frac{1}{n^s}=\prod_{p=2,3,5,\cdots}\left(1-\frac{1}{p^s}\right)^{-1}.$$

ここで n は整数，p は素数である．Euler はこれを，代数的に，全く形式的な等式として扱ったのだが，とにかくも彼はこれが，Euclid に由来する一つの定理，すなわちすべての整数は素数の積に一意的に分解される，という定理と同等であることを認識していたのである．

この，Euclid の定理は，数学における最初の重要な発見の一つで，また上の等式も，簡単ではあるが，数学の発達史上一つの本質的な step である．

また Euclid は，素数の個数が無限であるという重要な定理を発見している．これは上の等式から考えてもわかることで，この式の両辺は $s>1$ で収束するが，素数が有限個であるとして両辺の order を比べれば矛盾が出る．

ところでこの動物について，Euler の最も重要な発見は，ζ-函数の函数等式である．つまり $\zeta(s)$ と $\zeta(1-s)$ との間に或る関係が成立つことをいうものであるが，この函数等式は次の型である．ただしここでは詳しく書く必要は全くない．

$$\begin{aligned}&\text{exp.}\times\text{gamma}\times\zeta(s)\\&\qquad=\text{exp.}\times\text{gamma}\times\zeta(1-s).\end{aligned}$$

さて，この函数等式の最も驚くべき点は，$\zeta(s)$ を表わす級数が，$s>1$ では収束するが，$s<1$ では発散することである．$\zeta(1-s)$ は，$s>1$ では一体何を表わしているものだろうか？　大抵の歴史の本を読むと，Euler は，級数の発散収束など意に介せず，全く形式的に考えたと書いてあるが，Euler の書いたものを読めば，このような見方は完全にウソであることがわかる．彼が実際にしたことというのは次のことである．彼はまず $\zeta(s)$ を級数で書き，1より大きい整数に対して，函数等式の右辺を，現在 Poisson の総和法とよばれている方法で加えている．s が整数のときはこの和が explicit に計算できるのであるが，それを使って s が整数のときこの等式を証明し，そうでないときは，これがやはり成立つことは極めて本当らしいといっているのである．Euler の所論を取り上げて，それに，函数論の現在の知識を応用して，彼の証明を完全なものにすることは面白い練習問題だろう．

われわれが ζ-函数とよんでいるが，この種属の動物の，本質的な特色を考えてみよう．第一にそれは，或る種の級数で定義されて，それがまた或る型の無限積に等しいことで，第二にそれが函数等式を満たすことである：

$$\begin{cases}\text{級数}=\text{積}\\ \text{函数等式}.\end{cases}$$

Riemann はこの話題を取り上げた．彼は

Euler について何もいっていないが，私は，彼は Euler の仕事を知っていたのではないかと思う．一方 Riemann から 15 年程前 (1840 年) に，或るドイツ人が，たしか Schlömilch だったと思うが，ζ-函数に似た，別な函数を考えた：

$$\frac{1}{1^s}-\frac{1}{3^s}+\frac{1}{5^s}-\frac{1}{7^s}+\cdots.$$

これは $s>0$ で収束するから，ζ-函数よりも都合が良い．つまり，0 と 1 との間に対しては，函数等式の両辺に意味があるからである．Riemann は，彼の論文には書いていないが，この函数や Euler のことを知っていたと思われる．とにかくも彼は ζ-函数を考えて，決定的な進歩をもたらしたのである．第一に，その頃函数論は，解析接続の概念にまで発展していた．この発展はもちろん Riemann 自身によるものであるが，Riemann は Euler の ζ-函数を再び取上げた．第一に彼は，ζ-函数の積分公式を与えた．それにより $\zeta(s)$ は，s のすべての実数値だけでなく，すべての複素数値に対して意味を持つ式で表わせるので，函数等式の意味が明かになった．第二に，彼は，この積分公式を使って，函数等式の証明を与えた．これは，Jacobi のテータ函数の理論を本質的に利用する証明である．それはテータ函数と ζ-函数との関係をつけることにより成し遂げられる．ついでにいえば，Jacobi が，現在 Jacobi のテータ函数と呼ばれている函数に ϑ という文字を使ったのに示唆されて Riemann は ζ という文字を使ったのではないかと思われる．Jacobi のテータ函数は Jacobi により発見され，楕円函数の理論で，疑いもなく重要な役割を果している．尤も歴史についていうと，Jacobi 以前 Gauss がすでにこのテータを発明し，そのいくつかの基本的な公式を知っていたが，彼はそれを発表しなかったので，後になるまで知られなかった．それゆえこれが Jacobi のテータと呼ばれているのである．

次の step をなし遂げたのは Dedekind である．これが，bigger and better zeta-function の育成の始まりである．私自身も近頃は，その育成に，私の数学的な努力の一部を捧げているのである．とにかくも Dedekind は，Riemann の定義を，任意の数体に拡張した．次にその定義を書くが詳しいことは専門的になるから略す．非専門家にとっては，これが，有理数体における Riemann の ζ とちょうど同じことを，数体に対してやっているということさえわかればよいのである．記号は詳しく説明しないが，Dedekind の $\zeta(s)$ は次の式で定義される：

$$\sum_{\mathfrak{a}}\frac{1}{N\mathfrak{a}^s}=\prod_{\mathfrak{p}}\left(1-\frac{1}{N\mathfrak{p}^s}\right)^{-1}.$$

Riemann の ζ とこの Dedekind の ζ の定義との主要なちがいは，ローマ字とドイツ文字とのちがいであることがわかるであろう．ここでドイツ文字が使われているのは，長い間ドイツ人だけしか数論をやらなかったので，数論の記号にはドイツ文字を使うのが習慣になっているからである．この等式は同時に，Dedekind の主要な発見の一つを表わしている．それは代数体におけるイデアルの，素イデアルへの一意分解性の定理で，Kronecker も同じことを証明したが，この定理はちょうど，前の Euclid の定理と同じ意味をもっているのである．

この函数に対しては，始めは長い間，函数等式がなかった．遂に Hecke が，この函数等式を証明したのだが，当時これは非常に難かしい問題だと考えられていたので，Hecke が最初できたといったとき，Landau はその証明を信用しなかった．これは，かの有名な Landau が，数論に対しいかに深い洞察を持っていたかを示したいくつかの例の一つである．第二の例は，Artin が一般相互法則に達したときで，Landau はそのニュースを，軽蔑をもって取扱ったと私は聞いている．

第三の例は，Siegelが或る重要な定理を証明したとき，Landauはやはり，その知らせを軽蔑し，証明を信用しなかった．

さてHeckeのこの証明は1917年だが，それ以後，Riemannのζ-函数と，前に述べた性質を共有する函数が数多くあることが知られた．その性質というのは，それがまず或る種のDirichlet級数で定義され，第二に函数等式を持ち，第三にこの級数が，或る型の無限積に等しくなることである．専門家にとっては，これらすべての函数は，明白な特性を共有している．このような函数はいろいろあるが，その中のいくつかに対しては，今の所函数等式を証明できず，それは単に推測されるだけである．

このような函数全部を挙げることはできないので，私がそのような函数を育成するときに使って来た一つの特別なトリックをお話しよう．

普通のζ-函数についてRiemannは，現在Riemann予想と呼ばれている有名な予想を述べた．この予想というのは次のものである．$\zeta(s)$は，自明な零点を除けば，すべての零点を実数部が0, 1である二直線の間に持っているが，それらがすべて，実数部1/2なる直線の上にあるであろうというのである．証明について余り厳密さを要求しなければ，確率の計算で簡単にできる．それは次のようなものである．$\zeta(s)$の逆数は，容易に，

$$\frac{1}{\zeta(s)}=\sum_{n=1}^{\infty}\frac{\mu(n)}{n^s}$$

と表わせる．ここに$\mu(n)$は，nの素因子によって，

$$\begin{aligned}\mu(n) &= +1 \qquad n=p_1\cdots p_{2n},\\ &= -1 \qquad n=p_1\cdots p_{2n+1},\\ &= 0 \qquad \text{その他}\end{aligned}$$

($p_1, \cdots$は相異る素数)．

さて，勝手な数nが，偶数個の素因子を持つ確率も，奇数個の素因子を持つ確率も，ともに1/2であることは大変確からしい．ところで確率論の，よく知られた定理がある．それによれば，確率1/2の遊戯があるとき，N回やってみて，$N/2+r_N$回勝ち，$N/2-r_N$回負けるとすれば，Nが十分大きくなるとき，r_Nの大きさは$\sqrt{N}$のorderである．これを使えばRiemann予想が正しいことはほとんど直ちにわかる．

最近，次のような遊戯がArtinにより考えられた．

$$F(x, y)=0$$

という等式を考える．Fは整係数の多項式とし，これをmod. pで試みて，0に合同なら勝ち，合同でなければ負けとする．この式の取り得る値は，$0, 1, \cdots, p-1$のp個であるから，勝つ確率は$1/p$である．x, yの，$0, 1, \cdots, p-1$という値の組合せはp^2個あるから，p回勝ち，p^2-p回負けることが予想される．実際に勝った回数を$p+r$，負けた回数をp^2-p-rとすれば，rは誤差を表わすわけである．Artinは，遊戯の専門家だから確率論的に，rが$\sqrt{p}$のorderであると考えた．後に，この直覚が正しいことがわかったのである．実際この予想を，より一般的な場合に，厳密に証明できるのである．それを説明しよう．技術的な点は説明できないが，それは重要ではない．

そのために，Galoisにより始めて考えられた，一般の有限体で考える．

$$F(x, y)\equiv 0$$

を，p^ν個の元を持つ体で考える．νの値をきめたとき，この方程式の解の個数をN_νとし，

$$\sum_{\nu=1}^{\infty}\frac{N_\nu}{(p^\nu)^s}$$

という級数を導入する．(これを少し修正しなければならない場合もあるが，細かいことは略す．) それはArtinにより，この方程式のζ-函数とよばれた．私はそれをζの大文字で，

$$\boldsymbol{Z}_p(s)$$

と書くことにしている．これに対し，上に述べた様々な点をテストしてみよう．まずこれは級数であるが，前にいった形の無限積で書けるか？　答は簡単に Yes である．次は函数等式を満たすかどうかだが，初等的な代数幾何学の或る定理，つまり Riemann-Roch の定理を用いて考えれば，この答も簡単に Yes と出る．

次の step：私が思うに，Artin が興味をひかれた点は，この函数は Riemann の $\zeta(s)$ よりも非常に簡単だから，これに対し'Riemann 予想'が成立つことを確かめることができるだろうという点である．先程の遊戯に関する'Artin の直覚'は実はこの'Riemann 予想'と同等なのである．Trivial でない最初の重要な場合，それは種数 1 の場合と呼ばれるが，そのとき実際'Riemann 予想'の成りたつことは Hasse により証明された．一般の場合の証明は，'Riemann 予想'が，Castelnuovo に由来し，彼や他のイタリー人によって盛に使われた，代数幾何学の或る定理と同等なことに注意すれば，函数等式の証明と類似のやり方でできる．唯一の困難は，イタリーの代数幾何に比べて，極度に抽象的な場合を扱うので，事情が甚だ複雑になることだけである．

次の step は，非常に面白いものであることがわかり，現在，数学的植物学者に対し，非常に大きな研究分野を繰り拡げているものであるが，それは次のようなものである：整係数の多項式，

$$F(x, y)$$

から出発し，素数 p に対して上のようにして $\boldsymbol{Z}_p(s)$ を得る．そしてそれをすべての素数 p に関して掛け合わせ無限積

$$\boldsymbol{Z}(s) = \prod_p \boldsymbol{Z}_p(s)$$

を作る．最初は，このような無限積を書く必然性はないようにみえる．私の知っている限り，このような無限積は，特別な場合に，Hasse が書いたのが最初である．Hasse はこの $\boldsymbol{Z}(s)$ が函数等式を満たすことの証明を，或る学生に，問題として提出したのである．その学生は，この問題を暫くやっていたが手が着かず，Siegel により与えられた，全く違った系統の問題をやって割合に成功したが，数年後脳腫瘍でなくなった．

最近いくつかの場合に，この積が explicit に計算された．一つの例は $F(x, y) = 0$ が

$$y^2 = 1 - x^4$$

となる場合で，この場合には Gauss がすでに関連した問題を考えている（もっともそれはこの見地からなされたものではないが）．これは有望であるようにみえたので，私はその研究から出発し，体系的に考え，また彼のアイディアを，2 変数以上の場合にも拡張した．この話を終えるに当ってこの方向でなされた最近の結果を簡単に紹介しよう．

まず私は，

$$ax^m + by^n + c = 0$$

から定義される ζ-函数を考えた．これは Hecke により考えられた L-函数で表わされる．L-函数は ζ-函数の仲間であるが，Hecke はそれをこの名前で呼んだのである．さらに最近，Deuring は虚数乗法を持つ楕円曲線の場合に，同じような結果を得，この結果は，全く最近，谷山により，最も一般的な Abel 函数で，適当な虚数乗法を持つものの場合に拡張された．

一方 Eichler も重要な結果を得た．それは，

$$y^2 = 4x^3 - g_2 x - g_3$$

なる曲線で，モジュラー函数から生じたものを扱ったので詳しくいえばそれは，モジュラー函数についての Hecke の理論と関連している．その場合も $\boldsymbol{Z}(s)$ が前から知られている ζ-函数の積として表わされ，函数等式を満足することがわかる．

このように，今までわかっている場合はすべて，この新しい ζ-函数が実は新しくないことを示し，すでに知られているものについての結果から函数等式が得られたのである．

だからこの場合，bigger, better という言葉は，厳密にはあてはまらない．

しかしすでに，その次の最も簡単な場合，つまり虚数乗法を持たない楕円曲線で，Eichler のようにしてモジュラー函数に結びつけることのできないものに対してさえ，この函数 $\boldsymbol{Z}(s)$ の性質は完全に神秘的である．その函数等式を証明することは，これまでの場合に比べて非常に深い問題であるようにみえる．この問題は非常に難しくまた非常に重要である．私は，この函数が，これまで知られている何物にも似ていないという印象をもっている．だから私はこれを，bigger and better zeta function と呼ぶのである．

（数学，7 巻 4 号，1956 年）

A. Weil「イデール類群の或る指標について」

整数論における当面の最も重要な問題は，イデール類群の連結成分に関するものである．代数体 k のイデール類群の指標は，k の量指標に他ならない[1]．これはイデアル群の指標であるから，それを用いて L 函数を，

$$L(s,\chi)=\sum_{(\mathfrak{a},\mathfrak{f})=1}\frac{\chi(\mathfrak{a})}{N\mathfrak{a}^s}=\prod_{\mathfrak{p}\nmid\mathfrak{f}}\left(1-\frac{\chi(\mathfrak{p})}{N\mathfrak{p}^s}\right)^{-1}$$

により定義できる．ここで $\mathfrak{f}$ は量指標 χ の導手である．

イデアル指標 $\chi(\mathfrak{p})$ からイデール類群の指標を作ることは，やさしい練習問題にすぎない．一方，導手 $\mathfrak{f}$ と素な素イデアル $\mathfrak{p}$ の函数 $\chi(\mathfrak{p})$ が量指標を定義するための条件は，それを乗法的に，$\mathfrak{f}$ と素なイデアル $\mathfrak{a}$ の函数 $\chi(\mathfrak{a})$ に拡張したとき，単項イデアル (α) で，

$$\alpha\equiv 1 \quad \mathrm{mod.}\,\mathfrak{f}, \qquad \alpha \text{ は総正}$$

なる数 α で代表されるものに対し，$\chi((\alpha))$ が α の種々の共軛数の積で表わされること，即ち α_ρ を実，α_ι を虚の共軛とするとき，次の形の式が成立つことである：

$$(*)\quad \chi((\alpha))=\prod_\rho \alpha_\rho^{\eta_\rho}\cdot\prod_\iota|\alpha_\iota|^{\xi_\iota}\cdot\prod\alpha_\iota^{n_\iota},$$

ここに，ξ_ι,η_ρ は複素数，一方 n_ι は整数を意味する．そのときこの冪指数 ξ,η,n には，単数と関連する自明な条件がある．すなわち α として，総正で，mod. $\mathfrak{f}$ で 1 に合同な単数を取るとき $\chi((\alpha))=1$ でなければならないから，それらは任意ではあり得ない[2]．

さて，L-級数 $L(s,\chi)$ の算術的説明という意味は正確にはいえないが，例えば，χ を位数有限，即ちイデール類群の連結成分の上で 1 となる指標とすれば，この L-級数は，k の或る巡回拡大から生じ，k の素イデアルの Frobenius 置換を使って定義される，もう一つの L-級数と同一視できる[3]．これが，Artin により発見され，類体論，特に相互法則により示された，$L(s,\chi)$ の算術的説明である．

私はここで，狭義の，即ち連結成分の上で 1 でない，量指標により作られる Dirichlet-級数の，或る意味での算術的説明を与えよう．Γ を量指標全体の作る群，Γ_0 を位数有限な，すなわち連結成分の上で 1 になり類体論で説明できる指標の作る部分群とすれば，Γ/Γ_0 は，rank n の Abel 群である，ここで n は体 k の次数[4]．その生成元の中には一つ trivial なもの，すなわちノルムがあるから，$n-1$ 個の独立な量指標の L-級数を説明できれば大体満足すべきであろう．現在はまだその段階には到っていないが，虚数乗法が使える場合，即ち k が，総実な体 k_0 の，総虚な 2 次の拡大であるときには[5]，その一部が算術的に説明できるのである．

ここで量指標の間に一つの区別が現れる．これは，今までどの文献にも見られなかったものである．すなわちその指標から作った Dirichlet-級数の係数がすべて代数的数であるか否かによって鋭い区別が生ずるのである．所で式（$*$）で，すべての ξ_ι,η_ρ が 0 であれば，$\chi((\alpha))$ は明かに代数的数であるが，又 mod. $\mathfrak{f}$ での類数が有限だから，すべての値 $\chi(\mathfrak{a})$，即ち $L(s,\chi)$ の係数，が代数的数であることもやはり明かである．一方この条件は又必要でもあるように思われる．これが第一の予想である．

先に進む前に，二つの典型的な例を挙げよう．虚 2 次体では，すべての量指標は上の意味で代数的値を持つ．一方実 2 次体ではそう

でなく，超越的値の指標が存在する[6]．このことは，虚２次体には虚数乗法論があるが，実２次体にはそれに類するものが何もないということと密接に関連しているのである．

さて，代数的値を持つ量指標につき何かお話ししよう．それはイデール類群の，何か或る構造を指示しているのだが，今までどこにも見られなかったもので，現在の所かなり神秘的に見える．

最も都合の良い場合は，k が総実な体 k_0 の，総虚な２次の拡大の時で，この時は，k の単数は本質的には k_0 の単数である．すなわち後者は前者の中で，指数有限な部分群を作っているから，k の単数の偏角（argument）は 2π の有理数倍である．一般の場合にはそれは単に正の実数倍であるのだが．それ故私の予想が正しいとすれば，代数的値を持つ独立な量指標は丁度 $n_0 = n/2 = [k_0 : Q]$ 個ある．そこで，一般の型の代数体の場合に私の予想が証明できたとして，代数的値を持つ量指標が丁度どれだけあるかを見出すことが次の問題になる．これは，Artin が指摘したように，容易に，群環の問題に帰着される．それは単数の，虚の共軛数の偏角の間に，どれだけ一次独立な関係があるかということである．

少くとも，k が，総実な体 k_0 の，総虚な２次の拡大であるときには，上にいった型の量指標は丁度 $n/2$ 個ある．それ故，幾分ルーズだが，このような $n/2$ 個の指標の L-級数が，k を虚数乗法の体に持つ Abel 多様体のゼータ函数により説明される[7]といえば，大雑把について，イデール類群の連結成分の半分が算術的に説明されたことになる．これだけでは不十分であるが，少くとも何も無いよりはましである．

次の論点は，代数的値を持つ量指標に，位数有限な指標を対応させることであるが，その方法はかなり奇妙なものである．

今問題にしている型の量指標が，値 $\chi(\mathfrak{p})$ により与えられているとする，ここで $\mathfrak{p}$ は或イデアル $\mathfrak{m}$（それは導手 $\mathfrak{f}$ より大きくてもよい）を割らない素イデアルすべてとする．この値 $\chi(\mathfrak{p})$ からイデール指標を作ることは，やさしい練習問題だが[8]，そのためには，値 $\chi(\mathfrak{p})$ の属する代数体を複素数体の中に embed する，なぜなら量指標は複素数値を取る函数だから．そして χ を自由に定めうる所，即ち $\mathfrak{m}$ を割る素点と無限遠素点での値を，χ が principal イデールの上で１になる様に調整して定めるのである．所で今やり方を少し変えて，$\chi(\mathfrak{p})$の属する代数体を，その $\mathfrak{q}$-進完備体の中に embed して見る．ここに $\mathfrak{q}$ はこの代数体の素イデアルである．そして $\chi(\mathfrak{p})$ を，イデール群の，$\mathfrak{q}$-進体の中での表現に拡張するのである．所が $\mathfrak{q}$-進体は非連結だから，この表現は無限遠素点の連結成分で１にならなければならない．そこで χ を自由に定めうる所を少し変える必要がある．すなわち，$\mathfrak{q}$ で割れる有理数を q とする．…………色々な celebrations が余り沢山あって，考え直す暇がなかったので，この話に間違いがないと保証できないが，全体かなりルーズに話しているのだから，それは大して問題ではない[9]，……そこで $\mathfrak{q}$ で割り切れる素数 q を取り，$\mathfrak{m}$ を割る素点と，q を割る素点での χ の値を自由に定める．これは χ が連続で，principal イデールの上で１になるように調整するためで，このようにして，イデール類群の指標にもなる χ が一つ，ただ一つ定まる．所が $\mathfrak{q}$-進体の乗法群は非常に不連続で，有限群の projective limit として表わされるから[10]，その中で何かを表現することは，無限に多くの，有限群の中への表現を与えることと同等なのである．

このようにして，連結成分の上で１ではない量指標とイデール類群の，有限群の中への表現の無限列との間の関連が見出された．所が，有限群への表現は普通の類体論により説明されるもので，k の有限次 Abel 拡大を定める．これをまとめれば，k の無限次 Abel 拡大が得られ，これは χ と $\mathfrak{q}$ とで定まるものである．

これで話はほとんど終りだが，もう一つ予想が残っている．この予想は，最近志村，谷山及び私自身により虚数乗法に適用されたのと[11]丁度同じ方法によって証明されることはほとんど確かであるが，先ず手始めに，普通の虚数乗法の場合に試して見ることもできる．そのときには，既に Deuring により得られている結果の中でこれが解決されることは，極めて確かである．

そこで k を虚2次体，E を楕円函数体で，その虚数乗法の環が k の order であるものとする．E のゼータ函数は，Deuring によって計算され[12]，Hecke の L-函数により表わされるが，その指標 χ から，位数有限の指標を法として，k のすべての量指標が得られる[13]．今いったように，素イデアル $\mathfrak{q}$ を一つ取れば，この χ には，k の無限次 Abel 拡大が対応するが，私の予想というのはこれが，E の週期の $\mathfrak{q}^n$ 等分により生ずる無限次 Abel 拡大と丁度一致するであろうということで，これは詳細に証明することができるであろう．次に，より一般的な高次元の場合にも，対応する予想が正しいことはかなり確かである．このことは，虚数乗法論の中での，Abel 多様体と量指標との関連が，偶然的なものではなくて，実質的なものであることを確証しているのである[14]．

註

1）これは岩沢-Tate の定理であるが，まとまった文献としては未だ発表されていないようである．尤もこれだけのことならば，Weil［6］の始めにも簡単に説明してある．又 Weil［5］参照．

2）（＊）の具体的な形については Hecke［4］を見よ．

3）即ち Artin の L-函数である．このことは，類体論の分解法則，相互法則と本質的に同等である．Artin［1］,［2］参照．

4）Hecke［4］参照．

5）この間の事情については，Weil の三日目の講演に詳しく説明されている．

6）Hecke［4］の最後の2節参照．

7）谷山の講演参照．

8）$\mathfrak{m}$ を割らない素イデアル $\mathfrak{p}$ に対し，$\mathfrak{p}$-進体 $k_{\mathfrak{p}}$ の乗法群での局所指標を，$\mathfrak{p}$-単数で1，$\mathfrak{p}$-素元で $\chi(\mathfrak{p})$ なる条件により定める．それ以外の素点に対しては Weil のいう通りにやればよい．

9）以下の話にも本質的には間違いはない．ただ全体としてかなりルーズなことは Weil のいう通りで，定式化には注意を要する．尤もこの講演の意義はアイディアを述べることにあるので，それがわかれば定式化は自然にできる．

10）これは $\mathfrak{q}$-進単数の群への表現であるから mod. $\mathfrak{q}^n$, $n\to\infty$ によりこの有限群を定めるのが最も都合がよい．

11）それぞれの三日目の講演参照．

12）Deuring［3］参照．

13）これはかなりルーズな表現である，χ は E の定義体 k'（k の類体に取れる）の指標だから．然し χ は，位数有限な指標を法とすれば，k の量指標 χ_0 の co-norm として表わされ，この χ_0 から，k の指標がすべて得られるのである．従って以下の $\mathfrak{q}^n$-等分云々も，χ_0 を基本に取って考えれば，Weber の τ 函数の等分の意味に解さなければならない．

14）この関係は，或る意味で，より一般的な体の指標にも拡張できる．谷山の近刊の論文参照．

文献

［1］ E. Artin, Über eine neue Art von L-Reihen, Abh. Math. Sem. Univ. Hamburg, **3** (1923), 89-108.

［2］ E. Artin, Beweis des allgemeines Reziprozitätsgesetzes, Abh. Math. Sem. Univ. Hamburg, **5** (1927), 353-363.

［3］ M. Deuring, Die Zetafunktionen einer algebraischen Kurven vom Geschlecht

Eins, Nachr. Akad. Wiss. Göttingen, 1953, 85-94.

[4] E. Hecke, Eine neue Art von Zetafunktionen und ihre Beziehung zur Verteilung der Primzahlen, II, Math. Z., **6**(1920), 11-51.

[5] A. Weil, Sur la théorie du corps de classes, J. Math. Soc. Japan, **3**(1951), 1-35.

[6] A. Weil, Sur les "formules explicites" de la théorie des nombre premiers, Communications du sém. math. univ. de Lund, tome supplimentaire (1952), dédié à M. Riesz.

(数学, 7巻4号, 1956年)

『近代的整数論』(志村五郎との共著)

第1章　歴史

1・1　楕円函数の虚数乗法

$f(u)$ が周期 ω_1, ω_2 をもつ楕円函数であるとき，有理整数 n に対して $f(nu)$ も同じ周期に属する楕円函数であるから，$f(u)$ と $f(nu)$ との間に代数的な関係があり，その関係は加法公式によって具体的な形に書き表わされる．周期の比 ω_1/ω_2 が虚二次体 Φ に属する場合には，Φ の数 μ に対して $f(u)$ と $f(\mu u)$ との間にも代数的な関係が存在することが知られる．このとき楕円函数 $f(u)$ は虚数乗法をもつというのである．虚数乗法の歴史は，殆んど楕円函数の発生とともにはじまり，それははじめ代数的に，後に整数論の問題として研究されてきた．次に時代を追ってその理論の発展してきた跡をたどってみよう．

Gauss における lemniscate の研究は，今日の言葉でいえば，modulus $\sqrt{-1}$ に対する sn 函数の研究であり，それはかれの楕円函数研究の出発点であったが，その中にもわれわれは虚数乗法の具体的な例を見出すことができる．この函数に対して成り立つ関係 $\mathrm{sn}(\sqrt{-1}u, \sqrt{-1}) = \sqrt{-1}\,\mathrm{sn}(u, \sqrt{-1})$ は，上の $f(u)$ と $f(\mu u)$ との間の関係の最も簡単な一例であって，Gauss はこれによって複素変数の函数としての $\mathrm{sn}(u, \sqrt{-1})$ を定義したのであった．そしてまたかれは lemniscate の5等分が作図できることを示しているが，これは虚数乗法をもつ楕円函数の等分値が Abel 体を生成するという事実の最初の例といえる．この Gauss の研究に続いて，任意の modulus を扱い，問題を一般的に打ちだしたのが Abel である．いわゆる楕円函数の等分に関するかれの研究(1827〜29)を簡単に述べれば次のとおりである．

$$u = \int_0^z \frac{dz}{\sqrt{(1-z^2)(1-k^2z^2)}}$$

とおくと $z = \mathrm{sn}(u, k)$ であるが，いま簡単のために $z = f(u)$ と書くと，奇数 n に対して

$$f(nu) = \frac{P_n[f(u)]}{Q_n[f(u)]}$$

となる．ここで $P_n(X), Q_n(X)$ はそれぞれ n^2 次，n^2-1 次の多項式である．$f(nu) = y$ とおくと，$f(u) = x$ は n^2 次の方程式

$$P_n(X) - yQ_n(X) = 0$$

の根である．この方程式を sn 函数の一般等分方程式という．次にこの楕円函数の一組の基本周期 ω_1, ω_2 をとれば

$$P_n(X) = \prod\left(X - f\left(\frac{\alpha\omega_1+\beta\omega_2}{n}\right)\right)$$

となる．ここで右辺は $0 \leqq \alpha \leqq n-1, 0 \leqq \beta \leqq n-1$ である整数 α, β の組すべてについての積である．従って $P_n(X) = 0$ は $f\left(\dfrac{\alpha\omega_1+\beta\omega_2}{n}\right)$ を根とする n^2 次の方程式であり，それを sn 函数の特殊等分方程式という．$f(0) = 0, f(-u) = -f(u)$ であるから，$P_n(X) = XP_n^*(X^2)$ と書かれる．さて Abel は，一般等分方程式 $P_n(X) - yQ_n(X) = 0$ が，$\boldsymbol{C}(y)$ の上に代数的に解けることを示し，特殊等分方程式については，n が奇素数の場合，$\dfrac{1}{2}(n^2-1)$ 次の方程式 $P_n{}^*(X) = 0$ を $\boldsymbol{Q}(k)$ の上で解くことを，まず $n+1$ 次の方程式を解いてそれから $\dfrac{1}{2}(n-1)$ 次の方程式を解くことに帰着させ，その後者が代数的に解ける

ことを示した．これらの代数的に解ける方程式は，今日使う言葉でabelianであって，そのabelianという名も実はこのAbelの等分方程式の研究に因んでつけられたものである．Abelはその$n+1$次の部分(いわゆるmodular equation)が，$n>3$ならば，一般には代数的に解けないであろうと予想したのであるが，それが実際そのとおりであることはGaloisによって示された(1831)．これはGalois群の理論によって解決された問題のうち，最もきわだったものの一つである．さらにまたAbelは，その$n+1$次の方程式は，周期の比が二次無理数であるときには，すなわちその楕円函数が虚数乗法をもっているときには，代数的に解けることを示している．この認識が，明確な形における虚数乗法論の起原といえるであろう．

これらの理論はすべて代数的なものであるが，Kroneckerはこれを整数論的に取りあげたのである．かれは，一つの代数方程式が代数的に解かれ得るか否かを判定するという問題のほかに，代数的に解かれ得る方程式を"構成"するという問題を提出した．そして，具体的な形においては，"有理整係数の任意のAbel方程式の根は1のベキ根の有理式として表わされる"という言明をし，また同様な関係が$\sqrt{-1}$を係数に含むAbel方程式の根とlemniscateの等分方程式の根との間に存在するとして，さらに一般の数体の上のAbel方程式へ問題を拡張している(1853)．とくに虚二次体の上の方程式について，いわゆる"Kronecker青春の夢"を抱いていたのである．かれはDedekindへの手紙(1880)の中で，"ちょうど整係数のAbel方程式が円分方程式によって尽くされるように，有理数の平方根を係数にもつAbel方程式はsingulärenmodulnをもつ楕円函数の変換方程式によって尽くされるという私の最愛のJugendtraum"についての多くの困難の最後のものを克服したといっている．そしてその長大な論文"Zur Theorie der elliptischen Funktionen"(1883〜1890)のうちで，かれはJacobiの影響の下に，sn函数の特殊等分方程式を精細に研究し，奇素数nに対するsn函数の合同関係式

$$\begin{aligned}&(-1)^{\frac{1}{2}(n-1)}\sqrt{\lambda}\sin\operatorname{am}(\mu u,\lambda)\\&\equiv(\sqrt{x}\sin\operatorname{am}(u,x))^{n}\pmod{\mu}\end{aligned}$$

を与えている．これは，上の言明の解決への志向を示すものであったが，この研究はかれの死とともに未完に終った[1]．われわれはこのKroneckerの研究の中に，sn函数の等分値の生成する体における素因子の分解，分岐，イデアルの単項化の問題など，後の代数的整数論を形づくった具体的要因のいくつかを見出すことができるのである．

さて，これらのKroneckerの残した結果や予想は，整数論において多大の興味をひき起した．著名な研究者としてはWeber, Fueterらがある．なかでもWeberは，その大著Lehrbuch der Algebraの第III巻の大部分をこの問題に捧げたのであるが，具体的事実と定式化の上において多くの進歩はありながらも，またついにこの理論の頂上をきわめることはできなかった．そして完全な解決は高木の類体論の完成(1920)によってなされたのである．すなわち，虚二次体の上のAbel体はmodular函数$j(\tau)$の虚二次無理数における値，いわゆる特異値と，適当な楕円函数の等分値とによって生成され得るということが確定した．そして，それら楕円函数の特殊値の示す整数論的性質は類体論の特別な場合として認識されたのである．

以上が古典的虚数乗法論の簡単な歴史であ

1) Kroneckerが虚二次体上のAbel体について誤った予想をしたかのごとくいわれることがあるが，それは不当であって，おそらくはHilbertあたりの誤解であろうと思われる．Kronecker, Werke V巻末のHasseの註34参照．

るが，その中においてわれわれは次のことを見出し得るであろう．今日われわれは代数学ないし整数論を一般的な理論として有しているのであるが，その重要な部分の母胎ともなるべきものを以上述べてきたより具体的な理論のうちに認めるということである．

さて，Kronecker青春の夢は，類体論の完成とともに解決されたのであるが，もとより楕円函数の虚数乗法は，類体論の一例をもって目すべきものではない．それは，虚二次体の場合において，一般論を超えたものをわれわれが保持していることを意味する．そしてKroneckerが問題としたところのAbel方程式を一般の代数的数体の上に構成するという問題は，類体論の完成以後は，いわゆる類体の構成の問題として一つの重要な目標となったのである．Kroneckerの後，高木の前においてHilbertは，1900年パリの国際数学者会議の講演の中に数学の重要な問題を23挙げ，その第12番の問題としてこのAbel体構成の問題を取りあげた．すなわち，"有理数体に対する指数函数，虚二次体に対する楕円Modul函数に当る役割を果たす(解析)函数を，任意の代数的数体に対して見出し"て，その適当な点においてとる値によって，その代数体のAbel拡大をすべて構成しようという問題である．

1·2 Heckeによる拡張

すでに1893～94年に，Hilbertは虚数乗法論の一般化を試みていた．かれにあっては，modular函数の変換論が中心課題と考えられていたから，虚二次体に対する楕円modular函数と同様な意義を高次の数体に対してもつような函数の発見が問題であった．それは現在，Hilbertのmodular函数とよばれている函数で，総実な代数体の各々に対し定義されるものである．かれは，この函数の函数論的な基本性質についての短いnoteを書いたが，未発表のままその完成を弟子のBlumenthalに委ねた．この完成の仕事は，多変数函数の一般論があまり発達していなかった頃のこととて，かなり困難であったらしく，われわれはBlumenthal［1］において，その悪戦苦闘ぶりを見ることができる．一方，Picardその他のフランスの解析学者達は，純粋に函数論的な見地から，やはりHilbertのmodular函数に到達した．またHumbertは虚数乗法をもつ種数2の超楕円函数体についての一連の研究を発表したが，ある総実な2次体を虚数乗法の体としてもつような，種数2の超楕円曲線の不変量が，この2次体に対応するHilbertのmodular函数になるのである．Hecke［1］はこの観点から出発し，この2次体の上の総虚な4次体を生成するある数とその共役数に対応する上記の不変量の特殊値[1)]により生成される体の代数的性質を解明し，次に［2］において，その整数論的性質を研究した．その際の基本的な手段はもちろん，Hilbertのmodular函数の変換の理論である．ところでその結果は，はじめの予想とは異なり，これらの不変量の生成する体は，問題の4次体の，ある定まった共役体の上のAbel体となり，それも一般には絶対類体でなく，その部分体となるのである．このHeckeの研究は，"重要な美しい理論"[2)]として敬遠され，その一般化を試みる人の現われないまま，整数論の主流は，類体論を経て，抽象的な方向へと移ってゆくのである．

1·3 虚数乗法の抽象化・ドイツ学派の研究

抽象化のさきがけは，すでに早くArtinの学位論文［1］に見られる(1924)．かれは，有限体の上で定義された超楕円函数体の理論をつくり，さらに2次代数体のζ-函数の類似として，この超楕円函数体のζ-函数を定義し，これに対しRiemann予想の類似が成

1) すなわち，この4次体を虚数乗法の体にもつような超楕円曲線の不変量．

2) Hasse［5］参照．

り立つことを，特別な場合に確かめたのである．その後(1930) F. K. Schmidt は，有限体の上の代数函数体の一般論，およびその ζ-函数(合同 ζ-函数)の理論を展開し，合同 ζ-函数に対する Riemann 予想の類似を証明することが，ここに一つの中心問題となるに至った．この面での最初の成果は Hasse [2] により挙げられた．かれは，ある有限次代数体の上で定義され，虚数乗法をもつ楕円函数体から出発し，それの，常数体の素イデアルを法とする reduction により得られる，有限体の上の函数体の ζ-函数を問題にしたが，ここで Kronecker の合同関係式を利用して，この ζ-函数の零点が，はじめの曲線の虚数乗法の体に属することを証明した．楕円函数体の合同 ζ-函数に対しては，このことから直ちに Riemann 予想の類似が導かれるのである．しかしながら，Hasse のこの方法では，虚数乗法論における技術的な困難のために，楕円函数の合同 ζ-函数すべてを扱うわけにはゆかず，例外が現われる．Hasse はこの点を克服するために，上記の方法を逆行することを考えた．すなわち，有限体の上の楕円函数体の一般論を展開して，ζ-函数の零点が，ある虚2次体に含まれることを直接に証明することにより，楕円函数体の合同 ζ-函数に対する Riemann 予想の類似を一般に証明したのであるが，その際，虚数乗法による類体の構成の理論もまた，この方法により逆に代数的に構成できることに注意している(Hasse [3])．しかしこの計画に従って類体の構成の理論をつくりあげるためには，多くの技術的な準備が必要であって，後に Deuring の一連の研究 [1], [5], [2], [3] によりはじめて完成されたのである．

また Davenport - Hasse [1] は，$y^n = 1 - x^m$，あるいは $y^p - y = x^m$ (p は標数) という形の方程式で定義される，ある有限体の上の一変数代数函数体の ζ-函数に対しても，Riemann 予想の類似を証明した．これは本質的には，レムニスケート函数を定義する方程式の，有限体における解の個数についての Gauss の研究の発展であると見ることができる．Gauss はこれを，いわゆる Gauss の和の研究に利用したが，上記の一般の形の方程式に対しても，その ζ-函数の零点は，Gauss の和と類似な，Jacobi の和により表わされる．そして Jacobi の和の性質から直ちに Riemann 予想の類似が証明されるのである．後に Weil もまた，Jacobi の和等を含む，いわゆる exponential sum を取りあげているが，はじめかれは，Davenport - Hasse のこの研究を知らなかったように思われる．とにかく Weil も，このような和と，Riemann 予想の類似との関係に注目しているが，このときはすでに彼自身により，Riemann 予想の類似が証明された後であったので，かれの関心は逆に，この予想を用いて，このような和，とくに Kloostermann の和の絶対値の評価の限界を改良することにあった．

さて，Hasse は，有限次代数体の上の代数函数体に対して，今日 Hasse の ζ-函数とよばれている函数を定義し，それについていわゆる Hasse の予想を述べている．Weil は Jacobi の和の研究を続けるうちにこの予想に注目しはじめた．かれはまず [5] において，高次元の代数多様体に対する合同 ζ-函数を定義し，Davenport-Hasse の結果を，類似の形の方程式で定義される高次元の代数多様体に拡張し，いわゆる Weil の予想を提出したが，その後，$y^n = 1 - x^m$ という形の方程式で定義される代数曲線の，Hasse の ζ-函数が，Hecke の量指標の L-函数により表わされることを証明した([9])．これは，Jacobi の和の整数論的な性質から導かれるものであるが，Hasse 自身このことに気づかなかったのは不思議なことである．その後 Deuring [4] は，その抽象的な虚数乗法論の応用として，虚数乗法をもつ楕円曲線の

Hasseのζ-函数が，やはりHeckeの量指標のL-函数により表わされることを証明した．事実，合同ζ-函数の零点と，虚数乗法による類体の構成の理論との間には，すでに見たとおり非常に密接な関係があるのである．

1·4 代数幾何学の応用

一般の合同ζ-函数に対するRiemann予想の類似が，本質的には，代数幾何学においてCastelnuovoのlemmaとよばれるもの[1)]に基いていることを発見したのはWeilであった．その証明の方針は簡明であったが，それを具体的に遂行するためには，このlemmaを含む，代数曲線の代数的対応の理論を，厳密に，しかも抽象的につくりあげなければならなかった．Weilはそのために，代数幾何学の基礎づけ［1］から出発し，Riemann予想を証明し［2］，さらにAbel多様体の一般論を構成した［3］．代数幾何学の厳密な基礎づけはまた，その当時の代数幾何学の基本的な課題でもあって，Chevalley, Zari-ski等によっても，それぞれ異なる方法で取扱われている．

抽象的なAbel多様体の理論の完成によって，虚数乗法論の，高次の代数体への拡張が再び現実の問題となるに至った．抽象的な取扱いの原型としてはDeuringの理論［2］がすでに存在していた．一般化に当って，まず楕円函数の場合を代数幾何的に再構成する必要があるか否かは，人によって意見の分かれるところであろうが，Weilは，1950年Cambridgeにおける国際数学会議での講演［17］で，Kroneckerの合同関係式の重要性を強調し，その，代数幾何学的な意義を説明している．因みにHasseのζ-函数という概念がはじめて公表されたのも，この講演においてである[2)]．また，補助手段として，一般の代数多様体に対する常数体のreductionの理論が必要であったが，志村はこの理論を構成した(Shimura ［1］)．このような気運の下で，虚数乗法論の拡張は，Weil，志村，谷山によりそれぞれ独立して行われ，1955年，東京－日光における，代数的整数論に関する国際シンポジウムで報告されたのである(Weil ［14］, Shimura ［2］, Taniyama ［1］)．この三者の理論には，実質的に重複する所も少なくないのであるが，どちらかといえば，Weilのそれは，偏極多様体，Kummer多様体等の概念に，志村のそれはArtinの相互律に，また谷山のそれはFrobenius自己準同型の整数論的特性づけに，それぞれ重点がおかれていた．(その大要は本書の3章－6章で紹介される．)

さて，虚数乗法をもつ楕円函数体のHasseのζ-函数と量指標のL-函数との関係についてのDeuringの理論は，一般的な虚数乗法論を応用することにより，CM型に属するAbel多様体に対して，谷山により拡張された（Taniyama ［1］)．Hasseのζ-函数と量指標のL-函数との間のこのような関連はきわめて注目すべきものであるが，量指標はまたすでに他の関連の中に現われていた．すなわち，岩沢およびTateは，独立に，量指標がイデール類群の指標として説明され，量指標のL-函数の函数等式が，イデール類群の上のFourier解析を使うことにより証明されることを見出していたのである[1)]．このイデールは，Chevalley ［1］が，類体論を整理するために導入した概念であるから，虚数乗法論と，Hasseのζ-函数と，量指標との間の上記の関係が，イデールの言葉により表現されるであろうと考えるのはきわめて自然である．事実Weilは，同じシンポジウムにおいてこの問題に取組んでいる(Weil ［13］)．かれは，イデール類群の(A_0)-型の表現という概念を導入し，この表現に対して

1) 実質的には，［原著］51頁の不等式$\sigma(\alpha'\alpha)>0$と同じものである．

2) Hasseはこの概念を公表しなかった（数学7巻4号198ページを見よ)．

イデール類群の局所表現を対応させ得ることを示し，とくにAbel多様体のζ-函数を表わす量指標に対応する局所表現により定まる基礎体のAbel拡大と，その多様体の虚数乗法により得られる体との間に，ある密接な関係があるであろうと予想した．まもなく谷山はこの予想を解決したが，その際，Hasseのζ-函数と，量指標と，基礎体のAbel拡大との間に存在する関連は，一般の(A_0)-型の表現を公理的に特性づけることにより説明されている(Taniyama [2])．(この大要は本書の7章の後半で紹介される．)

最後に，虚数乗法およびHasseのζ-函数の理論と関係のあるもう一つの問題に注目しよう．1916年に，Ramanujan [1] は，Weierstrassの楕円函数の判別式（-12次元のmodular形式になる）の展開係数についての予想をたてているが，その中で，この展開係数の間のある種の乗法的な関係について，およびその絶対値についての予想がある．この第一の予想はその後Mordell [1] により解決され，さらにHeckeにより，一般的なmodular形式にまで拡張されて，いわゆるHeckeの作用素の理論となって結実した(Hecke [4]，[5])．一方第二の予想は現在でも解決されていないが，この予想も，一般のmodular形式に対し拡張できる．Eichler [1]，[2] は，-2次元のある種の形式に対しこの予想を取りあげた．かれはまず，modular形式をϑ-函数により表わし，常数体のreductionの理論を応用することにより，Heckeの作用素と，reductionされた函数体のp-乗準同型との間の関係式を見出した．この関係式により，問題にしている型のmodular函数体のHasseのζ-函数に対するHasseの予想が証明され，さらに，合同ζ-函数に対するRiemann予想の類似を利用して，問題のmodular形式の展開係数に対するRamanujanの第二の予想の類似が証明される．ここで，上記の関係式は，古典的な虚数乗法論における変換方程式と同様な方法で証明されるのである．志村はこの見地を取りあげ，modular函数体が，連続なパラメーターによる楕円函数の等分値により生成されることに注意して，この楕円函数に対するKroneckerの合同関係式を利用することにより，上記Eichlerの結果を，より一般的な型のmodular形式に対し証明した(Shimura [3])．(その大要は本書の9章で紹介される．) また，Eichlerは，1956年，Bombeyにおける，ζ-函数に関する国際シンポジウムで，Heckeの結果を他の面から拡張し，Heckeの作用素の跡に関する関係式を見出している．ここで，この関係式がまた，同じシンポジウムにおけるSelbergの講演で，弱対称Riemann空間の上の調和解析を応用することにより，より一般的な形で導かれていることは注目に値することである．

このように一般化された形における虚数乗法論は，単なる類体の構成の理論として，類体論の枠の中でのみ把握されるべきものではない．事実，以上に見てきたとおり，それは，整数論の幾つかの分野に新しい知見を加え，かくされた深い関連をあきらかにしながら，新しい理論の発展の原動力となりつつあるのである．前世紀において古典的な虚数乗法論が類体論に対してそうであったように．それ故，Hilbertにならって，あえて次のようにいうことが許されるであろう．新しい虚数乗法論の志向するこの方向をさらに追求し，発展させることは，現代の整数論および代数幾何学における最も重要な問題の一つであると．

1) ともに未発表．

参考文献

秋月康夫 [1] 調和積分論上下，岩波書店，1956.

E. Artin [1] Quadratische Körper im Gebiet der höheren Kongruenzen I, II, Math.

Zeitschr. 19 (1924), 153-206, 207-246.

I. Barsotti [1] Abelian varieties over fields of positive characteristic, Rend. Circ. Mat. Palermo 5 (1956), 1-25.

O. Blumenthal [1] Über Modulfunktionen von mehreren Veränderlichen I, II, Math. Ann. 56 (1903), 509-548, 58 (1904), 497-527.

C. Chevalley [1] La théorie du corps de classes, Ann. Math. 41 (1940), 394-418.

W. L. Chow [1] The jacobian variety of an algebraic curve, Amer. Journ. Math. 76 (1954), 453-476.

H. Davenport, H. Hasse [1] Die Nullstellen der Kongruenzzetafunktionen im gewissen zyklischen Fälle, Journ. f. R. u. A. Math. 172 (1935), 151-182.

M. Deuring [1] Die Typen der Multiplikatorenringe elliptischer Funktionenkörper, Abh. Math. Sem. Univ. Hamburg 14 (1941), 197-272.

—— [2] Algebraische Begründung der komplexen Multiplikation, Abh. Math. Sem. Univ. Hamburg 16 (1949), 32-47.

—— [3] Die Struktur der elliptischen Funktionenkörper und Klassenkörper der imaginären quadratischen Zahlkörper, Math. Ann. 124 (1952), 393-426.

—— [4] Die Zetafunktionen einer algebraischen Kurven vom Geschlecht Eins I, II, III, Nachr. Akad. Wiss. Göttingen (1953) 85-94, (1955) 13-42, (1956) 37-76.

—— [5] Reduktion algebraischer Funktionkörper nach Primdivisoren des Konstantenkörpers, Math. Zeitschr. 47 (1942) 643-654.

M. Eichler [1] Quaternäre quadratische Formen und die Riemannsche Vermutung für die Kongruenzzetafunktion, Arch. Math. 5 (1954), 355-366.

—— [2] La théorie des correspondances des corps des fonctions algébriques et leurs application dans l'arithmétique. (lecture note), Nancy, 1954.

R. Fricke [1] Die elliptischen Funktionen und ihre Anwendungen II, Leipzig u. Berlin, 1922.

—— [2] Lehrbuch der Algebra III, Braunschweig, 1928.

R. Fueter [1] Vorlesungen über die singulären Moduln und die komplexe Multiplikation der elliptischen Funktionen I, II Leipzig 1924, 1927.

H. Hasse [1] Neue Begründung der komplexen Multiplikation I, II, Journ. f. R. u. A. Math. 157 (1927), 115-139, 165 (1931), 64-88.

—— [2] Beweis des Analogons der Riemannschen Vermutung für die Artinschen und F. K. Schmidtschen Kongruenzzetafunktionen in gewissen elliptischen Fälle, Göttingen Nachr. (1933), 253-262.

—— [3] Abstrakte Begrüdung der komplexen Multiplikation und Riemannsche Vermutung in Funktionenkörpern, Abh. Math. Sem. Univ. Hamburg 10 (1934), 325-348.

—— [4] Über die Kongruenzzetafunktionen, Sitzungsber. Preuss. Akad. Phys. Math. 1934 XVII.

—— [5] Bericht über neuere Untersuchungen und Probleme aus der Theorie der algebraischen Zahlkörper I, Iα, II, Jahresber. d. D. M. V. (1926, 27, 30).

A. Hurwitz [1] Grundlagen einer independenten Theorie der elliptischen Mo- dulfunktionen und Theorie der Multiplikator-Gleichungen erster Stufe, Werke I.

—— [2] Über die Klassenzahlrelationen und Modularkorrespondenzen primzahliger Stufe, Werke II.

E. Hecke [1] Höhere Modulfunktionen und ihre Anwendung auf die Zahlentheorie, Math. Ann. 71 (1912), 1-37.

—— [2] Über die Konstruktion relativ Abelscher Zahlkörper durch Modulfunktionen von zwei Variabeln, Math. Ann. 74 (1913), 465-510.

—— [3] Eine neue Art von Zetafunktionen und ihre Beziehung zur Verteilung der Primzahlen II, Math. Zeitschr. 6 (1920), 11-51.

—— [4] Über die Modulfunktionen und die Dirichletschen Reihen mit Eulerschen Produktenwicklung I, II, Math. Ann. 114 (1937), 1-28, 316-351.

—— [5] Analytische Arithmetik der Positiven Quadratischen Formen, Kgl. Danske Vidensk. Selskab. Math.-Fys. Med. XVII, 12. Kopenhagen, 1940.

D. Hilbert [1] Mathematische Probleme, Vortrag, Intern. Math. Kong., Paris, 1900, Gesammelte Abh. III.

岩沢健吉 [1] 代数函数論，岩波書店，1952.

F. Klein [1] Vorlesungen über die Theorie der elliptischen Modulfunktionen I, II, Leipzig, 1890, 1982.

S. Koizumi [1] On the differential forms of the first kind on algebraic varieties, Journ. Math. Soc. Japan 1 (1949), 273-280.

L. Kronecker [1] Zur Theorie der ellipts-chen Funktionen, 1883-1889 Werke IV.

S Lang, A. Weil [1] Number of points of varieties in finite fields, Amer. Journ. Math. 76 (1945), 819-827.

S. Lefschetz [1] On certain numerical invariants of algebraic varieties with application to Abelian varieties, Trans. Amer. Math. Soc. 22 (1921) 327-482.

T. Matsusaka [1] Polarized varieties, the fields of moduli and generalized Kummer varieties of Abelian varieties, Proc. Japan. Akad. 32 (1956) 367-372.

L. J. Mordell [1] On Mr. Ramanujan's Empirical Expansions of Modular Functions, Proc. Cambridge Philos. Soc. 19 (1917).

Y. Nakai [1] On the divisors of differential forms on algebraic varieties, Journ. Math. Soc. Japan 5 (1953), 184-199.

中井喜和，永田雅宜 [1] 現代数学講座，代数幾何学，共立出版，1957.

H. Petersson [1] Konstruktion der sämtlichen Lösungen einer Riemannschen Funktionalgleichung durch Dirichlet-Reihen mit Eulerscher Produktentwicklung II. Math. Ann. 117 (1940), 39-64.

S. Ramanujan [1] On certain arithmetical functions, Trans. Cambr. Philos. Soc. 22. No. IX, 1916.

B. Riemann [1] Über die Anzahl der Primzahlen unter einer gegebenen Grösse, Werke.

F. Severi [1] Funzioni Quasi Abeliane, Rome, 1947.

G. Shimura [1] Reduction of algebraic varieties with respect to a discrete valuation of the basic field. Amer. Journ. Math. 77 (1955), 134-176.

—— [2] On complex multiplications, Proc. Int. Symp. Alg. Nb. Th. Tokyo-Nikko 1955, 23-30.

—— [3] La fonction zeta du corps des fonctions modulaires elliptiques. Comptes Rendus 224 (1957), 2127-2130.

末綱恕一 [1] 解析的整数論，岩波書店，1950.

菅原正夫 [1] 虚数乗法論，I，II，岩波講座数学，1930.

T. Takagi [1] Über eine Theorie des relativ-Abelschen Zahlkörpers, Journ. Coll. Sci., Tokyo, 41 (9), (1920), 1-132.

高木貞治 [2] 代数的整数論，岩波書店，1948.

Y. Taniyama [1] Jacobian varieties and number fields. Proc. Int. Symp. Alg. Nb. Th. Tokyo-Nikko, 1955, 31-45.

—— [2] L-functions of number fields and zeta-functions of abelian varieties. (forthcoming in Journ. Math, Soc. Japan).

H. Weber [1] Lehrbuch der Algebra III, Braunschweig, 2 Auflage, 1908.

A. Weil [1] Foundations of algebraic geometry, New York, 1946.

—— [2] Sur les courbes algébriques et les variétés qui s'en déduisent, Paris, 1948.

—— [3] Variétés abéliennes et courbes algébriques, Paris, 1948.

—— [4] Théorèmes fondamentaux de la théorie des fonctions thêta, Seminaire Bourbaki, Paris, 1949.

—— [5] Number of solutions in finite

fields, Bull, Amer. Math. Soc. 55 (1949), 497-508.

—— [6] Sur la théorie du corps de classes, Journ. Math. Soc. Japan 3 (1951), 1-35.

—— [7] Arithmetic on algebraic varieties, Ann. Math, 53 (1951), 412-444.

—— [8] On Picard varieties, Amer. Journ. Math. 74 (1952), 865-894.

—— [9] Jacobi sums as "Grössencharactere", Trans. Amer. Math. Soc. 73 (1952), 487-495.

—— [10] On algebraic groups of transformations, Amer. Journ. Math. 77 (1955), 355-391.

—— [11] On algebraic groups and homogeneous spaces, Amer. Journ. Math. 77 (1955), 493-512.

—— [12] Sur les critères d'équivalence en géometrie algébrique, Math. Ann. 128 (1954), 95-127.

—— [13] On a certain type of characters of the idèle-class group of an algebraic number-field, Proc. Int. Symp. Alg. Nb. Th. Tokyo-Nikko, 1955, 1-7.

—— [14] On the theory of complex multiplication, ibid. 9-22.

—— [15] The field of definition of a variety, Amer. Journ. Math. 78 (1956), 509-524.

—— [16] On the projective embedding of abelian varieties, Algebraic geometry and topology, a symposium in honor of S. Lefschetz, Princeton, 1957.

—— [17] Number theory and algebraic geometry, Proc. Int. Cong. Math. 1950.

ある人の話

僕の友人に，Ｓといふ人が居た．彼は幾分体の弱い，おとなしい秀才であったが，此の冬頃から何となく態度が変になって來て，僕ともあまり口を利かなくなってしまった．六月の初めの或る日，彼に偶然會った時，彼はすべてを僕に語つて呉れた．此れがその話である．以下僕とあるのはＳの事と思はれたい．

一

僕がまだ勤労動員に行ってゐる頃の事だ．秋の末の，曇った，うら寒い日であつた．僕は仕事が嫌になって，ブラブラしながら池の邊まで來た．君は知ってゐるね，工場の裏の，あの大きい池だ．生ひ茂ってゐた草は大方枯れ果てて，弱い北風に戰いてゐる．あゝ寒いなあと思ひながらふと見ると，向岸にＫが腰を下して，ぢっと何かを見つめてゐる．僕が彼のそばに行くと，彼は靜かに話し出した．

「僕は工場へ來てから，毎日此處へ來て，水の色を眺めてゐるが，不思議に，いつも少しづつ変ってゐるね．季節によって，天候によって，温度によって．一寸見ただけでは解らないさ．擬っと見てゐ給へ．今日は曇ってゐるので，此んなに白っぽく，灰色に近い様な緑色をしてゐるが，夏の曇天には此んな色ぢゃない．とても繪具などでは，描けないよ．然し，よく晴れた秋の日は好いね．水が，底まで澄んで深い空色の中に，水草がゆらいでゐる　のまで見える．それに，強い秋の日が照り附けて，水の色を一層明るく，透明にしてゐる．僕は何回か，あの色を描かうとして失敗したが，實に神祕的な色だね…．」

彼の話は，盡きようとわ（ママ）しなかった．僕はＫの顔をつくづくと見直した．今まで，ろくに口も利かなかったＫではあるが，急に親しくなった様な気がした．僕の知らない，別の大きな世界の空気を呼吸してゐる様に思はれるＫは，僕にとって一つの驚異であり空想の的となってしまった．

二

その翌日，僕がＫと顔を合せると彼は言った．

「おい．池に行かう．」

僕等は，昨日の様に，池の岸に腰を下した．然し，彼は全然物を言はずに，吸ひ込まれる様に池に見入ってゐた．そして時々，何か考へる様にうなだれるかと思ふと，急に目を輝かせて，顔を上げたりした．僕も池を見たが別に昨日と変ってゐる様にも見えなかった．唯だ，その寒さうな水の色や，岸の枯れた草などが，冬が來たといふ感じを強く表してゐる様に思はれた．然し，彼が毎日毎日飽きもせずに此處へ來てゐるのが不思議でならなかった．暫くの無言の後，彼は思ひ出した様に言った．

「君は純眞で好いなあ．僕なんかもう駄目だ．」

「何故？」

然し彼は再び無言にかへってしまった．僕はその意味が全然解らなかった．

その後も彼はよく僕をさそった．そして，いろいろ話し合つてゐる中に彼は，僕の知らない，或る大きなものを持ってゐる様に思はれた．彼は畫家であると同時に詩人でもあった．僕は彼と語るのが，樂しかった．大なるもの，或ひは未知なるものに対する憧れの気持ちであらう．又彼は何のわだかまりもなく僕に話した．今までろくに話もしなかった彼が何故あんな親しい態度を取ったか，今だにわからない．たゞ，あの頃の事を考へると，運命といふものの力を，つくづくと感じるよ．

三

その頃の或日久喜の驛で，四年から松本高校に行つたTに遇った．彼は四年から入ったせゐか，あまり蠻カラな風を好まない様に見えた．今まで，同じく勉强して來たTの白線を見て，僕は堪まらなく思った．そして，今年どうしても，高校を突破しようといふ気持ちに驅られた．が，彼と話しをして行く中に，彼はあまり中學校時代と変りない様にも思はれた．然しそんな事よりも僕は彼の白線に夢中になってゐた．やがて，彼が汽車に乘ってしまふとKは言つた．

「彼は眞面目だね．善いにしろ悪いにしろ，あまり大きいことはしさうもない．」

四

實際Kはサボリの名人だった．彼は，入所してから，仕事をした日は三日しかないとか言ってゐた．その後は，全然職場へも行かず，例の池へ行ったり，裏の林の中をさまよったり，工場内を散歩したり，目に附かない様な所で本を讀んだり，二三人の同志と喋り合ったりしてゐた．その癖，缺席は人一倍するので，僕は，屢々彼を待ってゐて失望する事があった．が，此んな事にも関らず，僕と彼とは，段々親しくなって行った．

彼は好く突飛な事をして僕達を驚かしたが，それも，極く自然にやって，少しもわざとらしい奇をてらふ様な所はなかった．

或る時は辨當を忘れたから何とかして呉れと，所長の所へ頼みに行った．此れには所長も呆れた様だったが，結局彼は御馳走にありついた．又ある時は，サボリ場所に困って，總務部長室に行って遊んでゐる内はまだ好かったが，誰も居ないのに附け込んで引出しの中を掻き廻してゐる所へ總務部長が入って來て，ひどく油をしぼられたこともあった．しかし彼は，いつものんきなものだった．

そんな風だから，職場主任や組長はよく彼に注意したが，彼は全然感じなかった．しまひに組長が業を煮して言った．

「君は一体何をしに來てゐるんだ．」

「飯を食ひに來てゐるんですよ．」

彼の聲は普通と変りなかった．後で彼は言った．

「職工の言ふ事を一々気にかけてはゐられないよ．」

五

その頃，僕は三太郎の日記を讀み出した．著者の激しい自己分析の気持は，僕をもその中に捲き込まずにはゐられなかった．自はそれをむさぼる様に讀んだ．

「いやに熱心だね．」とKは言った．

「僕は何か始めると，それに熱中して終るまで，止められないのだ．」

「やはり，君は純眞で好いね．然し，あまり多くの事に熱中し過ぎると，結局何も出來なくなってしまふ事もあるよ．」

六

僕は，自分の気持について，つくづく考へずには居られなかった．

僕はKの様に心臓强くサボル事は出來ない．と言って，今までの様に，何の考へもなく働く気にもなれない．僕は今まで，どんな気持で働いて來たのだろう．

他人に悪く思はれるのが嫌で，他人の心を乱すのが心苦しいから，出來るだけ事を起さない様にして來たのではなかったか．然しすべての人に好く思はれるのは不可能だ．それに，他人は遠慮なく僕の心を乱すではないか．

それに，他人に好く思はれようとするのは，自己を僞って，何等かの利益を得ようとするためではないか．他人の心を乱すのが心苦しいと言っても，さうしないのは，自分に対して自己を僞って，一時の気休めをしてゐるのではないか．僕がさうしなくても，他の人或ひは社會的，自然的事象が，多くの人の心を乱してゐるではないか．それを，自分一人，他人の心を乱さないからと，安心するのは現實よりの逃避ではないか．

又，そのすべての裏には，自分を偉く見せようとする高慢な心が働いてゐはしなかったか．だから徹底的に働くのは何だか嫌な気もする．此んな風だから，俺のすることが，皆中途半端な，あいまいなものになってしまふのではないか．さう思ふと僕は，心にもないことを續けるのが嫌になって來た．然も，僕には，思った事を行ふだけの決断力と實行力とが缺けてゐた．

又，眞面目に働いていゐるＡやＨ等は，何の考へもなく，命ぜられる事に従ってゐる様に見える．そんな人は幸福だ．

然し，盲目的にさうする事の出來ない僕はどうすれば好いのだ．自己に従ふか．調圍（ママ）と調和するか．僕は中途半端な態度はいやになった．が結局，どうして好いのか僕には解らなかった．そしてつくづくＫが，うらやましくなった．

七

僕は，心の中に，割り切れぬものを殘しつつも，段々とサボリになって行き，Ｋと行動を共にするようになった．工場をエスケイプして方々へ遊びに行ったり，工事場のトロッコを押したり乘ったりして一日遊んだり，半日で終して，早く家へ歸ったり，池の邊で話し合ったり，工場の品物を持ち出したり，一々言っては居られない．然し，前から教師に好く思はれて居なかった僕は，完全に教師に，にらまれてしまった．尤もＫには，教師も呆れ返って居た様であるが．とに角，僕は一日も仕事をする所か職場に行った日さへなかった．

或る時，僕はスピンドル油を持ち出さうとして，つかまった事があった．そして防衛課長に説教されたがそれを，正門の脇に大きく掲示された．僕は憤慨した．わざわざ掲示しなくても好いぢゃないか．何のためにあんな事をするのだ．

Ｋは僕の所へ來て言った．

「とうとうつかまつたね」と．が僕があまり憤慨してゐるのに多少驚いたらしく

「いやに怒ってゐるね．が，どうでも好い事ぢゃないか．」

どうでも好い事ぢゃないか．僕は兄の手紙の一節を思ひ出した．

「他人が，お前の悪口を言ふとか，皮肉な態度を取るとか，自分一人でうまい事をやってゐるとか，気にくわぬ様子をしてゐるとかそんな事を気にしたらきりがない．又日常の細かいことに気を使ってゐたら，それだけ愚だ．教師や工場側で何と言っても好いではないか．そんな事を気にしてゐたら，何も出來やしない．」大行は細謹をかへりみずか．

が一部の人々は更に憤慨して，僕の所へ來て，一緒に行って防衛課長に文句を言はうぢゃないかと言った．僕がいやだと言ふと彼等は覇気がなさすぎると批難した．結局，僕には，つまらない事の様に思はれて來た．

然し，此れは一種の逃避ではなかったか．そして，実際，覇気がなさ過ぎたのではなかったか．それがつまらぬ事ならば，つかまった時，防衛課長に呼ばれた時何故あんなにビクビクし，おどおどしたのか僕は都合の好い様に理屈を附けて居るのではないか．と言って結局，むきになって憤慨するのも馬鹿らしくなって來た．

が此の事は，僕に対する教師の信用を根底から破壊してしまつた．

八

その中に，受驗期になった．皆，希望に輝きつつも，どこか心配さうな目附で，お互ひ

の志望を語り合ったりしてゐた．入学願書を取り寄せたり，内申書を頼みに行ったり，忙しかった．僕は水高（編集部注：水戸高校，茨城大学文理学部の前身）を選び，Ａと一緒に，教師の所へ志願名票を持って行った．が，Ａには親切に，いろいろと注意したり話しをしたりする教師は，僕には甚だ冷淡で，殆んど口も利かなかつた．普段から憎まれてゐた上に受験校の撰擇を全然相談しなかったので，無視されたと思ったのであらう．その目には，「後で後悔するな」と言ふ様な，陰険な光があった．

僕は多少失望した．然し，成績を書き変へることは出來まいと高をくくってゐた．

一月の十二三日頃になると，僕は不安になり出した．工場に居ても，今日あたり一次合格通知が來てゐはしまいかと，居ても立っても居られない思ひであった．そして，いつも，家に歸って失望するのであった．

その中にＡに合格通知が來た．僕より成績の悪いＡに來て，僕に來ないとは．僕は始めて，工場の勤勞成績を，あまり軽視し過ぎた事に気が附いて，多少焦燥し出した．

Ｋはその少し前から缺席してゐた．そして美術学校（編集部注：東京美術学校，東京芸術大学の前身）を受けること，必勝を期してゐること等を手紙に書いて寄越した．然し僕より遙かにサボリでにらまれてゐたＫが，一次に通るだらうかと思った．

僕は第二期は横浜工專（編集部注：横浜工業専門学校，横浜国立大学工学部の前身）を受けた．そして，二十日間の不安と焦燥の後に横浜まで，發表を見に行って一次に落ちた事を確認して歸って來た．

所がＫは一次に通ったのだ．受驗者が意外に少かったのだ．さもなければ通る筈がない．彼は締切の翌日，内申書を書いてもらひに教師の所へ行き，さんざん油を搾られてやっと出来上つた内申書を持って郵便局へ行き，前日のスタンプを押して呉と言って，局員を呆れさせた程なのだ．そのＫが入ったとは．然もその後，彼は二次にもパスした．僕は取り殘された様な，何とも言へない淋しい感じがした．

三期には臨教（編集部注：臨時教員養成所）に出したが勿論振られた．そして遂に，今の様な代用教員に落ちぶれたのさ．

あの頃の僕の気持は，君には倒底解るまい．前途は全く暗くなってしまって，すべての希望を失ってしまった．たゞ，惰性的に生きてゐた様なものだった．Ｋと交わったために，此んな風になったのかと思ふと，つくづくＫがうらめしかった．而も彼は，何となく去り難い魅力を持ってゐた．

九

三月の中頃といふのに，何となく寒い日だった．僕はすっかり失望して，悲哀の底に沈んでゐた．僕とＫとは，例の如く，池の所に行ったが，Ｋは，僕に，いろいろな歌を聞かせて呉れた．その中でも，殊に僕の心にしみじみと感じたのは，童謡であった．その空想的な文句，夢幻的な節，それは僕を，子供の頃の思ひ出の世界へと誘った．僕は，その思ひ出の世界に長い間さまよってゐた．

あゝ，僕にも子供の頃はあったのだ．あんな歌を聞きながら，限りない夢を見てゐた頃もあったのだ．あの頃は，すべてが美しく，すべてが大きく見えた．

あの頃から，僕は段々大きくなり，自己に目ざめ思想も発達して來たが，結局それは，僕にいろいろな悪を見せただけで，それを解決することは出來なかつた．

もう一度あの無邪気な気持にかへりたい．いつまでも子供でゐたい．あの童謡の様な気持ちの中に，一生，ひたってゐたい．あの時はつくづく子供の頃がなつかしく思はれた．

十

Ｋとは卒業式の日に別れたきり，全然會は

ない．今，Kと交わってゐた頃を思ひ出すと，暗い，厚ぼったい雲が，心を一面に覆ってゐた様な気がする．

Kの様に，すべてを自由に爲し得る人は，つくづく幸福だと思ふ．僕はKの様にしようとすると，すぐ，周圍の圧迫を感じ，それに対し努力せずには居られない．そして，どうでも好い事に努力するのは，つまらぬことではなかったか．

運命が僕達を奔弄し，交友や環境が僕達を左右する様に見えても，性格そのものを根本的に覆へす様な強力なものではない限り，人は皆，それぞれの性格に従って生きて行くものだね．僕も，いろいろな迂路曲折を経たが，結局僕自身の性質に依つて生活し，考へる様になった事には変りない．結局僕は僕自身でしかあり得ない．

僕は入学試験に落ちた頃，随分落膽したが，今考へて見ると，上級学校へ行つて，又工場へ行き，前の様な苦しい気持で生活するよりも，今の様に，割合に自由な生活の方が，好かったかも知れない．

それに僕は，社会的に成功しようとか，偉くならうとかいふ考へを捨ててしまった．そんな事は，つまらない事の様に思はれて來たのだ．そして自分自身に従って生き，自己を完成する事の方が，遙かに大事な事の様に見えて來たのだ．

僕は今まで，あまりにも自分の事ばかり考へ過ぎた．自分の他にも，いろいろ物を考へる他の多くの人人が居ること，そして，その人々から孤立しては生きて行けない事に無関心であり過ぎた．

彼はかう話し終って淋しく笑った．それは，すべてをあきらめた様な笑ひであつた．

『プラトンの自叙傳』書き込み

五月三日（金曜日）
予は本日此の書を、求むる時此の書が記念すべき物にならうとは誰が考へたであらうか．
予は愛の光を見失へり．然れども失いし汝よ汝には學問の苷界あるにあらずや．
行け唯眞直に見向きもするな．唯學問の愛情の中に．

遠き別れに堪えかねて
此の高樓に登るかな
悲しむなかれ吾が友よ
旅の衣をととのへ

別れといへば昔より
此の人の世の常なれど
流るる水を　眺めれば
夢はづかしき涙かな

君が清けき目の色も
君が紅の唇も
君が清けき黒髪も
又何時か見ん　此の別れ

君が行くべき山川は
落つる涙に見え分かず
袖のしぐれの冬の日に
君に送らんはなもがな

初恋の歌（石川啄木）
砂山の砂に腹這ひ初恋の
いたみを遠く思い出る日

●島崎藤村『若菜集』所収「高樓」より．『若菜集』の詩はひらがなで表記されている．
●写し違いもあるが，谷山のメモをそのまま再現した．
●啄木の歌は歌集『一握の砂』の一首．
●「五月三日（金曜日）」より昭和32年（1957年）5月3日と思われる．（編集部）

書簡

杉浦光夫　へ

1953年7月20日
大塚　より

今日工学部の人の持って来た問題，帰りに次の様な証明を考へました．

最初の間隔を $\dfrac{l}{n}+k_i \quad (i=0,\cdots,n-1)$ とする（n は点の個数，l は円周の長さ）．$\sum_{i=0}^{n-1} k_i=0$ である．

N 回平均を取ったときの間隔は

$$\frac{l}{n}+\frac{1}{2^N}\sum_{j=0}^{N}\binom{N}{j}k_{i+j}$$

$$(i=0,1,\cdots,n-1)$$

但し，$m\equiv i\ (\mathrm{mod}\, n)$ のとき $k_m=k_i$ とする．

従って

$$\left|\frac{1}{2^N}\sum_{j=0}^{N}\binom{N}{j}k_{i+j}\right|\to 0 \qquad (N\to\infty)$$

を証明すればよい．ここまでは今日云った通り．

所で，k_{i+j} の係数を比較する．$n>h>j\geqq 0$ として，

(1) $\binom{N}{j}+\binom{N}{j+n}+\binom{N}{j+2n}+\cdots\cdots$

と，

(2) $\binom{N}{h}+\binom{N}{h+n}+\binom{N}{h+2n}+\cdots\cdots$

とを比較する．$\binom{N}{i}$ は，$i=\left(\begin{smallmatrix}N\\ \left[\frac{N}{2}\right]\end{smallmatrix}\right)$ まで単調増大（[　] は Gauss 記号），以下単調減少．

故に，(1)の中央に $\binom{N}{\left[\frac{N}{2}\right]}$ を 2 つ入れて，(1)，(2)の対応する項を，中央から両端へと比較して行けば，(1)$+2\binom{N}{\left[\frac{N}{2}\right]}$ が(2)の和より大なことがわかる．

註，例へば $\left(N=11, n=3, j=0, h=1\right.$ として，$\left.\binom{N}{\left[\frac{N}{2}\right]}=\binom{11}{5}\right)$

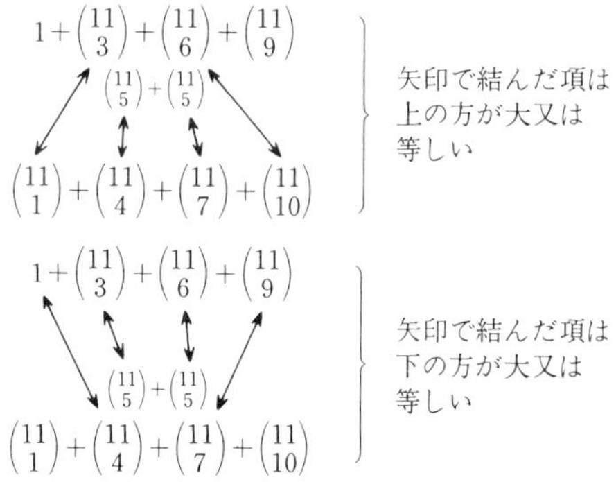
矢印で結んだ項は
上の方が大又は
等しい
矢印で結んだ項は
下の方が大又は
等しい

次に，(2)の中央に $\binom{N}{\left[\frac{N}{2}\right]}$ を 2 つ入れて同時に比較すれば(2)$+2\binom{N}{\left[\frac{N}{2}\right]}$ は(1)の和より大．

従って，(1)の和と(2)の和との差は $4\binom{N}{\left[\frac{N}{2}\right]}$ より大でない．

所が，Stirling の公式より

$$\frac{1}{2N}\cdot\binom{N}{\left[\frac{N}{2}\right]}$$

は，$N \to \infty$ のとき，$\frac{1}{N^{\frac{1}{2}}}$ の order である．

故に，$\sum k_i = 0$ を考慮に入れて，

$$\left|\frac{1}{2^N}\sum_{j=0}^{N}\binom{N}{j}k_{i+j}\right| < \frac{C}{N^{\frac{1}{2}}}\sum_{i=0}^{n-1}|k_i| = \frac{C'}{N^{\frac{1}{2}}}$$

故に

$$\frac{l}{n} + \frac{1}{2^N}\sum_{j=0}^{N}\binom{N}{j}k_{i+j} = \frac{l}{n} + \varepsilon_i,$$

$$|\varepsilon_i| < \frac{C'}{N^{\frac{1}{2}}}$$

此れが証明すべきことでした．

此の証明は，最も簡単ではないとしても，最も簡明でせう．あの人が又やって来て，君達を悩ましたりしない様に，一応お知らせしておきます．

杉浦光夫　へ

1953年8月11日
巣鴨　より（はがき）

ご無沙汰しました．……と云っても，此れが僕から君への最初の手紙になるわけですが，慣習上さう云って置きます．

君の事とて，相変らず，昼寝をして，夜は勉強したり，悲憤慷慨(大げさですか)したり……そんな所でせう．

今年の夏は，僕に取って，一寸変な夏になりさうです．……変と云ふのは，此れまでそんな事は無かったと云ふ意味であり，又，十分に怠けているにも拘らず，相当勉強した様な気持になっていられると云ふ意味でもあります．……と云っても君には良くわからないでせうが，僕にも良くわからないのだから，仕方がない……だから，此んなおしゃべりは止めませう．

つまり，数学に於ける一つの流れの中に身を置く様になったと云ふことです．それは，独りでボヤボヤしているより，ずっと有意義だと云ふ事です．

そんなわけで，代数幾何の分科会があるので，京都に出かけます．その帰りに，君の家へ遊びに行かうと云ふわけです．(勿論，僕独りで) 18日か19日，その頃君は家に戻るでせうね？

出来たら道順を教へて下さい．「京都市左京区北白川　京都大学　数学教室　谷山豊」宛で僕の手に入ると思ひます．入らなくても，交番ででも聞けば道順はわかるでせう．

色々お話ししたいことがあります．尤も，無くても，退屈しのぎにはなるでせう．

京都へ着いてから，宿所をお知らせします．もし18日か19日頃，君の方の都合が悪い様なら，そこ宛に電報でも打って下さい．

すべて夏枯れで，頭の中まで空っぽになりさうです．それを見越してか，馬鹿馬鹿しいことばかり起ります．ワンマンの保安隊員への訓示，南鮮の選挙，イギリス水兵事件 etc. ……数へ上げたら切りがありません．

「野火」を読みました．何か物足りない気持……．

10円切手が手許にないため……郵便局で買ってなど云ったら，何時投函出来るかわからないので……有り合せのハガキで失礼．おしゃべりは又後で，……

杉浦光夫　へ

1956年11月29日
池袋　静山荘　より（はがき）

前略

その後，お風邪はどうですか．案じて居ます．実際，悪いときに風邪を引いたものですね．

さて，御承知の通り，今月30日(金)午後5時半から，秋月さんとの座談会を開きます．そこで，余り御健康に差し障らない様でしたら，是非出席して，熱弁を振って下さい．期待しています．……（今日，委員，有志が相談して，大体の方針を決めました）．

尚「数学の歩み」先号の原稿の残り，あづかっています．金曜日に，君か長野君かにお渡しします．

お大事に，奥さんにもよろしく．

谷山清司　へ

1950年10月17日
上野櫻木町　より（はがき）

鈴文堂と云ふ本屋から書物及び雑誌が届いて居ります．持つて行くのは大変ですから，取りに来て下さい．

定期は買ひました．今度の日曜に家に歸りませんか？

何も変つたことはありません．毎日忙がしくて，消耗しています．

谷山清司　へ

1951年４月4日
上野櫻木町　より（はがき）

今日，小包デ本ト時計ヲ送リマス．“産科手術学”又ハソレニ類似シタ名前ノ本ハ見当リマセンデシタ．

時計ハ包装ニ注意シマシタガ，ソチラニ着イテ動クカドウカハ一寸保証デキマセン．途中デホウリ投ゲタリ，取扱ヒハカナリ乱暴ラシイカラ．

今日，家ニ歸ツテ四五日泊ツテ来ル予定デス．

谷山清司　へ

1951年７月26日
大塚坂下町　より（はがき）

御無沙汰シマシタ．

今度表記ニ引越シマシタ．３疊デ1500円．今ハ此ノ位ガ普通デス．交通ハ便利デスガ，アマリ良イ所デハアリマセン．

近イ中ニ家ニ歸ツテ１ケ月位居ルツモリデス．ソノ間ニ一度ソチラニ遊ビニ行カウト思ヒマス．詳シイコトハ又後カラ．大体ドンナ風ニ行タノカオ知ラセ下サイ．

暑イノデ閉口デス．何モ出来マセン．

何モ他ニ変ツタコトハアリマセン．

以上，オ知ラセマデ

（文京区ニハ駒込坂下町モアリマスカラ．大塚坂下町ノ大塚ハ抜カサナイデ下サイ.）

谷山清司　へ

1953年４月８日
大塚坂下町　より（はがき）

土曜日（11日）ハ都合ガ悪イノデ日曜日（12日）ニシマス．

午前中ハ適当ナ時刻ニコチラニ来テイタダケレバ幸デス．（朝寝坊デスカラソノ方ガ時間ノ節約ニナルデセウ）

以上，オ願ヒマデ

谷山清司　へ

1953年４月５日
大塚坂下町　より（はがき）

今日，家から出て来ました．近い中に，いつか，洋服を見立てていたゞきたいと思ひます．適当な日時，場所をお知らせ下さい．

今，胃を悪くして一寸消耗しています．

以上，お願ひまで

谷山清司　へ

1953年４月15日
大塚坂下町　より（はがき）

先日ハゴチサウサマ．

本立ハ，来週デナケレバ（ツマリ20日以後）入ラナイトノコト，ソレ故持ツテ行クノハ，ソノ後ニナリマス．杉製デ700円ラワン製デ1100円，ラワンノハ，小シ巾ガ狭クナル様デスガ，杉ノヨリハ，ズツト丈夫デセウ

（大キサト見比ベテ適当ナノヲ買ツテ行キマス）

以上

谷山清司　へ

1953年７月19日
大塚坂下町　より（はがき）

御無沙汰しました

本棚も来たし，休みにもなつたので，一度，そちらに伺はうと思ひます．いつお暇なのかお知らせ下さい．

但し，こちらは，25日以後でないと，一寸暇はありません．忙しいこと，驚くべきものがあります．

来月始めに家に帰ります．

ではいづれ又．

成田正雄　へ

1953年10月25日
大塚坂下町　より（はがき）

昨日は，折角いらして下さつたのに，居なくて失礼しました．文理大での輪講がなくなつたと聞いて，君が帰り途に，僕の下宿へ寄るんぢやないかなと思つたのですが，学校で，杉浦とおしやべりしている中に，ついおそくなつてしまつたので….僕は28日の水曜日に帰るつもりです．それまでも何やかとあつて，もうお会ひする機会がないかと思ひますが，暇を見て，家の方に，遊びにいらつしやい（その前に一寸，ハガキででもいつ来るか知らせて下さると好都合です．勿論，僕の都合の悪い日など，ありませんが）

僕は夢を気にしませんし，憶えておくことにもしていませんがそれにも拘らず，僕の記憶に残る夢があれば，それは非常に面白いものであるからに違ひないと思つています．その面白さが一見わからないとしても．

以上，取敢へず．

追伸　学会に行くとき，君の都合がよかつたら，一緒に行かうと杉浦が云つています．そして一緒に，山菅の下宿に泊らうと．28日の朝の汽車で行くさうですが，彼と連絡して御覧なさい．一人旅は退屈ですからね．

再伸　僕の机の引出しやノートなど，かきまはさなかつたでせうね．

谷山義泰　へ

1954年９月１日
埼玉県　騎西　より（はがき）

今度来る時に，机の引出し（多分右の）に入つている昔の写眞を持つて来て下さい

手紙及び葉書二枚，確かに受取りました．

サカタから水仙，チューリツプなど，到着．

以上

谷山欣隆　へ

1955年５月２日
池袋　靜山荘　より（はがき）

御葉書，いただきました．又，先日の書留も確かに，受取りました．少し勉強が忙しかつたので，面倒になり，つい，御返事を忘れてしまつたのです．まだ忙しいため，当分は家には帰りません．

おばあさんによろしく，どんな具合なのですか．

義泰がそちらに帰つた時に，次の本を小包にして，東大の方に，送る様に話して下さい．

1．「和獨辞典」（２階の，北向きの窓際の箱の上にあり）
2．関口存男著「髙等独文典」（東の家か，義泰の下宿かにある筈）

谷山義泰　へ

1955年5月10日
池袋　静山荘　より（はがき）

御葉書拝見.
東大の方の住所は,
東京都文京区本富士町一
東京大学理学部数学教室
谷山豊
尚, 高等独文典は, 濃いミドリ色の紙の表紙で, 脊に薄いドイツ文字で題名が入つているだけなので, わかりずらいのかと思ふが, 或はどこか別の所に行つていてそちらになかつたのかもしれない. 文法書は一冊買つたから, 無ければ勿論送る必要はない.
以上

成田正雄　へ

1955年6月21日
池袋　静山荘　より（はがき）

前略
今度の木曜日に, 弥永さんの所で, 君がセミナリーをやる様にしましたから, よろしく.
以上.

成田正雄　へ

1955年7月1日
池袋　静山荘　より（はがき）

昨年, 立教から借りて来てお貸ししたzariskiのHolomorphic functions, 返す様に催促されていますので至急お持ち下さい. 出来るなら明日の土曜日に. どうせ岩沢, 玉河さんの談話会を聞きに来るのでせう.
以上取敢へず

谷山欣隆　へ

1955年8月31日
池袋　静山荘　より（はがき）

もう, 国際数学者会議が目前に迫つているので, 忙しいため, 当分帰りません. 此の会議で, 50分位, 話すことになりました. 日本人で, 発表するのは全部で7人か8人です.
所で, やはりいろいろと金がかかるので, 少し足りなくなり, 出来れば3000円位, 送つて貰へませんか. (宛先はアパートの方にして下さい), 無ければもう少し少くても結構です.
此の夏は休み無しでした. その他は変つたことはありません. “ビーチオニン” あつたら少し送つて下さい. 此れは良く効きます.
以上

谷山欣隆　へ

1955年9月13日
日光　より（はがき）

先日の小包及び書留受取りました.
昨日こちらに着き, 今日は一寸見物し, 明日講演の予定です. 一寸疲れましたが, それ程のこともない様です. 帰りに一寸寄るつもりでしたが, やはり, 終わつてからしばらく用事があり, 家にはもう少し帰りません.
以上

谷山欣隆　へ

1955年11月13日
池袋　静山荘　より（はがき）

薬受取りました. 又日暮里に行つて柿ももらつて来ました（少しシブい所があつた）その外, 変りありません.

谷山欣隆　へ

1956年４月16日
池袋　靜山荘　より（はがき）

前略
都立大学の時間割，受取りました．４月から，９月まで，内職に，週１回都立大学で講義することになつたので送つて来たのでせうか．昔の住所録でも使つたのでせう．
学校は今日から始まりました．その他，格別のこともありません．

谷山欣隆　へ

1956年６月13日
池袋　靜山荘　より（はがき）

前略
都立大からの通知並びにハガキ受取りました．こちらの住所は知らせてあるのですが，教務課の方には傳つていないのでせう．
大部前にイチゴがどうのこうのと云つていましたが，今年は雨ばかりで，余り取れなかつただらうと思つています．
夏休みは大部おそくなります．
以上

谷山清司　へ

1956年９月２日
池袋　靜山荘　より（はがき）

此の間御葉書をもらつたまゝ御無沙汰しました．
前から余り工合は良くなかつたのですが，この間の暑さですつかり参つてしまひ，まだ參つています．それに少し忙しいので，つい返事を書かなかつたのですが．
明日から又学校が始まりますが，二週間で試験休みになるので，その頃伺はうと思ひます．都合を知らせて下さい．

谷山義泰　へ

1957年１月10日

謹賀新年
大久保兼之
　　文恵
（日暮里で，くじ附きの葉書がないというので，代筆を頼まれた．当ればせめてものなぐさみか）
豊

谷山欣隆　へ

1957年３月15日
池袋　靜山荘　より（はがき）

小包受取りました．４角のモチは全部たべてしまひました．
月末に一寸京都に行きます．その時何か買つて来て送ります．
その他は変りありません
以上

谷山欣隆　へ

1957年９月13日
池袋　靜山荘　より（はがき）

御ハガキ，何度も受取りましたが，面倒なのでお返事しませんでした．
不動岡高校の件
卒業年次　昭和20年（回数は不明）
目下の職業，東京大学教養学部講師（東京都目黒区駒場）
最終学校部科　東京大学理学部数学科
以上の通りです．学校の住所忘れましたので，そちらから知らせて下さい．
特に変つたこともありません．もう学校は始つています．夏にはずつとこちらにいました
以上

第五部

谷山豊の生涯

杉浦光夫

1. 高校卒業まで

谷山豊は，1927年(昭和2年)11月12日，埼玉県騎西町で，医師の父佐兵衛(通称欣隆)と母格の第六子(姉三人(内一人死亡)，兄二人)として生れた．後弟妹が一人ずつ生れた．豊と名づけられたが，家族以外の人達は皆「ユタカ」と呼ぶので，後には本人もそう称していた．大学を卒業した年の夏休みに，帰省していた谷山から，遊びに来いという手紙をもらって，騎西の家を尋ねたことがあった．その時，お祖母さんが「豊は小学校へ入る前から，算術が出来ましてな，買物へ行くと，店の人がそろばんでするより先に，いくらいくらいくらと言うんですよ．そしてお金を出すと，またお釣りはいくらいくらと言ったものでした」と話された．小学校は地元の騎西小学校．中学(旧制)は騎西にないので，隣の加須市の県立不動岡中学へ4キロ程の道を自転車通学した．この年(1940年)中学の入試選抜の方法が変って学科試験がなくなり，体育の試験と面接だけになった．谷山は補欠でやっと入学できた．年譜によると中学時代は体が弱く，欠席がちであったという．そして中学時代の後半は，勤労動員で授業はほとんどなかった．

1945年中学を卒業，浦和高等学校(旧制)理科甲類(英語が第一外国語，ドイツ語が第二外国語のクラス)に入学した．

翌年4月発熱，肺浸潤と診断される．一年間休学の後復学したが授業には出ず，試験だけ受けて第二学年を修了した．翌47年8月頸部淋巴腺炎の手術を受けた．48年4月から学校に出るようになったが，夏休みに病気が再発して自宅で療養した．病状が好転したので，49年4月からは学校に出て翌年3月浦和高校を卒業した．そして4月に東大理学部数学科に入学した．こうして谷山は高校時代を闘病に費した．回復できたのはまったく幸運であった．

後に谷山が書いた「自己紹介」によると，「浦高時代に読んだ『近世数学史談』(高木貞治著)が病みつきで数学に非常に興味を感じるようになったが，本格的に始めたのは大学へ入ってからである．」

2. 大学時代

さて1950年に谷山と私が入学したのは，いわゆる旧制大学であった．現在の制度とは異なるので若干説明が必要であろう．旧制の学制では，小学校6年，中学5年，高校3年，大学3年で，中学4年でも高校は受験できた．

英独仏のうち二箇国語の修得を含む一般教育は高校までで，大学は専門科目だけであった．東大数学科では，3年間で9科目に合格すれば卒業できた．9科目のうち6科目(1年の微積分，代数，幾何，2年の函数論と微分方程式，3年のセミナー)が必修で，他に3科目以上の選択科目を履修することになっていた．

また現在の東大では，数学の教官は大学院が本務で学部は併任であり，大学院の講義がたくさん開講されていて，学部の上級生はそれを聴講できるようになっている．これに対して旧制の大学院では講義はまったくなかったから，結局私達が聞くことができたのは，数学のごく基礎的な部分の講義だけであった．

その代り，自由な時間は相当あったので，学生同士でする輪講がいくつも行われた．年譜によると，谷山は1年の時に，バーコフ『束論』，ヴェイユ『位相群上の積分とその応用』を5月頃から輪講し，11月からはワイル『典型群—その不変式と表現』，12月からはトゥーキイ『トポロジーにおける収束と一様性』，翌年4月からはシュヴァレー『リー群論Ⅰ』の輪講に参加している．これらは皆当時よく読まれた本である．

こうして輪講をしたり，一緒に演習にでたりしている内に，仲間の得意の方向や苦手の面などもわかって来る．一年の代数の講義で，菅原正夫教授が，ジーゲルのシンプレクティック幾何の話をして，ジーゲル空間の測地線を求める問題を出したことがあった．このような計算は，誰もやったことがなかったのだが，これを解いたのが谷山だった．こうして谷山は，同級生の誰からも一目置かれるようになった．

二年になると辻正次教授の函数論が始まった．毎週2時間の講義が2回と演習が1回あった．辻先生の講義は，きわめて内容豊富で，一変数函数論の多くの定理を網羅していた．

いつだったかクラスでピクニックに行こうという話が出て，秋川渓谷に行くことになった．日取はいつにするかと話し合っていると，細越康暢が「辻さんの講義の日にしよう．オレが行って話してくる」と言い，結局そうすることになった．私共はあまりよい学生ではなかったようである．

年譜によるとこの年，谷山はワイルの『リーマン面』，ストイロフの『解析函教論の位相的原理』等を輪講している．

またこの年谷山は，同級の小野貴生と淡中忠郎『代数的整数論』を読んだ．これはシュヴァレーによる類体論の算術的証明を紹介した本である．小野からの来信によると，この本の最後の章は単項化定理を扱っており，当然そこまで読むつもりだったが，谷山はその章には興味を示さず，類体論の証明までやったところで輪講は終りになったという．

辻教授の講義は非常に早い速度で進むので完全なノートをとることは困難であったが，函数論を専攻することにきめていた及川廣太郎は，一年間で十数冊のノートをとった．谷山は夏休みにこのノートを借りて勉強した．「函数論ノ Notes ハ直接森本ニ渡シテ良イデセウネ．…講義ヲ聞クヨリ，君ノ Notes ヲ読ンダ方ガ，良クワカル様デス」という，旧カナの及川宛書簡がこの『全集』に収められている．学年末の函数論の試験の結果が発表されて見ると，我我のクラスで合格したのは谷山一人であった．

しかし，谷山の数学にとって一番重要なこの年の出来事は，ヴェイユの著書三冊(『代数幾何学の基礎』ほか二冊)を読んだことだと思われる．上述の及川宛書簡の中で，谷山は「van der Waerden ノ代数幾何ヲ大部読ミマシタ．Weil ノ本デ，直観的ニ摑ミ難カッタ概念モ，此ノ本デハッキリシテ，思ッタヨリ有効デシタ」とある(1951年9月4日付)．

ヴェイユのこの本は，標数 p の場合を含む代数幾何学を，始めて体系的に展開した歴史的な本で高度の専門書であり，初学者のための入門書ではない．したがって，論理的には完全であるが，始めてこの方面を学ぶ者にとっては，大変抽象的に感じられる本である．谷山はこのハードルを，ヴァン・デル・ヴァルデンの『代数幾何学入門』を読むことで克服しようとしたわけである．この年の暮から翌年始めにかけて，ヴェイユのこの本と，それに続く『代数曲線とそれに関連した多様体』『アーベル多様体と代数曲線』を，谷山は玉河恒夫講師と読んだ(年譜)．1952年の始め頃谷山は杉浦に「僕はアーベル函数をやることにしたよ」と語った．彼はこの頃自分の研究の方向を見出したのである．こうしてヴェイユの三部作は，谷山の数学の方法的基礎となった．

1952年に谷山は，藤田輝昭，立川三郎，宮原克美等と回覧雑誌『自信ある仲間』を作った．これはノートに好きなことを書くものであった．谷山が書いた数篇の文章は，この『全集』に収録した．

1952年4月，3年生になった谷山は，玉河講師の下で数論のセミナーを行った．そこではヴェイユの学位論文などを読んだ．この年玉河講師は，代数的整数論の講義をした．付値論から始って，ヒルベルトの分岐の理論までが内容だった．谷山はこの講義の試験には出席しなかった．その代り代数体上定義されたアーベル多様体の mod $\mathfrak{p}$ での reduction に関するレポートを提出して単位を得ている．このレポートは失われたが，もう一つ彌永昌吉教授に提出したレポートが全集に収められている．これは，ヴェイユの学位論文の主結果であるモーデル-ヴェイユの定理(有限次代数体 k 上定義されたアーベル多様体の k 有理点の作る群は有限生成)を，ヴェイユの後年のアーベル多様体論を用いて代数的に証明したものであった．

元の学位論文では，テータ函数等解析的手段が用いられていたのである．このレポートは，この時すでに谷山が，ヴェイユを使いこなすようになっていたことを示している．(後に谷山はもう一つの証明を得た.)

こうして53年3月谷山は数学科を卒業した．

3. 虚数乗法

卒業して大学院(旧制)に入った谷山は，特別研究生に採用された．特研生になれば，何年か研究に専念する経済的保証が与えられるので，研究者の卵にとって有難い制度であった．もっとも給費は今日の目から見れば決して多くはないが，当時としてはとにかくそれで暮して行けたのである．この年助手になった私の初任給は確か手取り八千二百円位だった．特研生の給費はそれより少し良かったが賞与はなかった．5月の数学会で，谷山は「Abelian varietyの常数体のreductionについて」,「Abel函数体のTeilungについて」という二つの研究発表を行った．翌年志村五郎のより一般なreduction理論が公表されたので，谷山は前者についての論文は発表しなかった．後者は上述のモーデル・ヴェイユ定理の代数的証明が内容である．こうして谷山は，順調に研究生活を開始したのであった．

この年の始め，谷山の活躍の場の一つとなる新数学人集団(SSS)が発足した．これは，都内の大学の数学科学生・卒業生の集りで，毎週一回集って，メンバーの一人が講演した後，自由に数学について語り合う会であった．この新数学人集団は，『月報』(実際は年5回刊行，後『数学の歩み』と改題)という雑誌を発行した．谷山は，この雑誌のもっとも有力な書き手であった．この『月報』は谷山の生涯をたどろうとする本稿にとっては，谷山の論文及び書簡と共にもっとも重要な情報源である．以下主としてこの三つと谷山を知る人達の談話[1)]によって記すことにする[2)]．

53年7月の杉浦宛書簡で，谷山は次のように述べている.「modular函数は，当分一休みです．abelian varietyの虚数乗法，つまりそのendomorphismのringの構造を，LefschetzやWeilに従ってまとめて見ようと思っているのです.」とある．谷山はこの時考えたことを，『月報』2号(53年10月)に,「類体の構成について」という題で書いている．またそこで谷山は，ヘッケの学位論文の結果を，自分の研究の立場から見直している．

谷山が『月報』1号(53年7月)に書いた「A. Weilをめぐって」は，興味深い人物論である．『数学原論』によって，抽象数学の権化のように思われることの多いブルバキの代表的存在であるヴェイユに，谷山が見出すのは,「伝統に培われた見識」と「腕力の強さと息の長さ」に由来する「単なる抽象以上に出た彼の業績の深遠さ」である．そして最後に谷山は「SiegelはWeilより遙かに独創的」であると断言している．

54年10月の『月報』第2巻2号に「モヅル函数と合同ζ-函数」という題で，谷山はこの年アムステルダムで開かれた国際数学者会議(ICM)でのアイヒラーの講演を紹介した.「それは，楕円モヅル形式に対応するヂリクレ級数のオイラー積展開に関するHeckeの理論を，所謂HasseのL函数と結びつけたもので，整数論の，二つの離れた分野の間に一つの架け橋を作る点から，又Heckeの理論の意味の探求のために，HasseのL函数の理論の進歩のためにも喜ばしい貢献である」と谷山は述べている．さらに次の3号の「補遺」で，アイヒラーが，ヘッケ作用素の固有函数である尖点形式のフーリエ係数の絶対値に対するいわゆるペーターソン・ラマヌジャン予想を-2次元の尖点形式に対して解決したことを説明している．当時の谷山の関心がどの辺にあったかを示す資料である．またこの文章によって，谷山はこの頃にはすでに，ヘッケの後期の保型形式の仕事に親しんでいたことが伺われる．

54年10月の『月報』2巻2号に，谷山は投書をした．ここでは,「純粋数学をなぜやるのか」についての種々の主張を，批判している.「純粋数学の研究も人類の幸福に役立つ」という主張を,「空空しく響く」といい,「国民の名誉のために」とか「科学的合理的精神の涵養」という目標についても疑問を呈している．

54年9月谷山は理学部の助手となった．特研生の給費は，大学に就職したので返さなくてもよいと考えていた谷山は，特研生だった期間が短いので返済する必要があると知り，驚いて交渉したが，返済を一時猶予してもらうことしかできなかった．

SSSが企画した分野別の「日本における数学の発展」というシリーズの第一回として，谷山は日本における整数論の研究史を書いた（『月報』2巻5号，1955年5月）．そこで谷山は，日本における整数論の研究が十分強固な伝統を形成できず，数論から他の分野への転向が多いことを指摘している．唯一つ高木貞治の類体論は，世界の学界をリードするものであり，「殆んど充足した伝統を作るに至ったかに見える」が，結局この方面でもアルチン，ハッセ，シュヴァレー，エルブランなどの影響の下に我が国の研究者は研究を進めて行くことになったと谷山はいう．そして「戦後の整数論も，主としてコホモロジーと云う，極めて代数的な理論に代表され，而も再び代数学の方向に転化しつつある」．また「抽象的代数函数論は，アルチン以来，合同 ζ-函数論を軸として発展して来たものであるにも拘らず，我が国では，主として賦値論及び類体論の立場から取扱われていたのである．」という指摘をしている．このように，谷山は，それまでの日本の数論の研究の方向に批判的であり，彼がより数論的と考える新しい道を自覚して進んだのである．

それは谷山一人だけの選択ではなかった．1946－52年に東大数学科を卒業して，数論の研究者となった玉河恒夫，佐武一郎，久賀道郎，志村五郎，小野孝等の人人も，皆それぞれ新しい方向の数論を探究していたのである．

1955年9月8日－13日に，東京および日光で，彌永教授等が計画した代数的整数論に関する国際会議が開かれた．ヴェイユ，アルチン，シュヴァレー，ドイリンク，セール等外国人数学者10名が来日した．この会議で，谷山はかねてから研究していたアーベル多様体の虚数乗法に関する研究を発表した．ところがヴェイユおよび志村五郎も同じテーマの発表を行ったので，この三つの研究は，この会議の中心的な議題となったのである．これら三人の研究は，いずれもヴェイユのアーベル多様体論を基礎としており共通の結果も相当にあったが，三人共それぞれ独自の結果を得ていたのであった．

谷山の研究意図を，『月報』第3巻1号(1955年7月)の「虚数乗法と合同 ζ-函数」という彼の文章によって紹介しよう．アーベルの発見した楕円函数の虚数乗法は，クロネッカーによって，その数論的意義が見出された．

クロネッカーは，虚2次体上のすべてのアーベル方程式が，これによって得られるのではないかと予想した．この「クロネッカーの青春の夢」は，後に高木貞治によって「虚2次体 k のすべてのアーベル拡大体は，楕円モヂュラー函数 j の特異値(虚数乗法を持つ楕円函数の周期の比 τ に対する $j(\tau)$ の値)及びこのような周期の適当な楕円函数の等分値を k に添加した体に含まれる」という形で解決された．

谷山の注目したのは，青春の夢以後の諸研究である．その主なものは次の三つである．

(1) 2変数のヒルベルト・モヂュラー函数を用いて，ある4次体上の類体の構成を行ったヘッケの研究．

(2) 有限体上の楕円函数体における虚数乗法とその合同ゼータ函数に対するリーマン予想の関係を扱ったハッセの研究．

(3) 代数体 k 上で定義された代数多様体 V のハッセのゼータ函数(k の素イデアルに関する簡約によって得られる合同ゼータ函数の積)がヘッケの量指標の L 函数で表わされる場合があることを発見したヴェイユとドイリンクの研究．

特にヘッケの結果は，古典的な虚2次体の場合と大きく食い違っていた．「此の様な食い違いの意味は何であろうか」と谷山は疑問を抱く．これが彼のこの方面の研究の出発点の一つであった．ヘッケ自身も他の方面の研究に転じ，この「食い違い」の意義を解明する人も現われないままに40年が経過していた．谷山は「然し，楕円函数の虚数乗法論の新展開，代数幾何学の整備，特にアーベル多様体とその乗法子環の理論の発展などは，Heckeの研究を，新たな角度から見直すことを，現実の問題に戻した様に思われた．筆者は以前から，その様な意図を抱いていたが，その後志村による，代数多様体の，常数体のreduction［簡約］の理論が出たので，必要な道具立てはすべて整った．後は，第一歩をどの方向に踏み出すかと云うことだけが問題であった．然しその方向は明白である様に思われなかったのである．筆者は多少戸惑っていたのである．然しHeckeの扱った函数体に対するHasseの ζ-函数を考えている内に，その証明のキー・ポイントから，上記Heckeの結果が，自然に出て来ることがわかった．今から考えれば此れは当然であって，上で説明した様に，中心的な問題は常に，合同 ζ-函数の零点の算術的性質を決定することにあったのである．一方では此れから，Hasse の ζ-函数の理論が出，他方では類体の構成の理論が可能になるのである．此の様な応用の契機は，合同 ζ-函数の零点と，変換 $(x, y)\to(x^q, y^q)$ との間の，既に説明された，Hasse-Weil による関連である.」と述べている．

こうして谷山の得た結果は，「ヤコービ多様体と数体」という題で発表された．有限次代数体 k 上で定義された g 次元アーベル多様体 A の乗法子環(endomorphism ring) $\mathscr{A}(A)$ が，$2g$ 次代数体 K の整環 R を含むと仮定し，K を含む最小のガロア体 $\bar{K}$ が k に含まれるとする．このとき谷山はアーベル多様体 A のハッセのゼータ函数 $\zeta_A(s)$ は，k の量指標の L 函数いくつかの積と有理函数の積となることを証明した．系として，C が k 上定義された代数曲線で，そのヤコービ多様体 A が上記の仮定をみたせば，C のゼータ函数 $\zeta_C(s)$ は，k のゼータ函数と k の量指標の L 函数と有理函数で表わされることを示した．谷山は最初この系を証明して，国際会議で発表することにし，題を「ヤコービ多様体と数体」とした．すぐ後で谷山は，一般のアーベル多様体の場合の証明にも成功したので，この題は内容と若干食い違うこととなった．さらに谷山は類体の構成についての定理を証明しているが，技術的に複雑なので，ここではその結果は述べない．この結果を国際会議で発表して，谷山はヴェイユを喜ばせ，この方面の研究の第一線に登場したのであった．

4. 谷山の問題

1955 年の国際会議の際に，出席者から未解決問題が集められ，そのプリント(英文)が配布された．後雑誌『数学』7 巻 4 号，8 巻 1 号にその日本語版が掲載された．(両者の間に若干の出入りがある)．谷山はいくつかの問題を提出している．その内問題 11 は，代数体の量指標の L 函数とメリン変換によって対応する多変数の保型形式を見出し，それに対しヘッケ作用素の理論を拡張せよというものである．問題 12，13 の原文を次に掲げよう．

問題 12　C を代数体 k 上で定義された楕円曲線とし，k 上 C の L 函数を $L_C(s)$ とかく：

$$\zeta_C(s) = \zeta_k(s)\,\zeta_k(s-1)/L_C(s)$$

は k 上 C の zeta 函数である．もし Hasse の予想［ζ_C が C 上有理型で函数等式をみたす］が $\zeta_C(s)$ に対し正しいとすれば，$L_C(s)$ により Mellin 逆変換で得られる Fourier 級数は特別な形の -2 次元の automorphic form でなければならない(cf. Hecke)．もしそうであれば，この形式はその automorphic function の体の楕円微分となることは非常に確からしい．

さて C に対する Hasse の予想の証明は上のような考察を逆にたどって，$L_C(s)$ が得られるような適当な automorphic form を見出すことによって可能であろうか．

問題 13　問題 12 に関連して，次のことが考えられる．“Stufe” N の楕円モジュラー函数体を特性づけること，特にこの函数体の Jacobi 多様体 J を isogeny の意味で単純成分に分解すること．また $N=q=$ 素数，かつ $q\equiv 3\,(\mathrm{mod.}\,4)$ ならば，J が虚数乗法をもつ楕円曲線をふくむことはよく知られているが，一般の N についてはどうであろうか．

この問題に関連して，55 年 10 月の『月報』3 巻 2 号の「代数幾何学と整数論」の中で，谷山は次のように述べている：「前に述べた様に，モヅル函数又はそれに類似な性質を持つ函数で，一意化出来る代数函数に対しては，Hasse の ζ-函数を研究する現実的な道が開けるので，その意味でもこの Eichler の試みは注目に値する．然しながらいずれにせよ，どの様な代数函数が，此の様な函数により一意化出来るかと云う疑問が当然起るであろう．此れは逆に云えば，モヅル函数体のヤコビ多様体を，その単純成分に分解する問題に外ならず，それは又第 1 種積分の周期を決定する問題である．(下略)」．

この谷山の問題は，数論的なディリクレ級数と保型形式の間のメリン変換による対応を発見したヘッケの研究の流れに属するが，ここで注目すべきことは，谷山が当時その研究が始まったばかりだった数体上の楕円曲線の L 函数と保型形式の関連をはっきりと問題にした点である．1955 年という時点における谷山のこの目のつけ所は，高く評価さるべきだと思われる．谷山はこれを問題として提出したのであり，その定式化についても十分精密な形をとってはいない．谷山はこの問題をさらに追求する前に世を去ってしまった．谷山の問題は，後に志村によって，研究され精密化された．

正整数Nに対して，モヂュラー群$\Gamma = SL_2(\boldsymbol{Z})$の階数$N$の合同部分群を

$$\Gamma_0(N) = \left\{ \begin{pmatrix} a & b \\ c & d \end{pmatrix} \in \Gamma \;\middle|\; c \equiv 0 \mod N \right\}$$

とする．このとき，谷山の問題は，現在次の形の予想として定式化されている．

予想 $\boldsymbol{Q}$上定義された任意の楕円曲線EのL函数$L(E, s) = \sum_{n=1}^{\infty} a_n n^{-s}$に対して，$E$の導手が$N$のとき，$\Gamma_0(N)$に対する重さ2(次元−2)の尖点形式$f(z)$で，$p \nmid N$となる素数$p$に対するヘッケ作用素$T_p$の同時固有函数となるものが存在し，$f(z) = \sum_{n=1}^{\infty} a_n e^{2\pi inz}$となる．

さらに上半平面Hの商空間$H/\Gamma_0(N)$のコンパクト化$X_0(N)$から，Eの上への型射ϕが存在して，E上の不変微分ωの引きもどし$\phi^*(\omega)$は$f(z)dz$の定数倍となる．

志村は1962—64年頃に，その予想をセール，ヴェイユ等に話した．ヴェイユはその論文(全集III［1967 a］)の最後で，これに触れているが，当時ヴェイユは，まだこの命題を予想と呼ぶだけの根拠がないという意見であった．その後志村は，Eが虚数乗法を持つ場合には，この予想が成立つことを証明した(Nagoya Math. J., vol. 44 (1971))．このような経緯から，上の予想は現在「谷山・志村予想」と呼ばれている．[3)]

1985年にフライ(Gerhard Frey)が，「フェルマの大定理」と「谷山・志村予想」の関連を発見した．彼はこの関連を完全に証明することはできなかったが，この結果は，この方面の研究者の関心の的となった．間もなくセールが証明に必要な命題を予想として示した．そしてリベットが，準安定な楕円曲線に対してこのセールの予想を証明し，谷山・志村予想からフェルマーの大定理が導かれることを確定するのに成功した．これは，次のような関係に基づく．

$p \geqq 5$を素数とし，フェルマの大定理が誤りで，0でない整数a, b, cで，$a^p + b^p = c^p$をみたすものがあったとする．一般性を失うことなく，$\gcd(a, b, c) = 1, a \equiv -1 \mod 4, b \equiv 0 \mod 2$としてよい．このとき楕円曲線

(1) $E \;:\; y^2 = x(x - a^p)(y + b^p)$

は，谷山・志村予想をみたさない．谷山・志村予想が正しければ，これは矛盾であり，フェルマーの大定理が成立つ．

その後1993年6月に，ケムブリッヂ大学で開かれた研究会で，A. ワイルズは，$\boldsymbol{Q}$上定義された楕円曲線で準安定なもの(各素数pで簡約したとき尖点の生じない楕円曲線)に対して，谷山・志村予想が成立つことを証明したと発表した．上の(1)は準安定であるから，フェルマーの大定理が証明されたことになり，大きな反響を呼んだ．その後ワイルズの証明は不完全な所があることがわかったので，現在の所まだ証明は完成していない．しかしワイルズの仕事は谷山・志村予想が，手に届かない所にある予想ではないことを示した点で非常に重要である．以上挙げた多くの人人の研究によって，谷山の問題が見直され，正確に定式化され，その重要性が明らかになって来たのである[4)]．

5. 1956〜58年

以上谷山の数学について記して来たが，もちろん彼は数学にしか興味がなかったのではない．杉浦宛の二つの書簡で谷山は，大岡昇平『野火』，森鷗外『澁江抽斎』の読後感を語っている．この二つはあまり谷山の趣味に合わなかったようである．彼の好きな作家としては，チェーホフがある．彼の書棚には何冊かのチェーホフがあったし，片山孝次には「チェーホフが好きだ」と言ったそうである．

また級友の藤田輝昭が立派なLP再生装置を持っていたのでよく聞きに行っていた．卒業前後に，これからダミアのシャンソンを聞きに行くのだと言っていたこともあった．後に谷山は小さなプレイヤーを買いレコードを楽しんだ．58年9月22日の志村宛の手紙では，「最近はBeethovenの第8ばかり聞いています」とある．

谷山は1955年12月東大教養学部講師となった．56年度か57年度の微積分の講義で，微分より積分を先にやることにしたと言っていたことがある．

なるほどこうすると，対数函数や逆三角函数を積分で定義することができて，初等函数の理論を解析的に展開することが可能になるのである．

谷山は55年の論文で扱ったアーベル多様体のゼータ函数に関する結果を，新しい視点に立って見直し，「代数体の L 函数とアーベル多様体のゼータ函数」という論文にまとめ，数学会の Journal 9 (1957) に発表した．この論文は谷山の学位論文ともなった．この研究の一つのきっかけは，ヴェイユの国際会議での講演の一つ「イデール類群の或る指標について」であった．この講演の谷山による記録は本全集に収めた．

有限次代数体 k のヘッケの量指標は，1950年岩澤・テイトによって k のイデール類群 $C=J/k^*$ の指標と一対一に対応することが示された．ヴェイユは，特別な量指標として，(A_0) 型量指標 χ を導入した．このような χ の，k の素イデアル $\mathfrak{p}$ における値 $\chi(\mathfrak{p})$ はすべて，一つの有限次代数体 K に含まれる．K の素イデアル $\mathfrak{q}$ を一つ定めるとき，$\mathfrak{p}$ が $\mathfrak{f}\cdot N(\mathfrak{q})$（$\mathfrak{f}$ は χ の導手）と互いに素ならば，$\chi(\mathfrak{p})$ は $\mathfrak{q}$ 進単数群 $U_{\mathfrak{q}}$ に含まれる．$\tilde{\pi}_{\mathfrak{p}}$ を，$\mathfrak{p}$ 成分が丁度 $\mathfrak{p}$ で割り切れる $k_{\mathfrak{p}}$ の元で，他の成分がすべて1であるイデールとする．写像 $\tilde{\pi}_{\mathfrak{p}} \mapsto \chi(\mathfrak{p})$ は，$C \to U_{\mathfrak{q}}$ の連続準同型に一意的に拡張できる．$U_{\mathfrak{q}}$ は完全不連結だから，この準同型は C の連結成分 D 上で1に等しく，$C'=C/D$ から $U_{\mathfrak{q}}$ への表現を定義する．このヴェイユの考察に対して谷山は「代数的整数論における ζ-函数」で，次のように述べている：「所で，これは未だ量指標を説明したことにならない．なぜならば［k の最大アーベル拡大体の k 上のガロア群］$\mathfrak{G} \cong C'$ から，$U_{\mathfrak{q}}$ への C' の表現は沢山あって，そのすべてが量指標に対応するわけではないから，どの様な表現が量指標に対応するかは，改めて決定されなければならない．」そして谷山は実際に，C' から $U_{\mathfrak{q}}$ への表現が，(A_0) 型量指標に対応するための必要十分条件を与えることに成功した．これが学位論文の定理1である．

一方代数体 k 上で定義された n 次元アーベル多様体のゼータ函数 $\zeta_A(s)$ は，ガロア群 $G(\overline{k}/k)$ の l 進表現を用いて計算される．特に A が十分多くの虚数乗法を持つとき（すなわち A の乗法子環が $\boldsymbol{Q}$ 上 $2n$ 次元の可換半単純部分環を含むとき），定理1を用いて，$\zeta_A(s)$ が k の (A_0) 型量指標の L 函数達を用いて表わされることを谷山は証明した．これから $\zeta_A(s)$ がハッセの予想をみたすこともわかる（定理4）．こうして谷山は，前論文の主定理の，数論的に見通しのよい証明を得た．

谷山の死後ヴェイユは「Y. Taniyama」という文章（著作集II［1959 b］）において，若くしてなくなったエルブランと谷山を対比させながら，谷山の数学を語った．その一節でヴェイユはこの論文について次のように述べている：「谷山は，自ら設定した枠組の中で申し分なく完成され，しかも現代の整数論のいくつかの重要な問題に対し広大な展望を開く第一級の一つの著作を我我に残すことができた点において，少くともエルブランより幸運であった．」

57年7月，志村五郎・谷山豊共著の『近代的整数論』（共立出版）が出版された．ヴェイユ・志村・谷山によるアーベル多様体の虚数乗法論とその数論への応用を詳しく述べた著書である．さらにこの本は，谷山の学位論文の内容（第7章），及びアイヒラーと志村によるモヂュラー函数体（モヂュラー曲線）のゼータ函数とペーターソン・ラマヌジャン予想についての結果（第9章）をも含んでいる．この本の英語版は，谷山の死後志村の手によって日本数学会から出版された（1961年）．この英語版で除かれている「第1章　歴史」を，本全集では収録した．

57年から数学研究連絡委員会で検討されていた数理科学研究所設立の動きが具体化する中で，谷山は何人かの人達とこの計画に若い研究者の声を反映させるための活動を行った．また谷山は「大学院の問題について」（『数学の歩み』6巻1号）という文章を発表して，新制大学院の実態が貧弱であることを批判し，改善を求めた．谷山が正式に大学院担当となったのは，1958年度からであるが，実質的には57年度から谷山は二人の大学院生清水英男，上林達治の指導教官となっていた．東大の学部学生で谷山のすすめで数学科へ進学がきまっていた草場敏夫に対する1958年3月6日付の書簡がこの全集に含まれている（319ページ）．この書簡で谷山はこれから数学を勉強しようとする草場に対して，勉強の仕方を懇切に説いている．

58年4月に助教授に昇任した谷山は，活発な研究活動を行った．10月に出た「ヒルベルトのモヂュラー函数（I）」(『数学の歩み』6巻2号)は，$GL(2)$のアデール群を用いてヘッケ型の理論を扱おうと試みたものであった．（I）という文字が示すように，これを用いて次に数論的な研究に進むつもりだったのである．総実な代数体は，(A_0)型の量指標を一つも持たない．従ってこのような代数体の数論は全く新しい立場からの研究が必要である．谷山はこの方面の研究を開始する決心をし，総実な代数体に対して定義されるヒルベルトのモヂュラー函数の組織的な研究を始めたのであったが，その早すぎた死によって，この研究は中絶してしまった．また10月16日受理の論文「有限定数体上の代数多様体の絶対類における正の0-サイクルの分布」(教養学部自然科学科紀要(英文)）では，合同ゼータ函数に対するラングのある予想が，ヴェイユ予想の一部を用いると，谷山の得た定理から導かれることが示されている．

一方プリンストンの高等研究所の教授となったばかりのヴェイユから，谷山は翌年9月から研究所に来るように招かれ，熟慮の末これを受ける決心をした．そして10月に谷山は婚約した．

こうして充実した活動を行っていた谷山は，58年11月17日自らの命を絶った．彼の遺書の冒頭は次の通りである．

「昨日まで，自殺しようという明確な意志があったわけではない．ただ，最近僕がかなり疲れて居，また神経もかなり参っていることに気付いていた人は少なくないと思う．自殺の原因について，明確なことは自分でも良くわからないが，何かある特定の事件乃至事柄の結果ではない．ただ気分的に云えることは，将来に対する自信を失ったということ．僕の自殺が，或る程度の迷惑あるいは打撃となるような人も居るかも知れない．このことが，その将来に暗いかげを落すことにならないようにと，心から願うほかない．いずれにせよ，これが一種の背信行為であることは否定できないが，今までわがままを通して来たついでに，最後のわがままとして許してほしい．」12月2日婚約者は谷山の後を追って亡くなった．こうして谷山は突然我我から立去ったのである．

註

1） 本稿を記すに当って次の方方から貴重な情報を提供して頂いた．厚く御礼申し上げる：小野貴生，片山孝次，上林達治，草場公邦，佐武一郎，清水英男，谷山清司，藤田輝昭，堀田良之，山崎圭次郎．

なお谷山のノートが谷山清司氏によって保管されている．

2）谷山の人と数学及びその時代を知るには，次の文献は見逃せない．

G. Shimura, Yutaka Taniyama and his times. Very personal recollections, Bull. London math. Soc. 21 (1989), 186-196.

また『数学の歩み』6巻4号（1959）は，谷山追悼号であり，友人達の追悼記事が収められている．

3） 谷山・志村予想の由来については，次の文献を見よ．

飯高茂・吉田敬之，谷山-志村予想の由来，数学46-2（1994)，177-180.

4）その後ワイルズは，問題であった点の証明を完成したとして，論文「モヂュラー楕円曲線とフェルマの最終定理」およびR．テイラーとの共著の「ある種のヘッケ環の環論的性質」のプレプリントを，94年10月に発表した．

これに対し専門家の検討が進められているが，現在まで問題点は発見されていない．間もなく証明の正しさが最終的に確認されるものと期待される．こうしてフェルマの大定理が確定すると共に，谷山・志村予想が準安定な場合に証明され，谷山が「代数幾何学と整数論」（本書 p.197 右）で述べた「ハッセのゼータ函数を研究する現実的な道」が，この場合に実現されることになったのである．

谷山豊とSSS

高瀬正仁

1．福富節男先生の話

平成28年（2016年）の秋10月，８日と９日の二日間にわたり国分寺の津田塾大学で数学史シンポジウムが開催されたとき，９日の午後，SSSの往時を知る福富節男先生にお会いして少時お話をうかがったことがある．おりしも谷山さんの没後60年の節目が２年後に迫り，SSSに寄せる関心が高まって，SSSの消息を今に伝える定期刊行物『月報』『数学のあゆみ』のバックナンバーを読み始めていたころであった．

谷山さんはどのような人でしたかと率直におたずねすると，先生は即座に，なんとも名状しがたい人物だったと応じられた．立派には違いないが，立派というのはあてはまらないような気がする．偉いには違いないが，偉いというのはあてはまらない．すぐれた数学者には違いないが，秀才天才というのもあてはまらない．普通の誉め言葉はみな陳腐に思えてしまうというほどのことを先生は口にして，言葉のない言葉を幾重にも重ねてこのうえもなく高く称賛した．大正８年（1919年）の10月末日に樺太に生れた福富先生は谷山さんより８歳の年長で，このとき満96歳である．若い日に出会った谷山さんの印象が今も鮮明に生きている様子がしのばれて，感慨が深かった．

2．数学方法論研究会（仮称）

『月報』の「３巻５号」に「新数学人集団　例会の歩み」という一覧表が掲載されていて，SSSのはじまりのころの状況が綴られている．昭和27年（1952年）６月23日の例会の報告から始まっているが，この日は遠山啓先生の著作『無限と連続』の合評会であった．岩波新書の一冊で，昭和27年５月10日付で発行されたばかりの新刊の書籍である．出席者は約15名．三日後の６月26日に開催された２回目の例会は遠山先生を囲む会で，この日も15名ほどの出席者が参集した．SSSの結成へと向う最初の一歩がこうして踏み出された．約15名の出席者たちの氏名は明らかではないが，谷山さんもそこにいたと見てさしつかえないであろう．

1952年６月の２回の集まりはSSSの淵源ではあるが，この時点ではまだ正式に発足したわけではなく，SSSという名称も定まっていなかった．遠山啓先生を囲む会から３箇月がすぎて，９月14日の例会は「準備会」と銘打たれている．新たに「数学方法論研究会（仮称）」という会が企画されて，そのための準備会が開かれたのである．出席者数は記録されていない．10日後の９月24日，目黒区高木町の宮原克実さんの下宿を会場にして，この新たに発足した会の第１回目の例会が開催され，立川三郎という人が「仮説検定論」の話をした．宮原さんは数学者ではなく，どのような人物なのか長らく不明だったが，福富先生にうかがったところ，谷山さんの友人で絵描きとのことであった．

数学方法論研究会（仮称）の例会は10月に４回，11月に３回，12月に３回と，週に１度ほどのペースで順調に継続した．年が明けて昭和28年になり，１月に３回，２月はやや変則で５回の例会があった．３月は２回で，14日の２回目の例会は「回顧と展望の会」とされた．出席者は16名．ここにいたるまでに通算して21回の例会がもたれたが，次の４月15日の第22回目の例会から会の呼称が変更されて，「新数学人集団」となった．略称はSSS.「新」のS,「数学人」のS,「集団」のSを合わせて「エスエスエス」と読む．数学方法論研究会はどこまでも仮称だったのである．

3．数学方法論研究会（仮称）から SSS へ

SSS が発足した1953年は春３月に谷山さんが大学を卒業した年であった．同期の卒業生に杉浦光夫，銀林浩，山﨑圭次郎という諸先生がいて，翌1954年３月の卒業生の中には倉田令二朗先生と斎藤正彦先生がいた．1953年３月の時点で大学院に在籍していた清水達雄先生が SSS の団長になった．これらの諸先生はみな SSS の創立同人である．団長の清水先生は清水建設にゆかりの人で，建設の世界には1947年６月に結成された新日本建築家集団があり，これにならって清水先生が新数学人集団という呼び名を提案した．

清水団長に対し，副団長を名乗ったのは倉田先生であった．もっとも団長や副団長の位置づけが組織的に規定されたというわけではない．

1953年７月15日付で，SSS の機関誌の役割を果たすことになる小冊子『月報』が創刊された．定価10円．わずかに８頁の小冊子で，清水先生の「創刊号に寄せて」という一文が巻頭に掲載された．

「東大新数学人集団」という記事があり，概略ではあるが，SSS の誕生の経緯が綴られている．何よりも先に目を引くのは「東大新数学人集団」という名称で，「新数学人集団」に「東大」の２文字が冠せられて，東京大学の数学科の学生を中心にしたグループとして発足したことが明示されている．冒頭に「東大の新数学人集団は今年三月，本年度卒業生を中心に結成された研究団体」と明記され，「略称 SSS」と書き添えられ，「協同研究，教室運営の民主化，五月祭等に活発に活動」して，現在の会員は40名と伝えられた．

この時期の情報をもう少し拾うと，東大の数学教室の1949年当時の主任は彌永昌吉先生であった．福富先生は助手で，学生の中に佐武一郎先生がいて，この３人が中心になって尽力して教室会議が設立された．これで新時代の運営体制が整ったが，わずか１年後の1950年に主任が交代して末綱恕一先生になると有名無実になってしまった．ところが谷山さんたち1953年３月の卒業生が数学科に進んだのはまさにその1950年であった．これに加えて岩澤健吉先生と矢野健太郎先生が渡米してそのまま帰ってこなかったため，学生や若い研究者に対する指導がまったくなおざりにされるという状況が現れたという．SSS の結成が要請される理由はこのあたりの消息に認められる．指導者の不在を協同研究により補おうという切実な心情が共有されていたのである．

このような事情はフランスの数学者集団ブルバキの場合と似通っている．フランスでは第１次大戦で多くの若い数学者が戦死したため，老大家たちは生存していたものの，アンドレ・ヴェイユやアンリ・カルタンなど，戦後のエコール・ノルマルの卒業生たちの指導にあたるべき少し上の世代の数学者が不在になっていた．かろうじてガストン・ジュリアが重篤な戦傷を負ったものの戦死を免れていたため，ヴェイユとカルタンは同世代の仲間に呼びかけて，ジュリアのもとでセミナーを行うという名目を立てて集結した．これがブルバキのはじまりである．SSS の場合，ブルバキにおけるジュリアに相当する人物として，もっともよく該当するのは遠山啓先生である．

遠山先生は（旧制の）福岡高等学校を経て東大数学科に進んだが，旧態依然とした微積分の講義をする坂井英太郎教授との折り合いが悪かったため，退学を余儀なくされたという体験をもつ人物である．退学したのちに決意を新たにして東北大学の数学科に入学し，1928年，28歳のとき卒業した．数学の姿が大きく変容しつつある時期であった．『無限と連続』は新しい数学への手引きとして書かれた作品であり，それだけに SSS の創立同人たちの心情に訴えるものがあったのである．『無限と連続』の時期の遠山先生は東京工業大学の教授であった．倉田先生や銀林先生のように，東大卒業後，東工大の大学院に進み遠山研究室に所属した人も多く，遠山研究室は SSS の有力な活動拠点のひとつになった．

4．拡大する SSS

「東大新数学人集団」として発足した SSS だが，当初から東大の枠をこえて全国的な広がりをめざそうとする傾向が現れていた．1953年10月15日付で発行された『月報』第２号を見ると，連絡先が「東大数学教室内　新数学人集団」と明記され，「東大新数学人集団」から「東大」の２文字が消えている．第３号の発行は同年12月８日．注目に値するのは「東大新数学人集団，都立大民科，東京工大有志，京大有志，岡山大有志，九大民科連合機関誌」（註．「民科」は民主主義科学者協会

の略称）と各大学の学生団体名が記されていることである.「東大新数学人集団」は消滅したわけではないが，これらの諸団体のひとつとして位置を占めている．この連合組織の呼び名が「新数学人集団」で，その機関誌が『月報』である．実際の編集作業を担当したのは東大新数学人集団で，原稿の投稿先として清水先生が指定された．各地の大学にそれぞれ新数学人集団を設立し，全国に展開する組織を作ろうとする気配が感じられる．

1955年12月の時点で作成された会員名簿には，95名の会員の氏名と住所と所属先が記載されている．東大の学生（学部生と大学院生）や東大の出身者が大半を占めているが，東京教育大学やお茶の水女子大学，津田塾大学の学生もいる．発足時の会員数40名に比べると，SSS は急速に拡大したのである．

『月報』という誌名は1956年５月発行の「３巻５号」までで終焉し，1956年７月に発行された「４巻１号」から改題して『数学の歩み』になった．末尾の「改題の辯」を見ると改題にいたるまでのいきさつが略記されているが，加えて『数学の歩み』の位置づけが変更されたことも報告された．これまで『月報』は SSS，都内数学科学生集合（都数集），各大学有志で構成された全国連絡会連合機関誌だったが，この年の春の全国連絡会で，編集その他いっさいの責任が SSS と都数集にまかされることが決った．この決定を受けて，今後は SSS と都数集の共同による「数学の歩み刊行会」が発行の主体となることになったという．刊行会への連絡先は「東京大学理学部数学教室気付」と指定された．

5．投書「手帳より」

谷山さんは『月報』と『数学の歩み』に多くの原稿を書き続けた．アンドレ・ヴェイユを論じ，数論と代数幾何学を語ったが，ときおり数学研究の根源に触れるような不思議な文章を寄せることがあり，会員たちの間に波紋を呼んだ．際立った一例として，1954年10月15日発行の『月報』の「２巻２号」に掲載された一通の投書を紹介したいと思う．著者名はなく，ただ「T」とのみ記入されたが，谷山さんの投書である．「手帖より」と題されて，「数学研究は何のためか」という問いをめぐってさまざまな所見が披歴され，しかもみなことごとく否定されてしまう．この時点で谷山さんは満26歳である．

谷山さんの語る「数学を研究する理由」を摘記する．

・空々しい理由

工業技術は人間の生活を支え，豊かにする．その基盤は自然科学．そのまた一部，もしくは科学そのものの基礎として数学がある．数学を進歩発展させるのが有意義である理由はそこにある．数学は統一ある有機体である．それゆえ，純粋数学の研究も人類の幸福に役立つのである．これは立派な見解だが，なぜか空々しく響く．

・無邪気な理由

日本では科学と工業が乖離していることを説く人もいる．科学は空に蒸散し，技術は輸入に頼る．科学は尊重されず，科学者は国家の支持を感じない．戦時中はともあれ，今や科学は無用の長物と化した感がある．ましてあまりにも迂遠な数学など，やりたい人がやればいいさというわけである．現在はすべてが植民地化されてしまった．だからこそ，科学の殿堂だけでも毅然として独立を維持し，民族的誇りのひとつの拠り所をそこに築くのはすばらしいことだ．日本国民の名誉のために．古橋と橋爪の水泳の世界記録樹立，湯川秀樹のノーベル賞，小平邦彦のフィールズ賞．これらは劣等感に打ち挫かれた国民精神を感奮喚起させる．

自国の高い文化を誇り，受賞してさらに高める努力をするのは自然でもある美しくもある．ただし，そんなふうに言えるのは，その文化が真に国民の中に根をおろしている場合のことである．普段は見向きもせず考えもしないくせに，輸入した問題と取り組んで外国で業績を挙げた人びとを，何かの賞をもらったというだけでかつぎまわり，誇りとするのはどうか．なるほど十二歳の少年（引用者註．マッカーサーが日本人を評した言葉）に相応しい無邪気さかもしれない．

・合理的精神の涵養

日本人のもつ非合理性，日本社会における資本主義と前近代性との奇妙な混淆，その上に重なる植民地政策．このジャングルを切り抜けるには科学的合理的精神の涵養以外に道はない．その目的にもっともよくかなうのはきちんとした理科，数学教育である．数学をちゃんとやっている研究室があり，その雰囲気の中で育った教師が教育に当る．これが理想である．

もし求められているならば，これはそのとおり．だが，教育も馬に水を飲ませることはできない．かつて合理主義の担い手であった資本主義は，日本では合理主義の徹底を恐れ，時には植民地主義に順応することによりみずからを守ろうとしている．勤労大衆もまた合理主義に救いを求めようとはしていない．合理主義は表立って求められていない．それをたたき込むには人は啓蒙家にならなければならない．だが，フェルマの問題を研究しながら啓蒙家になることはできない．純粋数学が啓蒙の光となりえた時代ははるか昔のことなのだ．

・趣味としての数学

趣味として数学をやるという人がいる．たいへんけっこうな趣味だが，人は単なる趣味に一生を捧げるようなことはしない．この言葉にはさまざまなニュアンスが伴っている．数学のもつ純粋さ．曖昧を許容しない明確さ．一種の精神的な高み．そのようなものに精神の拠り所を求めようとする人．学問に対する漠然とした尊敬．何かに徹する生活の魅力から，学の中の学たる数学に没頭しようとする人もあるだろう．人間精神の名誉のために透明で希薄な空気を呼吸しつつ，高い氷河につかまるのもよい．数学者は詩人であり，その論文は芸術作品であるとも考えられる．だれかが言ったように，詩を理解しない者は真の数学者ではない（引用者註．ヒルベルトの言葉）．

あるいはまたアカデミズムへのあこがれから象牙の塔に立て籠もり，暇つぶしのためでも，収集癖のためでも，骨董趣味のためでも，とにかく何でもよい．暇と余裕のある人はどんな生活を選ぼうと自由ではないか．

どれもみな「何のための数学か」という問いに対する解答ではありえないと，谷山さんは言いたそうである．趣味としての数学というのはよさそうな感じもないではないが，語り口は皮肉めいている．

・類体論の読者はマラルメ（引用者註．フランスの詩人）の読者の何千分の一かもしれない．独特な記号のもとに，著者にしか理解できない「詩」をものしようとして煩雑な計算，記号的形式的推論により試作というよりは大量生産の名にふさわしい論文が山積みしようと，浮世の俗物のことなど顧慮する必要がどこにあるだろう．その俗物どものあなた方に対してはきわめて寛容である．実際，あなた方の生活はすばらしい．合理主義的啓蒙とか，資本主義の矛盾とか喚きたてて，現在の秩序を乱し，社会を転覆しかねない不逞の輩の尻馬に乗るおそれもなく，おまけに，求めようともせず，考えてもみなかった何かすばらしい名誉を，人間精神に与えてくれるかもしれないのだから．

「数学は何のためか」と問うよりもむしろ，数学研究者を揶揄して，無害無益の無用の長物と言われているかのような印象がある．マラルメ詩集をひもとく人も多いとは言えないかもしれないが，類体論の書物を読む人となるとさらにその何千分の一．数学の研究などは自己満足にすぎないではないかと，ひとまず指摘してみせたのであろう．

・自分の生きていたしるしを世の中に残すために，死の床にあって，私はこれだけのことをしてきた，私の生活は無意味ではなかったこと，満足の中に目を閉じられるように，数学の殿堂に自分の鑿（のみ）の跡を残そうとする人もいる．山小屋の壁に落書きをし，木の幹に自分の名を彫りつけるのも同じ心理．死の床にあって，守銭奴は子孫に遺しうる金高に満足の笑みを浮かべ，政治家は自分の当選回数を思い浮かべ，将軍は自分の灰にした町の数を思い浮かべる．ある人は自分の書き写した経典の山を，などなど，何とすばらしい「しるし」が世に遺ることだろう．人間の事業がすべてむなしいというのではないが，それに

「価値」を与えるのは何か別の原則によるのだ.

・何も死の床に限るわけではない. 私はこんなに頭がいいんだ. 私はこんなに能力がある. 私はこんなにいろんなことをした. それは私が前に証明しました. それよりこのほうが簡単な証明だ. これは私の発見した最大の定理だ. 私はどんな論文でも理解できる. 私はこれだけ本を書いた. 私は学位を6個取った. 私は何とか大学の教授だ. 私は大物だ. …… 別に数学でなくても……私はこんなに将棋が強い. 私は柔道6段だ. 私はこんなにお酒が飲める. 私は金庫破りの名人だ. 泥棒の親分だ. 私は神の使いだ. 私は神様だ！

わかった, わかった, わかった. ただ, よくわからないこともある. なんでも証明できるあなたにとって, あなたの能力の証明だけはむずかしいらしく, その証明にあなたの一生を要すると思っているらしいのはなぜなのだろう.

「数学は何のためか」という問いが提示され, いろいろな答が試みられたものの, 答のないままに終ってしまった. だれにも答えることのできない難問を提示して,「月報」の読者にあえて議論を吹きかけたような印象もあるが, 案の定, いくつかの感想が寄せられて小さな誌上討論会が出現した.

谷山さんは「なぜ数学を学ぶのか」「なんのために数学を学ぶのか」「数学は何のためか」という, だれも答えることのできない問いを問うことのできる人であった.「人はなぜ生きるのか」「人は何のために生きるのか」と問うのと同じことで, 正解はありえないが, その代わりひとたび思索に踏み込めばどこまでも果てしなく深まっていきそうである. 深遠な魅力を秘めていて, しかもあまりにも危険な問いというほかはないが, 数学研究に携わろうとする以上, 問わなければならない問いであることもまたまちがいない.

谷山さんの発言は読む人の内省を誘い, SSS の会員相互の心の交流の契機をもたらして, SSS を1個の生きた有機体にする力を備えていたのである.

谷山豊著作目録

（ ）は全集掲載ページ，（未）は全集（初版）未収録を示す．

『自信ある仲間』は谷山の参加していた同人誌．
『月報』（第4巻から）『数学の歩み』は新数学人集団が全国の諸大学の数学科関係有志と発行していた連絡機関誌．後数学の歩み刊行会発行となる．

1．Quelques Questions，自信ある仲間，1952年6月．（pp. 262-263）
2．無題，自信ある仲間，1952．（pp. 264-266）
3．或る映画を見てある人がどう考へるかと云うことについて――一つの実例――，自信ある仲間，1952．（pp. 266-267）
4．四行詩――Q君に――，自信ある仲間，1952．（p. 267）
5．屋上屋を架す，自信ある仲間，1952．（pp. 267-269）
6．ABEL 函数体ノ n-分割ニツイテ，Weil ノ有限性定理ノ代数的証明，彌永昌吉教授「幾何学続論」（1952年度）リポート．（pp. 151-159）
7．正雄さんの話（草稿），1953．5．（pp. 270-271）
8．A. Weil をめぐって，月報1巻1号，5-6ページ，1953．7．（pp. 176-177）
9．新人紹介 Peter Roquette，月報1巻1号，5ページ，1953．7．（未）
10．類体の構成について，月報1巻2号，11-12ページ，1953.10．（pp. 177-178）
11．W. Maak，教師としての Erich Hecke（翻訳），月報1巻3号，11-14，1953.12．（未）
12．“類体の構成について”の訂正，月報1巻3号，15，1953.12．（pp. 178）
13．A. Weil の Fibre Spaces in Algebraic Geometry，月報2巻1号，24-32，1954．7．（未）
14．モヅル函数と合同 ζ 函数，月報2巻2号，8-9，1954.10．（pp. 179-180）
15．投書　手帖より，月報2巻2号，28-29，1954.10．（pp. 181-182）
16．モヅル函数と合同 ζ 函数（補遺），月報2巻4号，24，1955．3．（pp. 179-180）
17．日本における数学の発展1，整数論，月報2巻5号，19-23，1955．5．（pp. 183-187）
18．投書　空しさということ，月報2巻5号，42，1955．5．（p. 187）
19．虚数乗法と合同 ζ 函数，月報3巻1号，31-35，1955．5．（pp. 188-192）
20．シンポジウムについて，月報3巻1号，45，1955．7．（p. 192）
21．Jacobian Varieties and Number Fields（abstract），Sept. 1955．（pp. 3-6）
22．Jacobian Varieties and Number Fields（mimeographed notes），Sept. 1955．（pp. 7-56）
23．1955年国際数学会議，新初等数学講座月報4，1955．9．（pp. 259-261）
24．代数幾何学と整数論，月報3巻2号，60-64，1955.10．（pp. 193-198）
25．名著紹介　吉田洋一著『零の発見』，科学読売7巻12号，1955.12．（pp. 252-255）
26．On a certain relation between L-functions（manuscript），1956．2．（pp. 71-84）
27．Regular Local System for an Abelian Extension（manuscript），1956．2．（pp. 85-98）
27′．A Letter to Andre Weil, 1956．2．（pp. 99-100）
28．A. Weil に接して，月報3巻3号，37-46，1956．2．（pp. 199-208）
29．Weyl と整数論，月報3巻4号，33-34，1956．3．（pp. 208-209）
30．数論グループ，月報3巻5号，4-5，1956．5．（pp. 210-211）
31．A. Weil，ゼータ函数の育成について（翻訳），数学7巻4号，4-7，1956．5．（未）

32. A. Weil の印象，数学 7 巻 4 号， 7 - 8 ，1956. 5 .（p. 175）
33. A. Weil，イデール類群の或る指標について（記録），数学 7 巻 4 号，205-207.（未）
34. Jacobi 多様体と数体，数学 7 巻 4 号，218-220，1956. 5 .（pp. 162-165）
35. 虚数乗法に関する非公式討論会（記録），数学 7 巻 4 号，227-231.（pp. 166-173）
36. 問題 11,12,13,29，数学 7 巻 4 号，269，271 ページ，1956. 5 .（p. 174）
37. 名著紹介　ポアンカレ著　吉田洋一訳『科学と方法』，科学読売 8 巻 6 号，1956. 6 .（pp. 256-258）
38. 代数的整数論における ζ 函数，数学の歩み 4 巻 1 号， 1 - 8 ，1956. 7 .（pp. 212-220）
39. Jacobian Varieties and Number Fields, Proc. Int. Symp. on Algebraic Number Theory, Tokyo-Nikko 1955, 31-45, 1956.10（pp. 57-70）
40. 入門講座に望む（杉浦光夫と共著），数学の歩み 4 巻 2 号，76-77，1956.10.
41. A. ヴェイユ，テータ函数の理論（翻訳），数学の歩み 4 巻 5 号，45-51，1957. 5 .（未）
42. L-functions of Number Fields and Zeta Functions of Abelian Varieties, J. Math. Soc. Japan 9, 330-366, 1957.7.（pp. 101-132）
43. 巻頭に寄せて，数学の歩み 5 巻 1 号， 1 ページ，1957. 7 .（pp. 221-222）
44. 少数精鋭主義について，数学の歩み 5 巻 1 号，24-27，1957. 7 .（pp. 222-226）
45. 『近代的整数論』（志村五郎と共著），共立出版，1957. 7 .（未）
46. 正夫さんの話，駒場の四季，1957.11.（pp. 271-274）
47. 整数論展望（久賀道郎と共著），数学の歩み 5 巻 3 号，10-16，1957.12.（pp. 227-233）
48. 自己紹介，駒場，1958. 3 .（p. 252）
49. 現代数学の性格について，駒場，1958. 3 .（pp. 274-281）
50. 数理研設立の問題について――全国の研究者各位に訴える――，数学の歩み 5 巻 4 号， 1 - 6 ，1958. 4 .（pp. 233-239）
51. 米国留学について，二つの意見，数学の歩み 5 巻 4 号，94，1958. 4 .（pp. 239-240）
52. 大学院問題について，数学の歩み 6 巻 1 号，35-39，1958. 7 .（pp. 241-245）
53. 応用数学小委員会への質問状――数理研の問題について，数学の歩み 6 巻 2 号， 1 - 2 ，1958.10.（pp. 245-246）
54. ヒルベルトのモヂュラー函数について 1 ，数学の歩み 6 巻 2 号， 3 - 6 ，1958.10.（pp. 247-251）
55. スミルノフ『高等数学教程』II 巻 3 章，重積分と線積分（翻訳），共立出版，1958.10.（未）
56. Distributions of Positive 0-Cycles in Absolute Classes of an Algebraic Variety with Finite Constant Field, Sci. Papers of the College of Gen. Education. Univ. Tokyo, vol. 8. 123-137, 1958.3.（pp. 133-146）
57. M. Eichler 教授講演記録，数学 10 巻 3 号，50-58，1959. 4 .（未）
58. Complex Multiplication of Abelian Varieties and its Application to Number Theory（with G. Shimura），1961, Math. Soc. Japan, iv＋159 pages（未）

年譜

昭和2年 (1927年)	11月12日　埼玉県騎西(キサイ)町大字騎西1287番地において，谷山佐兵衛(通称は欣隆，医師)と格との間に生まれる．(姉3人うち1人死亡，兄2人，のちに弟1人，妹1人) 豊(トヨ)と名づけられたが，ユタカと呼ぶもの多く，のちに本人もユタカと称するようになる．
昭和7年頃 (1932年)	幼稚園に入園したが，人となじまず，すぐに退園．
昭和9年 (1934年)	4月　騎西尋常小学校に入学，よろこんで通学する．
昭和15年 (1940年)	3月　騎西尋常小学校を卒業． 4月　埼玉県立不動岡中学校(旧制)に入学． (入試は体育が主であり，補欠入学となる) 中学時代を通じ，体が弱く，欠席がちであった．(なお当時は，軍事教練や武道などが盛んで，学業は重んじられなかった．また中学後半は勤労動員が多く，授業はほとんど行われなかった．)
昭和20年 (1945年)	3月　埼玉県立不動岡中学校を卒業． 4月　浦和高等学校(旧制)理科甲類に入学，共済部に入る．寮(武原寮)においては，共済部のもの数人と同室，親しく交わる． 高校時代は，一般的な社会事情に加え，家庭事情による経済難に見舞われた．(当時は，第2次大戦末期，敗戦の混乱の時代である．)
昭和21年 (1946年)	4月　春休みで帰省中に発熱し，肺浸潤と診断される．第2学年を休学．
昭和22年 (1947年)	2度目の第2学年を欠席の形で病気療養する．なお，8月に頸部リンパ腺炎で入院し，手術を受ける．
昭和23年 (1948年)	3月　試験だけを受けて第2学年を修了． 4月より7月まで学校に出席，夏休みも寮で頑張ったところ，病気が再発し，自宅で療養する．以後ほとんど出席せず，試験も受けなかった．
昭和24年 (1949年)	4月　ようやく病気が回復し，2度目の第3学年は大部分出席する．
昭和25年 (1950年)	3月　浦和高等学校を卒業． 4月　東京大学理学部数学科(旧制)に入学． この頃から，浦和市郊外の百姓家に間借りして東京に通学する．

5 月頃より始まった前期(第 1 学年)学生のセミナーのほとんど全部に出席する．G. Birkhoff “Lattice theory,” A. Weil “L'integration dans les groupes topologiques” の輪講をする．

7 月頃より，文京区坂下町に間借りする．

10 月　数学科前期学生自治委員となる．また，レッド・パージ反対闘争委員に選出される．(当時，連合軍総司令部教育担当者イールズらによって，大学教官のレッド・パージが企図された．)

11 月より，化学教室を借りて，H. Weyl “Classical groups” の輪講をする．

12 月より級友，宮川典代の家で，J. Tukey “Convergence and uniformity in topology” の輪講をする．

昭和 26 年 (1951 年)

4 月より，C. Chevalley “Theory of Lie groups I,” 淡中 “代数的整数論” などの輪講をする．

10 月より，H. Weyl “Die Idee der Riemannschen Fläche” の輪講をする．

11 月より，Stoilow “Leçons sur les principes topologique de la théorie des fonctions analytiques”を読む．

中期(第 2 学年)の頃から，虚数乗法論に関心をもち始め，E. Hecke の仕事に注目する．この年の末頃より，玉河講師と共に A. Weil “Foundation of algebraic geometry” を読む．

昭和 27 年 (1952 年)

この年の初め，玉河講師と共に，A. Weil “Sur les courbes algébriques et les variétés qui s'en déduisent,” “Variétés abéliennes et courbes algébriques” を読む．

4 月より，玉河講師のセミナーにて，A. Weil の Thèse “L'arithmétique sur les courbes algébriques, Acta Math. 52 (1929) pp. 281-315” を読む．彌永教授のセミナーにも，しばしば出席する．H. Minkowski “Geometrie der Zahlen” を読む．

4 月頃より，一部の後期(第 3 学年)学生の間で，回覧書き込み式の雑誌 “自信ある仲間” が出され，その執筆者の一人であった．

[Quelques questions]*，[無題]，[或る映画を見てある人がどう考えるかと云ふことについて…一つの実例]，[四行詩]，[屋上屋を架す]，

この頃から，菅原教授その他の先輩を歴訪する．また，友人宮原克美の下宿で定期的に開かれた小会合に参加する．(この会合は，その後 “数学方法論研究会” と称され，9 月 14 日新橋の石膏会館で第 1 回例会が開かれた．)

11 月，数学方法論研究会の第 7 回例会にて講演：“Weilについて”．

昭和 28 年 (1953 年)

3 月　[Abel 函数体の n 分割について] (幾何学続論レポート)

東京大学理学部数学科を卒業．

新数学人集団(略称 SSS)結成に尽力する．(数学方法論研究会は SSS に発展的解消する．団長清水達雄) また，SSS の研究組織の一つとして，数論グループ発足の中心になる．

＊ 以下著作物を [ ] で示す．

[正雄さんの話]

4 月　大学院研究奨学生となる．

5 月　日本数学会春季例会にて講演：“Abelian variety の常数体の reduction について”“Abel 函数体の Teilung について”

7 月　[A. Weil をめぐって；月報 1 巻 1 号]，[新人紹介：P. Roquette；月報 1 巻 1 号]（SSS の機関紙を“月報”と題した．）この頃，文京区大塚坂下町に間借する．

9 月頃より，病気のため帰省して療養する．（のちに大したことはないことがわかった）

10 月　[類体の構成について；月報 1 巻 2 号]

12 月　[W. Maak の講演：教師としての Erich Hecke（翻訳）；月報 1 巻 3 号]，[“類体の構成について”の訂正；月報 1 巻 3 号]

昭和 29 年（1954 年）

2 月　病気が回復して上京する．

4 月頃より，世田谷区松原町に間借りする．

5 月　日本数学会春季例会にて講演：“Abel 函数体の Teilung について II”

この頃，Hecke の扱った超楕円曲線の合同 ζ 函数について考察している．

7 月　[A. Weil : Fibre spaces in algebraic geometry（論文紹介）；月報 2 巻 1 号]

9 月　東京大学理学部数学科助手となる．

10 月頃より，北区西ケ原町に間借りする．[モヅル函数と合同 ζ 函数（Eichler の講演紹介）；月報 2 巻 2 号]，[手帖より；月報 2 巻 2 号]（署名 T. 多くの反響をよんだ．）

12 月　SSS 例会で講演：“代数体と代数函数体”

年末，豊島区池袋のアパート静山荘に転居する．

この年の末より翌年にかけて，虚数乗法をもつ Abel 多様体の上の合同 ζ 函数，および虚数乗法論の拡張に関する研究をする．

昭和 30 年（1955 年）

この年の初めに，志村五郎と個人的に知り合い，数学上の交流があった．

3 月　[補遺：モヅル函数と合同 ζ 函数；月報 2 巻 4 号]

5 月　[日本における数学の発展（1）：整数論；月報 2 巻 5 号]，[空々しさということ；月報 2 巻 5 号]（署名T.）

日本数学会春季例会にて講演：“或る種の特異週期の 1 位超楕円函数体について”“簡単な代数的多様体の ζ 函数について”

[Hyperelliptische Kurve und algebraische Zahlen ; Abstracts]（9 月に予定されているシンポジウムにおける講演の簡単な予稿集）

7 月　[虚数乗法と合同 ζ 函数；月報 3 巻 1 号]，[シンポジウムについて；3 巻 1 号]（署名U.）

8 月　A. Weil が来日，大きな影響を受ける．（10 月に離日）

[Jacobian varieties and number fields ; Abstract]（シンポジウムに当っての講演アブストラクト集）

［Jacobian varieties and number fields］（シンポジウムの講演の詳しい内容を謄写版にしたもの）
9 月　国際数学連合(IMU)の代数的整数論シンポジウムにて講演：“Jacobian varieties and number fields”
［1955 年国際数学会議；小山書店数学月報 4 ］
10 月　［代数幾何学と整数論；月報 3 巻 2 号］
12 月　東京大学教養学部講師となる．
［‘零の発見’(紹介)；科学読売 7 巻 12 号］

昭和 31 年（1956 年）

［Jacobian varieties and number fields ; Proceedings of the International Symposium on Algebraic Number Theory, Tokyo-Nikko, 1955, pp. 31-45］（前年 9 月に行われた IMU シンポジウムの報告集）
2 月　［On a certain relation between L-functions ; an unpublished note］，［Regular local system for an abelian extension ; an unpublished note］（上記二論文は，はじめ日本数学会の Journal に出すつもりで書かれたが，結局発表せず．翌年ほぼ同じ内容を書き改めて発表した．なお，上記二論文を Weil に送ったときの手紙が，この全集に収録されている．）
［A. Weil に接して；月報 3 巻 3 号］（署名 SSS 会員）
3 月　［Weyl と整数論；月報 3 巻 4 号］
この頃から英語のサークル“シェイクスピアを読む会”に出席する．
4 月　東京都立大学非常勤講師を委嘱される(任期は 9 月まで)．講義“Galois 理論から整数論まで”を行う．この頃［知性パズル；知性（河出書房)］
5 月　SSS 大会にて特別演説．
［研究グループの活動と状勢分析－数論グループ；月報 3 巻 5 号］（署名数論グループ運営委員会）
日本数学会春季例会にて特別講演：“代数的整数論における ζ 函数”
［A. Weil の印象］，［Jacobi 多様体と数体］，［虚数乗法に関する非公式討論会］，［来日数学者と接触して(及川広太郎，杉浦光夫と共著)］，［問題 11, 12, 13, 29］以上，数学 7 巻 4 号(国際数学会議特集号)．
6 月　［‘科学と方法’(紹介)；科学読売 8 巻 6 号］
7 月　［代数的整数論における ζ 函数；数学の歩み 4 巻 1 号］（SSS 月報を“数学の歩み”と改題．）
10 月　［入門講座に望む(杉浦光夫と共著)；数学の歩み 4 巻 2 号］
12 月　［Zariski 氏を囲む座談会(記事)；数学の歩み 4 巻 3 号］

昭和 32 年（1957 年）

4 月　東京都立大学非常勤講師を委嘱される(任期は翌年 3 月まで)．講義“Faisceaux algébriques cohérents”を行う．
5 月　［A. Weil：テータ函数の基本定理（翻訳)；数学の歩み 4 巻 5 号］
日本数学会春季例会にて講演“十分多くはない虚数乗法をもつ Abel 多様体の ζ 函数”，“Weil 群の表現について”
7 月　［近代的整数論(志村五郎と共著)：共立出版］

[巻頭に寄せて；数学の歩み5巻1号]（署名編集委員会），[少数精鋭主義について；数学の歩み5巻1号]

[SSSについての提案；数学の歩み5巻1号]（渡辺毅，上野正と共著）．

10月 [SSSについての提案の前書；数学の歩み5巻2号]（渡辺毅，上野正と共著）．

この頃，数理科学研究所(仮称)設立の計画が表面化したが，そのあり方をめぐって同志と共に活躍をはじめる．

11月 駒場祭にて講演："現代数学の性格について"．[正夫さんの話（童話）；駒場の四季]（担任のクラス雑誌，昭和31年度入学理科1類8組）

12月 [整数論展望；数学の歩み5巻3号]（久賀道郎と共著）．

[L-function of number fields and zeta function of abelian verieties ; Journal of the Mathematical Society of Japan. vol. 9, pp. 332-336]．

昭和33年（1958年）

[自己紹介，駒場]（東大教養学部学友会刊行の小冊子）

4月 東京大学教養学部助教授となり，あわせて大学院数物系研究科数学課程担当を命ぜられる．

[数理科学研究所についての要望；数学の歩み5巻4号]（署名SSS数理研対策委員会，Y.S.と共著），[数理科学研究所設立の問題について——全国の研究者に訴える——数学の歩み5巻4号]，（署名SSS, 上野正と共著），[米国留学について二つの意見；数学の歩み5巻4号]（無署名）．

5月 東京大学理学部より理学博士の学位を受ける．

[大学院の問題について；数学の歩み6巻1号]，[数理科学研究所に関するシンポジウム(記事)；数学の歩み6巻1号]

日本数学会春季例会にて講演："代数体の0次のcycleの同値類について"，"リーマン面の週期のHasse-Wittの行列について"，"Hilbertのモジュラー函数について"．

10月 [ヒルベルトのモジュラー函数I；数学の歩み6巻2号]

この頃，鈴木美佐子と婚約する．

東京大学理学部講師併任を命ぜられる(任期は翌年3月まで)．講義"Abel多様体"を行う．

プリンストン研究所より招きを受ける．

11月17日，東京都豊島区池袋1丁目539番地静山荘20号室(自室)にて自殺する．満31歳．

同日付で内閣総理大臣岸信介より従七位に叙せられる．

23日，埼玉県騎西町にて葬儀が行われる．同町善応寺に葬られる．戒名理顕明豊居士．

(12月2日婚約者あとを追う．)

[Distribution on positive 0-cycles in absolute classes on an algebraic variety with finite constant field; Scientific Papers of the College of General Education, University of Tokyo, vol. 8, pp. 123-137]

昭和34年（1959年）

1月25日 埼玉県騎西町において，婚約者美佐子との葬婚式が行われる．

初版あとがき（谷山豊全集刊行会版）

この全集は，不幸にも若くして亡くなった谷山 豊君の業績とその人柄を記念するため家族及び友人其他生前故人と親しかった人々を中心として編まれたものである．

内容は次の三部からなる．

第一部は，欧文で書かれた数学に関する論文の中，次の一篇をのぞくすべてを収めた．

Hyperelliptische Kurve und algebraishen Zahlen.

（これは欧文の推敲が不十分であるので収録をみあわせた．内容は本全集冒頭の論文の原型である．）

第二部は，和文で書かれた数学その他に関する論文，評論，随筆等からなり，学生時代のリポート一篇および日本数学会発行の雑誌“数学”，新数学人集団(S.S.S.)機関紙“月報”，“数学の歩み”その他の雑誌にのせられた記事の中から主なものを集めたものである．この中には，委員会名のものではあるが本人が中心となって執筆したと思われるものも含まれている．

第三部は，書簡および遺書からなる．書簡は死後，故人の知己から集めることのできたものの中ごく少数をのぞきそのすべてを収めた．

末尾に故人の略歴と著作目録をかねた年譜を掲げておいた．本書に収録されていない著作物についてはこれを参照されたい．

最後に本全集の出版のため寄附をおよせ下さった多くの方々，および快く故人の書簡をお貸し下さった方々の御厚意と御助力に心からの謝意を表する．

谷山豊全集刊行会編集委員

久賀道郎，佐武一郎，清水達雄，志村五郎，杉浦光夫，谷山清司

中村得之，成田正雄，山崎圭次郎

なお有馬 哲，石田 信，上野 正，齋藤正彦，片山孝次，清水英男の諸氏には校正その他で御尽力いただいたことを記して感謝したい．

増補版あとがき

この増補版では，初版に次の谷山の文章を第四部として追加した．

1．新人紹介 Peter Roqutte,『月報』1巻1号 p. 5，1953.

2．A. Weil「ゼータ函数の育成について」,『数学』7巻4号 p. 4 - 7，1956.

3．A. Weil「イデール類群の或る指標について」,『数学』7巻4号 p.205-207，1956.

4．『近代的整数論』第1章　歴史，共立出版，1957.（志村五郎と共著）

5．Problems, 1955.

6．書簡三通

一方初版にあった「入門講座に望む」は，谷山が共著者であることを承諾していた文章であるが，全文が共著者杉浦の執筆であるので，今回除くことにした．上記2,3はヴェイユの講演の谷山による紹介であり，内容も谷山の仕事に密接に関係しているので，今回収録することにした．

また4は，現在絶版中なので，その全体の復刻を考えたが，内容としては，谷山の死後出版された英語版の方が改善された部分があるので，この英語版の復刻を出版者である日本数学会に希望することとし，英語版に含まれていない第1章を，この全集に収録することとした．

5は55年の国際会議で配布された未解決問題集(英語版)の谷山提出分である．その日本語版はp. 174にある．ただし英語版の問題11は日本語版になく，英語版の問題10が日本語版では問題11となっている．問題28，29は当時の英文が入手できなかったので，今回佐武が日本語版から英訳した．

また初版掲載の文章についても，今回新たに校正を行い，誤植，誤字を訂正した．また谷山著作目録と略伝「谷山豊の生涯」を付記した．

この増補版全集出版に際し，多くの方々からの御好意，御支援を頂いた．『近代的整数論』の共著者である志村五郎氏は，その第1章の本書への収録をおすすめ下さった．また A. Weil 氏は，講演の谷山による翻訳紹介の収録を許可して下さった．また，日本数学会は，『数学』および Journal of Math. Soc. Japan に掲載された谷山執筆の文章を本書に収めることを許可して下さった．また共立出版株式会社は，『近代的整数論』第1章を復刻収録することを許可して下さった．また，片山孝次，清水英男，草場公邦の三氏は，谷山論文の校正をして下さった．また日本評論社と亀井哲治郎氏は，この増補版全集の出版を企画，実現して下さった．また谷山清司氏は，著作権者として，この全集の出版を許可され，支援して下さった．谷山氏は，この全集発行に際し，次の言葉を寄せられた．

「「何もしないのは罪悪である」と彼は常日頃言っていた．之は彼の公理である．

或る日海岸に出てみると，沖の方から「海はいつでも日曜……」という歌がきこえてきた．こういう夢をみたと彼は私に話した．

この二つを私は結びつけてみたい．つまり「何もしないのは罪悪である．しかし日曜日はその限りにあらず」

肉体的に或は精神的に何も出来なくなった時，犯罪者にならない為には，いつでも日曜である海に入って行く他なかったのである．」

以上の方々に，編集委員会は，厚く御礼申し上げる．

1994年4月

谷山豊全集(増補版)編集委員会
佐武一郎　清水達雄
杉浦光夫　山﨑圭次郎

新版あとがき

『数学の歩み』の「7巻4－5合併号」(1960年5月発行)に,「「谷山豊全集」発刊準備会のこと」という記事があり,『谷山豊全集』の刊行が企画された当時の消息が伝えられている.ここに全文を引きたいと思う.

「谷山豊全集」発刊準備会のこと

故　谷山豊君の全集の発刊が計画されています.

谷山豊君は,御承知のように一昨年11月に夭折した数学者で,僕等の誇るべき仲間でした.

彼の業績は,今世紀の整数論を画期的に飛躍させ,将来の整数論学者に豊かな瞑想の手がかりを与えています.

全集には,それらの発表された論文の他,発表された論文の原型となった幾つかの未発表原稿ものせます.それらは,完成され,発表された論文よりも,よりよく,彼の生のアイデアを伝えていますので,歩みにのせた多くの数学的記事(それももちろん全集にのせます)と共に彼の発想の跡や思考の型を裏づけるための絶好の資料になるでしょう.

僕達が,彼を誇りに思うのは,そればかりではありません.彼はSSSの創立者の一人であり,活動の中心の一つであり,SSSの精神的なシンボルですらありました.

僕らは,全集のために資料を集めながら,主として「歩み」にのせられた彼の主張をよみ,それが主張された当時の事を思い返して,ひるがえって,今の我々自身を思うとき,いかに彼が僕達にとって必要な人であったかを知って,口惜しくなります.

彼は,現代整数論の創り手であったと同時に,SSSのリーダーでありましたけれども,それよりも何よりも,僕は彼を一人の友と呼びたい.それは僕ばかりでなく,SSSの殆どすべての人の気持と思います.彼はそのように,SSSのすべての若者達に信頼された先輩であり友人でした.多くの若い人達が嬉しいにつけ,悲しいにつけ,彼のアパートに押しかけたものでした.

彼が,いろんな友人にあてた書簡が,ずい分沢山——全集の頁にして50頁分程も——あつまりました.受取人の方達の御協力を感謝します.まだ外にも書簡をお持ちの方がありましたら,差支えのない限り発表させていただきたいと思いますが,よろしかったら,僕等に拝借させて下さい.

また，彼が「歩み」や「数学」や，「Journal」に書いたものや手紙の他に，青年時代に書いた同人ノート「自信ある仲間」や，講演の記録や，クラス雑誌にのせた童話などものせます．

僕達の手に集った資料は下記の通りですが，尚この他にお心当りの方がありましたら，どうぞ下記へお知らせください．

尚，現在は資料収集と整理の段階で，発刊の方法等に関しては，何も具体的には決っておりません．

東京都目黒区駒場　東大教養学部
数学教室気付
谷山豊全集発刊準備会

尚，当会のメンバーは下の通りです．
久賀道郎，清水達雄，志村五郎
杉浦光夫，谷山清司，中村得之
成田正夫，山﨑圭次郎
（漢字表記，送り仮名等，原文のまま）

この企画は実現し，1962年，最初の『谷山豊全集』が刊行された．これを土台として，37回忌にあたる1994年の秋10月に，日本評論社から『[増補版] 谷山豊全集』が刊行された．編集委員として，杉浦光夫，清水達雄，佐武一郎，山﨑圭次郎の諸先生の名が挙げられている．

新版の刊行にあたり，谷山豊の遺品の中から若干の文書と写真が選定され，新たに収録された．谷山豊の遺品は長く谷山家に保管されていたが，兄の清司氏による委託を受け，現在は二つの段ボール箱におさめられて，朝日新聞記者の木脇みのりさんのもとにある．この間の消息は木脇さんのエッセイ「谷山豊との25年」（『数学文化』第30号，2018年）に詳述されている．遺品の一部を列記すると次のとおりである．

・騎西尋常小学校卒業証書（昭和15年３月29日）
・不動岡中学校卒業証書（昭和20年３月27日）
・褒状（昭和20年３月27日）「本学年成績ノ優等をヲ賞ス」 不動岡中学校より
・学位記（東京大学より　昭和33年５月２日）
・写真　東大理学部数学科卒業式の日に（昭和28年３月）同級生一同
・写真　同上，卒業式の日に　彌永昌吉宅にて　有志とともに
・浦和高等学校に在籍中に書かれた小説風のエッセイ「ある人の話」
・写真数葉「代数的整数論に関する国際会議」（東京－日光整数論シンポジウム．昭和30年９月）にて
・蔵書『プラトンの自叙傳』（和田堀書店，青木巌訳，昭和22年）．書き込みあり
・写真数葉　昭和33年８月．婚約者（鈴木美佐子さん）と鍾乳洞に遊ぶ
・遺書　原稿用紙３枚

ほかに多くのはがきも遺されている．また，谷山豊の遺品ではないが，
・婚約者の遺書　谷山清司氏宛て

・写真１葉　葬婚式（昭和34年１月ころ）のときの集合写真

などもある．

これらのうち，新版に収録された文書と写真は次のとおりである．

文書

１．はがき22通

２．「ある人の話」

写真（口絵に掲載）

１．東大理学部数学科卒業式の日の写真２葉

２．東京―日光整数論シンポジウムの写真２葉

３．学位記

４．蔵書『プラトンの自叙伝』のカバーと書き込み

５．はがき３通

６．遺書

『月報』に掲載された論攷「Weylと整数論」「代数幾何学と整数論」「モヅル函数と合同ζ-函数」の自筆原稿の写真も収録された．これらの原稿は杉浦光夫先生が所蔵していたものである．

2018年10月

高瀬正仁（元九州大学教授，数学者・数学史家）

「谷山豊全集（増補版）編集委員会」の清水達雄氏から，『[増補版] 谷山豊全集』の新版化に際し，次の言葉を寄せられた．（編集部）

たしか久賀道郎君からの訃報を受けて，あの下宿部屋に行ったのだと記憶しています．

行ってみたら，弥永先生（原文ママ）がいらしていて，ひかえ目に座りました．意外だったのですが，本人の書いたものを見るとややわかってゆきました．

ガスがたちこめていくのを何とか出したところだったのですが，いやなことでした．

全集をまた出すのは賛成です．婚約者があったと知りました．墓前であと追いされたのでしたね．

2018年８月

［新版］谷山豊全集

1994年10月30日　第1版第1刷発行
2018年12月15日　新　版第1刷発行

著　者　谷　山　豊
編　者　杉浦光夫・佐武一郎
清水達雄・山﨑圭次郎
発行者　串　崎　浩
発行所　株式会社 日　本　評　論　社
〒170-8474 東京都豊島区南大塚 3-12-4
電話 (03)3987-8621 販売・(03)3987-8599 編集
印　刷　藤原印刷株式会社
製　本　牧製本印刷株式会社

ISBN 978-4-535-78886-2